Interaction Models

The radical interdependence between humans who live together makes virtually all human behavior conditional. The behavior of individuals is conditional upon the expectations of those around them, and those expectations are conditional upon the rules (institutions) and norms (culture) constructed to monitor, reward, and punish different behaviors. As a result, nearly all hypotheses about humans are conditional – conditional upon the resources they possess, the institutions they inhabit, or the cultural practices that tell them how to behave. *Interaction Models* provides a standalone, accessible overview of how interaction models, which are frequently used across the social and natural sciences, capture the intuition behind conditional claims and context dependence. It also addresses the simple specification and interpretation errors that are, unfortunately, commonplace. By providing a comprehensive and unified introduction to the use and critical evaluation of interaction models, this book shows how they can be used to test theoretically derived claims of conditionality.

WILLIAM ROBERTS CLARK is the author of *Capitalism, Not Globalism, Principles of Comparative Politics*, and numerous journal articles. With Sona and Matt Golder he was awarded the Brian Barry Prize by the British Academy. He has taught at six leading research universities and is currently President of the European Political Science Association.

MATT GOLDER is in the top 2 per cent of the most cited scientists worldwide, and his article with William Clark, "Understanding Interaction Models: Improving Empirical Analyses," is in the top 10 most cited articles in political science (http://charlesbreton.ca/assets/PS_Top10_2020.pdf). He is the winner of the GESIS Klingemann Prize and the Brian Barry Prize from the British Academy.

Methodological Tools in the Social Sciences

SERIES EDITORS
Paul M. Kellstedt, *Professor of Political Science, Texas A&M University*
Guy D. Whitten, *Professor of Political Science and Director of the European Union Center at Texas A&M University*

The Methodological Tools in the Social Sciences series is comprised of accessible, stand-alone treatments of methodological topics encountered by social science researchers. The focus is on practical instruction for applying methods, for getting the methods right. The authors are leading researchers able to provide extensive examples of applications of the methods covered in each book. The books in the series strike a balance between the theory underlying and the implementation of the methods. They are accessible and discursive, and make technical code and data available to aid in replication and extension of the results, as well as enabling scholars to apply these methods to their own substantive problems. They also provide accessible advice on how to present results obtained from using the relevant methods.

OTHER BOOKS IN THE SERIES
Eric Neumayer and Thomas Plümper, *Robustness Tests for Quantitative Research*

Interaction Models
Specification and Interpretation

William Roberts Clark
Texas A&M University

Matt Golder
Penn sylvania State University

CAMBRIDGE
UNIVERSITY PRESS

CAMBRIDGE
UNIVERSITY PRESS

Shaftesbury Road, Cambridge CB2 8EA, United Kingdom

One Liberty Plaza, 20th Floor, New York, NY 10006, USA

477 Williamstown Road, Port Melbourne, VIC 3207, Australia

314–321, 3rd Floor, Plot 3, Splendor Forum, Jasola District Centre, New Delhi – 110025, India

103 Penang Road, #05–06/07, Visioncrest Commercial, Singapore 238467

Cambridge University Press is part of Cambridge University Press & Assessment, a department of the University of Cambridge.

We share the University's mission to contribute to society through the pursuit of education, learning and research at the highest international levels of excellence.

www.cambridge.org
Information on this title: www.cambridge.org/9781108416719

DOI: 10.1017/9781108241762

First published 2023

A catalogue record for this publication is available from the British Library.

Library of Congress Cataloging-in-Publication Data
Names: Clark, William Roberts, 1962– author. | Golder, Matt, author.
Title: An introduction to interaction models / William Roberts Clark, Matt Golder.
Description: New York : Cambridge University Press, [2023] |
    Series: MTSS methodological tools in the social sciences
Identifiers: LCCN 2023006720 (print) | LCCN 2023006721 (ebook) |
    ISBN 9781108416719 (hardback) | ISBN 9781108404082 (paperback) |
    ISBN 9781108241762 (epub)
Subjects: LCSH: Social sciences–Statistical methods. | Political science–Statistical methods. |
    Regression analysis. | Qualitative research–Methodology.
Classification: LCC HA29 .C596 2023 (print) | LCC HA29 (ebook) |
    DDC 519.2–dc23/eng/20230505
LC record available at https://lccn.loc.gov/2023006720
LC ebook record available at https://lccn.loc.gov/2023006721

ISBN 978-1-108-41671-9 Hardback
ISBN 978-1-108-40408-2 Paperback

To our most important students:
Meaghan, Brian, Liam, Cameron, and Sean.

Contents

Figures

Tables

Preface

This book is about how to use interaction models to test the conditional implications of our social science theories. As we argue in more detail in Chapter 1, the radical interdependence of humans who live together means that many of the hypotheses that we can derive from our theories across a broad array of topics throughout the social sciences are likely to be context dependent. This explains why conditional claims such as "an increase in X is associated with an increase in Y when condition Z is present, but not otherwise" are so ubiquitous across the various fields in political science, economics, sociology, psychology, and other social science disciplines. It's well established that the intuition behind conditional claims and context dependence can be captured by interaction models (Wright, 1976; Friedrich, 1982; Aiken and West, 1991). Unfortunately, the implementation of interaction models is often flawed, and inferential errors are common.

Our mutual interest in interaction models is long-standing. We've been discussing the appropriate specification and interpretation of interaction models with each other for over two decades now. Our continued interest in interaction models is often met with bemusement by many of our colleagues, especially those trained in more computationally complex or "cutting-edge" quantitative methods. We suspect that their bemusement stems from the widespread belief, probably arising from the fact that interaction models are often introduced early on in someone's quantitative methods training, that interaction models are easy and well-understood. When we started writing our first methods paper on interaction models in the early 2000s, a senior methodology colleague openly questioned its utility. "Everybody already knows how to deal with interaction models" was the basic response. We were just wasting our time. He was, of course, trying to be helpful. Our own experience reviewing papers and reading published articles, though, made us much less sanguine about the quality of social science research based on the use of interaction models. We ignored our colleague's advice and went ahead with our paper anyway.

To motivate it, we conducted a survey of the literature in our home discipline of political science where we systematically examined the use of interaction models in articles published between 1998 and 2002 by the three leading non-specialized political science journals: the *American Journal of Political Science*, the *American Political Science Review*, and the *Journal of Politics*. Contrary to the beliefs of our senior methodology colleague, we found that just ten percent of the articles that we identified as using an interaction model followed all four of the basic "best practice" recommendations that we outlined in our paper. The results from our survey suggested that there was considerable potential for inferential errors in the articles we examined, something that was troubling as many had been written by the discipline's leading figures and had gone on to generate substantial research agendas. The message was clear. Although we as a discipline may well have thought we knew how to use interaction models, we obviously didn't. At least when it came to common practice. Our experience as reviewers and communication with others told us that political science was not unique in this regard.

Our paper, which contained a simple checklist of dos and don'ts for using interaction models, was published in *Political Analysis* in 2006 (Brambor, Clark and Golder, 2006). It quickly became clear that there was huge pent-up demand across the social sciences for advice on how to improve empirical analyses involving interaction models. Evidence for this comes from the fact that our paper is the third most cited article in political science published in the 2000s and, according to some measures, in the top ten of all political science articles ever published (http://charlesbreton.ca/assets/PS_Top10_2020.pdf). Other recent publications that offer additional advice on interaction models have also proven to be very influential (Ai and Norton, 2003; Kam and Franzese, 2007; Berry, DeMeritt and Esarey, 2010; Berry, Golder and Milton, 2012; Hainmueller, Mummolo and Zu, 2019). The result is that the quality of social science research that uses interaction models has improved significantly in recent years. For example, we're now much less likely to see scholars inappropriately omit "constitutive terms" from their interaction models or incorrectly interpret these terms as capturing unconditional effects. And scholars are much more likely to employ graphical techniques such as "marginal effect plots" to evaluate their conditional claims and provide substantively meaningful information about the effects of their variables.

While this progress is substantial and obviously welcome, simple specification and inferential errors with interaction models remain stubbornly commonplace. One thing we've noticed in recent years is that scholars will often cite our paper or one of the others offering methodological advice on the use of interaction models (perhaps because they feel that reviewers require this) but fail to actually implement the recommended practices. Indeed, it's not unusual to see scholars cite our 2006 paper to support practices that we explicitly note are unnecessary or, worse, inappropriate. This is disappointing.

In general, widespread confusion persists about certain aspects of interaction models. As an example, there are, as we'll show, alternative ways of specifying the exact same interaction model when one or more of the interacting variables is discrete. Scholars are often unaware of this equivalence or what it means for interpretation. The result can be that some researchers aren't aware that they're estimating an interaction model (Reingold, Haynie and Widner, 2020). As another example, few scholars appreciate the fact that there are two distinct sources of interaction when we estimate an interaction model with a limited dependent variable and the issues that this poses for interpretation and hypothesis testing. As we'll demonstrate, these confounding sources of interaction relate to the *variable-specific* interaction that arises from the inclusion of an explicit interaction term and the *compression-based* interaction that results from the inherent non-linearity that links the outcome variable to the independent variables in these types of models. Considerable uncertainty also surrounds exactly how to specify and interpret interaction models when we move beyond evaluating the simple two-variable interactions typically examined in pedagogical pieces on the use of interaction models to test more complex claims of conditionality. What's the best way to evaluate conditional claims when more than two variables interact or when a variable interacts with itself? What's the best way to present the results in these settings? It's also the case, in our experience, that scholars frequently fail to present all of the quantities of interest necessary to test their conditional claims or expose their theories to as strong an empirical test as is possible given the available data. This often has to do with a failure to think through all of the key predictions that can be derived from their underlying theory.

Our Approach

Our goal in writing this book is to provide a comprehensive and unified introduction to the use of interaction models and how they can be used to test theoretically-derived claims of conditionality. We take an "empirical implications of theoretical models (EITM) style" approach where we emphasize the importance of closely integrating the theoretical and empirical components of social science research. Throughout the book, for example, we always try to make a strong connection between the conditional implications that can be derived from our theories and the interaction model specification and quantities of interest necessary to fully evaluate our conditional claims. A consequence is that we discuss theory and its implications for empirical analysis much more than is usually the case in methods pieces dealing with interaction models.

The fact that the book is designed to help scholars to better use interaction models in situations where the primary goal is theory or hypothesis testing rather than, say, prediction is reflected in the types of interaction models and estimation techniques that we cover. To be specific, our book deliberately focuses on the types of regression-based parametric models that are most commonly used by applied researchers interested in theory testing and drawing inferences. There are, of course, other more sophisticated estimation techniques, such as neural networks (Zeng, 1999; Beck, King and Zeng, 2000), generalized additive models (Hastie and Tibshirani, 1986; Beck and Jackman, 1998), kernell regularized least squares (Hainmueller and Hazlett, 2014), tree-based models (Quinlan, 1986; Green and Kern, 2012; Montgomery and Olivella, 2018), and support vector machines (Vapnik, 1995, 1998; D'Orazio et al., 2014), that can be used to capture highly complex forms of interaction and conditionality. These methods, though, many of which come out of the statistical and machine learning literature (Hastie, Tibshirani and Friedman, 2017) and depart from the familiar regression framework, are arguably more suited to model-fitting problems related to prediction and classification rather than problems related to theory testing and inference. It can often be challenging to meaningfully interpret exactly what's going on in these models and derive the types of quantities of interest that applied scholars require to test their theoretical claims.

Who Is This Book For?

Our book is written for people who are interested in formulating contextual theories and testing conditional or "context-dependent" hypotheses using quantitative methods. Given the ubiquity of conditional relationships in the study of human behavior, we suspect that scholars from across the social sciences will find something of value in reading this book – or, at least, that's our hope. We've assumed that readers have some grounding in the basics of statistical inference and ordinary least squares (OLS) regression but not much beyond that. This means that our book could be used by most graduate students, faculty, and researchers in the social sciences. We suspect that it could also be used by some advanced undergraduate students who've taken at least one class in quantitative methods.

We've attempted to make the material in the book as accessible as possible. We've tried, for example, to build our approach to thinking about interaction models from the ground up, starting with what we call "the fundamentals" before introducing more "complex" forms of interaction in an incremental and systematic fashion. In contrast to many methods books, we try to show more of the intermediary steps that are involved in deriving and calculating particular quantities of interest and not just the final equations themselves. Where possible, we also try to provide the intuition behind the math. In addition, we offer practical advice on how to state conditional hypotheses and present quantities of interest in the most effective manner. Call-out boxes keep track of particularly important points. Throughout the book, we provide detailed substantive applications showing how each of the techniques and cases that we cover can be implemented in practice. The replication code, written in both Stata and R, for all of the substantive applications and exercises is available online at http://mattgolder.com/. We've tried to minimize pre-written packages and commands in our code in order to make the calculations underlying the methods clearer.[1]

How to Use This Book

We believe that our book would be ideal for a class that focuses on how to formulate contextual theories and test context-dependent hypotheses. Of course, we recognize that such a class, while likely to be valuable and interesting, is rarely taught, even in our own departments and institutions. Given this, we suspect that our book will most likely be used as supplementary reading in various quantitative methods classes and as a professional reference for applied researchers. In terms of quantitative methods classes, the material covered in *Part I: The Fundamentals* is particularly well-suited to an introductory class on regression analysis dealing with ordinary least squares estimation. Depending on the level of this class, the material introduced in *Part II: More Complex Forms of Conditionality* may also be appropriate. This additional material would certainly fit well in an advanced regression analysis class where students start to move beyond the simple examples typically covered in an introductory class. The material examined in *Part III: Interactions and Limited Dependent Variables* will be of greatest value in a standard maximum likelihood estimation class that introduces students to a variety of limited dependent variable models or in a class that specifically focuses on discrete choice models. As you can see, we think that our book will be of use to students in several of the classes that have usually made up a significant part of the traditional "methods sequence" in graduate programs in the social sciences. Given this,

---

[1] The figures that appear in this book were created using the PGFPLOTS package in LATEX (Feuersänger, 2010). The code for them can also be found online at http://mattgolder.com/.

we suspect that students may want to obtain a copy of our book in the first year of graduate school so that they can come back to it again and again as they progress through their required quantitative methods classes.

The book can also be used as a professional reference for applied researchers who seek advice on how to appropriately specify and interpret an interaction model for a particular substantive application. Over the years, we've been contacted by researchers from around the world who've been looking for advice on how to test specific types of conditional claims. They're usually reaching out to us because their particular conditional claim involves some kind of deviation from the scenario covered in the typical methods piece on interaction models where there's an interaction between two continuous independent variables and we have a continuous dependent variable. Perhaps one or more of their independent variables is discrete, there are more than two variables interacting, or there's some kind of limited dependent variable involved. If our book had been completed earlier, we could have simply referred these researchers to the relevant sections. Indeed, it's partly because so many scholars have contacted us with these sorts of requests over the years that we thought a book like this would be useful. We certainly hope that it will be.

Acknowledgments

We've had productive conversations about the use and misuse of interaction models with many scholars over the years. These conversations have been important for clarifying our thinking about how to appropriately test claims of conditionality. In this regard, we're particularly grateful for the discussions we've had with Neal Beck, William Berry, Rob Franzese, and Jonathan Nagler. Numerous individuals have provided comments and useful insights or responded to one of our many queries at various points while writing this book. These include Ray Block, Emma Cohen, Scott Cook, Charles Crabtree, Yaoyao Dai, Kostanca Dhima, Ben Ferland, Gilles Godefroy, Jerg Guttman, Boyoon Lee, Howard Liu, Eric Plutzer, and Chris Zorn. We're especially grateful to Garrett Glasgow for being such a valuable sounding board for thinking about interactions in the context of discrete choice models, to Ali Kagalwala for translating our Stata code into R, and to Anil Kuleli for help with the index. Special thanks must also go to Sean Golder for patiently working on various bits of the math in the book with his Dad and to Sona Nadenichek Golder for reading more drafts of the chapters than anyone should ever have to read.

We're indebted to Paul Kellstedt and Guy Whitten, the editors of the Methodological Tools in the Social Sciences book series at Cambridge University Press, for encouraging us to write this book in the first place and their enormous patience as we made slow progress in bringing it to fruition. We're also grateful to them for organizing a book conference at Texas A&M University where we received incredibly detailed and thoughtful feedback from our reviewers, Justin Esarey and Vera Troeger, and other participants, especially Dave Armstrong.

Our greatest debt is to our families. Sona and Sean have shown incredible support and patience throughout the entire writing (and non-writing) process. They make life much, much better. Laurie Clark's support during this project's long gestation (much of it during a global pandemic) has been unwavering and unconditional. Thank you for reinforcing and facilitating the positive aspects of life, while inhibiting and mitigating the negative.

The art on the cover is "Linear Space 359" by Amie Adelman (http://www.amieadelman.com/)

1 Introduction

Aristotle (1998) famously said that "man is by nature a political animal." In other words, the natural environment of women and men is the *polis*. The book of *Exodus* (2:18) quotes no lesser authority than God in saying that it's not good for people to "be alone." Humans are social beings, and, as a result, their lives are marked by a radical interdependence. One person's goal attainment depends on the behavior of others. My ability to sleep in on a Saturday morning depends on my neighbors' willingness to forebear from cutting their lawns. My neighbors' ability to cut their lawns before the heat of the day sets in without provoking my wrath depends on my willingness to get out of bed at a decent hour. We are social animals in the sense that our behaviors affect each other's well being. In fact, we have a word for people whose behavior demonstrates a callous disregard for their effect on others: anti-social.

This book starts from the premise that the radical interdependence that exists between humans who live together makes virtually all of human behavior conditional. The behavior of individuals is conditional upon the expectations of those around them, and those expectations are conditional upon the rules (institutions) and norms (culture) constructed to monitor, reward, and punish different behaviors. As a result, virtually all hypotheses about humans are conditional – conditional upon the resources they possess, the institutions they inhabit, or the cultural practices that tell them how one "ought to behave." Of course, we can, and often times should, simplify a situation by comparing behaviors at a particular resource level, within a particular institutional context, or among individuals who share a set of cultural practices. But when we do so, we lose the ability to understand how resource endowments, institutions, or culture influence behavior. We must also either give up on generalizing beyond particular contexts or take as a matter of faith that the relationships we have uncovered are invariant with respect to the contexts that we hold constant.

If, instead, we want to understand how resource endowments, institutions, or culture influence human behavior, we must observe human behavior in contexts where those factors vary. Further, if we have reason to believe that people's behavior depends on these contextual factors in the sense that their responses to changes in their environment depend on the context they find themselves in, we must account for this context dependence in our empirical analyses. This book is about one way to capture and evaluate this context dependence in statistical analyses: multiplicative interaction effects.

An example of a multiplicative interaction effect is the conditional relationship between foreign aid and economic growth. In 2000, two World Bank economists, Craig Burnside and David Dollar, published an influential study in the *American Economic Review* arguing that foreign aid has a positive effect on economic growth in recipient countries but only when those countries adopt "good policies." This study was so influential that it led the administration of President George W. Bush to begin conditioning the giving of foreign aid on the policies of recipient countries (Eviatar, 2003). To test their conditional claim about the relationship between foreign aid and the quality of the policy environment on economic growth, Burnside and Dollar (2000) employed a multiplicative interaction model similar to the one shown here:

$$Growth = \beta_0 + \beta_1 Aid + \beta_2 Good\ Policy + \beta_3 Aid \times Good\ Policy + \epsilon.$$
$$(1.1)$$

In the next chapter, we show why a multiplicative interaction model like this is a reasonable way to examine how the statistical relationship between two variables, such as economic growth and foreign aid, depends on the value of a third variable, such as the quality of the policy environment.

Before we proceed to the next chapter, we'll attempt to motivate your interest in interaction effects by pointing out the ubiquity of conditional relationships in the study of human behavior.

1.1 RESOURCE ENDOWMENTS

Many political economists believe that the assets actors hold influence the way they respond to changes in their environment. Some individuals possess stores of capital – either financial instruments, like stocks, bonds, and stockpiles of monetary assets, or physical capital, such as homes, factories, and machines and equipment, like textile looms or oil derricks. Others possess only their labor. There are many different political economy models and many ways to classify the assets that individuals own, but this class of models shares the idea that individuals will assess policy alternatives by predicting how those policies will influence the value of the assets they hold.

For example, according to a political economy perspective, the effect of an exogenous change in the flow of relatively low skilled immigrants across a nation's borders (perhaps due to war, crime, or economic dislocation in nearby countries) on citizen preferences for legislation restricting immigration will depend on the type of assets that citizens possess. Owners of businesses that employ relatively low skilled workers are likely, all else equal (holding factors such as any non-economic related animus or affection for foreigners constant), to be more welcoming of new immigrants and more resistant to legislation seeking to restrict immigration. In contrast, citizens who own nothing but their relatively low skilled labor will be worried about the increased competition from new immigrants either driving their wages down or causing them to lose their jobs and, as a result, will be in favor of legislation seeking to restrict immigration. We see from this that whether an increase in immigration leads to an increase or decrease in someone's enthusiasm for restrictive legislation on immigration depends on the type of assets that they hold. In effect, the type of assets held by an individual "moderates" or "modifies" the effect of an increase in immigration on attitudes towards immigration restrictions. If we could classify citizens as either workers or capitalists, our theory would make the following prediction:

Resource Endowment Hypothesis: An exogenous increase in immigration is likely to elicit increased support for restrictive immigration policy among workers but decreased support among owners of capital.

We can test this conditional hypothesis with a multiplicative interaction model similar to the one shown here:

$$\text{Support for Immigration Restrictions} = \beta_0 + \beta_1 \text{Immigration} + \beta_2 \text{Worker} + \beta_3 \text{Immigration} \times \text{Worker} + \epsilon,$$
(1.2)

where *Support for Immigration Restrictions* is a measure of an individual's support for restrictions on immigration, *Immigration* captures the level of immigration, and *Worker* is a dichotomous variable that equals 1 if an individual is a worker and 0 if they're a capitalist. Once again, we'll see exactly how this specification is able to capture the conditionality in our theory in Chapter 2.

The above example is just one of many where resource endowments (in this case, capital ownership) might "moderate" or "modify" the relationship between two other factors (an increase in immigration and support for immigration restrictions). A closely related example can be found in work that relies on the Stolper–Samuelson (1941) model of international trade. According to this model, an exogenous change in trade will affect citizens' income by influencing the value of the assets they hold. However, the precise manner in which it does this will depend on the factor endowments present in the country in question. Because countries can be expected to export goods that use their abundant factor intensively, an exogenous increase in trade, perhaps as a result of the development of containerized shipping, will lead to an increase in the income of workers where labor is the abundant factor but a decline in income where labor is the scarce factor. This suggests that the effect of trade on the policy preferences of workers should depend on whether labor or capital is the abundant factor in a society.

The point here is that according to a broad set of theories, an individual's policy preferences (and by extension, perhaps, their political behavior) are expected to depend on a combination of their individual characteristics (asset ownership) and characteristics of the economy in which they find themselves (resource endowments). Attempts to test explanations about how economic interests influence political behavior that rely only on attributes of the individual are likely, therefore, to be misspecified.

1.2 INSTITUTIONS

In August of 1992, renowned political scientist Theodore Lowi (1992, 363) wrote in the *New York Times* that

"[W]hatever the outcome of this year's Presidential race, historians will undoubtedly focus on 1992 as the beginning of the end of America's two-party system. The extraordinary rise of Ross Perot and the remarkable outburst of enthusiasm for his ill-defined alternative to the established parties removed all doubt about the viability of a broad-based third party."

This statement, made by a preeminent scholar of American politics, was astonishing because it seemed to fly in the face of almost a half century of comparative politics research summarized as "Duverger's Law" (Duverger, 1954) showing that single-member district electoral systems like the one in the United States tend to produce two-party systems. This particular line of comparative politics research had been recognized a decade earlier for demonstrating that the accumulation of knowledge was possible in political science (Riker, 1982).

In the years following Theodore Lowi's prediction, we came to see that much of the evidence for Duverger's Law was presented in a confusing manner that both obscured Duverger's theoretical insights and invited people to regard some observations as being more anomalous than they actually were (Clark and Golder, 2006). In this respect, it was perhaps understandable that Professor Lowi might have been confused about the possibility of the emergence of an electorally successful third party in the United States.

Maurice Duverger (1954) argued that political parties have their foundations in societal divisions that cause people to place different types of demands on the state. Parties can be thought of as teams of citizens and/or representatives who share policy goals and compete against other teams with different policy goals. Duverger thought that societies differed in the number of latent groups that might form parties and that the way these latent groups were translated into parties, in either the electorate or the legislature, depended on the nature of the electoral system. In proportional representation (PR) systems, where votes are proportionally translated into legislative seats, societal divisions are translated into electoral and legislative parties in a rather frictionless fashion. This means that socially diverse countries with a PR electoral system can expect to have many political parties, while socially homogeneous countries can expect to have only a small number of them. In contrast, majoritarian systems, such as the single-member district plurality systems used to elect representatives in the United States and United Kingdom, where only the largest party can win a seat, act as a brake on the translation of societal cleavages into political parties and thus constrain party systems to always be small. Majoritarian electoral rules constrain the size of the party system for two reasons. First, the mechanical way in which votes are translated into seats in majoritarian systems means that large parties that come first win legislative representation whereas smaller parties that come second or worse don't. Second, this mechanical effect of the electoral system favoring large parties creates incentives for both candidates and voters to act strategically in ways that benefit a small number of large parties even more. Supporters of small parties who don't think that their party will come first have an incentive to vote for the "lesser of two evils" among the two largest parties who can realistically win. The anticipation of this strategic behavior among voters, along with the mechanical effect of the electoral system, creates incentives for strategic entry on the part of political candidates. All other things equal, candidates in majoritarian systems have an incentive to run under the banner of one of the two larger parties that are going to be advantaged by the electoral system even if a smaller party is a better ideological fit. The end result is that countries with majoritarian electoral systems tend to be dominated by a small number of political parties, usually two, irrespective of their degree of social diversity.

As should be clear, the essence of Duverger's theory is that electoral institutions modify the relationship between societal divisions and the number of parties (Clark and Golder, 2006). For Duverger, then, the question was not whether social divisions or electoral laws were the key determinant of party system size. Rather, he was interested in the *interaction* between these two aspects of a polity. Failure to recognize the centrality of this interaction could lead no less of a scholar than Gary Cox, arguably one of the most important scholars in comparative politics, to become confused about Duverger's argument. For example, in his magisterial monograph, *Making Votes Count*, Cox (1997, 23) says that Duverger

"**took social structure more or less as a residual error,** something that might perturb a party system away from *its central tendency defined by electoral law*" [italics and bold added].

In fact, Duverger argued that

"the influence of ballot systems could be compared to that of a brake or an accelerator. The multiplication of parties, which arises as a result of other factors, is facilitated by one type of electoral system and hindered by another. *Ballot procedure,* however, *has no real driving power.* The **most decisive influences** in this respect **are** aspects of the life of the nation such as **ideologies and particularly the socio-economic structure**" [italics and bold added].

By comparing the italicized text across the two quotes, we see that the argument Cox attributes to Duverger about the centrality of electoral systems is pretty close to exactly the opposite of what Duverger actually said. Similarly, by comparing the bold text across the two quotes, we see that Cox misses the fact that Duverger thought of social structure as the primary driver behind the creation of parties.

We do not bring this up to poke a great scholar in the eye. Rather, we'd like to highlight that a failure to think clearly about the conditional nature of the arguments we encounter can cause even great scholars to become confused. We believe that Cox's confusion about Duverger's argument is caused by his failure to recognize that Duverger was making an argument involving interaction effects. Evidence of this fact is found in Cox's own words. After discussing Duverger, Cox (1997, 23) says that "Later scholars, however, have considered the possibility that cleavage and electoral structures may interact. For example, two recent papers take this tack ... both come to the conclusion that Duverger's institutionalist claims are conditioned by the nature of social cleavages." But clearly, from the passage we just discussed, Duverger had been making the argument that cleavage and electoral structures interact to shape party systems all along.

Cox is not alone. William Riker, another giant of political science, also fails to fully appreciate the conditional nature of Duverger's argument in his history of science essay looking at Duverger's Law (Riker, 1982). Riker proposes a distinction between what he calls "Duverger's *Law*" (the claim that majoritarian single-member district plurality systems encourage two-party systems) and "Duverger's *Hypothesis*" (the claim that proportional representation electoral systems favor multi-party systems) because Duverger appeared to treat the former relationship deterministically and the latter relationship as "at best probabilistic" (754). Duverger presumably chose to view the former claim as "law-like" and the latter claim as "probabilistic" because there appeared to be a larger number of anomalies for the latter claim (countries with proportional electoral laws but few parties) than the former claim (countries with majoritarian single-member district plurality systems but many parties). But the conditional nature of Duverger's argument actually predicts this outcome. If single-member district plurality systems act as a brake on the translation of social cleavages into parties, then we'd expect countries with these electoral rules to always have a small number of parties irrespective of their level of social diversity. But if proportional representation electoral systems permit social divisions to be accurately translated into parties, then we'd expect countries with these electoral rules to exhibit greater variance in their party system size due to the variation in their level of social heterogeneity.

Our point here is simply that thinking carefully about the effects of institutions, such as electoral rules, often requires clear thinking about conditional arguments. In fact, it may be the case that institutional arguments are intrinsically arguments about modifying effects. If institutions determine how political inputs are translated into political outputs, then it follows that in different institutional contexts, the mapping of inputs to outputs will be different. From this perspective, it's hard to imagine how institutions would be causally important if they didn't act as modifying variables. We believe that it follows, therefore, that good theoretical and empirical work on institutions is unlikely to occur in the absence of clear thinking about arguments involving moderating or modifying variables.

1.3 CULTURE/IDENTITY

Cultural arguments also produce hypotheses about the social and political world that are likely to be conditional. Cultures involve shared sets of understandings that help people interpret events that occur in their environment. As such, like institutions, culture moderates the way that political and social inputs get translated into political and social behaviors.

Emile Durkheim (2003/1895) argues that the discipline of sociology should be seen as the empirical study of what he called "social facts" – the "beliefs, tendencies and practices of the group taken collectively." Durkheim believed that these social facts influence individual behavior because they determine the consequences of individual choices, whether those choices are deliberate or driven by unconscious perceptions of what behaviors are socially acceptable. These facts constitute "currents of opinion, whose intensity varies according to the time and country in which they occur" and "impel us, for example, toward marriage or suicide, toward higher or lower birth rates, etc." These social facts exist outside individuals. Importantly, the "forms these collectives take when they are 'refracted' through individuals are things of a different kind" (2003/1895, 77). In effect, Durkheim saw *social* behavior as the product of individual characteristics and agency on the one hand and the societal context that individuals inhabit on the other. For Durkheim, if individual behavior is unconstrained by such social facts, it would fall in the province of psychology or biology rather than sociology.

One example of how social context shapes individual behavior can be found in the sociology of religion literature. In two initial studies, scholars were surprised to find that church attendance had no statistically discernible effect on the delinquency of teenagers (Hirschi and Stark, 1969; Burkett and White, 1974). Subsequent attempts at replicating these null results, however, were unsuccessful (Rhodes and Reiss, 1970; Albrecht, Chadwick and Alcorn, 1977; Higgins and Albrecht, 1977). Instead, these later studies found a strong negative correlation between church attendance and delinquency: teenagers who went to church more frequently were less likely to engage in delinquency than those who didn't.

Later, Stark, Kent and Doyle (1982) noted that the initial two studies on religiosity and delinquency were conducted in relatively secular communities in Redmond, California, while the latter three studies were conducted in highly religious communities in Atlanta and Mormon-dominated communities in Southern Idaho and Utah, respectively. Perhaps, they speculated, the effect of church attendance on delinquency was moderated by the religious behavior of others in one's community. Specifically, they argued that if we take a more social view of human affairs, it becomes plausible to argue that religion only serves to bind people to the moral order if religious influence permeates the culture and the social interactions of the individuals in question (1982, 7). Where the religious sanctioning system isn't pervasive, the effects of an individual's religious commitment will be muffled and curtailed. This is clearly a conditional claim: the consequences of an individual's religiosity on that person's delinquency depends on their social context. Consistent with their conditional hypothesis, Stark, Kent and Doyle (1982) found that attending church reduces delinquency among youths whose classmates are frequent church attenders but has no discernible effect on youths whose classmates don't attend church.

Despite the fact that Stark, Kent and Doyle (1982) were clearly following Durkheim's dictate to study social behavior (delinquency) as a product of individual behavior (church attendance) and social facts (the level of piety in their surrounding community), studies in subsequent decades repeatedly attempted to challenge their conditional claim by conducting empirical analyses that attempted to account for the importance of social context by additively including various independent variables to capture an individual's location, school, peer group, religious denomination, and level of alcohol and drug use. These studies produced a variety of findings that obscured the conditional effect that Stark, Kent and Doyle (1982) had hypothesized because they failed to evaluate the effect of these additional social facts with an interactive model specification. Stark (1996) responded by showing that the negative correlation between an individual's church attendance and subsequent "troubles with the law" was strong in regions of the country (East, Midwest, and South) where church membership was high (about 60%), non-existent in the Pacific region where it was low (36%), and modest in the Mountain region, where church membership was moderate (48%).

Given the tremendous influence of Durkheim on the discipline, the fact that social context matters should not be surprising for sociologists. Despite this, studies that examine the modifying effects of social context on the behavior of individuals are actually quite rare in sociology and, perhaps, even more so in social psychology and behavioral political science. Consider voting studies. It's commonplace to consider the effect of demographic information such as ethnic group membership on vote choice or political attitudes. But this is typically done in nationally representative samples where individuals are abstracted from their social context. This is surprising since, surely, it means something different to be a Korean American, for example, in Tuscaloosa, Alabama than it does in Queens, New York or Los Angeles, California, or a Cuban American in College Station, Texas rather than Union City, New Jersey or Miami, Florida.

We have argued that many of the hypotheses we can derive from our theories across a broad array of topics throughout the social sciences will be context-dependent. In economics and political economy, actors' policy preferences are likely to be the product of the types of assets they own and how abundant those assets are in an economy. Institutional arguments common in both political science and economics are likely to point to the way that relationships between variables differ across institutional contexts. Finally, arguments about culture, and indeed, if one follows Durkheim, perhaps all sociological arguments, are likely to involve claims that are context-dependent.

This book recommends best practices in formulating contextual theories and testing context-dependent hypotheses. Our overarching argument is that social scientists should work hard to make the contextual aspects of their theories as clear as possible, they should deduce as many implications from those theories as possible, be as clear as possible regarding the quantities of interest about which their theories make predictions, and present their findings in a manner that clearly captures the degree of uncertainty we have about those quantities of interest. To provide concrete examples of the practices we recommend, we'll provide myriad examples across a broad range of research questions, many of which present new empirical findings.

1.4 PLAN OF THE BOOK

The book is arranged in three parts. The first part of the book looks at a number of fundamental issues that arise when testing conditional claims involving two interacting variables in the context of a continuous dependent variable. In Chapter 2, we provide guidance on how to derive context-dependent hypotheses from social scientific theories and present them in ways to capture as many falsifiable predictions as possible. We also explain why multiplicative interaction models are well-suited to test conditional claims. Chapter 3 provides recommendations for the specification of interaction models, while Chapter 4 indicates best practices when it comes to interpreting and communicating the results of interaction models. We end the first part of the book on the fundamentals of interaction models with three substantive applications in Chapter 5 that show how to put our recommendations into practice. The substantive applications cover interaction effects involving different combinations of dichotomous and continuous independent variables. The first application looks at how race and gender interacted to affect support for the Republican Party in the 2016 presidential elections in the United States. The second examines how ideology and race combined to affect support for President Barack Obama during the 2012 US presidential elections. And the third application investigates how supply-side and demand-side factors interact to influence women's legislative representation around the world.

The first part of the book focuses on theories that posit interaction between two independent variables on a continuous dependent variable. Not all of the theories in which we're interested, though, are as simple as these. In the second part of the book, we begin to look at some more theoretically complex forms of conditionality, still in the context of a continuous dependent variable. In Chapter 6, we turn our attention to theories that imply that the effect of an independent variable depends on the value of more than one other modifying variable. As we'll see, much depends on whether the modifying effects of these other variables are "independent" or "dependent." To illustrate the case where the modifying effects are independent, we employ a substantive application looking at how gender, education, and age interact to affect support for feminism. And to illustrate the case where the modifying effects are dependent, we revisit our substantive application looking at the determinants of female legislative representation by examining how the interactive effect of supply-side and demand-side factors is modified by a country's regime type. In Chapter 7, we look at theories that imply that the effect of a variable depends not on the value of another variable but on its own value. We might refer to this type of interaction as a "self-interaction." In effect, we examine theories that predict a non-linear relationship between an independent variable and a dependent variable. As we explain, a common approach to testing the conditional implications of these types of theories involves using some kind of polynomial regression. Our substantive application looks at how a party's ideological position affects its use of emotive language during election campaigns in Europe. The theories that we've examined up to this point have all assumed a linear interaction effect where the effect of some independent variable varies in a *linear* way with the value of some modifying variable. As we demonstrate, polynomial regression models allow us to relax this assumption.

The book has so far focused on theories with conditional implications for a continuous dependent variable. However, not all of our theories deal with a continuous dependent variable. In the third part of the book, we begin to look at interaction models in the context of limited dependent variables. A limited dependent variable is one where the range of values is restricted in some important way. In Chapter 8, we look at how to evaluate the conditional implications of our theories when we have a dichotomous or binary dependent variable. Our substantive application examines the factors that influence the formation of pre-electoral coalitions at election time. In Chapters 9 and 10, we extend our coverage of how to evaluate conditional claims to other discrete choice models. In Chapter 9, we look at the case where we have an ordered dependent variable. Building on an earlier substantive application, we examine how ideology and race interacted to affect the presidential approval of Barack Obama in the 2012 US presidential elections. In Chapter 10, we look at the case where we have an unordered dependent variable. Our substantive application this time examines how ideology and gender combined to affect vote choice in the 1992 legislative elections in the United Kingdom.

PART I

The Fundamentals

2 Theories and Their Conditional Implications

Interaction models allow us to test the conditional implications of our theories. In the next few chapters of this book, we're going to talk about how to specify and interpret interaction models. In this chapter, though, we're going to start by discussing theories that have conditional implications. How do we know when we have a theory with conditional implications? How do interaction models help us test these conditional implications? What are the best practices for stating conditional hypotheses? As we'll see later in the book, a firm understanding of what's going on in a theory with conditional implications is key to correctly specifying and interpreting interaction models. For example, it's only by carefully thinking through the logic of our theories and their conditional implications that we can identify the key quantities of interest that would falsify our claims.

2.1 HOW DO I KNOW WHEN MY THEORY POSITS A CONDITIONAL RELATIONSHIP?

Suppose you have an idea about how some aspect of the world works. How do you know if you have a theory with conditional implications? Put differently, how do you know if you should use an interaction model to test your idea? When you find yourself in this position, it's often a good idea to think about what words or phrases you're using to convey your theoretical claims about the world – they can often tell you whether your theory posits a conditional relationship and therefore whether the use of an interaction model is appropriate.

Consider the following theoretical claims, all of which posit a conditional relationship:

Gender Pay Gap: On average, women earn less than men for doing the same job. However, women are not a homogenous group, and there are reasons to believe that some women perform better relative to men when it comes to pay than others. Race scholars, for example, argue that the gender pay gap is larger for Black women than White women. In other words, they argue that race <u>modifies</u> or <u>moderates</u> the effect of gender on pay. In this particular case, being Black is expected to <u>strengthen</u> the negative effect of being female on an individual's earnings.

Party System Size: According to Duverger's theory, the number of political parties in a country is <u>jointly</u> driven by the level of social diversity in a country and the permissiveness of the electoral system. The more (cross-cutting) cleavages there are in a country, the greater the demand for political parties. The extent to which this social "demand" for political parties is translated into actual political parties <u>depends on</u> the permissiveness of the electoral system. When the electoral system is permissive, social demand translates freely into political parties. When the electoral system isn't permissive, the extent to which social demand is translated into political parties is <u>inhibited</u>.

Democratization: There are many factors that affect whether a country will become a democracy. However, the development of a strong middle class is a <u>necessary condition</u> for the successful emergence of democracy.

Clinical Trial: A medical research company has developed a seratonin-based drug to help reduce the symptoms of depression. The company is beginning a randomized clinical trial to evaluate the effectiveness of its new drug. Due to biological differences between men and women – women naturally produce less seratonin than men – the company expects to find evidence of <u>treatment heterogeneity</u>. Specifically, the company expects that the drug will produce a positive treatment effect for women but that this treatment effect will be <u>attenuated</u> for men.

The words and phrases that we have underlined signal that a conditional claim is being made and that some kind of interaction model is appropriate. If you find yourself using these sorts of words and phrases in your own theory, or see them being used in the theory of others, then you should immediately think of interaction models.

> ⚠ **Important:** You can often identify a theory with conditional implications by the words that it uses. If you see words like "depends on," "strengthen," "conditions," "augment," "only if," "attenuate," "weaken," "modify," "inhibit," "moderate," "necessary condition," "sufficient condition," "jointly," or "treatment heterogeneity," then there is a good chance that the theory has conditional implications.

In addition to thinking about the words and phrases that are used when discussing our theoretical ideas, it can also be useful to think about how we might graphically represent our theoretical story. In Figure 2.1a, we graphically present the basic structure of a simple theory with conditional implications. There's some factor, X, that we think affects some outcome of interest, Y. This is indicated by the horizontal arrow from X to Y. There's also some other factor, Z, that we think changes *how* – the direction or degree to which – X affects Y. This is indicated by the vertical arrow

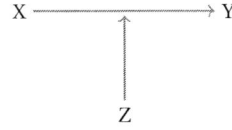

(a) Simple Conditional Relationship

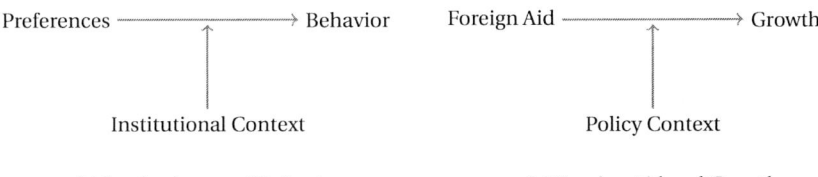

(b) Institutions and Behavior   (c) Foreign Aid and Growth

Figure 2.1 Possible causal relationships: *Moderation*

Note: Figure 2.1 shows three possible causal relationships that involve moderation. Panel (a) shows a causal relationship in which the effect of X on Y is "moderated by" or "depends on" Z. Panel (b) shows a causal relationship in which the effect of preferences on behavior is moderated by the institutional context. Panel (c) shows a causal relationship in which the effect of foreign aid on economic growth is moderated by the policy context in a country.

from Z to the horizontal arrow going from X to Y. We often refer to Z as a "moderating," "modifying," or "conditioning" variable. In some ways, we can think of a modifying variable like Z as being similar to the volume control on a radio in that it changes the "volume" of the effect of X on Y – it can turn the volume up, turn it down, turn it on, turn it off, and so on. This is why words like "strengthen," "augment," "attenuate," "weaken," "modify," "inhibit," and "moderate" are often used when describing conditional relationships. Unlike a volume control, though, a modifying variable can also change the direction of the effect of X on Y from negative to positive or positive to negative.

In Figure 2.1b and 2.1c, we graphically present two stories that involve conditional claims. The generic story captured in Figure 2.1b highlights the importance of combining institutional and behavioral approaches when trying to explain the social world. The basic idea is that human behavior is driven by our preferences. However, the precise way in which our preferences are translated into actual behavior *depends on* the institutional context in which we live. For example, a voter who has a preference for environmental protection may choose to vote sincerely for a small Green party if the electoral institutions are proportional but strategically for a large center-left party if the electoral institutions are majoritarian. This is because a vote for a small party is less likely to be wasted in a proportional system where parties are rewarded in proportion to the votes they receive than it is in a winner-take-all majoritarian system where an additional vote leads to more representation only if it grants the party a plurality. As another example, consider an individual who holds anti-government preferences. The extent to which this person reveals their true preferences and engages in anti-government behavior is likely to *depend on* whether the institutional environment in which they live is open and permissive rather than closed and repressive. The causal story presented in Figure 2.1c suggests that the extent to which foreign aid promotes economic growth *depends on* the policy context in the recipient country. It may be the case, for example, that foreign aid promotes economic growth only if the government adopts "good" policies. If the government does not adopt good policies, then the foreign aid may be wasted, and there'll be no improvement in economic growth.

If your idea about how some aspect of the world works involves a causal relationship like those shown in Figure 2.1, then your theory has conditional implications that can be tested with an interaction model. Before we explain why interaction models can be used to test conditional claims about the world, we want to first clear up one area of common confusion.

2.2 THE DIFFERENCE BETWEEN MODERATION AND MEDIATION

All of the causal relationships shown in Figure 2.1 involve some kind of *moderation*. Some variable Z "moderates" the effect of some other variable X on some outcome of interest Y. It's this moderation that signals that we have a conditional relationship. In our experience, people often get confused when trying to distinguish between conditional relationships that involve *moderation* and unconditional relationships that involve *mediation*. This is quite important. While interaction models are appropriate for testing claims of moderation, they're not appropriate for testing claims of mediation.

> ⚠ **Important:** If your theory involves mediation, rather than moderation, then interaction models are not appropriate for testing your theoretical claims. Interaction models are appropriate for testing theoretical claims that involve *moderation*.

So, what's the difference between claims of moderation and claims of mediation? Just as there are certain words and phrases that signal

(a) Ethnic Diversity and Civil War: No Mediation

(b) Ethnic Diversity and Civil War: Complete Mediation

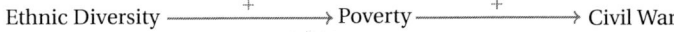

(c) Ethnic Diversity and Civil War: Partial Mediation

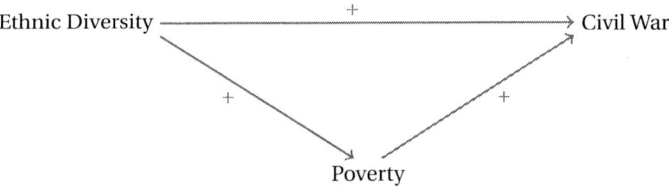

Figure 2.2 Possible causal relationships: *Mediation*

Note: The arrows in Figure 2.2 indicate three types of causal relationships involving mediation. The '+' indicates that the causal effects are positive. Figure 2.2a shows a causal relationship in which ethnic diversity has an unmediated effect on civil war onset. Figure 2.2b shows a causal relationship in which the effect of ethnic diversity on civil wars is *completely mediated* by the level of poverty in a country. Figure 2.2c shows a causal relationship in which the effect of ethnic diversity on civil wars is *partially mediated* by the level of poverty in a country.

that we're making claims of moderation, there are also certain words and phrases that signal that we're making claims of mediation. If you find yourself talking about "causal mechanisms," "causal paths," "channels," "conduits," "proximal and distal causes," "direct and indirect effects," "intervening" factors, or "mediating" variables, then you're talking about claims of mediation. It's very important to make sure that you don't use a word or phrase that signals mediation, like "intervening" or "mediating," if you're really trying to posit a causal claim involving moderation.

Causal relationships that involve mediation look quite different from those that involve moderation. In Figure 2.2, we graphically examine different types of claims involving mediation. In claims of mediation, we typically have some factor, X, whose effect on some outcome of interest, Y, is partially or completely mediated by some other variable, Z. We often refer to Z as a "mediating" or "intervening" variable. A mediating variable indicates how or why a particular relationship exists between X and Y; it provides a mechanism or path by which X affects Y. In Figure 2.2, we look at different ways of thinking about how the impact of ethnic diversity on the probability of civil war onset might be partially or completely mediated by the level of poverty in a country. Figure 2.2a represents a story in which ethnic diversity has a direct and unmediated positive effect on civil wars. This is indicated by the positive sign "+" and the unbroken horizontal arrow from ethnic diversity to civil war. Figure 2.2b proposes a different story in which ethnic diversity has a positive effect on civil wars but only through its positive impact on poverty, which positively affects civil wars. This is indicated by the horizontal arrow from ethnic diversity to poverty and the horizontal arrow from poverty to civil war. In this particular story, we say that the positive effect of ethnic diversity on civil wars is completely mediated by the level of poverty in a country. We might also say in this story that ethnic diversity has only an indirect effect on civil wars through its impact on poverty, or that ethnic diversity is a distal cause of civil wars while poverty is a proximal cause of civil wars. Figure 2.2c offers a third story, in which ethnic diversity has a positive direct effect on civil wars *and* a positive indirect effect on civil wars by increasing poverty. The direct effect is indicated by the horizontal arrow from ethnic diversity to civil wars, while the indirect effect is indicated by the sequence of arrows that reach civil war via poverty. In this particular story, the impact of ethnic diversity on civil wars is only partially mediated by the level of poverty in a country. We might also say in this story that ethnic diversity is both a distal and a proximal cause of civil wars.

⚠ **Important:** A moderating or modifying variable changes the strength or direction of the relationship between an independent variable and a dependent variable. A mediating or intervening variable provides a mechanism or path by which an independent variable and a dependent variable are related.

If you compare the causal relationships in Figure 2.2 with those we saw earlier in Figure 2.1, it should be clear that arguments involving mediation are quite different from arguments involving moderation. If you determine that the implications of your theory involve mediation, rather than moderation, then interaction models are not appropriate for testing your theoretical claims. Instead, you should consult the literature that focuses on testing claims of mediation (Baron and Kenny, 1986; MacKinnon, 2008; Imai et al., 2011). If, however, you determine that the implications of your theory involve moderation, then you're making conditional claims, and interaction models are an appropriate tool for your empirical analyses. In the next section, we look at precisely why interaction models are appropriate for testing theories with conditional implications.

2.3 HOW DOES A MULTIPLICATIVE TERM CAPTURE CONDITIONALITY?

To provide some substantive context for our discussion, let's consider the relationship between foreign aid and economic growth. For a long time, observers argued that foreign aid should have a positive effect on economic growth in recipient countries. This claim is captured in the *Unconditional Aid Hypothesis*.

Unconditional Aid Hypothesis: An increase in foreign aid boosts economic growth in the recipient country.

We can test this hypothesis with a standard linear regression model,

$$Growth = \alpha_0 + \alpha_1 Aid + \upsilon, \tag{2.1}$$

where *Growth* is a measure of economic growth, *Aid* is a measure of foreign aid, α_0 is the constant term, α_1 is a coefficient that captures the effect of an additional dollar in foreign aid on economic growth, and υ is the error term.[1] According to the *Unconditional Aid Hypothesis*, we expect that $\alpha_1 > 0$. In this model specification, the effect of foreign aid on economic growth is unconditional in the sense that it doesn't depend on the value of any other variable. Figure 2.3 graphically illustrates a linear model that's consistent with the claim that foreign aid boosts economic growth; we've assumed that $\alpha_0 > 0$. As you can see, α_1 captures the slope relationship between foreign aid and economic growth.

In an influential paper, Burnside and Dollar (2000) challenged this prevailing wisdom regarding the expected relationship between foreign aid and economic growth. In line with the conditional causal story we saw earlier in Figure 2.1c, Burnside and Dollar argue that foreign aid doesn't have an unconditional effect on economic growth. Instead, they claim that the effect of foreign aid *depends on* the policy environment in the recipient

---

[1] For simplicity, we'll assume that there are no other independent variables in the model.

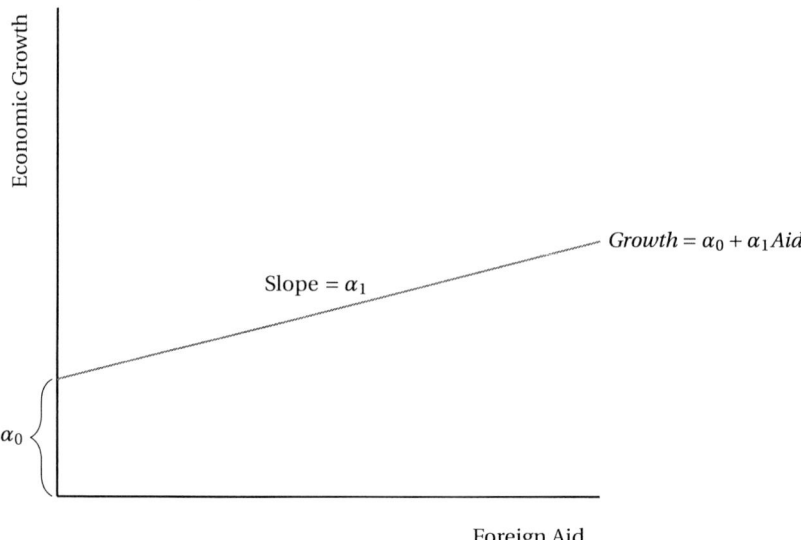

Figure 2.3 Graphical illustration of a linear-additive model consistent with the *Unconditional Aid Hypothesis*

country. Specifically, they believe that foreign aid will have a positive impact on economic growth only when the recipient country adopts the kinds of "good" policies recommended by economists at the World Bank (like them). If the recipient country adopts "bad" policies, then foreign aid will be wasted and have no effect on economic growth. Burnside and Dollar's claim is captured in the *Conditional Aid Hypothesis*.

Conditional Aid Hypothesis: An increase in foreign aid increases economic growth in the recipient country if the policy environment is good but has no effect if the policy environment is bad.

What model specification can we use to test the *Conditional Aid Hypothesis*? You might think that we just need to add a measure of the policy environment in the recipient country to our original regression model,

$$Growth = \gamma_0 + \gamma_1 Aid + \gamma_2 Good\ Policy + v, \qquad (2.2)$$

where *Good Policy* is a measure of the extent to which the recipient country has adopted the types of policies recommended by the World Bank. In this way, we could "control" for the policy environment in the recipient country. Adding a variable to a linear model in an additive manner like this, though, is never an appropriate strategy for testing a conditional claim about the effect of a variable. The model specification shown in Eq. 2.2

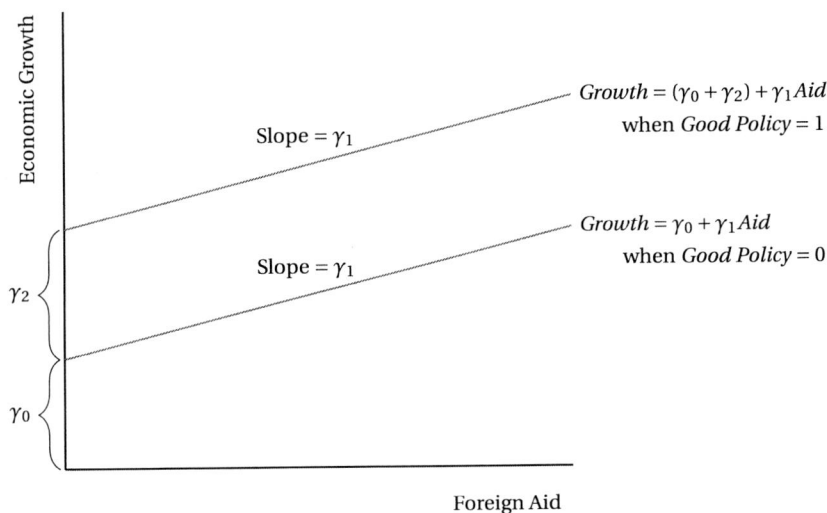

Figure 2.4 Graphical illustration of a linear-additive model consistent with Eq. 2.2 where $\gamma_0, \gamma_1, \gamma_2 > 0$

can't test the *Conditional Aid Hypothesis* because the effect of foreign aid on economic growth is constant, γ_1, and doesn't allow for the possibility that the effect of foreign aid may vary with the policy environment in the recipient country.[2]

To see what's going on visually, consider Figure 2.4 in which we present a graphical illustration consistent with the model specification shown in Eq. 2.2 where $\gamma_0, \gamma_1, \gamma_2 > 0$. For simplicity, we've assumed that our measure of *Good Policy* is dichotomous and takes the value of 1 when the policy environment in the recipient country is good and 0 otherwise. As you can see, economic growth is certainly higher when the policy environment is good (*Good Policy* = 1) than when the policy environment is bad (*Good Policy* = 0). This is because we assumed that $\gamma_2 > 0$. However, the key thing to note is that although the intercept of the two regression lines is allowed to vary with the quality of the policy environment, this isn't the case for the slope relationship between economic growth and foreign aid. The slope relationship is always the same, γ_1, and doesn't (can't) vary depending on whether the policy environment is good or bad. Entering

---

[2] The estimated effect of foreign aid on economic growth may well be different in the model where we control for the policy environment in the recipient country (Eq. 2.2) than in the model where we don't do this (Eq. 2.1). This is why we changed the Greek letter used to denote our parameters across the two models. In each model, though, the effect of foreign aid on economic growth is constant and doesn't depend on the value of any other independent variable.

Good Policy into our model in an additive manner "controls" for the policy environment in the recipient country but doesn't allow us to test whether the effect of foreign aid depends on, or changes with, the policy environment. The bottom line is that the linear-additive specification shown in Eq. 2.2 isn't appropriate for testing the *Conditional Aid Hypothesis*. The problem is that while "controlling" for *Good Policy* makes the intercept conditional on the quality of the policy environment, it doesn't make the slope conditional on it (which is what the *Conditional Aid Hypothesis* implies). The broader point here is that you can't test conditional claims about the effects of variables using linear models in which the variables of interest have been entered in an additive manner.

> ⚠ **Important:** Linear-additive models in which the independent variables have been entered in a purely additive manner cannot test conditional claims about their effects.

To test the *Conditional Aid Hypothesis*, we must enter the variables *Aid* and *Good Policy* into our model in an interactive, rather than additive, manner. This means adding an interaction term, *Aid* × *Good Policy*, to our model specification. The interaction term is created by multiplying the *Aid* and *Good Policy* variables together. This explains why the new variable, *Aid* × *Good Policy*, is sometimes referred to as a multiplicative interaction term, a multiplicative term, or a product term. The variables we multiply together, *Aid* and *Good Policy*, are referred to as the constitutive terms, as they are used to create or "constitute" the interaction term. As we'll explain in the next chapter, scholars should include the interaction term and all of its constitutive terms in any interactive model specification:

$$Growth = \beta_0 + \beta_1 Aid + \beta_2 Good\ Policy + \beta_3 Aid \times Good\ Policy + \epsilon.$$
$$(2.3)$$

But how does the inclusion of a multiplicative term help us get at conditionality? Why is the linear-interactive specification shown in Eq. 2.3 appropriate for testing the *Conditional Aid Hypothesis*? Let's continue to examine the case where the policy environment in the recipient country is either good or bad. When the policy environment is bad (*Good Policy* = 0), our model in Eq. 2.3 becomes

$$Growth = \beta_0 + \beta_1 Aid + \beta_2[0] + \beta_3 Aid \times [0] + \epsilon$$
$$= \beta_0 + \beta_1 Aid + \epsilon. \qquad (2.4)$$

This is a linear equation where the intercept is β_0, and the slope with respect to foreign aid is β_1. In other words, the expected effect of an additional

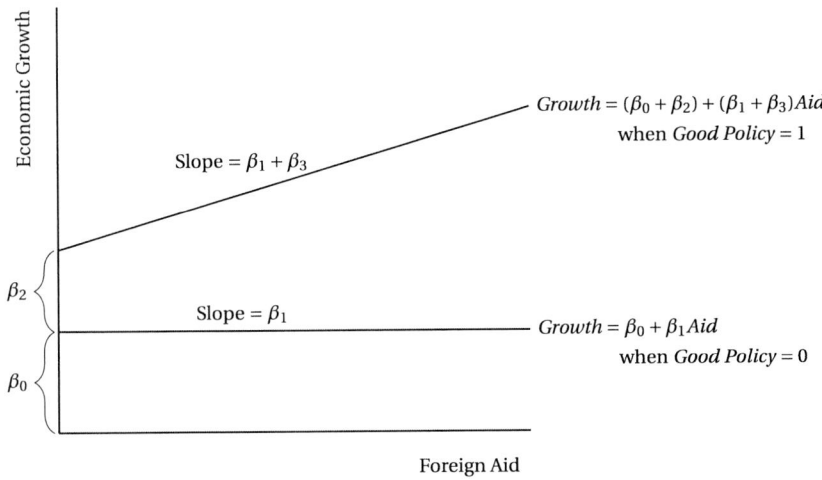

Figure 2.5 Graphical illustration of a linear-interactive model that's consistent with the *Conditional Aid Hypothesis*

dollar of foreign aid on economic growth is β_1 when the policy environment is bad. In contrast, when the policy environment is good (*Good Policy* = 1), our model in Eq. 2.3 becomes

$$Growth = \beta_0 + \beta_1 Aid + \beta_2[1] + \beta_3 Aid \times [1] + \epsilon. \tag{2.5}$$

With a little manipulation, this can be rewritten as

$$Growth = (\beta_0 + \beta_2) + (\beta_1 + \beta_3)Aid + \epsilon. \tag{2.6}$$

This is also a linear equation. Now, though, the intercept is $\beta_0 + \beta_2$, and the slope with respect to foreign aid is $\beta_1 + \beta_3$. In other words, the expected effect of an additional dollar of foreign aid on economic growth is $\beta_1 + \beta_3$ when the policy environment is good. In Figure 2.5, we graphically illustrate a linear-interactive model that's consistent with the *Conditional Aid Hypothesis*; to ease presentation, we've assumed that $\beta_2 > 0$.

It should now be clear that the inclusion of a multiplicative term, *Aid* × *Good Policy*, leads to a model specification that allows for the possibility that the effect of foreign aid on economic growth might be different when the policy environment is good compared to when the policy environment is bad. This is exactly what we need to test the *Conditional Aid Hypothesis*. The claim in the *Conditional Aid Hypothesis* that foreign aid boosts economic growth if and only if the policy environment is good amounts to the claim that $\beta_1 = 0$ and that $\beta_1 + \beta_3 > 0$. Put differently, the claim that the effect of foreign aid on economic growth depends on the policy environment in the recipient country implies that the slope describing the relationship between foreign aid and economic growth varies depending on the policy environment. In the simple case we've been examining where policies are either "good" or "bad," we can see the difference between the two slopes by subtracting one from the other, $[\beta_1 + \beta_3] - [\beta_1] = \beta_3$. A simple t-test on the interaction term coefficient, β_3, can be used to test the null hypothesis that the slope relationship between foreign aid and economic growth doesn't depend on the policy environment in the recipient country.

⚠ **Important:** Analysts should include multiplicative interaction terms in their statistical models whenever they are testing conditional claims about the world.

The basic intuition that comes from looking at a case where we have a dichotomous modifying variable extends easily to the case where we have a continuous modifying variable. Rather than treat *Good Policy* as a dichotomous variable that takes the value 1 when the policy environment in the recipient country is good and 0 otherwise, let's now treat *Good Policy* as a continuous variable that captures the extent to which the policy environment in the recipient country is good. We'll assume that the lowest value of *Good Policy* is 0 and that higher values indicate a better policy environment. Recall that we're still using the linear-interactive model shown in Eq. 2.3.

Rather than use algebra to find the slope relationship between foreign aid and economic growth for different values of *Good Policy* as we did before, we now use a little calculus. Specifically, we find the slope relationship by taking the derivative of *Growth* with respect to *Aid*,

$$ME(Aid|Good\ Policy) = \frac{\partial Growth}{\partial Aid} = \beta_1 + \beta_3 Good\ Policy. \qquad (2.7)$$

This is a linear equation that tells us the marginal effect (ME) of foreign aid on economic growth for every possible value of *Good Policy*.[3] A marginal effect describes a slope relationship; in this case, the slope relationship between foreign aid and economic growth. Since we're dealing with a

---

[3] As we discuss in more detail in Chapter 4, thinking about the effects of our independent variables in terms of derivatives, and hence marginal effects, is incredibly useful. Those who don't know much calculus shouldn't be too concerned, though, as the discussion in Chapter 4 also demonstrates how we can use algebra to think about the effects of our variables in terms of differences. Thinking in terms of a difference involves looking at how the predicted value of Y changes in response to a given change in X. As we explain, thinking in terms of derivatives or differences is essentially equivalent in the context of a linear-interactive model in which the interaction is itself linear. Since the calculation of differences tends to be a little more tedious, we often prefer, as we do here, to think in terms of derivatives and marginal effects.

linear-interactive model in which the interaction is itself linear, the marginal effect shown in Eq. 2.7 tells us how an additional dollar – a one-unit increase in *Aid* – affects economic growth at every possible value of *Good Policy*.[4] It's easy to see that so long as the coefficient on the interaction term isn't zero, $\beta_3 \neq 0$, then the marginal effect of foreign aid depends on, or changes with, the value of *Good Policy*. When *Good Policy* is 0, the marginal effect of foreign aid is $\beta_1 + \beta_3[0] = \beta_1$. It should now be clear that β_1, the coefficient on *Aid* in Eq. 2.3, tells us the marginal effect of foreign aid on economic growth when *Good Policy* is 0. When *Good Policy* is *1*, the marginal effect of foreign aid is $\beta_1 + \beta_3[1] = \beta_1 + \beta_3$. When *Good Policy* is 2, the marginal effect of foreign aid is $\beta_1 + \beta_3[2] = \beta_1 + 2\beta_3$. When *Good Policy* is 3, the marginal effect of foreign aid is $\beta_1 + \beta_3[3] = \beta_1 + 3\beta_3$. And so on. As you can see, the marginal effect of foreign aid on economic growth starts at β_1 when *Good Policy* is 0 and increases by a constant amount β_3, the coefficient on the interaction term *Aid* $\times$ *Good Policy*, for every unit increase in *Good Policy*.

Another way to see how the policy environment modifies the effect of foreign aid on economic growth is simply to take the derivative of the marginal effect of foreign aid on economic growth with respect to *Good Policy*,

$$\frac{\partial \left(\frac{\partial Growth}{\partial Aid} \right)}{\partial Good\ Policy} = \frac{\partial^2 Growth}{\partial Good\ Policy\ \partial Aid} = \beta_3. \qquad (2.8)$$

This tells us the marginal effect of *Good Policy* on the marginal effect of foreign aid on economic growth. Put differently, it tells us the slope relationship between the marginal effect of *Aid* and *Good Policy*. Again, we see that each unit increase in *Good Policy* changes the marginal effect of *Aid* on *Growth* by β_3. The modifying effect of the policy environment on the marginal effect of foreign aid, β_3, is commonly referred to as the "interaction effect." As we noted earlier, we need only conduct a *t*-test on the interaction term coefficient, β_3, to test the null hypothesis of no conditionality or no interaction. If we were to find that $\beta_3 = 0$, then this would indicate that there's no conditionality or interaction. In the context of our ongoing example, this would indicate that the effect of foreign aid on economic growth doesn't depend on the policy environment in the recipient country.

As we've seen, the use of multiplicative interaction terms provides us with a way to test conditional claims. Before moving on to think about

---

[4] Technically, the marginal effect of an independent variable X on some outcome variable Y only tells us how Y changes when we increase X from some value by an infinitesimally small amount divided by the change in X. In linear-interactive models in which the interaction is itself linear, though, the marginal effect of X on Y can also be interpreted as the effect of a one-unit increase in X.

the issues that arise when specifying and interpreting interaction models, it's worthwhile spending some more time thinking about theories and their conditional implications.

2.4 MOVING BEYOND THE INTERACTION EFFECT

Over the years, we've noticed that scholars often focus on the sign of the coefficient on the interaction term when evaluating a theory with conditional implications. It's important to recognize, though, that the sign of the coefficient on an interaction term is not sufficient on its own to corroborate a conditional claim. This is because any observed "interaction effect" is always consistent with a wide variety of ways in which X and Z interact to determine Y, some of which may be inconsistent with one's underlying theory. Importantly, simply knowing the sign of the coefficient on the interaction term says nothing about whether X has a positive, negative, or zero effect on Y for any value of Z. In effect, proposing and testing only a prediction about the sign of the coefficient on an interaction term constitutes an extremely weak, and often substantively uninformative, test of a conditional claim. As a result, we recommend that scholars always supplement a prediction about the sign of the coefficient on their interaction term with predictions about the sign of the marginal effect of X on Y. Doing so significantly narrows the range of conditional relationships that are consistent with one's underlying theory, thereby strengthening any empirical test.

To illustrate this point, let's return once more to our foreign aid and economic growth example. As you'll recall, Burnside and Dollar (2000) posit a theory in which foreign aid and the policy environment in the recipient country interact to influence economic growth. The *Conditional Aid Hypothesis* speaks to the effect of foreign aid on economic growth. Specifically, it states that an increase in foreign aid increases economic growth in the recipient country if the policy environment is good but has no effect if the policy environment is bad. We can test this hypothesis with the linear-interactive model specification shown in Eq. 2.3. As we saw earlier, the marginal effect of foreign aid on economic growth is

$$ME(Aid|Good\ Policy) = \frac{\partial Growth}{\partial Aid} = \beta_1 + \beta_3 Good\ Policy. \qquad (2.9)$$

According to the *Conditional Aid Hypothesis*, β_1 should be 0 as foreign aid is predicted to have no effect on economic growth when the policy environment is bad. In contrast, foreign aid is predicted to have a positive effect on economic growth if the recipient country adopts the types of "good" policies that are expected to create an environment in which aid is catalytic for growth. The better the policy environment in the recipient country, the larger the expected positive effect of foreign aid on economic growth. This suggests that the marginal effect of *Aid* should increase with higher values of *Good Policy*; that is, the slope of the linear equation in Eq. 2.9 should be positive, $\beta_3 > 0$.

Since the marginal effect of foreign aid in Eq. 2.9 is a linear equation, we can easily plot it. In Figure 2.6, we present a "marginal effect plot" for foreign aid that's consistent with the *Conditional Aid Hypothesis*, $\beta_1 = 0$ and $\beta_3 > 0$. As we discuss in more detail in Chapter 5, scholars often use their estimated model parameters from an interaction model to produce a marginal effect plot similar to the one in Figure 2.6. The goal is to see if the estimated marginal effect plot looks like the one predicted by their theory. If we were to estimate our linear-interactive model and obtain a marginal effect plot for foreign aid that looks like the one in Figure 2.6, we'd claim support for the *Conditional Aid Hypothesis*.

Every conditional claim about how the effect of some variable X on Y is linearly conditioned by another variable Z will, if the conditioning variable Z is continuous, imply a marginal effect plot that's broadly similar to the one shown in Figure 2.6. By broadly similar, we mean that the predicted marginal effect of X on Y will always vary in a linear way with Z. The presence of an interaction implies that the slope of the marginal effect plot won't be flat or zero. In almost all cases, a theory positing interaction will make a prediction as to whether the slope of the marginal effect plot

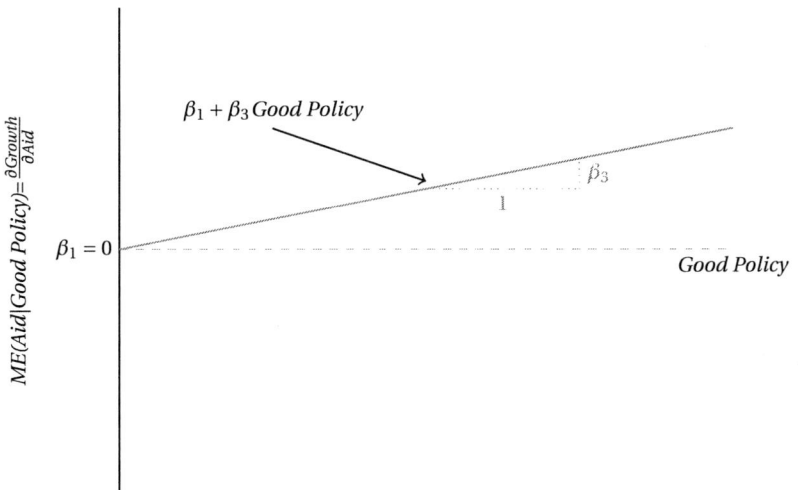

Figure 2.6 Marginal effect plot for *Aid* that's consistent with the *Conditional Aid Hypothesis*, $\beta_1 = 0$ and $\beta_3 > 0$

will be positive or negative. The *Conditional Aid Hypothesis*, for example, predicts that the slope of the marginal effect plot for *Aid* will be positive. As mentioned earlier, the slope of the marginal effect plot – the interaction effect – is determined by the coefficient on the interaction term, β_3.

Throughout our careers, we've noticed that scholars often focus their theoretical argument on predicting the sign of the "interaction effect" and that they claim support for their theory if they find that the estimated coefficient on the interaction term has the predicted sign and is statistically significant. This is problematic, though, because a conditional claim about how the effect of X on Y varies with Z can be wrong even if the coefficient on the interaction term has the predicted sign. The basic issue here is that knowing the sign of the coefficient on the interaction term – the slope of the marginal effect plot for X – isn't sufficient to fully characterize the effect of X on Y. In particular, seeing that the slope, β_3, is positive or negative provides no information about the value of the intercept, β_1, in the linear equation depicting the marginal effect of X. It follows that knowing the sign of the interaction effect establishes neither the sign (positive or negative) nor the magnitude of the marginal effect of X on Y at any value of Z. This is extremely important as different values for the intercept β_1 imply quite different ways in which the marginal effect of X is conditional on Z and thus quite different ways in which X and Z interact to determine Y.

In their study, Burnside and Dollar (2000) find that the coefficient on the interaction term between *Aid* and *Good Policy* has the predicted positive sign. The important thing to recognize here is that it would be a mistake to conclude from this that their theory is corroborated. In Figure 2.7, we show three possible marginal effect plots for *Aid* that each have the same positive slope, $\beta_3 = 0.04$. We've assumed, arbitrarily, that the observed values of *Aid* and *Good Policy* range from 0 to 10 in the population of interest. Although the three marginal effect plots all have the same slope, you can easily see that each plot tells a very different story about the conditional relationship between *Aid* and *Growth*. These different stories arise because the value of the intercept, β_1, is different in each plot.

Only the marginal effect plot shown in Figure 2.7a, where the intercept $\beta_1 = 0$, is consistent with the *Conditional Aid Hypothesis*. In this plot, the marginal effect of *Aid* on *Growth* is 0 when *Good Policy* is at its lowest value and $0 + 0.04 \times 10 = 0.40$ when *Good Policy* is at its highest value. In this scenario, a good policy environment "facilitates" a positive relationship between foreign aid and growth. In contrast, the marginal effect plot shown in Figure 2.7b, where the intercept $\beta_1 = -0.20$, is inconsistent with the *Conditional Aid Hypothesis*. In this plot, the marginal effect of *Aid* on *Growth* is -0.20 when *Good Policy* is at its lowest value

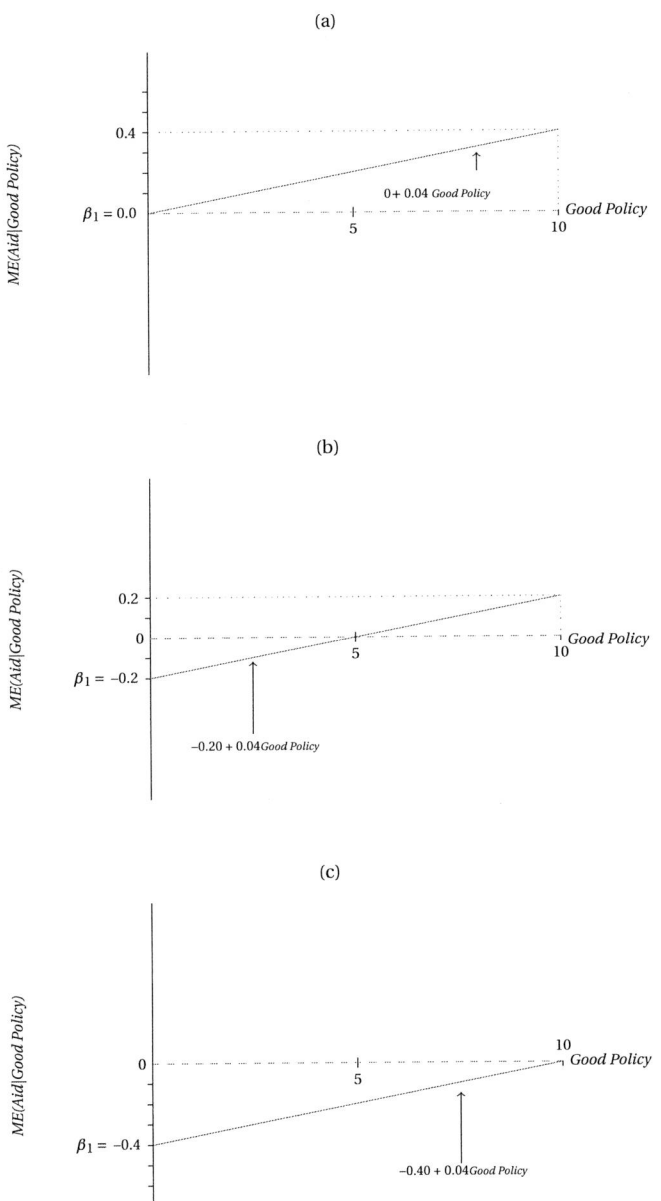

Figure 2.7 (a)–(c) Three very different marginal effect plots for *Aid* that each have the same slope, $\beta_3 = 0.04$

Note: The three plots show the marginal effect of *Aid* on *Growth*, $\beta_1 + \beta_3$ *Good Policy*, across the possible values of *Good Policy*. Although these marginal effect plots all share the same slope, $\beta_3 = 0.04$, they differ in the value of their intercept, β_1, and therefore represent quite different conditional relationships between *Aid* and *Growth*. Only the plot shown in panel (a) is consistent with the *Conditional Aid Hypothesis*.

and $-0.20 + 0.04 \times 10 = 0.20$ when *Good Policy* is at its highest value. In this scenario, a good policy environment "transforms" the relationship between foreign aid and economic growth from negative to positive. In effect, foreign aid reduces growth when the policy environment is bad but increases growth once the policy environment is sufficiently good. The marginal effect plot shown in Figure 2.7c, where the intercept $\beta_1 = -0.40$, is almost entirely at odds with the *Conditional Aid Hypothesis*. In this plot, the marginal effect of *Aid* on *Growth* is -0.40 when *Good Policy* is at its lowest value and $-0.40 + 0.04 \times 10 = 0$ when *Good Policy* is at its highest value. In this scenario, a good policy environment "inhibits" an otherwise negative relationship between foreign aid and growth.

⚠ **Important:** Scholars testing a theory with conditional implications should never focus solely on deriving and testing a prediction about the sign of the coefficient on an interaction term. This is because any observed interaction effect is always consistent with a wide variety of ways in which X and Z interact to determine Y, some of which may be inconsistent with one's underlying theory.

The issue we're raising here can be substantively important and have significant policy implications. As Figure 2.7 demonstrates, identifying that the coefficient on the interaction term is positive isn't sufficient to know whether the data support Burnside and Dollar's theory or some alternative story such as the one depicted in Figure 2.7c. This is critical because the plot shown in Figure 2.7c suggests that foreign aid never promotes economic growth and that the best good policies can do is inhibit foreign aid's negative effect on development. The policy implication to be drawn from Figure 2.7c is that donor governments should always reduce or eliminate foreign aid. Since economic growth can have a profound effect on important outcomes, such as infant mortality and life expectancy, knowing whether one is in the world where good policies facilitate the positive effects of foreign aid (Figure 2.7a) or the one in which good policies can only inhibit the negative effects of foreign aid (Figure 2.7c) can be a matter of life and death. We can't distinguish between these two worlds if we only know the sign of the coefficient on the interaction term. This is why we encourage scholars to always supplement any prediction about the sign of the coefficient on their interaction term with predictions about the sign of the marginal effect of X on Y.

While it would be ideal if our theories were strong enough to predict the precise magnitude of the effect of X on Y at every possible value of Z, this is rarely the case in the social sciences. Almost all theories with conditional implications, though, are strong enough to generate three predictions about the effect of X on Y and how this varies with Z. These predictions speak to the direction of the interaction between X and Z and the direction of the marginal effect of X on Y.

1. P_{XZ}: The marginal effect of X on Y is [positively, negatively] related to Z.
2. $P_{X|Z_{min}}$: The marginal effect of X on Y is [positive, negative, zero] when Z is at its *lowest* value.
3. $P_{X|Z_{max}}$: The marginal effect of X on Y is [positive, negative, zero] when Z is at its *highest* value.

By calling on scholars to make predictions about the marginal effect of X on Y when Z is at its lowest and highest values, we don't mean to imply that analysts should necessarily focus their greatest attention on the *estimated* marginal effects at these extreme values. Instead, we emphasize predictions at the extremes due to the fact that if one assumes linearity, as in Eq. 2.3, or at least monotonicity, such predictions automatically imply predictions at values between the extremes.

The three predictions that come from Burnside and Dollar's conditional theory about the relationship between foreign aid and economic growth are:

1. $P_{Aid \times Good\ Policy}$: The marginal effect of *Aid* on *Growth* is positively related to *Good Policy*.
2. $P_{Aid|Good\ Policy_{min}}$: The marginal effect of *Aid* on *Growth* is zero when *Good Policy* is at its *lowest* value.
3. $P_{Aid|Good\ Policy_{max}}$: The marginal effect of *Aid* on *Growth* is positive when *Good Policy* is at its *highest* value.

All three of these predictions are contained in our *Conditional Aid Hypothesis*.

It's important for scholars to make all three of the predictions we've suggested as these predictions are necessary to distinguish between the different possible conditional relationships that might exist between X and Y. In other words, scholars must make and evaluate all three predictions if they wish to fully corroborate a conditional claim and distinguish it from alternative competing stories. To see this, note that our proposed predictions can be represented as a set with three elements: $\{P_{XZ}, P_{X|Z_{min}}, P_{X|Z_{max}}\}$. For example, the *Conditional Aid Hypothesis* can be represented by the prediction set $P = \{$Positive, Zero, Positive$\}$. This set indicates that the marginal effect of *Aid* on *Growth* is predicted to be *positive* with respect to *Good Policy*, *zero* when *Good Policy* is at its lowest value, and *positive* when *Good Policy* is at its highest value.

Different elements in the prediction set describe different conditional relationships between X and Y. There are eighteen possible combinations of elements in the set of predictions when we have a theory positing interaction between X and Z. However, only ten of these combinations are logically consistent.[5] For example, the set {Positive, Positive, Negative} is contradictory. This is because if the effect of X is predicted to be increasing in Z, as the first element in the set indicates, then it's impossible for the predicted effect of X to be positive at Z's lowest value and negative at Z's highest value. We saw three of the possible conditional relationships between X and Y earlier in Figure 2.7. In Figure 2.8, we show all ten of the possible conditional relationships between X and Y. The panels on the left show the possible conditional relationships between X and Y that involve a positive interaction effect, while the panels on the right show the possible conditional relationships that involve a negative interaction effect.

Looking at the left column in Figure 2.8, panel (a) describes a conditional relationship in which the modifying variable Z "reinforces" the positive effect of X on Y, panel (b) describes a conditional relationship in which Z "facilitates" the positive effect of X on Y, panel (c) describes a conditional relationship in which Z "transforms" the effect of X on Y from negative to positive, panel (d) describes a conditional relationship in which Z "inhibits" the negative effect of X on Y, and panel (e) describes a conditional relationship in which Z "mitigates" the negative effect of X on Y. Similar language can be used to describe the conditional relationships shown in the right column. Each of the panels shown in Figure 2.8 depicts a distinctly different conditional relationship between X and Y. Only by deriving and testing all three of the predictions we've suggested is it possible for us to know whether the data support our particular conditional claim as opposed to one of the possible alternative conditional relationships shown in Figure 2.8.

In this section, we've argued that scholars should look beyond the interaction effect when testing the conditional implications of their theories. In particular, we've encouraged scholars to make and test three predictions that relate to the marginal effect of X on Y and how this varies with Z. In the next section, we highlight the symmetry of interaction and discuss its

---

[5] In a conditional hypothesis, the element P_{XZ} can take on two possible values (positive and negative), the element $P_{X|Z_{min}}$ can take on three possible values (positive, negative, zero), and the element $P_{X|Z_{min}}$ can also take on three possible values (positive, negative, zero). This means that there are $2 \times 3 \times 3 = 18$ possible combinations. Eight combinations, though, are logically inconsistent: {Positive, Positive, Negative}, {Positive, Positive, Zero}, {Positive, Zero, Zero}, {Positive, Zero, Negative}, {Negative, Negative, Positive}, {Negative, Negative, Zero}, {Negative, Zero, Zero}, and {Negative, Zero, Positive}.

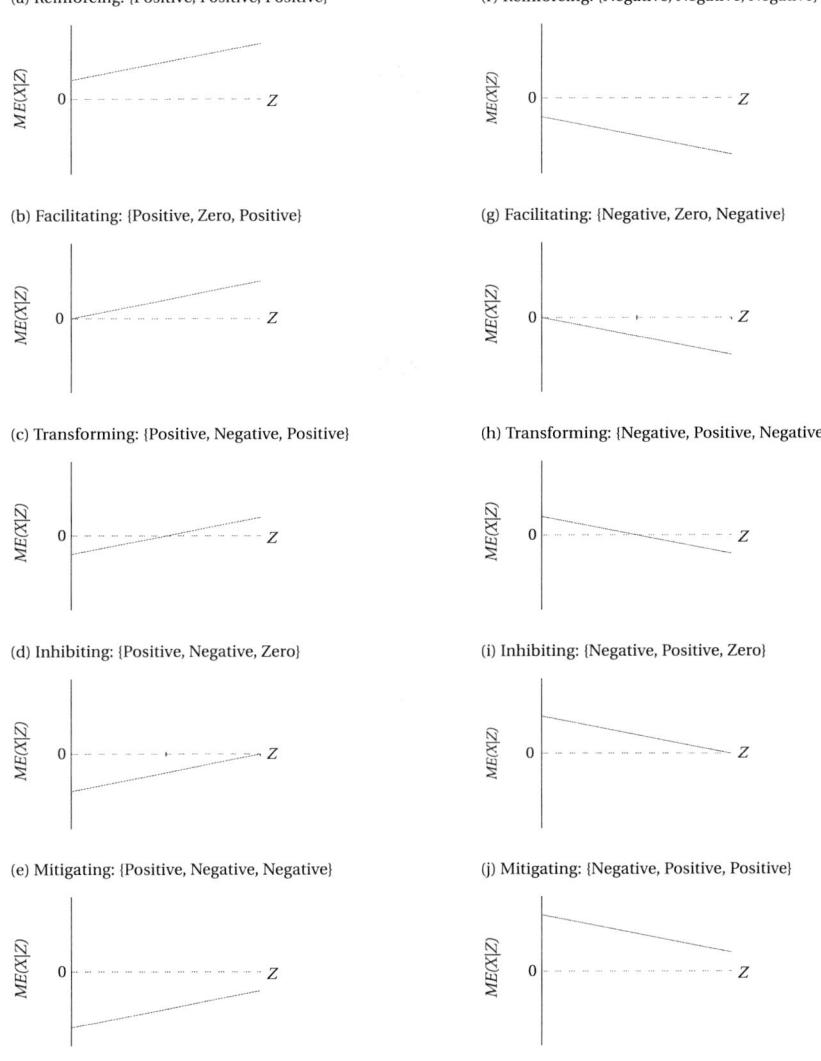

Figure 2.8 (a)–(j) Ten possible conditional relationships between X and Y in a theory positing interaction between X and Z

Note: Each plot indicates a possible conditional relationship between X and Y that could come from a theory positing interaction between X and Z. Each plot is based on a set of predictions with three elements: $\{P_{XZ}, P_{X|Z_{min}}, P_{X|Z_{max}}\}$.

implications for hypothesis testing. Perhaps the most important implication is that scholars should, when possible, also derive and test predictions about the marginal effect of Z on Y.

2.5 THE SYMMETRY OF INTERACTION AND ITS IMPLICATIONS FOR THEORY TESTING

Many scholars conceive of the two variables in a theory positing interaction as playing different roles. One variable, we'll call it Z, is typically thought of as the moderating or modifying variable, the role of which is to modify the impact of the other variable, X, on the dependent variable, Y. While it's perfectly reasonable when we have two variables that interact to think that Z modifies the effect of X on Y, the inherent symmetry of interactions means that it's a mathematical truism that X also modifies the effect of Z on Y (Brambor, Clark and Golder, 2006; Kam and Franzese, 2007; Berry, Golder and Milton, 2012). In this sense, Z and X can both be seen as moderating or modifying variables.

Let's think about what this means in terms of our foreign aid example. Burnside and Dollar argue that the policy context in a country modifies the effect of foreign aid on economic growth. This story is depicted in panel (a) of Figure 2.9. Due to the symmetry of interactions, Burnside and Dollar must also accept that foreign aid modifies the effect of a country's policy context on economic growth. This story is depicted in panel (b) of Figure 2.9. The two stories shown in Figure 2.9 might look different, but they logically imply each other.

To see why an interaction is necessarily symmetric, consider the following generic linear-interactive model positing an interaction between X and Z:

$$Y = \beta_0 + \beta_1 X + \beta_2 Z + \beta_3 XZ + \epsilon. \tag{2.10}$$

The marginal effect of X on Y is

$$ME(X|Z) = \frac{\partial Y}{\partial X} = \beta_1 + \beta_3 Z. \tag{2.11}$$

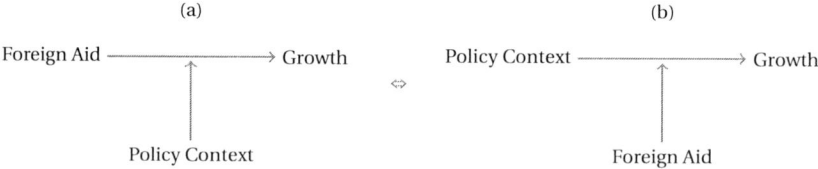

Figure 2.9 The symmetry of interaction

Note: If you accept the story in panel (a) where the policy context in a country modifies the effect of foreign aid on economic growth, then due to the symmetry of interactions you must also accept the story in panel (b) where foreign aid modifies the effect of a country's policy context on economic growth.

So long as the coefficient on the interaction term, β_3, isn't equal to 0, the marginal effect of X on Y depends on the value of Z. In other words, Z modifies the effect of X on Y. Similarly, the marginal effect of Z on Y is

$$ME(Z|X) = \frac{\partial Y}{\partial Z} = \beta_2 + \beta_3 X. \qquad (2.12)$$

Again, so long as the coefficient on the interaction term, β_3, isn't equal to 0, the marginal effect of Z on Y depends on the value of X. In other words, X modifies the effect of Z on Y. As you can see, both X and Z act as modifying variables.

You can also see from the two marginal effects shown in Eq. 2.11 and Eq. 2.12 that the coefficient on the interaction term, β_3, indicates the slope of the relationship between $ME(X|Z)$ and Z *and* the slope of the relationship between $ME(Z|X)$ and X. In other words, the modifying effect of Z on the marginal effect of X is exactly the same as the modifying effect of X on the marginal effect of Z. We can confirm this by taking the derivatives of the two marginal effects shown in Eq. 2.11 and Eq. 2.12 with respect to Z and X, respectively,

$$\frac{\partial \left(\frac{\partial Y}{\partial X} \right)}{\partial Z} = \frac{\partial \left(\frac{\partial Y}{\partial Z} \right)}{\partial X} = \beta_3. \qquad (2.13)$$

Thus, if Z has, say, a positive effect on the marginal effect of X on Y, it automatically follows that X has a positive effect on the marginal effect of Z on Y.[6] Similarly, if Z has a negative effect on the marginal effect of X on Y, it automatically follows that X has a negative effect on the marginal effect of Z on Y. It's logically impossible for a theory positing interaction to predict that the modifying effect of Z on the marginal effect of X on Y will be positive (negative) and that the modifying effect of X on the marginal effect of Z on Y will be negative (positive). An implication of this is that scholars who introduce a hypothesis about how the marginal effect of X on Y varies with Z are also implicitly introducing a hypothesis about how the marginal effect of Z on Y varies with X. This is something to keep in mind when deriving conditional predictions from a theory.

---

[6] The inherent symmetry of interactions, which in mathematics is often referred to as the symmetry of second derivatives or the equality of mixed partial derivatives, isn't specific to the linear-interactive specification in Eq. 2.10; *all* interactive specifications that we're likely to deal with in our empirical research will be symmetric (Kam and Franzese, 2007, 16). According to Schwarz's Theorem (1873), the symmetry of second derivatives always holds if our function has continuous second partial derivatives at the point at which the derivatives are being taken.

⚠ **Important:** All interactions are logically symmetric. If we posit that two variables X and Z interact to affect Y, then Z modifies the effect of X on Y *and* X modifies the effect of Z on Y. As a result, X and Z are both modifying variables. The modifying effect of Z on the marginal effect of X is exactly the same as the modifying effect of X on the marginal effect of Z. The two modifying effects are commonly known as the "interaction effect."

Many scholars fail to acknowledge the inherent symmetry of interactions when testing their theories. This is problematic as it frequently leads them to understate or, more worryingly, overstate the empirical support for their theories (Berry, Golder and Milton, 2012, 653). A common approach when it comes to testing theories that posit interaction between two variables X and Z is to focus on the effect of X on Y and how this varies with Z and to ignore the effect of Z on Y and how this changes with X. In effect, scholars often present and test a hypothesis about the marginal effect of X on Y but not a hypothesis about the marginal effect of Z on Y. In conversation, we've often heard the refrain, "But I'm only interested in the effect of X on Y." Ignoring the effect of Z on Y, though, often throws away potential information that may be useful for testing one's conditional theory.

In the last section, we saw that there are many different ways in which X can be related to Y that are consistent with a particular interaction effect, but that many of these conditional relationships may be inconsistent with one's specific theory. This was why we called on scholars to supplement any prediction about the sign of the coefficient on the interaction term with predictions about the marginal effect of X on Y. Along similar lines, any observed relationship between the marginal effect of X and Z is always consistent with a wide variety of ways in which the marginal effect of Z varies with X, some of which may be inconsistent with one's underlying conditional theory (Berry, Golder and Milton, 2012, 654). This means that just proposing and testing a hypothesis about how the effect of X on Y varies with Z without proposing and testing a hypothesis about how the effect of Z on Y varies with X passes up an opportunity to distinguish evidence consistent with one's own theory from evidence consistent with alternative theories. Thus, just as one should examine all of the implications of one's theory about the conditional relationship between X and Y, one should also examine all of the implications of one's theory about the conditional relationship between Z and Y. Even overlooking the symmetry of interactions, scholars will generally be interested in the substantive effects of both X and Z in most real-world applications.

To see what's going on here, let's return once more to our foreign aid example. As you'll recall, Burnside and Dollar (2000) posit a conditional theory in which foreign aid and the policy environment in the recipient country interact to influence economic growth. The *Conditional Aid Hypothesis* speaks to the marginal effect of foreign aid on economic growth. Specifically, it states that an increase in foreign aid increases economic growth in the recipient country if the policy environment is good but has no effect if the policy environment is bad. We can test this hypothesis with the linear-interactive model specification shown in Eq. 2.3. As we saw earlier, the marginal effect of foreign aid on economic growth is

$$ME(Aid|Good\ Policy) = \frac{\partial Growth}{\partial Aid} = \beta_1 + \beta_3 Good\ Policy. \qquad (2.14)$$

In the previous section, we presented a marginal effect plot for foreign aid in Figure 2.6 that's consistent with the *Conditional Aid Hypothesis*, $\beta_1 = 0$ and $\beta_3 > 0$. If we were to estimate our linear-interactive model and obtain a marginal effect plot for foreign aid that looks like the one in Figure 2.6, we'd claim support for the *Conditional Aid Hypothesis*. From this, we might conclude that the theory underlying this hypothesis – the one that posits a particular interaction between foreign aid and the policy environment in the recipient country – is also supported. It turns out, though, that a wide variety of conditional relationships among *Aid*, *Good Policy*, and *Growth* are still possible even after we've established that the marginal effect of *Aid* is consistent with our theory. Only some of these relationships may be consistent with our underlying theory.

The issue has to do with the marginal effect of *Good Policy* on *Growth* and how this varies with *Aid*. Note that we care about the marginal effect of *Good Policy* not just because of the inherent symmetry of interactions but also because it matters in the real world whether the "good" policies recommended by institutions such as the World Bank actually facilitate economic growth. We know from the fact that the slope of the marginal effect of *Aid* on *Growth* increases with *Good Policy* that the slope of the marginal effect of *Good Policy* on *Growth* increases with *Aid*. But simply knowing the slope of the marginal effect of *Good Policy* is not sufficient to fully characterize the effect of *Good Policy* on *Growth*. In particular, knowing that the slopes of the two marginal effects are the same, β_3, provides no information about the value of the intercept, β_2, in the linear equation depicting the marginal effect of *Good Policy*,

$$ME(Good\ Policy|Aid) = \frac{\partial Growth}{\partial Good\ Policy} = \beta_2 + \beta_3 Aid. \qquad (2.15)$$

If we were to plot the marginal effect of *Good Policy* on *Growth*, we'd produce a similar marginal effect plot to the one we produced for *Aid*.

This time the marginal effect of *Good Policy* would be on the vertical axis, and *Aid* would be on the horizontal axis. Due to the symmetry of interactions, the slope of the marginal effect plot for *Good Policy* will be identical to the slope of the marginal effect plot for *Aid*. In Figure 2.7, we assumed, in line with the *Conditional Aid Hypothesis*, that the slope of the marginal effect plot for *Aid* was 0.04. As a result, the slope of the marginal effect plot for *Good Policy* will also be 0.04. Knowing the slope of the marginal effect plot for *Good Policy*, though, doesn't establish the sign (positive or negative) or the magnitude of the marginal effect of *Good Policy* at any value of *Aid*. This will depend on the value of the intercept, β_2, in the linear equation depicting the marginal effect of *Good Policy*. This is extremely important because different values for this intercept imply quite different ways in which the marginal effect of *Good Policy* is conditional on *Aid* and thus quite different ways in which *Aid* and *Good Policy* interact to influence economic growth.

In Figure 2.10, we depict three quite different conditional relationships among *Growth*, *Aid*, and *Good Policy* that are all consistent with the marginal effect plot for *Aid* shown in Figure 2.7a.[7] Recall that this marginal effect plot has an intercept, β_1, of 0 and a slope, β_3, of 0.04, and that we assumed that the observed values of both *Aid* and *Good Policy* range from 0 to 10. On the left of Figure 2.10 are three-dimensional (3-D) predicted value plots of *Growth* against *Aid* and *Good Policy*. These plots permit us to visualize how the two independent variables jointly influence economic growth. The slope of each edge of the "surface" in the plots is indicated in small text. These slopes indicate the marginal effects of *Aid* and *Good Policy* at the minimum (0) and maximum value (10) of the other independent variable with which they interact. The slopes at the front right and back left edges, for example, indicate the marginal effect of *Aid* on *Growth* when *Good Policy* is 0 and when *Good Policy* is 10.[8] These two slopes take on the same values, 0 and 0.4, in all three of the 3-D predicted value plots. These slopes correspond to the lowest and highest values we found in our marginal effect plot for *Aid* in Figure 2.7a. As a result, all three 3-D predicted value plots are consistent with the marginal effect plot for *Aid* that we saw earlier.

As you can see, though, the three surfaces depicting the conditional relationship between *Growth*, *Aid*, and *Good Policy* all look quite different. This is because the marginal effect of *Good Policy* on *Growth* is

---

[7] We've assumed that $\beta_0 = 4$ in our linear-interactive model in Eq. 2.3.

[8] These slopes indicate that a one-unit increase in *Aid* – moving one unit to the right along the edge in the front right – has zero effect on *Growth* when *Good Policy* is 0, but that a one unit increase in *Aid* – moving one-unit to the right along the edge in the back left – leads to a 0.4 increase in *Growth* when *Good Policy* is 10.

different in each plot. To the right of each 3-D predicted value plot is the associated plot showing the marginal effect of *Good Policy* on *Growth* across the possible values of *Aid*. A key feature to note about these marginal effect plots is that although they share the same slope, $\beta_3 = 0.04$, the value of the intercept β_2 is different in each.

In Figure 2.10a, β_2 is 0.2, indicating that the marginal effect of *Good Policy* is 0.20 when *Aid* is at its lowest value. The fact that β_3 is positive means that the marginal effect of *Good Policy* is always positive and grows with *Aid*, reaching a maximum of 0.6 when *Aid* is at its highest value. This is reflected in the corresponding 3-D plot by the fact that the slopes of the edges in the front left (0.2) where $Aid = 0$ and back right (0.6) where $Aid = 10$ are both positive. In this scenario, good policies always increase growth, even in the absence of foreign aid.

In Figure 2.10b, β_2 is -0.2, indicating that the marginal effect of *Good Policy* is -0.2 when *Aid* is at its lowest value. This negative value is sufficiently small in magnitude that the marginal effect of *Good Policy* eventually becomes positive once *Aid* is large enough (> 5), reaching a maximum value of 0.2 when $Aid = 10$. This is reflected in the corresponding 3-D plot by the fact that the slope for the front left edge is negative (-0.2), and the slope for the back right edge is positive (0.2). In this scenario, good policies encourage economic growth only when foreign aid is sufficiently high and are detrimental otherwise.

In Figure 2.10c, β_2 is -0.6, indicating that the marginal effect of *Good Policy* is -0.6 when *Aid* is at its lowest value. This negative value is sufficiently large in magnitude that despite the positive value of β_3 the marginal effect of *Good Policy* remains negative for all values of *Aid*. In this scenario, the negative effect of *Good Policy* declines in strength with increases in *Aid*, reaching -0.2 when *Aid* obtains its maximum value. This is reflected in the corresponding 3-D plot by the fact that the slopes of the edges in the front left (-0.6) and back right (-0.2) are both negative. In this scenario, "good policies" are in fact bad for economic growth, and the best that foreign aid can do is mitigate their negative effects.

The three sets of plots shown in Figure 2.10 depict fundamentally different processes by which *Growth* is jointly determined by *Aid* and *Good Policy*.[9] In panel (a), *Growth* is maximized when *Aid* and *Good Policy* are at their maximum; it's minimized whenever *Good Policy* is at its minimum.

---

[9] Figure 2.10 shows three different marginal effect plots for *Good Policy* that are each consistent with a positive coefficient on the interaction term between *Aid* and *Good Policy*. As we saw in Figure 2.8, there are actually five possible conditional relationships between *Good Policy* and *Growth* when the coefficient on the interaction term is positive. There are a further five possible conditional relationships if the coefficient on the interaction term is negative.

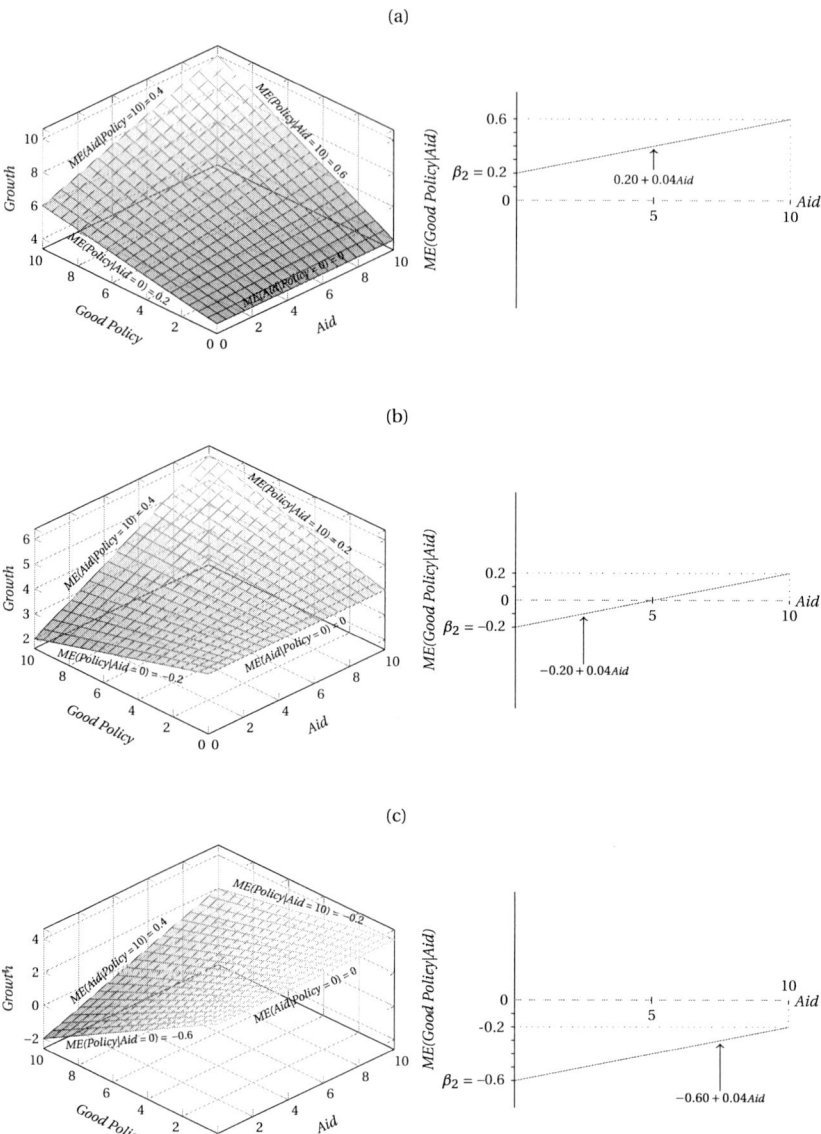

Figure 2.10 (a)–(c) Three conditional relationships among *Aid*, *Good Policy*, and *Growth* consistent with the marginal effect plot for *aid* in Figure 2.7a (assuming $\beta_0 = 4$, $\beta_1 = 0$, and $\beta_3 = 0.04$)

Note: The three-dimensional (3-D) predicted value plots on the left show different conditional relationships between *Growth*, *Aid*, and *Good Policy* that are all consistent with the information conveyed in the marginal effect plot for *Aid* shown in Figure 2.7a. To the right of each 3-D predicted value plot is an associated plot showing the marginal effect of *Good Policy*, $\beta_2 + \beta_3 Aid$, across the possible values of *Aid*. Although these marginal effect plots all share the same slope, 0.04, they differ in the value of their intercept, b_2.

In panel (b), *Growth* is maximized when *Aid* and *Good Policy* are at their maximum; it's minimized when *Good Policy* is at its maximum and *Aid* is at its minimum. In panel (c), *Growth* is maximized when *Aid* is at its maximum and *Good Policy* is at its minimum; it's minimized when *Good Policy* is at its maximum and *Aid* is at its minimum. If the underlying theory from which the *Conditional Aid Hypothesis* is derived predicts that the conditional relationship among *Growth*, *Aid*, and *Good Policy* should be like the one depicted in panel (a), which seems reasonable,[10] it's difficult to imagine anyone claiming empirical support for the theory if the estimated relationship ends up looking like that shown in either panel (b) or panel (c).

The point here is that if we limit ourselves to deriving and testing a hypothesis about the marginal effect of just one variable involved in an interaction, say X, then we might claim support for our theory ignorant of the possibly inconsistent evidence that would be apparent if we were to also derive and test a hypothesis about the marginal effect of the other variable involved in the interaction, Z. Thus, even when there's strong empirical support for a hypothesis about how the marginal effect of X on Y varies with Z, a failure to use one's theory to derive and test an additional hypothesis about how the marginal effect of Z on Y varies with X (beyond a prediction about the value of β_3) may mask either (1) additional evidence in support of one's theory or, of more concern, (2) evidence that's inconsistent with one's theory. This is why we encourage scholars, when possible, to derive and test hypotheses about both the marginal effect of X on Y and the marginal effect of Z on Y whenever they have a theory positing an interaction between X and Z.

It's important to note that we're not suggesting that researchers testing a hypothesis about how the marginal effect of X on Y varies with Z should "manufacture" a second hypothesis about how the marginal effect of Z on Y varies with X when their theory is silent on the matter. We're simply observing that many theories positing interaction generate more predictions than the one about the marginal effect of X and that if the researcher limits themself to testing just this prediction, then they are subjecting their theory to a weaker test than is possible given the data available. In practice, we strongly urge scholars to think hard before concluding that their theory makes no prediction about the marginal effect of Z on Y and how this varies with X. As we've seen, this is never strictly true. Any theory that

---

[10] We might expect, for example, that the adoption of the good policies (budget surpluses, low inflation, trade openness) recommended by the economists at places like the World Bank and the International Monetary Fund should always have a positive (or, at least, a non-negative) effect on economic growth irrespective of the level of foreign aid. This prediction is consistent only with the marginal effect plot and 3-D plot shown in Figure 2.10a.

produces a hypothesis about how the slope of the marginal effect of X varies with Z automatically produces a hypothesis about how the slope of the marginal effect of Z varies with X. This is because these slopes are the same. It doesn't demand much more from a theory that's specific enough to establish the signs of the intercept and slope of the marginal effect of X to generate additional predictions about the marginal effect of Z that would permit a stronger test of the theory.[11]

In terms of our foreign aid example, if the theory proposed by Burnside and Dollar were to additionally predict that *Good Policy* has a positive effect on *Growth* when *Aid* is at, say, its highest (or, indeed, any) value, then this would imply that the plots shown in Figure 2.10c, where the marginal effect of *Good Policy* is always negative, are inconsistent with the theory. If, in contrast, their theory were to predict that *Good Policy* has a positive effect on *Growth* when *Aid* is at its lowest value, which would seem to be implied by the policy environment being "good," then the plots in both Figures 2.10b and 2.10c would be inconsistent with the theory. In both cases, supplementing a hypothesis about the marginal effect of *Aid* on *Growth* with an additional one about the marginal effect of *Good Policy* would permit a stronger test of Burnside and Dollar's underlying theory.

> ⚠ **Important:** Scholars testing theories positing interaction between two independent variables should, whenever possible, derive and test predictions about how the marginal effect of each independent variable varies with the value of the other. If they do not do this, then they are subjecting their theory to a weaker test than is possible given the data available.

In the next section, we summarize our recommendations regarding theories positing interaction and briefly discuss ways that scholars can improve the clarity of their hypotheses.

2.6 FIVE KEY PREDICTIONS

In order to allow for the strongest possible test of a theory positing interaction, it's important to clearly present all of the predictions of the theory to the reader. In our experience, scholars rarely take the time to

---

[11] Even if it turns out not to be possible to derive a hypothesis about the effect of Z on Y from a theory, it's likely that scholars will still want to evaluate the marginal effect of Z on Y as doing so could provide substantively useful information about how the world works.

think through all of the implications of their theory and to present them in as clear a way as possible. It's common, for example, to see vague hypotheses predicting interaction or that the effect of some variable X on Y depends on some other variable Z. This often leaves the reader unsure as to whether the results from any ensuing empirical test are fully in line with the author's theory. After all, what kind of interaction is predicted? How exactly does the effect of X on Y depend on Z? Is the effect of X supposed to be positive or negative? Does the direction of the effect of X on Y change over the observed range of Z? What about the marginal effect of Z on Y and how this varies with X? Scholars can avoid this unfortunate situation by making more of an effort to clearly outline and explain their theory's predictions. This means carefully thinking through all of the implications of one's theory, writing clear hypotheses that capture all of the aspects of these implications, and indicating exactly what one expects to see in the empirical results. All of this should be done before moving on to the presentation of any empirical results. In what follows, we offer some practical advice on deriving and stating hypotheses from theories positing interaction that can be accurately specified with the type of linear-interactive model shown earlier in Eq. 2.3.[12]

Most theories that posit an interaction between X and Z on Y are strong enough to generate five basic predictions about the marginal effects of X and Z on Y (Berry, Golder and Milton, 2012, 6–7). These predictions speak to the direction of the interaction between X and Z, the direction of the marginal effect of X on Y, and the direction of the marginal effect of Z on Y.

1. P_{XZ}: The interaction effect between X and Z is [positive, negative].
2. $P_{X|Z_{min}}$: The marginal effect of X on Y is [positive, negative, zero] when Z is at its *lowest* value.
3. $P_{X|Z_{max}}$: The marginal effect of X on Y is [positive, negative, zero] when Z is at its *highest* value.
4. $P_{Z|X_{min}}$: The marginal effect of Z on Y is [positive, negative, zero] when X is at its *lowest* value.
5. $P_{Z|X_{max}}$: The marginal effect of Z on Y is [positive, negative, zero] when X is at its *highest* value.[13]

---

[12] More complex theories positing interaction produce testable implications of a different kind and require an alternative model specification. We'll look at some of these alternative model specifications in later chapters.

[13] When Z is dichotomous, the predictions $P_{Z|X_{min}}$ and $P_{Z|X_{max}}$ should be stated in terms of the response of Y to a *discrete change* in Z rather than in terms of the *marginal effect* of Z. This is because the concept of a marginal effect makes sense only when it's possible to conceive of an infinitesimally small, or "marginal," change in Z. The predictions $P_{X|Z_{min}}$ and $P_{X|Z_{max}}$ should be stated similarly when X is dichotomous.

> ⚠ **Important:** Theories positing interaction between two independent variables X and Z are typically strong enough to generate five key predictions about the marginal effects of X and Z on Y. These predictions speak to the direction of the interaction between X and Z, the direction of the marginal effect of X on Y, and the direction of the marginal effect of Z on Y.

While we encourage scholars to use their theory to make these five predictions, there's no need to present them as five separate hypotheses. It's usually the case that all five predictions can be incorporated into a single hypothesis about how the marginal effect of X on Y varies with Z and a single hypothesis about how the marginal effect of Z on Y varies with X. This is illustrated in the following pair of hypotheses:

- $H_{X|Z}$: The effect of an increase in X on Y is negative at all values of Z; this effect is strongest when Z is at its lowest value and decreases in magnitude as Z increases.
- $H_{Z|X}$: The effect of an increase in Z on Y is negative when X is at its lowest level. This effect declines in magnitude as X increases; at some value of X, Z has no effect on Y. As X rises further, the effect of Z becomes positive and strengthens in magnitude as X increases.

Hypothesis $H_{X|Z}$ implies that the marginal effect of X is negative at both the lowest and highest values of Z, thereby offering predictions $P_{X|Z_{min}}$ and $P_{X|Z_{max}}$. By stating that the marginal effect of Z is negative at X's lowest value and positive at X's highest value, hypothesis $H_{Z|X}$ offers predictions $P_{Z|X_{min}}$ and $P_{Z|X_{max}}$. There's no need to state a separate hypothesis that each variable is positively related with the marginal effect of the other because such a prediction, P_{XZ}, is already implicitly stated in both hypotheses $H_{X|Z}$ and $H_{Z|X}$. Together, our two hypotheses include all five of the predictions we recommend and offer as complete a description of the expected interaction between X and Z as we could offer for a linear-interactive model like the one in Eq. 2.3 without predicting specific magnitudes for the marginal effects at particular values of the independent variables.

Comprehensive hypotheses like the ones shown in $H_{X|Z}$ and $H_{Z|X}$ are important because they allow us to easily visualize the *predicted* marginal effect plots for X and Z. We can then produce *estimated* marginal effect plots based on our regression results and compare them to the predicted plots to help determine how much empirical support we have for our

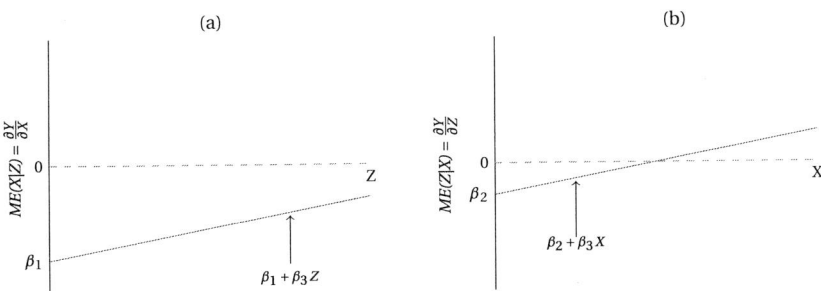

Figure 2.11 (a)–(b) Marginal effect plots consistent with hypotheses $H_{X|Z}$ and $H_{Z|X}$

Note: Panel (a) shows a marginal effect plot consistent with hypothesis $H_{X|Z}$, while panel (b) shows a marginal effect plot consistent with hypothesis $H_{Z|X}$.

theory. The predicted marginal effect plots for $H_{X|Z}$ and $H_{Z|X}$ are shown in Figure 2.11. In our experience, the ability to write clear and precise hypotheses is a strong indicator of how well someone understands the implications of their theory.

Scholars can often help to improve the clarity and utility of their hypotheses by adopting the following recommendations. Many of these recommendations apply to stating hypotheses in general and not just those derived from theories positing interaction. First, scholars should use names, instead of numbers, when labeling hypotheses. Labeling hypotheses with names helps the reader to remember the content of these hypotheses when they are referred to later. For example, think about how much easier it was throughout this chapter to remember that our *Conditional Aid Hypothesis* had to do with the effect of foreign aid than if we had simply labeled it as *Hypothesis 1* or H_1. Names also make it easier to distinguish between closely related hypotheses. For instance, we suspect that it was probably much easier to remember the distinction between the *Unconditional Aid Hypothesis* and the *Conditional Aid Hypothesis* because of our use of hypothesis names than if we had simply referred to these hypotheses as *Hypothesis 1* and *Hypothesis 2*. Reducing the mental tax burden on a reader, especially a reviewer, is almost always appreciated and makes it more likely they'll retain enough mental processing power to understand the argument you're making.

Second, hypotheses should be (i) declarative statements that (ii) make a relational claim between at least two variables that (iii) can be potentially falsified. As a declarative statement, a hypothesis shouldn't take the form of a question or command. A hypothesis should describe a relationship, such as positive, negative, or zero, between two variables, an independent variable X and a dependent variable Y. Hypotheses derived from a theory positing an interaction between X and Z should also indicate how this relationship between X and Y is modified or moderated by Z. Hypotheses should also be falsifiable, meaning that they're potentially testable. As a result, hypotheses must not include tautologies.

Third, the relationship described in a hypothesis should reference the relationship between theoretical concepts, not empirical measures. As an example, it's better to state a hypothesis in terms of the relationship between some outcome of interest and a country's income level (concept) as opposed to, say, a country's gross domestic product (measure). It's better to state a hypothesis in terms of the relationship between some outcome of interest and electoral system proportionality (concept) rather than district magnitude (measure). And it's better to state a hypothesis in terms of the relationship between some outcome of interest and social class (concept) as opposed to family income or employment status (measure). Stating hypotheses in terms of empirical measures often gives the impression that the researcher is allowing data availability to drive their theoretical argument and can make the important discussion as to the most appropriate way to operationalize one's theoretical concepts, which should appear in the empirical section of one's research, incorrectly appear redundant.

Fourth, hypotheses should, where theory allows, indicate the direction of any relational claim. It's common to see hypotheses predicting that "X and Z interact to influence Y," that "X and Z jointly influence Y," or that "the effect of X on Y depends on Z." While each of these hypotheses predicts interaction, none provides information as to the direction of the effect of the interaction or, indeed, the direction of the effect of any of the variables on Y. The reader is, therefore, left unsure as to the precise nature of any predicted interaction. Providing predictions about the direction of the effects of the variables in one's model, as we did in our five key predictions, provides clarity and allows for a stronger test of one's theory. Since the effects of the variables in a linear-interactive model vary depending on the values of the variables with which they interact, the hypotheses derived from a theory positing interaction are almost always longer and more complicated than those derived from an unconditional theory. Consider, for example, the hypothesis $H_{Z|X}$ that we saw earlier. While this hypothesis is quite long and complicated, it's not unnecessarily so given that it needs to convey information as to how the marginal effect of Z on Y changes from negative to zero to positive as X changes from its lowest value to its highest value. The directional clarity of hypotheses can also be improved by avoiding terms, such as "sex," "race," or "polity," that are ambiguous with respect to direction. For example, does a hypothesis mentioning the effect of an individual's sex on some outcome of interest refer to the effect of being male as opposed to female or the effect of being female as opposed to male? This sort of confusion can easily be avoided by using terms such as "men" and "women," "Black" and "White," and "autocracy" and "democracy" in our hypotheses instead of "sex," "race," and "polity."[14]

Fifth, the relational claims we wish to make in our hypotheses typically refer to what we expect to happen to some outcome of interest when we *change* the value of an independent variable (holding everything else constant). Explicitly adopting language in a hypothesis that signals that we're interested in "change" can be helpful in reminding both the reader and ourselves that we're engaging in a counterfactual form of reasoning. In this regard, it's often better for a hypothesis to state that "An increase in X has a positive effect on Y" than to simply say that "X has a positive effect on Y." In those cases where an independent variable is not continuous, it can be helpful to explicitly signal the counterfactual that we have in mind. For example, it's often better for a hypothesis to state, say, that "Men are more likely to vote than women" than it is to simply state that "Men are more likely to vote."

Sixth, scholars should generally avoid making references to statistical significance in their hypotheses. It's not uncommon to see hypotheses predicting that "X has a statistically significant effect on Y" or that "X only has a statistically significant effect on Y when the value of Z is sufficiently high." Statistical significance, though, depends on one's data, including one's sample size. Most theories in the social sciences are not sufficiently strong to make predictions about statistical significance for a particular sample size. Indeed, most social science theories can't even predict the precise magnitude of the effect of our variables. This is why our hypotheses typically only make predictions about the *direction* or *relative* magnitude, and not the statistical significance, of the effect of our independent variables.

In this chapter, we've looked at theories with conditional implications. In particular, we've examined how interaction models can be used to test the conditional claims derived from a theory. In the next chapter, we focus on how to correctly specify interaction models to test conditional claims about the world.

---

[14] This recommendation also applies to the variable names we use in our empirical analyses. If the estimated coefficient on a variable named *Race* is positive, does this mean that Black individuals have a higher value of Y than White individuals or that White individuals have a higher value of Y than Black individuals? This sort of confusion can be avoided by, say, naming the "race" variable *Black*. It now becomes obvious that the positive coefficient indicates that Black individuals have a higher value of Y than White individuals.

2.7 EXERCISES

1. Consider the following theoretical claims. Use a diagram to graphically depict them. Do these claims involve moderation or mediation?

   - *Education, Income, and Spending:* The more education people obtain, the more money they spend. This positive relationship has to do with the fact that more educated people tend to have higher levels of income.
   - *Drugs, Exercise, and Illness:* The effectiveness of some drugs at combating illness depends on the rate at which they are absorbed by the body. Exercise is a factor that influences the rate of drug absorption.
   - *Development, Culture, and Democracy:* According to cultural modernization theory, economic development helps promote democratization by producing certain cultural changes that are conducive to democracy.
   - *Federalism, Ethnic Diversity, and Conflict:* Federalism helps to mitigate the negative effects of ethnic diversity on conflict.
   - *Incumbency, Deterrence, and Electoral Advantage:* Incumbent politicians enjoy an electoral advantage over their opponents. One reason for this is their ability to deter high quality challengers from entering a race.
   - *Ethnicity Cues, Anxiety, Support for Immigration:* Ethnicity-based media cues lower support for immigration by altering levels of anxiety.
   - *Inequality, Asset Mobility, and Democracy:* Economic inequality only undermines democracy in countries where elites rely on immobile assets.

2. In Section 2.1, we presented several theories that posited interaction:

   - *Gender Pay Gap Theory*
   - *Party System Size Theory*
   - *Democratization Theory*
   - *Clinical Trial Theory*

   Choose one of these theories and see if you can generate each of the five key predictions outlined in Section 2.6; you may need to develop the theories a little further to generate all five predictions. Once you have generated the five key predictions, see if you can include them in two hypotheses that each refer to the effect of one of the variables involved in the interaction.

3. A scholar presents the following two hypotheses that they say are derived from their theory positing an interaction between X and Z on Y. Why do we immediately know that something is wrong?

- $H_{X|Z}$: The marginal effect of X on Y is always positive. This positive effect declines as Z gets larger.
- $H_{Z|X}$: The marginal effect of Z on Y is negative when X is at its lowest value. The magnitude of this negative effect declines as X increases; at some value of X, Z has no effect on Y. As X rises further, the effect of Z becomes positive and strengthens in magnitude as X increases.

4. Consider the following linear-interactive model:

$$Y = \beta_0 + \beta_1 X + \beta_2 Z + \beta_3 XZ + \epsilon. \qquad (2.16)$$

What's the marginal effect of X? What's the marginal effect of Z? What does β_1 tell us? What does β_2 tell us? What does β_3 tell us? And what does β_0 tell us?

5. Below are some hypotheses that we and our colleagues have come across while reviewing manuscripts. We've removed variable names to maintain anonymity. Why are they problematic or unclear? How could the authors have improved them?
   - X and Z are both important causal variables in determining Y.
   - The interaction between X and Z is positively related to Y.
   - X has less effect on Y when Z is high.
   - The negative effect of X on Y can be diminished through Z.
   - The effect of X on Y is greater when Z is absent than when Z is present.
   - X should be a stronger predictor of Y when Z is high.
   - The effect of X on Y is contingent on Z.
   - X has a negative effect on Y when Z is present.
   - Y should be viewed as a process, not an event.
   - X has an insignificant effect on Y.
   - X increases Y independent of Z.
   - The impact of X on Y is less when Z is present than when Z is absent.
   - X and Z interact to determine Y.
   - X has the most influence on Y.

6. Suppose we're interested in how consumer sentiment and periods of recession affect consumer spending. Existing studies argue that consumer spending will decrease in recessions as consumers have less to spend and that it will increase when consumer sentiment improves as consumers feel more optimistic about the future. However, we might believe that consumer sentiment and periods of recession interact to affect consumer spending. If this is the case, then we might want to employ the following linear-interactive model to test our claims:

$$Consumer\ Spending = \beta_0 + \beta_1 Consumer\ Sentiment + \beta_2 Recession$$
$$+ \beta_3 Consumer\ Sentiment \times Recession + \epsilon,$$
$$(2.17)$$

where *Consumer Sentiment* is, say, a continuous variable whose lowest value is 0 and whose highest value is 10, and where *Recession* is a dichotomous variable that equals 1 when we're in a recession and 0 otherwise.

(a) Think about how consumer sentiment and periods of recession might interact to affect consumer spending. Generate the five key predictions outlined in Section 2.6 and then include them in two hypotheses, the *Consumer Sentiment Hypothesis* and the *Recession Hypothesis*, that speak to the effects of *Consumer Sentiment* and *Recession* on *Consumer Spending*.

(b) What's the marginal effect of *Consumer Sentiment* on *Consumer Spending*? What's the effect of *Consumer Sentiment* when we're not in a recession? What's the effect of *Consumer Sentiment* when we're in a recession? Based on your *Consumer Sentiment Hypothesis*, what should be the sign of β_1? What should be the sign of β_3? What should the predicted marginal effect plot for *Consumer Sentiment* look like?

(c) What's the effect of a one-unit discrete change in *Recession* on *Consumer Spending*? Based on your *Recession Hypothesis*, what should the sign of β_2 be? What should be the sign of β_3 be? Draw a plot showing the predicted effect of a one-unit discrete change in *Recession* on *Consumer Spending* across the observed range of *Consumer Sentiment*.

3 Interaction Model Specification

So, you've decided that you have a theory positing interaction, and you've been careful in deriving your hypotheses. What's next? In this chapter, we'll take a look at how to appropriately specify an interaction model. We're going to focus on linear-interactive models that allow us to test theories positing interaction between two independent variables X and Z on a continuous dependent variable Y. This is the most common type of theory with conditional implications in the social sciences. In later chapters, we'll examine theories with more complicated conditional implications. The main piece of advice in this chapter is that scholars should almost always include all of the constitutive elements of a multiplicative interaction term when they specify their models. Failure to do this is likely to produce omitted variable bias that affects all of the estimated parameters. Typical excuses for the exclusion of constitutive terms are not valid.

Scholars who posit an interaction in which one of the modifying variables, say Z, is discrete *may* choose to adopt an "alternative" interactive specification in which *some* constitutive terms are omitted. This is fine because the alternative specification is exactly equivalent to the "standard" interactive specification that includes all of the constitutive terms. Scholars with a single discrete modifying variable Z sometimes choose to eschew an interactive model altogether by adopting a "split-sample" strategy in which they employ an additive model to examine the effect of X on Y using different sub-samples that are restricted to observations with a particular value of Z. Several important points need to be kept in mind when adopting such a strategy. As we indicate, it's almost always better to adopt a "pooled" interactive model than it is to adopt a split-sample strategy.

Scholars who posit an interaction in which both of the modifying variables X and Z are discrete *may* choose to adopt a different "alternative" interactive specification in which *all* of the constitutive terms are omitted. This is fine because this alternative specification is also exactly equivalent to the "standard" interactive specification that includes all of the constitutive terms. Since this alternative specification involves including a series of dichotomous variables that capture different categories of observations, some scholars are unaware that they're actually estimating an interaction model. This can be problematic when they come to interpret the results from such a model.

3.1 INCLUDE ALL CONSTITUTIVE TERMS

As we saw in Chapter 2, the following linear-interactive model specification allows us to test claims derived from theories positing interaction between two independent variables X and Z on a continuous dependent variable Y,

$$Y = \beta_0 + \beta_1 X + \beta_2 Z + \beta_3 XZ + \epsilon. \tag{3.1}$$

The inclusion of the multiplicative interaction term XZ is the key to testing conditional claims about the world, as it's this term that allows the marginal effect of X on Y to vary with Z and the marginal effect of Z on Y to vary with X. We focus on this particular interactive model specification because it's the most common one found in the social sciences and because it easily handles both discrete and continuous modifying variables.[1]

Constitutive terms refer to those variables that we multiply together to make an interaction term – they're the elements that "constitute" the interaction term. The constitutive terms for the interactive model in Eq. 3.1 are X and Z. As we'll see in Chapter 6, multiplicative interaction terms can take a variety of different forms and may involve quadratic terms such as X^2 or higher order interaction terms such as XZW. No matter what form the interaction term takes, all constitutive terms should almost always be included in our model specifications. Thus, X should be included when the interaction term is X^2 and X, Z, W, XZ, XW, and ZW should be included when the interaction term is XZW. The general rule that all constitutive terms should be included when we have an interaction model also applies in non-linear settings where the dependent variable is not continuous, such as logit and probit.

⚠ **Important:** Scholars should almost always include all of the constitutive elements of a multiplicative interaction term when they specify their interaction models.

---

[1] In most substantive applications, the interactive model specification shown in Eq. 3.1 will also include various control variables. Unless otherwise stated, we'll assume that our model is correctly specified with respect to these control variables.

Before discussing why it's important to include all of the constitutive elements of an interaction term in our model specifications, we note that scholars don't always appear to recognize when they've included an interaction term in their models. Consider race and gender scholars who are interested in the behavior or attitudes of, say, Black women. To evaluate their theoretical claims, it's not uncommon for these scholars to include in their models a dichotomous variable *Black Woman* that equals 1 if an individual is a Black woman and 0 otherwise. Notice, though, that *Black Woman* is really an interaction term created by multiplying the variables *Black* and *Woman* together.[2] To see this more clearly, think about how one would go about identifying the Black women in a study. Typically studies will identify the race and sex of a respondent. Treating race and sex as dichotomous, we'll usually have a dichotomous variable *Black* that equals 1 if an individual is Black and 0 otherwise, as well as a dichotomous variable *Woman* that equals 1 if the individual is a woman and 0 otherwise. Black women are those who are coded as *Black* $= 1$ and *Woman* $= 1$. Practically speaking, the *Black Woman* variable is created by multiplying the values of *Black* and *Woman* together. Only if *Black* and *Woman* are both 1 will *Black Woman* equal 1: Black women are coded as $1 \times 1 = 1$, White women are coded as $0 \times 1 = 0$, Black men are coded as $1 \times 0 = 0$, and White men are coded as $0 \times 0 = 0$. As you can now hopefully see, *Black Woman* is an interaction term. As such, scholars should include the variables *Black*, *Woman*, and *Black Woman* in their model specifications.[3] The same reasoning applies to scholars interested in more complicated forms of identity. For example, scholars who study race, gender, and sexuality might be interested in some aspect of Black female heterosexuals. If these scholars include a dichotomous variable *Black Female Heterosexual* in their models, they should recognize that this is an interaction term and also include all six of its constitutive terms: *Black*, *Female*, *Heterosexual*, *Black* $\times$ *Heterosexual*, *Black* $\times$ *Female*, and *Female* $\times$ *Heterosexual*. The key point here is that you should always be on the lookout for these "hidden" interaction terms.[4]

---

[2] A similar argument can be made for those who include variables such as *White Men*, *Black Democrats*, *Male Republicans*, *College Educated Women*, *Rural Poor*, and so on in their models.

[3] As we'll see a little later in the chapter, it's possible to omit some, or all, of the constitutive terms if one or more of the interacting variables in our model, like *Black* and *Woman*, are discrete. However, this is only because these "alternative" interactive specifications are exactly equivalent to the "standard" interactive specification in which all of the constitutive terms are included.

[4] Some interaction terms are particularly difficult to spot. Consider a variable measuring gross domestic product per capita, *GDP per capita*. Although we don't typically think of it this way, *GDP per capita* is actually an interaction term created by multiplying a country's gross domestic product with the reciprocal of its population

3.2 WHY IS IT IMPORTANT TO INCLUDE ALL OF THE CONSTITUTIVE TERMS?

Scholars often fall prey to the temptation to exclude one or more of the constitutive terms in their interaction models. To see what happens when they do this, consider again the simple interactive model specification shown in Eq. 3.1. Suppose that we omit the constitutive term Z. Omitting Z is equivalent to assuming that $\beta_2 = 0$. With the omission of Z, our model becomes

$$Y = \gamma_0 + \gamma_1 X + \gamma_3 XZ + v. \tag{3.2}$$

Scholars rarely indicate why they've omitted one or more of the constitutive terms in their interaction models. In those cases where analysts do actually say something, they typically provide one of the following two reasons for why they specify a model similar to Eq. 3.2 rather than the fully specified model shown in Eq. 3.1. First, some claim that they don't believe that Z has any effect on Y on average and that, as a result, they don't need to include it as a separate term in the model. Second, others claim that they don't believe that Z has an effect on Y when X is zero and that this means that they can exclude it as a separate variable from their model. The first of these claims is never justified, and the second is rarely, if ever, defensible (Brambor, Clark and Golder, 2006).

Note that both claims used to justify the omission of the constitutive term Z are based on the expectation that β_2 in Eq. 3.1 is zero. The first claim is relatively easy to refute. As we saw in Chapter 2, β_2 doesn't represent the average effect of Z on Y; it only indicates the effect of Z on Y when X is zero.[5] This is easy to see by remembering that the marginal effect of Z on Y is

$$\frac{\partial Y}{\partial Z} = \beta_2 + \beta_3 X. \tag{3.3}$$

size; that is, $GDP\ per\ capita = GDP \times \frac{1}{Population}$. In most cases, scholars who include $GDP\ per\ capita$ in their models probably don't have a theoretical claim that the effect of GDP on some outcome of interest Y varies with the value of $\frac{1}{Population}$. Instead, we suspect that they simply have a belief that they should control for the size of the population when examining the effect of GDP on Y. If this is the case, though, why not simply include GDP in the model and control for $Population$? If the scholar does have a claim that the effect of GDP on Y varies with $\frac{1}{Population}$, then they should include the constitutive terms of $GDP\ per\ capita$ in their model.

[5] It's true that β_2 indicates the average effect of Z on Y when there's no interaction effect present; that is, when $\beta_3 = 0$. However, if this is the case, then neither the model in Eq. 3.1 nor the model in Eq. 3.2 is appropriate.

Thus, even if the average effect of Z on Y is zero, it isn't necessarily the case that β_2 is zero. As a result, the assertion that the average effect of Z on Y is zero is *never* a justification for omitting the constitutive term.

The second claim is based on the theoretical expectation that Z has no effect on Y when X is zero; that is, $\beta_2 = 0$. Unfortunately, there's reason to believe that the omission of a constitutive term may still lead to inferential errors even when the analyst is armed with as strong a theory as this. The basic intuition is that the analyst's theory may be wrong and β_2 may in fact not be zero. If this is the case and Z is correlated with either XZ (or X), as will occur in virtually any social science application, then omitting the constitutive term Z will result in biased (and inconsistent) estimates of β_0, β_1, and β_3. Although not always recognized as such, this is a straightforward case of omitted variable bias.

To see exactly how and why the omission of a constitutive term may lead to biased estimates in an interaction model, consider Figure 3.1, which illustrates a scatterplot of 500 observations generated by the process implied in Eq. 3.1.[6] Just as we saw when we presented a graphical illustration of a linear-interactive model that was consistent with the *Conditional Aid Hypothesis* in Figure 2.5 in Chapter 2, the intercept for the thick solid line when $Z = 0$ is β_0, and the intercept of the thick solid line when $Z = 1$ is $\beta_0 + \beta_2$. In other words, β_2 captures the difference in the intercepts between the regression lines for the case where $Z = 1$ and the case where $Z = 0$.

As we've already stated, omitting the constitutive term Z is the same as assuming that $\beta_2 = 0$. It should now be clear that omitting Z amounts to constraining the two regression lines to meet on the vertical axis. Note that in some sense this is equivalent to specifying a model without a constant term and forcing the regression line to go through the origin. As one would expect, forcing the two lines to meet on the vertical axis can only happen if the slopes of the regression lines (and the angle between them) change. In effect, instead of estimating the slopes as β_1 and $\beta_1 + \beta_3$, the model omitting the constitutive term estimates them as γ_1 and $\gamma_1 + \gamma_3$. And instead of estimating two intercepts, β_0 and $\beta_0 + \beta_2$, the under-specified model only estimates one, γ_0. In other words, the estimates of the parameters of interest will all be biased whenever β_2 is not zero. More precisely, the coefficients estimated by the under-specified model in Eq. 3.2 will be $\gamma_0 = \beta_0 + \beta_2 \alpha_0$,

---

[6] X was generated as a uniform variable on the unit interval and then multiplied by four. Z was originally drawn from a uniform distribution on the unit interval; it was then recoded as 1 if $Z \geq 0.5$ and 0 if $Z < 0.5$. The error term was randomly drawn from a normal distribution with mean 0 and variance 0.5. The true parameters of this model indicate that X has no effect on Y when Z is absent, $\beta_1 = 0$, but that X increases Y when Z is present, $\beta_1 + \beta_3(1) = 2$.

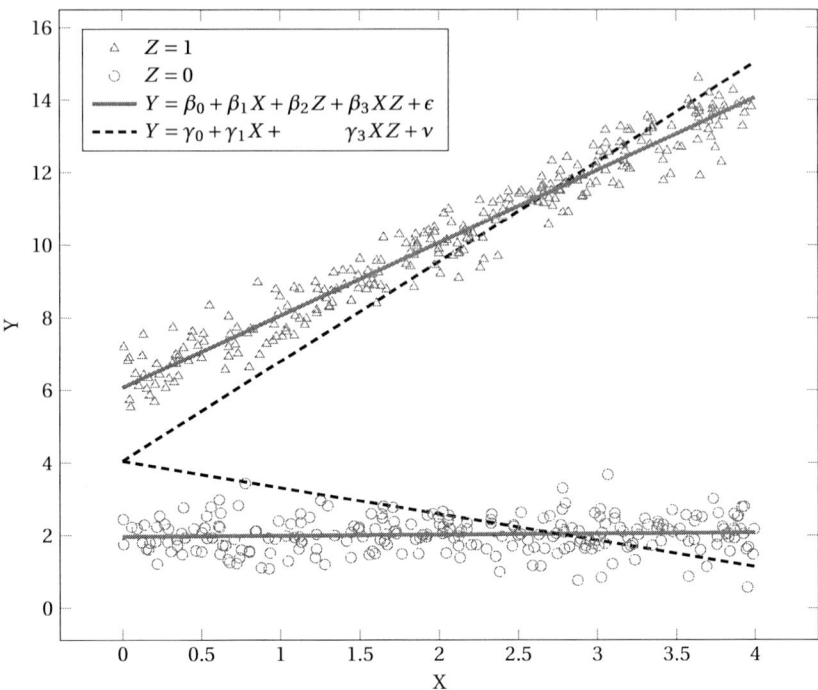

Figure 3.1 The consequences of omitting a constitutive term

Note: Figure 3.1 illustrates a scatter plot of 500 observations generated by the process implied in Eq. 3.1, where $\beta_0 = 2$, $\beta_1 = 0$, $\beta_2 = 4$, $\beta_3 = 2$, and Z is a dichotomous modifying variable. The open triangles indicate observations when $Z = 1$, while the open circles indicate observations when $Z = 0$. The thick solid lines indicate the predicted values of Y based on a fully specified model that includes Z; the thin dashed lines indicate the predicted values of Y from an under-specified model that omits Z. The intercepts for the two solid lines are β_0 and $\beta_0 + \beta_2$.

$\gamma_1 = \beta_1 + \beta_2\alpha_1$, and $\gamma_3 = \beta_3 + \beta_2\alpha_3$, where the βs are simply the coefficients from the fully specified model in Eq. 3.1, and the αs are the coefficients from the regression of Z on X and XZ; that is, $Z = \alpha_0 + \alpha_1 X + \alpha_3 XZ + \varepsilon$. The bias in each of the coefficients is shown in gray. Practically speaking, the extent of any bias will depend on (i) the degree to which β_2 differs from zero and (ii) the magnitude of β_2 relative to the magnitudes of β_0, β_1, and β_3. In addition, the extent to which each of the coefficients is individually biased depends on the distribution of the modifying variables.

Are there any circumstances in which omitting a constitutive term wouldn't lead to significant inferential errors? Possibly. However, there are at least two necessary conditions that must be met before we consider omitting constitutive terms. First, we must have a strong theoretical expectation that the omitted variable – Z in this case – has no effect on the dependent variable when the value of the other modifying variable – X in this case – is zero. Note, though, that the only situation in which this expectation can be justified *a priori* is if X is measured with a natural zero.[7] This is because the coefficients on constitutive terms depend on how an analyst scales these variables. For example, using a measure of democracy such as the familiar *Polity* score, which is scaled from -10 to $+10$, in an interactive model will generate a different coefficient on the variable that it interacts with than the same measure of democracy that's scaled from 0 to 20.

This can easily be shown (Braumoeller, 2004). Start with our fully specified model outlined in Eq. 3.1. Imagine that some arbitrary constant L, such as 10 in our *Polity* example, is now added to X to create X^*. The model now becomes

$$Y = \beta_0 + \beta_1(X^* - L) + \beta_2 Z + \beta_3(X^* - L)Z + \epsilon. \tag{3.4}$$

Multiplying through, we have

$$Y = \beta_0 + \beta_1 X^* - \beta_1 L + \beta_2 Z + \beta_3 X^* Z - \beta_3 L Z + \epsilon. \tag{3.5}$$

Collecting terms, we get

$$Y = (\beta_0 - \beta_1 L) + \beta_1 X^* + (\beta_2 - \beta_3 L)Z + \beta_3 X^* Z + \epsilon. \tag{3.6}$$

As you can see, rescaling X in this arbitrary way changes the coefficient on Z from β_2 to $\beta_2 - \beta_3 L$; the standard error of the coefficient on Z naturally changes as well.[8] The point here is that even if β_2 is truly zero, we have no way of knowing in practice whether we're estimating β_2 or $\beta_2 - \beta_3 L$ if our theory doesn't tell us which particular scale to use for X (something that's normally the case). In other words, we have no way of predicting *a priori* what the coefficient on Z will be before actually estimating our

---

[7] Loosely speaking, a variable has a natural (or absolute) zero if 0 indicates the absence of the thing that the variable is measuring. For example, height and weight have a natural zero since 0 indicates no height or no weight. In other variables, the 0 value is arbitrary and doesn't indicate the absence of the thing being measured. For example, centigrade as a measure of temperature doesn't have a natural zero as 0 degrees centigrade doesn't indicate the absence of temperature. Only variables measured on a ratio scale have a natural or absolute zero.

[8] Note that Eq. 3.6 also shows that the arbitrary rescaling of X, which results in a different zero point for this variable, has *no* effect on the coefficient on the interaction term. This reminds us that the magnitude and statistical significance of the coefficient on the interaction term doesn't depend on the zero point of the independent variables in an interactive model. It's also worth noting that the statistical significance of an interaction term coefficient doesn't depend on the units in which the variables are measured either. Although changing the units of the independent variables will change the size of the coefficient on the interaction term, the standard error will also change size proportionately, leaving the statistical significance unchanged.

model. As a result, it's *impossible* for us to have a theoretical justification for omitting Z in this case. Only if X has a natural zero can we be justified in having a theoretical prediction that the effect of Z on Y is zero when X is zero. This is because the zero point of a variable with a natural zero isn't subject to rescaling by definition. The widespread use in the social sciences of scales and indices, such as the *Polity* score, suggests that scholars are often working with variables that don't have a natural zero.

The second condition that must be met before omitting a constitutive term is that we should estimate the fully specified model outlined in Eq. 3.1 and find that β_2 is zero. In other words, we should test our theoretical claim that β_2 is, in fact, zero. Note that even if β_2 is statistically indistinguishable from zero, the other coefficients will still be estimated with bias to the extent that β_2 is not *exactly* zero if the constitutive term is dropped.[9] This may or may not be much of a problem in practice. Much will depend on whether β_2 is close to zero and whether β_2 is small relative to β_1 and β_3.

Thus, there may be a set of very limited conditions under which it's appropriate to estimate a standard interactive model that omits constitutive terms. However, given that the second condition requires us to estimate a fully specified model to check whether β_2 is in fact zero, we see no compelling reason why we shouldn't simply report these results.

> ⚠ **Important:** Model parameters are likely to be estimated with bias when one or more of the constitutive elements of an interaction term is omitted from an interaction model.

3.2.1 Multicollinearity

You might respond at this point that including all of the constitutive terms in an interaction model increases multicollinearity, thereby increasing the size of the standard errors and making it less likely that the interaction term coefficient will be statistically significant. However, this reasoning doesn't justify the omission of constitutive terms.

---

[9] This is particularly problematic for the coefficient on the interaction term. Earlier we showed that this coefficient in the under-specified model is estimated as $\gamma_3 = \beta_3 + \beta_2\alpha_3$, where β_3 represents the true parameter from the fully specified model, and $\beta_2\alpha_3$ represents the bias. Recall that α_3 comes from the regression of Z on X and XZ; that is, $Z = \alpha_0 + \alpha_1 X + \alpha_3 XZ + \varepsilon$. Given the obvious relationship between Z and XZ, it's implausible that α_3 would ever be zero. This raises the distinct possibility that the coefficient on the interaction term in the under-specified model, γ_3, could suffer from considerable bias even if β_2 is close to zero. We should, therefore, be extremely wary of omitting constitutive terms even if we find that β_2 is statistically insignificant or close to zero.

We start by noting that the problem with multicollinearity has been overstated in the context of interaction models (Friedrich, 1982). Much of the concern about multicollinearity arises when a researcher observes that the coefficients from a linear-additive model change when an interaction term is introduced. This is because the sensitivity of results to the inclusion of an additional variable is often taken as a sign of multicollinearity in the linear-additive world. However, this needn't be the case in the linear-interactive world. Consider the following two models, which are identical except that the second includes the interaction term XZ:

$$Y = \gamma_0 + \gamma_1 X + \gamma_2 Z + \varepsilon, \tag{3.7}$$

$$Y = \beta_0 + \beta_1 X + \beta_2 Z + \beta_3 XZ + \epsilon. \tag{3.8}$$

The key thing to note is that while γ_1 and γ_2 capture the average effects of X and Z on Y, this isn't the case for β_1 and β_2 when there's an interaction effect, $\beta_3 \neq 0$. As we've already seen, β_1 captures the effect of X on Y only when $Z = 0$, and β_2 captures the effect of Z on Y only when $X = 0$. In other words, the inclusion of the interaction term XZ in the second model completely changes the interpretation of the coefficients on X and Z. Since these coefficients now capture different quantities of interest, it's almost certainly the case that they'll be different from the coefficients on X and Z in the linear-additive model. Given this, we shouldn't interpret a change in the coefficients when we add an interaction term as necessarily being a sign of multicollinearity.

Even if there really is high multicollinearity and this leads to large standard errors in the model parameters, it's important to remember that these standard errors are never in any sense "too" large or "inflated" – they're always the "correct" standard errors. High multicollinearity simply means that there isn't enough information in the data to estimate the model parameters accurately, and the standard errors rightfully reflect this.

⚠ **Important:** High multicollinearity is never a reason to omit constitutive terms from an interaction model.

While *high* multicollinearity is never a reason to omit a constitutive term from an interaction model, it's important to recognize that scholars will *not be able* to include all of the constitutive elements of an interaction term when there's *perfect* multicollinearity. Perfect multicollinearity occasionally arises in the context of interactive models when one of the variables involved in the interaction is only observed for a subset of the cases being analyzed. As an example of this, consider a study by Crabtree et al. (2020) that claims that incumbent prime ministerial parties in a coalition government adopt more positive emotional language in their election campaign messages than their incumbent coalition partners. To evaluate their conditional claim, Crabtree et al. include an interaction term, *Incumbent Party* × *Prime Ministerial Party*, in their model specification. Note that all prime ministerial parties are, by definition, also incumbent parties. Put differently, it isn't possible to have a prime ministerial party that isn't also an incumbent party in the government. A consequence of this is that Crabtree et al. *cannot* include the constitutive term *Prime Ministerial Party* in their model due to perfect multicollinearity. If, for some reason, you're unsure as to whether you have perfect multicollinearity, you can try to estimate your interaction model with all of the constitutive terms. If it turns out that you have perfect multicollinearity, the statistical package you're using will automatically omit the 'problematic' variable or indicate that there's a problem. The bottom line is that with perfect multicollinearity, it won't be possible to include all of the constitutive terms. When multicollinearity is high but not perfect, though, scholars should include all of the constitutive terms.

While the problem of multicollinearity has been overstated in the context of interaction models and should never cause the analyst to omit constitutive terms, some scholars have argued that "centering" the relevant variables can help to mitigate any multicollinearity issues that might exist. Centering a variable typically involves subtracting its mean from all values of the variable.[10] There are two reasons why scholars sometimes think that centering the variables in an interaction model helps with multicollinearity. The first reason has to do with variance inflation factor (VIF) scores, which are often used as a diagnostic to measure the level of multicollinearity in a linear regression model. The VIF scores for an interaction model where the variables have been centered are almost always lower than the VIF scores for an interaction model where the variables have been left uncentered. The second reason has to do with the fact that scholars sometimes find that the coefficients on the constitutive terms achieve statistical significance when the variables have been centered but not when the variables have been left uncentered. Unfortunately, while it would be nice if we could reduce multicollinearity in our interaction models by centering the variables as these scholars claim, this is simply not the case. The basic intuition here is that multicollinearity issues arise because there's too little information in the data to accurately estimate our model parameters. Since centering our

[10] Centering a variable involves subtracting a constant from all values of the variable. A variable is "mean centered" when the constant we subtract is the variable's mean value.

variables doesn't provide us with any "new" or "more accurate" data, it can't help us to reduce the level of multicollinearity.

As Kam and Franzese (2007) note, centering "alters nothing important statistically and nothing at all substantively" in an interactive model. This is relatively easy to show. Compare our basic "uncentered" interaction model outlined earlier in Eq. 3.8 with the "centered" model shown below,

$$Y = \delta_0 + \delta_1 X_c + \delta_2 Z_c + \delta_3 X_c Z_c + \epsilon_c, \tag{3.9}$$

where the means have been subtracted from X and Z; that is, $X_c = X - \overline{X}$ and $Z_c = Z - \overline{Z}$. Some substitution and rearranging allows us to rewrite Eq. 3.9 as

$$
\begin{aligned}
Y &= \delta_0 + \delta_1 \left(X - \overline{X} \right) + \delta_2 \left(Z - \overline{Z} \right) + \delta_3 \left(X - \overline{X} \right) \left(Z - \overline{Z} \right) + \epsilon_c \\
&= \delta_0 + \delta_1 X - \delta_1 \overline{X} + \delta_2 Z - \delta_2 \overline{Z} + \delta_3 XZ - \delta_3 X\overline{Z} - \delta_3 \overline{X} Z + \delta_3 \overline{X}\ \overline{Z} + \epsilon_c \\
&= \left(\delta_0 - \delta_1 \overline{X} - \delta_2 \overline{Z} + \delta_3 \overline{X}\ \overline{Z} \right) + \left(\delta_1 - \delta_3 \overline{Z} \right) X + \left(\delta_2 - \delta_3 \overline{X} \right) Z + \delta_3 XZ + \epsilon_c.
\end{aligned}
\tag{3.10}
$$

It should now be clear that this centered model is just an algebraic transformation of our uncentered model from earlier, where $\beta_0 = \delta_0 - \delta_1 \overline{X} - \delta_2 \overline{Z} + \delta_3 \overline{X}\ \overline{Z}$, $\beta_1 = \delta_1 - \delta_3 \overline{Z}$, $\beta_2 = \delta_2 - \delta_3 \overline{X}$, and $\beta_3 = \delta_3$. In other words, the two models are, in fact, the same model. The fact that VIF scores are often lower in the centered version of the model simply reminds us that VIF scores shouldn't be employed in the context of interaction models. In particular, they should never be used to justify the omission of constitutive terms in an interaction model.

⚠ **Important:** Centering the variables in an interaction model does not reduce the level of multicollinearity.

Note that although the centered and uncentered models are effectively the same model, the algebraic transformation that results from centering the variables will almost certainly result in different coefficients and standard errors across the two models. While the coefficient and standard error on the interaction term will always be the same in the centered and uncentered models, this is unlikely to be the case for the coefficients and standard errors on the constitutive terms. This isn't because one model produces better estimates than the other but because the coefficients on the constitutive terms measure different substantive quantities in the two models. For example, β_1 in the uncentered model captures the marginal effect of a one unit increase in X *when Z is zero*, while the equivalent coefficient, δ_1, in the centered model captures the marginal effect of a one unit increase in X_c (the same as a unit increase in X) *when Z is at its mean*.[11] This difference in the substantive quantities captured by the coefficients on the constitutive terms in the centered and uncentered models explains why scholars who find that these coefficients are statistically significant in the centered model but not in the uncentered model shouldn't conclude that this is because multicollinearity is lower in the centered model. If a researcher were to actually calculate the marginal effect of a one unit increase in X at the same level of Z from the estimates of the centered and uncentered models, they'd obtain the exact same marginal effect and measure of uncertainty. Given that the centered and uncentered models are algebraically equivalent, we can unequivocally state that centering doesn't change the statistical certainty of the estimated effects and, therefore, can't mitigate any multicollinearity issues that might exist.[12]

To sum up, scholars should stop justifying the use of centered variables or the omission of constitutive terms in interaction models by claiming that this reduces multicollinearity. In the case of centered variables, this claim is simply unfounded. While it's true that the omission of constitutive terms may well reduce multicollinearity, it's almost certainly unwise to omit constitutive terms when estimating an interaction model due to induced omitted variable bias.

3.3 AN ALTERNATIVE MODEL SPECIFICATION WHEN ONE OF THE MODIFYING VARIABLES IS DISCRETE

All of the interaction models we've looked at so far to test claims from a theory positing an interaction between X and Z have the following specification:

$$Y = \beta_0 + \beta_1 X + \beta_2 Z + \beta_3 XZ + \epsilon. \tag{3.11}$$

As we've seen, it's important to include all of the constitutive elements of the interaction term in this type of setup. However, scholars can use an alternative model specification that omits *some* of the constitutive terms when one of the modifying variables, say Z, is discrete (Wright, 1976; Brambor, Clark and Golder, 2006). To estimate this alternative model, we must first break the discrete variable Z up into a set of logically exhaustive

---

[11] It should be clear that δ_1 will only equal β_1 if the mean value of Z is 0 and that δ_2 will only equal β_2 if the mean value of X is 0.

[12] Some scholars choose to center their variables in the belief that the effects of X and Z on Y are most likely to be statistically significant when the other variable with which they interact is at its mean value. As we discuss in Chapter 4 and show in Appendix C, this belief is incorrect.

and mutually exclusive dichotomous variables that each correspond to one of its K distinct values and then multiply each of these dichotomous variables by X to create K interaction terms. The alternative interactive model specification will include all of the K interactions and either (i) the K dichotomous variables and no constant or (ii) $K-1$ of the dichotomous variables and a constant; it won't include X.

Consider the case where Z is dichotomous and takes on the values 0 and 1. In this case, the alternative interactive model specification with a constant is

$$Y = \gamma_0 + \gamma_1 Z_1 + \gamma_2 X Z_0 + \gamma_3 X Z_1 + \epsilon, \tag{3.12}$$

where Z_0 is a dichotomous variable that equals 1 when $Z = 0$ and 0 otherwise, and Z_1 is a dichotomous variable that equals 1 when $Z = 1$ and 0 otherwise. Things can be simplified in the dichotomous case as Z_1 is just the same as Z. As a result, we can rewrite Eq. 3.12 as

$$Y = \gamma_0 + \gamma_1 Z + \gamma_2 X Z_0 + \gamma_3 X Z + \epsilon. \tag{3.13}$$

As you can see, this model specification omits some of the constitutive elements of the two interaction terms, XZ and XZ_0. Note that X *cannot* appear in the model as a separate term as there's perfect multicollinearity with XZ and XZ_0. The constitutive terms Z and Z_0 are perfectly collinear with each other and, as a result, only one of them can be included in the model specification if we have a constant. In Eq. 3.13, we chose to include Z, but a researcher could choose to include Z_0 instead. The important thing is that the researcher should include either Z or Z_0 in their model specification. This is because the omission of both of these terms constrains the two regression lines (one for $Z = 1$ and one for $Z = 0$) to have the same intercept, γ_0, and potentially biases the estimated parameters in exactly the same way as outlined earlier in Figure 3.1.

The reason why the omission of some of the constitutive elements of the interaction terms in the alternative model specification in Eq. 3.13 isn't problematic is that the model is exactly equivalent to the standard interactive model in Eq. 3.11 that includes all of the constitutive terms (Ferland, 2018). Note that Z_0 is just the opposite of Z; that is, $Z_0 = 1 - Z$. This means that we can create the interaction term XZ_0 simply by multiplying X by $1 - Z$; technically, we never need to create the dichotomous variable Z_0. It also means that we can rewrite Eq. 3.13 as

$$Y = \gamma_0 + \gamma_1 Z + \gamma_2 X(1 - Z) + \gamma_3 X Z + \epsilon. \tag{3.14}$$

Multiplying through, we have

$$Y = \gamma_0 + \gamma_1 Z + \gamma_2 X - \gamma_2 X Z + \gamma_3 X Z + \epsilon. \tag{3.15}$$

And collecting terms, we have

$$Y = \gamma_0 + \gamma_1 Z + \gamma_2 X + (\gamma_3 - \gamma_2)XZ + \epsilon. \tag{3.16}$$

It should now be clear that the alternative interactive model shown in Eq. 3.13 is just an algebraic transformation of the standard interactive model shown in Eq. 3.11, where $\beta_0 = \gamma_0$, $\beta_1 = \gamma_2$, $\beta_2 = \gamma_1$, and $\beta_3 = \gamma_3 - \gamma_2$. In effect, the two models are just different representations of the *same* model.

> ⚠ **Important:** Scholars who posit an interaction in which one of the two modifying variables, say Z, is discrete may choose to adopt an "alternative" interactive specification in which *some* constitutive terms are omitted. The alternative specification involves breaking Z up into a set of dichotomous variables that each correspond to one of its distinct values and then multiplying each of these dichotomous variables by X.

Is it better to use the alternative interactive model when one of our modifying variables is dichotomous? As we've just seen, the alternative interactive specification is algebraically equivalent to the standard interactive specification when one of the modifying variables is dichotomous. As a result, the exact same quantities of interest can be calculated from both model specifications. In this sense, it doesn't matter which specification we use. That said, each model specification makes it is easier to see particular quantities of interest directly from the regression output (Ferland, 2018).

The primary advantage of the alternative specification is that we can directly identify the marginal effect of X on Y when $Z = 0$ *and* when $Z = 1$ from the regression output. Using Eq. 3.13, we see that the marginal effect of X on Y is

$$\frac{\partial Y}{\partial X} = \gamma_2 Z_0 + \gamma_3 Z = \gamma_2 (1 - Z) + \gamma_3 Z. \tag{3.17}$$

As you can see, the marginal effect of X on Y is γ_2 when $Z = 0$ and γ_3 when $Z = 1$. In other words, the coefficient on XZ_0 in the alternative model specification shown in Eq. 3.13 tells us the marginal effect of X when $Z = 0$, and the coefficient on XZ tells us the marginal effect of X when $Z = 1$.

The main drawback of the alternative model specification is that we can't necessarily determine directly from the regression output whether these two marginal effects are statistically different. In other words, we can't always see whether there's any statistically significant interaction between X and Z on Y. When the confidence intervals for the two coefficients γ_2 and γ_3 don't overlap, we know for sure that the coefficients are significantly different, and this indicates that there's a statistically significant interaction. However, when the confidence intervals do overlap, we can't immediately tell whether the two coefficients are significantly different from each other or not (Schenker and Gentleman, 2001). In this situation, we must formally test whether $\gamma_2 = \gamma_3$ or, equivalently, whether $\gamma_3 - \gamma_2 = 0$, to know whether the coefficients are significantly different and hence whether there's any evidence of a statistically significant interaction.

A word of warning is in order at this point. It's not uncommon for scholars to claim evidence of interaction if they find that, say, γ_3 is significant and γ_2 is insignificant, or vice versa. The basis for such a claim, which focuses on the differing levels of statistical significance across the two coefficients, is never justified. The fact that γ_3 is statistically significant indicates only that the marginal effect of X on Y is statistically different from 0 when $Z = 1$. Similarly, the fact that γ_2 is statistically insignificant indicates only that the marginal effect of X on Y is not statistically different from 0 when $Z = 0$. Without additional information, we have no way of knowing whether the marginal effect of X when $Z = 1$, γ_3, is significantly different from the marginal effect of X when $Z = 0$, γ_2. In other words, we have no way of knowing whether Z significantly modifies the marginal effect of X on Y.[13] Paraphrasing the language of Gelman and Stern (2006, 328, 329), we need to look at the statistical significance of the difference in the effects of X across the different values of Z and not the difference between their significance levels. This is because the difference between "significant" and "not significant" may not itself be statistically significant.

Another potential drawback of the alternative specification comes when we think about the symmetry of interactions. So far, we've focused on the marginal effect of X on Y and how this varies depending on whether Z is 0 or 1. As we've seen, the alternative specification makes it especially easy to see the effect of X on Y when $Z = 0$ and when $Z = 1$. But all interactions are symmetric. This means that fully evaluating the predictions that come from a theory positing interaction will often mean also examining the effect of Z at different values of X. Given that the standard and alternative interactive models are equivalent, this can obviously be done with the alternative specification. However, analysts have to be especially careful. It's often appealing when using the alternative specification shown in Eq. 3.13, where Z explicitly appears twice, once as the constitutive term Z and once in the interaction term XZ, to think that the effect of Z on Y is calculated as

---

[13] Of course, if the confidence intervals for the two coefficients don't overlap, then we know that the two marginal effects are significantly different and that there's a statistically significant interaction. However, this has nothing necessarily to do with the differing levels of statistical significance across the two coefficients.

$$\frac{\partial Y}{\partial Z} = \gamma_1 + \gamma_3 X. \qquad (3.18)$$

However, this is incorrect. The problem arises in not recognizing that Z also appears in Eq. 3.13 due to the presence of Z_0 in the interaction term XZ_0. Recall that $Z_0 = 1 - Z$, and so the alternative specification in Eq. 3.13 can be rewritten as

$$
\begin{aligned}
Y &= \gamma_0 + \gamma_1 Z + \gamma_2 X(1 - Z) + \gamma_3 XZ + \epsilon \\
&= \gamma_0 + \gamma_1 Z + \gamma_2 X - \gamma_2 XZ + \gamma_3 XZ + \epsilon.
\end{aligned}
\qquad (3.19)
$$

Once we recognize this, we easily see that the effect of Z is actually calculated as

$$
\begin{aligned}
\frac{\partial Y}{\partial Z} &= \gamma_1 - \gamma_2 X + \gamma_3 X \\
&= \gamma_1 + (\gamma_3 - \gamma_2) X.
\end{aligned}
\qquad (3.20)
$$

The issue here may not strike you as too problematic. After all, it's perhaps somewhat obvious in the generic context that Z enters the alternative interactive specification through the inclusion of both Z and Z_0; in some sense, our variable names give it away. In most substantive applications, though, Z and Z_0 will have variable names that aren't so obviously related. For example, in the substantive application in the next chapter where we have an interaction between a continuous modifying variable and a discrete modifying variable, Z is a dichotomous variable indicating whether someone is Black or White. In the alternative interactive specification, Z is a dichotomous variable called *Black* that is 1 if someone is Black and 0 if otherwise, and Z_0 is a dichotomous variable called *White* that is 1 if someone is White and 0 otherwise. Given these variable names, it's much less obvious when we're calculating the effect of *Black* that *Black* enters the alternative model specification both through the constitutive term *Black* and through *White* in the interaction term $X \times$ *White*. In other words, we're not so primed to recognize that *White* $= 1 -$ *Black*. The point here is simply that extra care must be taken when evaluating the effect of Z in the alternative model specification.

The primary advantage of the standard interactive specification compared to the alternative specification is that we can always directly identify whether there's any statistically significant interaction between X and Z on Y. This is because the coefficient on the interaction term XZ indicates whether there's interaction or conditionality. The coefficient on the interaction term β_3 tells us how the slope of the relationship between X and Y changes with Z and how the slope of the relationship between Z and Y changes with X.

The main drawback of the standard interactive specification is that we can only identify directly from the regression output the marginal effect of X on Y when $Z = 0$ and not the marginal effect of X on Y when $Z = 1$. Recall that the marginal effect of X on Y from the standard interaction model in Eq. 3.11 is

$$\frac{\partial Y}{\partial X} = \beta_1 + \beta_3 Z. \tag{3.21}$$

The regression output provides us with β_1, which indicates the marginal effect of X on Y when $Z = 0$. To get the marginal effect of X on Y when $Z = 1$, though, we must add β_1 and β_3 and determine whether this sum is statistically different from 0; this isn't something that's directly provided in the regression output.

We summarize our comparison of the "standard" and "alternative" interaction models when one modifying variable, Z, is dichotomous in Table 3.1. The information contained in Table 3.1 also links our discussion back to the five key predictions that can typically be derived from a theory positing interaction between X and Z by indicating the quantities of interest necessary for evaluating each prediction. Whether these quantities of interest should be positive, negative, or zero will depend on the particular theory under consideration. As you can see, the standard and alternative interaction models differ in how easy they make it to see particular quantities of interest. No matter which model they employ, though, an analyst who wishes to fully evaluate a theory positing interaction and test all five of the key predictions we saw in Chapter 2 is going to have to make some post-estimation calculations. The regression output provided by both models isn't sufficient on its own to fully evaluate such a theory. Given this,

Table 3.1 Comparing the standard and alternative interaction models when one modifying variable, Z, is dichotomous: Five key predictions

Key Prediction	Standard Interaction Model $Y = \beta_0 + \beta_1 X + \beta_2 Z + \beta_3 XZ + \epsilon$	Alternative Interaction Model $Y = \gamma_0 + \gamma_1 Z_1 + \gamma_2 XZ_0 + \gamma_3 XZ_1 + \epsilon$
1. P_{XZ}	β_3	$\gamma_3 - \gamma_2$
2. $P_{X\mid Z_{\min}}$	β_1	γ_2
3. $P_{X\mid Z_{\max}}$	$\beta_1 + \beta_3$	γ_3
4. $P_{Z\mid X_{\min}}$	$\beta_2 + \beta_3 X_{\min}$	$\gamma_1 + (\gamma_3 - \gamma_2) X_{\min}$
5. $P_{Z\mid X_{\max}}$	$\beta_2 + \beta_3 X_{\max}$	$\gamma_1 + (\gamma_3 - \gamma_2) X_{\max}$

the choice of model specification when testing a theory in which one of the modifying variables is dichotomous is largely a matter of taste.

What about when the discrete modifying variable Z has multiple nominal or ordinal categories? In this situation, an interaction model in its standard form with a single interaction term XZ is rarely appropriate. This is because it *constrains* the modifying effect of Z on the slope relationship between X and Y to be constant across each of Z's categories.[14] To see what's going on here, suppose that our discrete modifying variable Z has three categories, 0, 1, and 2. We might think to estimate the following interactive specification:

$$Y = \beta_0 + \beta_1 X + \beta_2 Z + \beta_3 XZ + \epsilon, \tag{3.22}$$

with a single interaction term XZ. As we know from Eq. 3.21, the marginal effect of X changes by β_3 as we move up each category in Z. In other words, the marginal effect of X on Y is β_1 when $Z = 0$, $\beta_1 + \beta_3$ when $Z = 1$, and $\beta_1 + 2\beta_3$ when $Z = 2$. Put differently, the interactive model shown in Eq. 3.22 constrains the modifying effect of Z on the slope relationship between X and Y to be constant across each of Z's categories. This restriction is almost certainly never warranted in the case where Z is nominal and its categories are unordered. Consider the case where Z captures an individual's race or ethnicity, where 0 = Hispanic, 1 = Black, and 2 = Asian. It's hard to imagine a theory predicting that going from being Hispanic to Black would have the exact same modifying effect on the slope relationship between X and Y as going from Black to Asian. The restriction is, perhaps, slightly more plausible in the case where Z is ordered. Even here, though, we suspect that few theories are actually strong enough to predict that the marginal effect of X on Y changes by the exact same amount as we move up through each of Z's ordered categories. And even if such theories do exist, we might want to test this prediction, something that's not possible with the interactive model shown in Eq. 3.22.

However, this doesn't necessarily mean that we should adopt the alternative interactive specification in this setting. It simply means that

[14] It also constrains the effect of Z on Y to be constant across each of Z's categories. This particular issue is well-known to scholars who are thinking of including a discrete variable with multiple nominal or ordinal values in a linear-additive model. A linear-additive model that includes a discrete variable like this assumes that the effect of increasing this variable's value from, say, 0 to 1, is identical to the effect of increasing its value from 1 to 2, or 2 to 3, and so on. This assumption is likely to be unrealistic in many applications. The typical recommendation to allow these effects to vary is to break the discrete variable up into a series of dichotomous variables that each correspond to one of its K distinct values and include $K-1$ of these dichotomous variables in the model specification; we can't include all K of the dichotomous variables due to perfect multicollinearity. As we'll see, adopting a similar strategy is also useful in the context of interaction models.

we probably shouldn't use the *constrained* interactive specification shown in Eq. 3.22. It turns out that we can easily estimate an *unconstrained* interactive model in its standard form that allows the modifying effect of Z on the slope relationship between X and Y to vary across each of Z's categories. To do this, we must first break up the discrete variable Z into a series of dichotomous variables that each correspond to one of its K distinct values and then multiply $K-1$ of these dichotomous variables by X. We can then specify an unconstrained interaction model in its standard form by including the $K-1$ interactions terms along with all of their constitutive components. When Z has three discrete categories, the unconstrained interactive specification in its standard form is

$$Y = \beta_0 + \beta_1 X + \beta_2 Z_1 + \beta_3 Z_2 + \beta_4 X Z_1 + \beta_5 X Z_2 + \epsilon, \qquad (3.23)$$

where $Z_1 = 1$ when $Z = 1$ and 0 otherwise, and $Z_2 = 1$ when $Z = 2$ and 0 otherwise. We chose to omit the dichotomous variable Z_0 (and its interaction with X) to avoid perfect multicollinearity. The marginal effect of X on Y is now

$$\frac{\partial Y}{\partial X} = \beta_1 + \beta_4 Z_1 + \beta_5 Z_2. \qquad (3.24)$$

From this, we see that the marginal effect of X is β_1 when $Z = 0$ (Hispanics), $\beta_1 + \beta_4$ when $Z = 1$ (Blacks), and $\beta_1 + \beta_5$ when $Z = 2$ (Asians). Since there's no constraint that $\beta_4 = \beta_5$, we can immediately see that the modifying effect of Z on the marginal effect of X is allowed to vary across each of Z's categories. As expected, our unconstrained standard interaction model also allows the effect of Z on Y to vary across each of Z's categories. For example, the effect of increasing Z from 0 to 1 (Black vs. Hispanic) is $\beta_2 + \beta_4 X$, the effect of increasing Z from 0 to 2 (Asian vs. Hispanic) is $\beta_3 + \beta_5 X$, and the effect of increasing Z from 1 to 2 (Asian vs. Black) is $(\beta_3 - \beta_2) + (\beta_5 - \beta_4) X$.

We can, of course, choose instead to adopt the alternative interactive specification in this setting. When the discrete modifying variable Z has three categories, the equivalent alternative interactive specification is

$$Y = \gamma_0 + \gamma_1 Z_1 + \gamma_2 Z_2 + \gamma_3 X Z_0 + \gamma_4 X Z_1 + \gamma_5 X Z_2 + \epsilon. \qquad (3.25)$$

Recall that we can't include X as a separate term as there's perfect multicollinearity with the three interaction terms. And since the model includes a constant, we have to omit one of the K dichotomous variables, in this case Z_0, as these variables are perfectly collinear with each other. With this alternative setup, the marginal effect of X on Y is

$$\frac{\partial Y}{\partial X} = \gamma_3 Z_0 + \gamma_4 Z_1 + \gamma_5 Z_2. \qquad (3.26)$$

We see that the marginal effect of X on Y is γ_3 when $Z=0$ ($Z_0=1$, Hispanic), γ_4 when $Z = 1$ ($Z_1 = 1$, Black), and γ_5 when $Z = 2$ ($Z_2 = 1$, Asian). This means that we can identify the marginal effect of X on Y for each value of Z directly from the regression output by looking at the coefficients γ_3, γ_4, and γ_5. Simply looking at these coefficients, though, doesn't necessarily tell us whether these three effects are significantly different from each other. In other words, we can't always see whether there's any evidence of a statistically significant interaction between X and Z on Y from this alternative specification. When the confidence intervals for the coefficients overlap, we have to formally test whether $\gamma_3 = \gamma_4$, $\gamma_4 = \gamma_5$, and $\gamma_3 = \gamma_5$ to determine whether the coefficients are statistically different and hence whether there's any evidence of a statistically significant interaction. As you can see, this alternative setup allows the modifying effect of Z on the marginal effect of X to vary across Z's distinct categories. For example, the marginal effect of X increases by $\gamma_4 - \gamma_3$ when Z is 1 (Black) instead of 0 (Hispanic), by $\gamma_5 - \gamma_3$ when Z is 2 (Asian) instead of 0 (Hispanic), and by $\gamma_5 - \gamma_4$ when Z is 2 (Asian) instead of 1 (Black). As expected, the alternative setup also allows the effect of Z on Y to vary across each of Z's categories. For example, the effect of increasing Z from 0 to 1 (Black vs Hispanic) is $\gamma_1 + (\gamma_4 - \gamma_3)\, X$, the effect of increasing Z from 0 to 2 (Asian vs Hispanic) is $\gamma_2 + (\gamma_5 - \gamma_3)\, X$, and the effect of increasing Z from 1 to 2 (Asian vs Black) is $(\gamma_2 - \gamma_1) + (\gamma_5 - \gamma_4)\, X$.

The bottom line is that it's appropriate for scholars to adopt either of two exactly equivalent interaction models when one of their modifying variables is discrete. While it's important to include all of the constitutive terms when specifying the interaction model in its standard form, *some* of the constitutive terms will be omitted when specifying it in its alternative form. Given that the two interactive specifications are exactly equivalent, it's largely a matter of taste as to which one we choose to adopt.

3.3.1   Can I Just Split My Sample?

The main advantage of adopting the alternative interactive specification is that we can immediately see the marginal effects of X on Y for each value of the discrete modifying variable Z in the regression output. Some scholars who want to see the same quantities of interest in their regression output instead choose to "split" their sample and estimate a model that includes X on different sub-samples of the data that are restricted to include only those observations that have a particular value of Z. For example, if the discrete modifying variable Z is dichotomous, the scholar might split their sample so that they have a sub-sample of observations where $Z = 0$ and a sub-sample of observations where $Z = 1$. They'd then estimate

$$Y_{Z=0} = \kappa_0 + \kappa_1 X + \varepsilon \qquad (3.27)$$

on the sub-sample where $Z = 0$ to identify the marginal effect of X on Y when $Z = 0$ and

$$Y_{Z=1} = \delta_0 + \delta_1 X + \varepsilon \qquad (3.28)$$

on the sub-sample where $Z = 1$ to identify the marginal effect of X on Y when $Z = 1$.

Is this an appropriate strategy? As we'll see, much depends on the precise nature of one's theory and in particular on whether it makes sense to allow the effects of *all* of the independent variables in one's model to vary across the different values of Z. Critically, the "split-sample" strategy can introduce an often unrecognized asymmetry in how X and Z are allowed to affect the outcome of interest. Even in those cases where the split-sample strategy is theoretically appropriate, though, it's almost always better in our opinion to estimate an equivalent "pooled" interaction model instead. Despite its appearance, the split-sample strategy employs an inherently interactive research design, and we can always specify an equivalent pooled interaction model that provides us with the same results. Significantly, there are conditional claims that can easily be evaluated with a pooled interaction model that can't be so easily evaluated with the split-sample strategy. It's for this reason that we believe that a pooled interaction model is always weakly superior to the split-sample strategy.

Two cases are worth considering: (i) the case where we have a theory in which X and Z interact and there's no need to control for any other variables and (ii) the case where we have a theory in which X and Y interact and there is a need to control for other variables.

3.3.1.1 Case I: When There Are No Control Variables

We'll start with the case where we have a theory in which X and Z interact and there's no need to control for other variables. We'll assume that Z is a dichotomous modifying variable that takes on the value 0 or 1. In this case, the split sample strategy is equivalent to the "pooled" alternative or standard interaction models that we've already discussed in that it produces the exact same marginal effects of X when $Z = 0$ and when $Z = 1$.[15] In other words, κ_1 in the split sample model in Eq. 3.27, which indicates the marginal effect of X when $Z = 0$, will be the same as γ_2 in the alternative

---

[15] There's nothing special here about the fact that Z is a dichotomous variable. Even if Z has multiple nominal or ordinal categories, it's still the case that the split sample strategy will estimate the exact same marginal effects of X on Y for the different values of Z as we get from the alternative and *unconstrained* standard interactive specifications.

interaction model in Eq. 3.13 and the same as β_1 in the standard interaction model in Eq. 3.22. Similarly, δ_1 in the split sample model in Eq. 3.28, which indicates the marginal effect of X when $Z = 1$, will be the same as γ_3 in the alternative interaction model in Eq. 3.13 and the same as $\beta_1 + \beta_3$ in the standard interaction model in Eq. 3.22.[16]

While the estimated marginal effect of X on Y for different values of Z is the same across these different modeling strategies, this isn't the case for the standard errors. One reason for this has to do with the fact that the two regression models in the split-sample strategy necessarily use a smaller sample size than the "pooled" sample used in the alternative and standard interaction models. A second reason that contributes to the standard errors being different is that the observed variation in the independent variable X and the residual variation in the dependent variable Y may differ across the two sub-samples.

The appeal of the split-sample strategy for many scholars is that it makes it easy to see the marginal effects of X on Y both when $Z = 0$ and when $Z = 1$. One of the main drawbacks, as with the alternative interaction specification shown in Eq. 3.13, is that we can't directly determine from the regression output whether these two marginal effects are statistically different from each other. In other words, we can't read directly from the regression output whether there's any evidence of a statistically significant interaction between X and Z on Y.[17] To determine this, we'd need to test whether $\kappa_1 = \delta_1$. The fact that these two coefficients come from different models – Eq. 3.27 and Eq. 3.28 – means that this test, while possible, is not as straightforward as testing whether $\gamma_2 = \gamma_3$ in the alternative model specification shown in Eq. 3.13. Perhaps the easiest thing to do is simply estimate the standard interaction model in Eq. 3.22 and see if the coefficient on the interaction term is statistically significant.

A second drawback of the split sample strategy is that it's easy to overlook the inherent symmetry of interactions that's built into theories

---

[16] The constant term from the model estimated on the sub-sample where $Z = 0$, κ_0, will be exactly the same as the constant terms in the standard interaction model, β_0, and the alternative interaction model, γ_0. And the constant term from the model estimated on the sub-sample where $Z = 1$, δ_0, will be exactly the same as $\beta_0 + \beta_2$ in the standard interaction model and $\gamma_0 + \gamma_1$ in the alternative interaction model. As you'll recall, the intuition behind this is graphically shown in Figure 3.1.

[17] It's not uncommon for scholars to claim evidence of interaction if the coefficient on X in one of the sub-samples is statistically significant, and the coefficient on X in the other sub-sample is not statistically significant. As we noted earlier, though, this inference is never justified as the difference between "significant" and "not significant" may not itself be statistically significant (Gelman and Stern, 2006). In other words, a difference in significance levels doesn't necessarily indicate a statistically significant difference in the marginal effects of X on Y when $Z = 0$ and when $Z = 1$.

positing interaction. As we noted in the previous chapter, most theories positing interaction in the social sciences are strong enough to produce hypotheses about both the marginal effect of X on Y and the marginal effect of Z on Y. While the split sample strategy is nicely set up to provide information about the marginal effect of X on Y, it's not so well-designed to test claims about the effect of Z on Y. We suspect in many cases that scholars who adopt the split-sample strategy aren't even thinking about the effect of Z on Y beyond the way that it modifies the marginal effect of X. To the extent that this is true, these scholars are putting their underlying theory to a weaker test than is possible given the available data. This is not to say that we can't examine the effect of Z on Y with the split-sample strategy. We can. It's just not as straightforward as it is with one of the two pooled interaction models because we now have to compare sums of coefficients across different models. For example, the effect of Z in the split-sample strategy examined here is $(\delta_0 + \delta_1 X) - (\kappa_0 + \kappa_1 X)$ or $(\delta_0 - \kappa_0) + (\delta_1 - \kappa_1) X$.

Note that the alternative interaction model in particular offers all of the benefits associated with the split-sample strategy without most of the drawbacks. Like the split-sample strategy, the alternative interaction model provides us with the marginal effects of X on Y for all values of Z directly in the regression output. The alternative interaction model has the advantage that we need estimate only one regression model, and we can easily test whether the the marginal effects of X on Y for the different values of Z are statistically different without estimating any other models. It's also easier with the alternative interaction model to test any claims we might have about the effect of Z on Y and how this changes with X than it is with the split-sample strategy. Given this, we see no reason to ever prefer the split sample strategy to the alternative interaction model.

3.3.1.2 Case II: When There Are Control Variables

Let's now consider the case where we have a theory in which X and Z interact and we need to control for one or more other variables. We'll continue to assume that Z is a dichotomous modifying variable that takes on the value of 0 or 1. The key thing to recognize here is that the split-sample strategy allows the effects of *all* of the independent variables, including the control variables, to vary across the sub-sample of observations where $Z = 0$ and the sub-sample of observations where $Z = 1$. Many scholars seem unaware that this introduces an asymmetry in how X and Z are allowed to affect the outcome of interest, something that's especially relevant given that our theoretical concern is with how X and Z interact to influence Y. Scholars should think carefully about whether the possible conditional effects of X and Z on Y implied by the split-sample strategy are consistent with their underlying theory.

To see what's going on here, we'll start by considering the following pooled interaction model in its standard form,

$$Y = \beta_0 + \beta_1 X + \beta_2 Z + \beta_3 XZ + \beta_4 W + \epsilon, \tag{3.29}$$

and the same pooled interaction model in its alternative form,

$$Y = \gamma_0 + \gamma_1 Z_1 + \gamma_2 XZ_0 + \gamma_3 XZ_1 + \gamma_4 W + \varepsilon, \tag{3.30}$$

where W is some control variable. These two models, as we've indicated before, are equivalent and produce exactly the same quantities of interest. However, unlike when we had no control variables, they will now almost certainly produce different results from the split-sample strategy where we estimate

$$Y_{Z=0} = \kappa_0 + \kappa_1 X + \kappa_2 W + \epsilon \tag{3.31}$$

on the sub-sample where $Z = 0$ and

$$Y_{Z=1} = \delta_0 + \delta_1 X + \delta_2 W + \epsilon \tag{3.32}$$

on the sub-sample where $Z = 1$. This is because the split-sample strategy allows the effect of the control variable W to vary across the different sub-samples (the different values of Z), whereas the two pooled interaction models shown in Eq. 3.29 and Eq. 3.30 constrain the effect of W to be the same across the different values of Z. The "constrained" pooled interaction models in Eq. 3.29 and Eq. 3.30 are not equivalent to the split-sample strategy when we have control variables.

Note, though, that we can easily estimate an "unconstrained" pooled interaction model that *is* equivalent to the split-sample strategy simply by including additional interactions between each of the control variables and Z. With one control variable W, the *unconstrained* interaction model in its standard form would be

$$Y = \beta_0 + \beta_1 X + \beta_2 Z + \beta_3 XZ + \beta_4 W + \beta_5 WZ + \epsilon. \tag{3.33}$$

This model allows the effects of X *and* W to vary with the value of Z and, as a result, it produces identical point estimates for the quantities of interest to those obtained from the split-sample strategy.[18] As an example, κ_1 in the split-sample model in Eq. 3.31, which tells us the marginal effect of X when $Z = 0$, is the same as β_1 in Eq. 3.33. Similarly, δ_1 in the split-sample model in Eq. 3.32, which tells us the marginal effect of X when $Z = 1$, is the same

[18] The standard errors will be slightly different for the same reasons mentioned earlier when we discussed the case where there was no need to include control variables.

as $\beta_1 + \beta_3$.[19] The *unconstrained* interaction model in its alternative form would be

$$Y = \gamma_0 + \gamma_1 Z_1 + \gamma_2 X Z_0 + \gamma_3 X Z_1 + \gamma_4 W Z_0 + \gamma_5 W Z_1 + \varepsilon. \qquad (3.34)$$

This model again allows the effects of X *and* W to vary with Z and so produces identical point estimates for the quantities of interest to those obtained from the split-sample strategy. For example, κ_1 and κ_2 in the split-sample model in Eq. 3.31, which tell us the marginal effects of X and W when $Z = 0$, are the same as γ_2 and γ_4 respectively in Eq. 3.34. Similarly, δ_1 and δ_2 in the split-sample model in Eq. 3.32, which tell us the marginal effects of X and W when $Z = 1$, are the same as γ_3 and γ_5 in Eq. 3.34. The point here is that we can always write a pooled interaction model that's equivalent to the split-sample strategy.

Scholars who are considering using the split-sample strategy (or an equivalent pooled interaction model) should think carefully about whether doing so is consistent with their underlying theoretical argument. Does it make theoretical sense, for example, to allow the effects of *all* of the independent variables to vary with Z? The decision as to which variables to include in a model and how to include them is always first and foremost a theoretical issue. While allowing the effects of all of the variables to vary with Z increases the flexibility of the model, such an approach does place more demands on the data and can use up possibly valuable degrees of freedom. The inclusion of irrelevant variables in a model leads, as we know, to a loss of efficiency.

The more important issue, though, is that allowing all of the independent variables to vary with Z in the split-sample strategy introduces an asymmetry in how X and Z can affect Y. Due to the symmetry of interactions, allowing the effect of all of the independent variables to vary with Z logically implies that the effect of Z on Y is allowed to vary with each of the independent variables, including the control variables. In other words, we end up with a situation where the effect of X on Y is allowed to vary with Z, but the effect of Z on Y is allowed to vary not only with X but also with every other control variable in the model. It's not clear that this asymmetry is always theoretically appropriate. If this is considered appropriate, then it should be explicitly recognized in the theoretical discussion that generates our hypotheses about the effects of X and Z. In particular, our hypothesis about the effect of Z on Y should, where possible, specify how this effect varies not only with X but also with each of the other factors captured by the control variables.

---

[19] In terms of other quantities of interest, κ_2 in Eq. 3.31, which tells us the marginal effect of W when $Z = 0$, is the same as β_4, and δ_2 in Eq. 3.32, which tells us the marginal effect of W when $Z = 1$, is the same as $\beta_4 + \beta_5$.

Scholars who are thinking about using the split-sample strategy should also consider whether running regression models on different sub-samples changes the meaning of their independent variables. It's not uncommon, for example, for researchers with time-series cross-sectional data to have a theoretical reason to include unit fixed effects. The decision to adopt a split-sample strategy in this context is equivalent to allowing the effects of these unit fixed effects to vary across the different values of Z. In effect, we'll estimate as many fixed effects for the same unit as there are values in Z. Since each unit will have "multiple" fixed effects, it's not clear to us that we can continue to think of these effects as "unit" fixed effects.

For those who decide that their theory really does call for a model in which the effects of all of the independent variables are allowed to vary with Z, it still seems preferable to us to use an equivalent pooled interaction model rather than the split-sample strategy. Again, the pooled interaction model in its alternative form would seem to offer all of the benefits associated with the split-sample strategy without most of the drawbacks. Like the split-sample strategy, the alternative interaction model allows us to see the marginal effects of all of our independent variables for all values of Z directly from the regression output. Moreover, with the pooled alternative interactive specification, we need estimate only one regression model, and we can easily test whether the marginal effects of our variables for the different values of Z are statistically different without estimating additional models. We can also easily test any claims we might have about the marginal effect of Z on Y. In addition, a pooled interaction model has the advantage that we can easily allow some but not all of the effects of our independent variables to vary with Z if that's what our theory implies.

3.4 A DIFFERENT ALTERNATIVE MODEL SPECIFICATION WHEN BOTH MODIFYING VARIABLES ARE DISCRETE

So far, we've shown that we can use the following model specification to test claims from a theory positing an interaction between X and Z on a continuous dependent variable Y:

$$Y = \beta_0 + \beta_1 X + \beta_2 Z + \beta_3 XZ + \epsilon. \tag{3.35}$$

As we've seen, it's important to include all of the constitutive elements of the interaction term in this type of setup. We can use this interactive model specification irrespective of whether the modifying variables X and Z are continuous or discrete. In the previous section, we saw that researchers can use an alternative model specification that excludes *some* of the constitutive terms when one of the modifying variables, say Z, is discrete. In the case where Z is dichotomous, for example, researchers can employ the following alternative model specification:

$$Y = \gamma_0 + \gamma_1 Z_1 + \gamma_2 X Z_0 + \gamma_3 X Z_1 + \epsilon, \tag{3.36}$$

where Z_0 is a dichotomous variable that equals 1 when $Z = 0$ and 0 otherwise, and Z_1 is a dichotomous variable that equals 1 when $Z = 1$ and 0 otherwise. Although this model excludes some of the constitutive elements of the interaction terms, this isn't a problem as the specification is exactly equivalent to the standard interaction model in Eq. 3.35 that includes all of the constitutive terms.

It turns out that scholars can use a *different* alternative model specification that omits *all* of the constitutive terms when both of the modifying variables are discrete. To estimate this new alternative model, we must first break up X and Z into two sets of logically exhaustive and mutually exclusive dichotomous variables that correspond to one of the modifying variables' distinct values and then multiply each of the dichotomous variables obtained from X by each of the dichotomous variables obtained from Z to create a series of interaction terms. If X has m distinct values and Z has n distinct values, we'll end up with $K = m \times n$ interaction terms that are themselves dichotomous variables. The new alternative interactive specification will either include (i) all of the K dichotomous interaction terms and no constant or (ii) $K - 1$ of the dichotomous interaction terms and a constant;[20] none of the constitutive terms for the dichotomous interaction terms will be included. The reason why the omission of *all* of the constitutive terms isn't problematic in this new alternative model specification is that, as with the alternative model specification discussed in the previous section, it's exactly equivalent to the standard interaction model in Eq. 3.35 that includes all of the constitutive terms. The fact that this new alternative specification requires us to include a series of dichotomous variables into a model in an additive manner has led some researchers to claim that they're not actually estimating an interaction model (Reingold, Haynie and Widner, 2020, 13). This is incorrect. These researchers are failing to recognize that the dichotomous variables they're including in their model are, in fact, interaction terms and that the model setup is exactly equivalent to the standard interaction model shown in Eq. 3.35.

To see what's going on here, consider the case where X and Z are both dichotomous and each takes on the value 0 or 1. As an example, X may capture an individual's sex and indicate whether they're female (1) or

---

[20] The dichotomous interaction terms are collinear with each other, and this is why we can't include all of them in a model specification if there's a constant.

male (0), while Z may capture an individual's race and indicate whether they're Black (1) or White (0). In this case, the new alternative model specification with a constant is

$$Y = \gamma_0 + \gamma_1 X_1 Z_0 + \gamma_2 X_0 Z_1 + \gamma_3 X_1 Z_1 + \epsilon, \tag{3.37}$$

where X_0 is a dichotomous variable that equals 1 when $X = 0$ and 0 otherwise, X_1 is a dichotomous variable that equals 1 when $X = 1$ and 0 otherwise, Z_0 is a dichotomous variable that equals 1 when $Z = 0$ and 0 otherwise, Z_1 is a dichotomous variable that equals 1 when $Z = 1$ and 0 otherwise, and $X_0 Z_0$ is the omitted interaction term. All of the interaction terms in the model are, themselves, dichotomous variables. For example, and continuing with our sex and race example, $X_1 Z_0$ is a dichotomous variable that equals 1 if an individual is a White female and 0 otherwise, $X_0 Z_1$ is a dichotomous variable that equals 1 if an individual is a Black male and 0 otherwise, $X_1 Z_1$ is a dichotomous variable that equals 1 if an individual is a Black female and 0 otherwise, and the omitted interaction term $X_0 Z_0$ is a dichotomous variable that equals 1 if an individual is a White male and 0 otherwise. Each of the dichotomous interaction terms capture different categories of observations, with the omitted interaction term determining the "baseline" or "reference" category against which the other categories are compared. Things can be simplified in the case where both modifying variables are dichotomous as X_1 is the same as X, and Z_1 is the same as Z. As a result, we can rewrite Eq. 3.37 as

$$Y = \gamma_0 + \gamma_1 X Z_0 + \gamma_2 X_0 Z + \gamma_3 X Z + \epsilon. \tag{3.38}$$

As you can see, this alternative model specification omits all of the constitutive elements of the interaction terms. The key thing to note here is that this isn't a choice made by the analyst but is instead a simple result of the fact that there would be perfect multicollinearity between any of the constitutive terms and the included interaction terms. In other words, the constitutive terms *cannot* be included with this alternative model specification.

To see that this alternative model specification is exactly equivalent to the standard interaction model shown in Eq. 3.35, start by recognizing that X_0 is just the opposite of X and that Z_0 is just the opposite of Z. In other words, $X_0 = 1 - X$ and $Z_0 = 1 - Z$. This means that we can rewrite Eq. 3.38 as

$$Y = \gamma_0 + \gamma_1 X(1 - Z) + \gamma_2(1 - X)Z + \gamma_3 X Z + \epsilon. \tag{3.39}$$

Multiplying through, we have

$$Y = \gamma_0 + \gamma_1 X - \gamma_1 X Z + \gamma_2 Z - \gamma_2 X Z + \gamma_3 X Z + \epsilon. \tag{3.40}$$

And collecting terms, we have

$$Y = \gamma_0 + \gamma_1 X + \gamma_2 Z + (\gamma_3 - \gamma_1 - \gamma_2) XZ + \epsilon. \tag{3.41}$$

We can now see that the alternative interaction model shown in Eq. 3.37 is just an algebraic transformation of the standard interaction model shown in Eq. 3.35, where $\beta_0 = \gamma_0$, $\beta_1 = \gamma_1$, $\beta_2 = \gamma_2$, and $\beta_3 = \gamma_3 - \gamma_1 - \gamma_2$. In effect, the two models are just different representations of the *same* interaction model. Scholars who employ the alternative model specification shown in Eq. 3.37 should understand that they're estimating an interaction model even if their model includes a series of dichotomous variables that are included in an additive manner. The key comes in recognizing that the dichotomous variables included in the alternative model specification are *interaction terms*. Earlier in the chapter we referred to these types of variables as "hidden interaction terms" as they're so often overlooked and misunderstood by analysts.

⚠ **Important:** Scholars who posit an interaction in which both modifying variables X and Z are discrete may choose to adopt an "alternative" interactive specification in which *all* of the constitutive terms are omitted. The alternative specification involves breaking up X and Z into two sets of dichotomous variables that correspond to one of the modifying variables' distinct values and then multiplying each of the dichotomous variables from X by each of the dichotomous variables from Z to create a series of dichotomous interaction terms.

As an aside, it should now be obvious why it's inappropriate for scholars to estimate a model specification similar to the one shown below:

$$Y = \delta_0 + \delta_1 X_1 Z_1 + \varepsilon. \tag{3.42}$$

We bring this model up as we often come across some form of it in various literatures. In terms of our sex and race example, $X_1 Z_1$ is a dichotomous interaction term that equals 1 if an individual is a Black female and 0 otherwise. Note that this model is identical to the standard interaction model in Eq. 3.35 except that both constitutive terms have been omitted. As we discussed previously, this is problematic because the omission of constitutive terms is likely to produce omitted variable bias that affects all of the model parameters. Note also that in the alternative interactive specification shown in Eq. 3.37, we omitted the dichotomous interaction term $X_0 Z_0$, and so White men act as the reference category. In contrast, the model shown in Eq. 3.42 omits *three* dichotomous interaction terms, $X_0 Z_0$, $X_1 Z_0$, and $X_0 Z_1$. This means that White men, White women, and

Black men are lumped together into the reference category. Put differently, the model shown in Eq. 3.42 is trying to compare Black women against some weighted mixture of three different identity groups, where the weights are related to the proportions that these groups represent in the sample. By lumping these different identity groups together, the model is implicitly assuming that there are no differences between White men, White women, and Black men. While it's certainly possible that this is the case in some settings, the model specification in Eq. 3.42, unlike the standard interaction model in Eq. 3.35 and the alternative interaction model in Eq. 3.37, cannot test this possibility and simply leaves it as an assertion. We encourage scholars not to adopt the model specification shown in Eq. 3.42.

Is it better to use the alternative model specification in Eq. 3.37 when both modifying variables are dichotomous? As we've just seen, the alternative model specification is algebraically equivalent to the standard interaction model shown in Eq. 3.35 when both modifying variables are dichotomous. As a result, the exact same quantities of interest can be calculated from both model specifications. In this sense, it doesn't matter which specification we use. That said, each model specification makes it is easier to see particular quantities of interest directly from the regression output.

One advantage of the alternative model specification is that researchers can directly identify the effect of *jointly* changing X and Z on Y. Recall that the included interaction terms in the alternative specification capture different categories of observations, say White men $(X_0 Z_0)$, White women $(X_1 Z_0)$, Black men $(X_0 Z_1)$, and Black women $(X_1 Z_1)$. If we estimate the alternative model with a constant, then we have to omit one of these interaction terms. The omitted interaction term becomes the baseline or reference category against which the other categories are compared. In the model shown in Eq. 3.37, we omitted $X_0 Z_0$, and so White men act as the reference category. This means that γ_1 indicates the effect of being a White female as opposed to a White male, γ_2 indicates the effect of being a Black male as opposed to a White male, and γ_3 indicates the effect of being a Black female as opposed to a White male. Note that γ_3 captures the effect of changing *both* the sex and race of the individual in the reference category.[21] While the coefficients β_1 and β_2 in the standard interaction model shown in Eq. 3.35 are equivalent to the coefficients γ_1 and γ_2 in the

---

[21] It's important to recognize that γ_3 captures the *joint effect* of X and Z and not the *interaction effect* of X and Z. In other words, γ_3 indicates the joint effect of a one-unit change in X *and* a one-unit change in Z on Y. In contrast, the interaction effect of X and Z refers to how a one-unit change in Z modifies the effect of a one-unit change in X on Y and how a one-unit change in X modifies the effect of a one-unit change in Z on Y.

alternative model shown in Eq. 3.37, there's no equivalent coefficient in the standard interaction model for γ_3. Instead, $\gamma_3 = \beta_1 + \beta_2 + \beta_3$.[22] While it's certainly possible to calculate this quantity from the standard interaction model, it's not possible to read it directly from the regression output as it is with the alternative interaction model.

To take a closer look at exactly what we can read directly from the regression output when we estimate the alternative interaction model in Eq. 3.37 and the standard interaction model in Eq. 3.35, consider the predicted values and effects from the two models shown in Figure 3.2.[23] The predicted values of Y for the four categories of observations are shown in black.[24] The effects of changing X (sex) and Z (race), as well as the interaction effect between X and Z, are shown in gray in smaller font.[25] As you can see, both models allow us to see directly from the regression output the effect of X (being female instead of male) when $Z = 0$ (White) and the effect of Z (being Black instead of White) when $X = 0$ (male). These effects are captured by the coefficients $\gamma_1 = \beta_1$ and $\gamma_2 = \beta_2$.

Both models, though, require us to move beyond the regression output to examine the effect of X (being female instead of male) when $Z = 1$ (Black) and the effect of Z (being Black instead of White) when $X = 1$ (female). To determine whether the effect of being female instead of male when the individual is Black is statistically significant in the standard interaction model, we must formally test whether $\beta_1 + \beta_3 = 0$. To determine the same thing in the alternative interaction model, we must formally test whether $\gamma_3 = \gamma_2$, or equivalently whether $\gamma_3 - \gamma_2 = 0$. To determine whether the effect of being Black instead of White when the individual is female is statistically significant in the standard interaction model, we must formally test whether $\beta_2 + \beta_3 = 0$. To determine the same thing in the alternative interaction model, we must formally test whether $\gamma_3 = \gamma_1$, or equivalently whether $\gamma_3 - \gamma_1 = 0$. A different approach to calculate these quantities of interest in the case of the alternative interaction model is to simply reestimate the model with a different reference category. We would then

[22] Recall that we showed earlier that $\beta_3 = \gamma_3 - \gamma_1 - \gamma_2$, $\beta_1 = \gamma_1$, and $\beta_2 = \gamma_2$. It follows that $\gamma_3 = \beta_1 + \beta_2 + \beta_3$.

[23] Technically, the coefficients for the predicted values are estimated and should therefore have "hats." However, throughout the text, both here and elsewhere, we've chosen not to show "hats" wherever doing so doesn't lead to undue confusion in order to avoid unnecessary clutter.

[24] The predicted values are for an interaction model like those in Eq. 3.35 and Eq. 3.37 where there are no control variables or for similar interaction models where there are control variables but we're interested in knowing the predicted value when the control variables are all set to 0.

[25] The reported effects can be interpreted as the effects of changing X (sex) and Z (race) when there are no control variables or while holding the values of any control variables constant.

(a) Standard Interaction Model

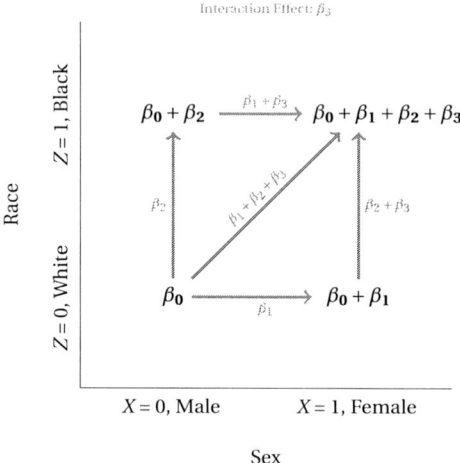

(b) Alternative Interaction Model

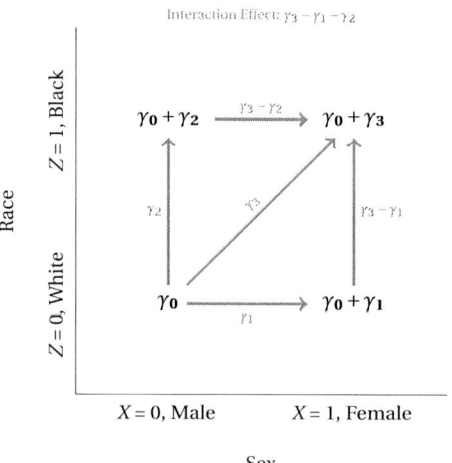

Figure 3.2 (a)–(b) Predicted values and effects from the standard and alternative interaction models

Note: Figure 3.2 shows the predicted values (in black) and effects (in gray) from the standard interaction model shown in Eq. 3.35 and the alternative interaction model shown in Eq. 3.37. The predicted values are for an interaction model like those in Eq. 3.35 and Eq. 3.37 where there are no control variables or for similar interaction models where there are control variables but we're interested in knowing the predicted value when the control variables are all set to 0. The reported effects can be interpreted as the effects of changing X (sex) and Z (race) when there are no control variables or while holding the values of any control variables constant.

be able to read off the desired quantity directly from the regression output. For example, if we omitted the interaction term X_0Z_1 (Black male) instead of X_0Z_0 (White male), then the coefficient on the included interaction term X_1Z_1 (Black female) would tell us the effect of being female instead of male for Blacks.

So far, we've framed our discussion in terms of evaluating the *effects* of X (sex) and Z (race). We did so because theoretical hypotheses typically focus on "effects." In other words, theories usually make predictions about how some dependent variable Y responds to a change in some independent variable such as X. We encourage scholars to think in terms of "effects" and use their empirical analyses as an opportunity to evaluate these effects. However, we can reframe our discussion in terms of whether the various categories of observations are significantly different when it comes to the *predicted value* of the dependent variable. In our sex and race example, for instance, we might be interested in knowing whether the predicted value of the dependent variable is different for White males, White females, Black males, and Black females. We can easily reframe our discussion in this way because evaluating the differences in predicted values across categories is exactly equivalent to examining the effects of sex and/or race.

In the alternative interaction model, we can immediately see whether the included categories are significantly different from the reference category by looking at γ_1, γ_2, and γ_3. In effect, we can immediately see whether White females, Black males, and Black females are significantly different from the omitted category, White males. However, we can't necessarily determine directly from the regression output whether the predicted values for the included categories are significantly different from each other. In other words, we can't necessarily tell whether White females, Black males, and Black females are significantly different from each other. If the confidence intervals for γ_1, γ_2, and γ_3 overlap, then we need to formally test whether $\gamma_3 - \gamma_1 = 0$ (Black female versus White female), whether $\gamma_3 - \gamma_2 = 0$ (Black female versus Black male), and whether $\gamma_2 - \gamma_1 = 0$ (Black male versus White female). Equivalent tests are that $\gamma_3 = \gamma_1$, $\gamma_3 = \gamma_2$, and $\gamma_2 = \gamma_1$. In the standard interaction model, we can immediately see whether White females are different from White males and whether Black males are different from White males by looking at the coefficients β_1 and β_2. To compare other groups, though, we have to formally test whether $\beta_2 + \beta_3 = 0$ (Black female versus White female), whether $\beta_1 + \beta_3 = 0$ (Black female versus Black male), whether $\beta_1 + \beta_2 + \beta_3 = 0$ (Black female versus White male), and whether $\beta_2 - \beta_1 = 0$ (Black male versus White female).

The key advantage of the standard interaction model is that we can directly identify whether there's any statistically significant interaction between X (sex) and Z (race) on Y. This is because the coefficient on the interaction term β_3 indicates whether there's interaction or conditionality. Recall that β_3 tells us how the relationship between X and Y changes with Z and how the relationship between Z and Y changes with X. In contrast, there's no way of knowing directly from the regression output with the alternative model specification whether X (sex) and Z (race) interact to determine Y. Instead, we must formally test whether $\gamma_3 - \gamma_1 - \gamma_2 = 0$ to determine if there's any evidence of interaction or conditionality. As an aside, those who fail to recognize that the alternative model specification is an *interactive* specification are likely to overlook the need to test this particular prediction as they won't be thinking to look for evidence of interaction. A consequence is that they won't know whether their empirical evidence is consistent or not with X (sex) and Z (race) simply having an additive effect on Y.

⚠ **Important:** It is not possible to determine directly from the regression output if there is an interaction effect of X and Z on Y when estimating the alternative interaction model

$$Y = \gamma_0 + \gamma_1 X_1 Z_0 + \gamma_2 X_0 Z_1 + \gamma_3 X_1 Z_1 + \epsilon.$$

Instead, the researcher must test whether $\gamma_3 - \gamma_1 - \gamma_2 = 0$.

We summarize our comparison of the standard and alternative inter-action models when both modifying variables X and Z are dichotomous in Table 3.2. The information contained in the table also links our discussion back to the five key predictions that can typically be derived from a theory positing interaction between X and Z by indicating the quantities of interest necessary for evaluating each prediction. Whether these quantities of interest should be positive, negative, or zero will depend on the particular theory under consideration. Table 3.2 also includes a prediction about the joint effect of changing X and Z. This is equivalent to seeing if the observations from the category $X_1 Z_1$ are different from the observations in the reference category $X_0 Z_0$. As you can see, the standard and alternative interaction models discussed in this section differ in how easy they make it to see particular quantities of interest. No matter which model they employ, though, an analyst who wishes to fully evaluate a theory positing interaction and test all five of the key predictions that we saw in Chapter 2 is going to have to make some post-estimation calculations. The regression output provided by both models isn't sufficient on its own to fully evaluate such a theory. Given this, the choice of model specification when testing a theory in which both of the modifying variables are dichotomous is largely a matter of taste.

Table 3.2 Comparing the standard and alternative interaction models when both modifying variables, X and Z, are dichotomous: Five key predictions

Key Prediction	Standard Interaction Model $Y = \beta_0 + \beta_1 X + \beta_2 Z + \beta_3 XZ + \epsilon$	Alternative Interaction Model $Y = \gamma_0 + \gamma_1 X_1 Z_0 + \gamma_2 X_0 Z_1 + \gamma_3 X_1 Z_1 + \varepsilon$
1. P_{XZ}	β_3	$\gamma_3 - \gamma_1 - \gamma_2$
2. $P_{X\mid Z_{\min}}$	β_1	γ_1
3. $P_{X\mid Z_{\max}}$	$\beta_1 + \beta_3$	$\gamma_3 - \gamma_2$
4. $P_{Z\mid X_{\min}}$	β_2	γ_2
5. $P_{Z\mid X_{\max}}$	$\beta_2 + \beta_3$	$\gamma_3 - \gamma_1$
Joint Effect of X and Z	$\beta_1 + \beta_2 + \beta_3$	γ_3

What about when one or both of the discrete modifying variables X and Z has multiple nominal or ordinal categories? The logic here is exactly the same as that presented in the previous section when we discussed having a single discrete modifying variable with multiple categories. An interaction model in its standard form with a single interaction term XZ is almost never appropriate because it (i) constrains the effect of X (Z) to be constant across each of its categories and (ii) constrains the modifying effect of Z (X) on the relationship between X (Z) and Y to be the same across each of Z's (X's) categories. Scholars should, instead, either adopt the *unconstrained* standard interaction model or the equivalent alternative interaction model.

To briefly see what this looks like, consider the case in which X captures an individual's sex, where $0 =$ male and $1 =$ female, and Z captures an individual's race or ethnicity, where $0 =$ Hispanic, $1 =$ Black, and $2 =$ Asian. In this particular case, the unconstrained interaction model in its standard form is

$$Y = \beta_0 + \beta_1 X + \beta_2 Z_1 + \beta_3 Z_2 + \beta_4 XZ_1 + \beta_5 XZ_2 + \epsilon, \qquad (3.43)$$

where $Z_1 = 1$ when $Z = 1$ and 0 otherwise, and $Z_2 = 1$ when $Z = 2$ and 0 otherwise. The observant among you will notice that this is the exact same unconstrained model in its standard form that we specified in Eq. 3.23 when X was continuous and Z had three discrete categories. We encourage you to look back at our earlier discussion to remind yourself of how this model works.

The equivalent alternative interactive specification is

$$Y = \gamma_0 + \gamma_1 X_0 Z_1 + \gamma_2 X_0 Z_2 + \gamma_3 X_1 Z_0 + \gamma_4 X_1 Z_1 + \gamma_5 X_1 Z_2 + \epsilon,$$

$$(3.44)$$

where $X_0 = 1$ when $X = 0$ and 0 otherwise, $X_1 = 1$ when $X = 1$ and 0 otherwise, $Z_0 = 1$ when $Z = 0$ and 0 otherwise, $Z_1 = 1$ when $Z = 1$ and 0 otherwise, and $Z_2 = 1$ when $Z = 2$ and 0 otherwise. As before, all of the interaction terms are dichotomous variables. For example, $X_0 Z_1$ is a dichotomous variable that equals 1 if an individual is a Black male and 0 otherwise, $X_0 Z_2$ is a dichotomous variable that equals 1 if an individual is an Asian male and 0 otherwise, $X_1 Z_0$ is a dichotomous variable that equals 1 if an individual is a Hispanic female and 0 otherwise, $X_1 Z_1$ is a dichotomous variable that equals 1 if an individual is a Black female and 0 otherwise, $X_1 Z_2$ is a dichotomous variable that equals 1 if an individual is an Asian female and 0 otherwise, and the omitted interaction term $X_0 Z_0$ is a dichotomous variable that equals 1 if an individual is a Hispanic male and 0 otherwise. As always, the omitted interaction term acts as the reference category against which the other categories are compared, and so γ_1 indicates the effect of being a Black male instead of a Hispanic male, γ_2 indicates the effect of being an Asian male instead of a Hispanic male, γ_3 indicates the effect of being a Hispanic female instead of a Hispanic male, γ_4 indicates the effect of being a Black female instead of a Hispanic male, and γ_5 indicates the effect of being an Asian female instead of a Hispanic male. As you can see, the coefficients indicate the effect of changing the sex and/or race of the reference category, Hispanic males.

Importantly, the alternative interaction model allows the interaction effect between X (sex) and Z (race) to vary in magnitude depending on an individual's particular sex and race category. For example, the modifying effect of X (being female) on the relationship between Z_0 (being Hispanic) and Y is γ_3, the modifying effect of X (being female) on the relationship between Z_1 (being Black) and Y is $\gamma_4 - \gamma_1$, and the modifying effect of X (being female) on the relationship between Z_2 (being Asian) and Y is $\gamma_5 - \gamma_2$. Due to the inherent symmetry of interactions, these interaction effects also describe how being Hispanic, Black, and Asian modifies the effect of X (being female) on Y. Scholars can test not only whether each of these interaction effects is statistically significant but also whether they're significantly different from one another.

In this chapter, we've looked at how to appropriately specify linear-interactive models with a continuous dependent variable. In particular, we emphasized the importance of including all of the constitutive elements of an interaction term when specifying an interaction model in its standard form. We've also seen how an alternative interactive model specification, in which *some* of the constitutive terms are omitted, can be useful when one of the modifying variables is discrete. This alternative model specification is exactly equivalent to a standard interactive model that includes all of the constitutive elements of an interaction term. Finally, we saw how a different alternative interactive model specification, in which *all* of the constitutive terms are omitted, can be appropriate when both of the modifying variables are discrete. This alternative model specification is also exactly equivalent to a standard interactive model that includes all of the constitutive elements of an interaction term. In the next chapter, we focus in more detail on how to calculate and interpret the key quantities of interest from an interaction model. Among other things, we introduce how to calculate measures of uncertainty for our quantities of interest and take a quick look at how to interpret some prototypical results from a linear-interactive model.

3.5 EXERCISES

1. Suppose that we have a theory in which X and Z interact to influence Y. The X variable is measured on a 0–10 scale, and Z is a continuous variable whose values range from 0 to 1. After estimating a standard linear-interactive model,

$$Y = \beta_0 + \beta_1 X + \beta_2 Z + \beta_3 XZ + \epsilon, \qquad (3.45)$$

we obtain the following results: $\beta_0 = 1.00$ (0.56), $\beta_1 = 0.20$ (0.08), $\beta_2 = -2.00$ (0.97), and $\beta_3 = 0.40$ (0.13), where the parentheses indicate standard errors.

   a. Suppose that we now rescale X so that its values run from -5 to $+5$ and interact this new variable with Z. Based on the results from the original model, indicate, where possible, the coefficients and standard errors that we'll obtain if we now estimate the same linear-interactive model specification shown in Eq. 3.45 with our new rescaled variables.

   b. Some scholars argue that it's appropriate to omit a constitutive term like Z from their interaction model if they have a theoretical prediction that the marginal effect of Z on Y is zero when $X = 0$. Under what conditions is it possible to have an *a priori* prediction that the marginal effect of Z on Y is zero when $X = 0$?

2. Why do some countries have many political parties, while others have few? A common explanation, as we saw in Chapter 1, points to the interaction between a country's social divisions and the permissiveness of its electoral system (Duverger, 1954; Clark and

Golder, 2006). According to this explanation, social divisions are the primary engine behind the formation of parties – the more divisions there are, the greater the demand for parties to form. The extent to which this latent demand for parties actually leads to the existence of new parties, though, depends on the permissiveness of the electoral system. Only if a country has high levels of social division and permissive electoral systems should we expect a country to have many political parties. To examine this explanation, we estimated the following interactive model using data from Amorim Neto and Cox (1997) on 51 countries:

Number of Parties
$$= \delta_0 + \delta_1 Social\ Divisions_c + \delta_2 Electoral\ Permissiveness_c$$
$$+ \delta_3 Social\ Divisions_c \times Electoral\ Permissiveness_c + \varepsilon, \qquad (3.46)$$

where the independent variables have been mean-centered. Our results, along with the mean variance inflation factor score, are shown in the first column of Table 3.3. The mean value for the uncentered *Social Divisions* variable is 1.57, and the mean value for the uncentered *Electoral Permissiveness* variable is 1.53.

a. How do we interpret the coefficient on the constitutive term *Social Divisions_c*?

b. How do we interpret the coefficient on the constitutive term *Electoral Permissiveness_c*?

c. What is the marginal effect of *Social Divisions* when *Electoral Permissiveness* is zero?

d. What is the marginal effect of *Electoral Permissiveness* when *Social Divisions* is zero?

We also estimated the same basic interactive model specification using the uncentered independent variables. We did not put the results from this regression in the second column of Table 3.3. However, we did report the mean variance inflation factor score from this uncentered regression along with the number of observations.

e. The variance inflation factor score is 6.0 in the uncentered model and 1.19 in the centered model. These scores suggest that multicollinearity is higher in the model with the uncentered variables. The coefficients on the constitutive terms, *Social Divisions* and *Electoral Permissiveness*, were not statistically significant in the uncentered model, but they are in the centered model. Does this information suggest that the model with the centered variables is better than the model with the uncentered variables? If so, why? If not, why not?

Table 3.3 Social divisions, electoral system permissiveness, and the number of parties

Dependent Variable: *Number of Political Parties*

	Mean-centered	Uncentered
Social Divisions	1.11***	
	(0.38)	
Electoral Permissiveness	0.61***	
	(0.16)	
Social Divisions ×	0.83***	
Electoral Permissiveness	(0.29)	
Constant	3.69***	
	(0.22)	
Observations	51	51
R^2	0.30	
Variance Inflation Factor	1.19	6.00

Standard errors in parentheses. $^*p < 0.10$; $^{**}p < 0.05$; $^{***}p < 0.01$ (two-tailed)

    f. Based on the information that we've provided, fill in as much information as you can in the second column of Table 3.3 about the coefficients, standard errors, and R^2 that we obtained when we estimated the uncentered model.

3. In a 2006 article in the *American Journal of Political Science*, Adams et al. (2006) look at how niche and mainstream political parties change their policy positions between elections in response to changes in public opinion. One of the claims in their article is that mainstream parties are more responsive to changes in public opinion than niche parties. In a replication of their study, Ferland (2018) estimates the following alternative interactive model:

$$\begin{aligned} Party\ Position\ Shift = \gamma_0 &+ \gamma_1 Niche \\ &+ \gamma_2 Mainstream \times Public\ Opinion\ Shift \\ &+ \gamma_3 Niche \times Public\ Opinion\ Shift \\ &+ \gamma\ Controls + \varepsilon, \end{aligned} \tag{3.47}$$

Table 3.4 Political responsiveness of niche and mainstream parties

	Alternative Specification	Standard Specification
Dependent Variable: *Party Position Shift*		
Niche	0.02	
	(0.12)	
Public Opinion Shift × *Mainstream*	0.97***	
	(0.19)	
Public Opinion Shift × *Niche*	−0.55**	
	(0.26)	
Public Opinion Shift		
Constant	0.09	
	(0.18)	
Controls	Yes	Yes
Observations	158	158
R^2	0.33	

Standard errors in parentheses. $^{*}p < 0.10$; $^{**}p < 0.05$; $^{***}p < 0.01$ (two-tailed)

where *Controls* refers to several control variables. *Party Position Shift* captures the change in a party's left-right policy position between the current election and the previous election; it's measured on a 0–10 scale where higher numbers indicate a shift to the right. *Public Opinion Shift* captures the change in public opinion between the current election and the previous election; it's also measured on a 0–10 scale where higher numbers indicate a shift to the right. *Niche* is a dichotomous variable that equals 1 if a party is a niche party and 0 if it's a mainstream party. *Mainstream* is the opposite of *Niche* in that it's a dichotomous variable that equals 1 if a party is a mainstream party and 0 if it's a niche party. The results from the replication analysis are shown in the first column of Table 3.4.

a. The model specification shown in Eq. 3.47 doesn't include all of the constitutive elements of the interaction terms. Is this problematic?

    b. How do we interpret the coefficient on the interaction term *Public Opinion Shift × Mainstream*?

    c. How do we interpret the coefficient on the interaction term *Public Opinion Shift × Niche*?

    d. How do we interpret the coefficient on *Niche*?

    e. The only constitutive term included in Eq. 3.47 is *Niche*. Would it be appropriate to omit *Niche* and replace it with *Mainstream*? Would this change how we interpret the coefficients on the interaction terms?

In their original article, Adams et al. didn't present results from the "alternative" interactive specification shown in Eq. 3.47. Instead, they presented results from the following "standard" interactive specification:

$$\textit{Party Position Shift} = \beta_0 + \beta_1 \textit{Public Opinion Shift} + \beta_2 \textit{Niche}$$
$$+ \beta_3 \textit{Public Opinion Shift} \times \textit{Niche}$$
$$+ \beta \textit{Controls} + \epsilon. \qquad (3.48)$$

    f. Use the results from the alternative model specification shown in the first column of Table 3.4 to identify as much information as you can about the coefficients, standard errors, and R^2 that Adams et al. obtained when they estimated the standard interactive model specification shown in Eq. 3.48. Put this information in the second column of Table 3.4.

    g. How do we interpret the coefficient on *Public Opinion Shift*?

    h. How do we interpret the coefficient on *Niche*?

    i. How do we interpret the coefficient on the interaction term, *Public Opinion Shift × Niche*?

4. In a 2012 *American Journal of Political Science* article, Banks and Valentino (2012) argue, among other things, that anger is the primary emotional trigger of Whites' negative racial attitudes in the contemporary period. Specifically, they claim that opposition by Whites toward race-conscious policies such as affirmative action depend on the interaction between an individual's level of symbolic racism (or racial resentment) and their level of anger. Using their data, we estimated the following linear-interactive model specification:

$$\textit{Oppose Race Policies} = \beta_0 + \beta_1 \textit{Symbolic Racism} + \beta_2 \textit{Anger}$$
$$+ \beta_3 \textit{Symbolic Racism} \times \textit{Anger} + \epsilon, \quad (3.49)$$

where *Oppose Race Policies* is a continuous variable on a 0–1 scale that captures an individual's level of opposition to race-conscious policies, *Anger* is a dichotomous variable that equals 1 when an

	Dependent Variable: *Oppose Race Policies*			
	Full Sample		**Split Sample**	
	Standard Specification	Alternative Specification	Calm	Angry
Symbolic Racism	0.52*** (0.05)			
Anger	−0.11 (0.10)			
Symbolic Racism × Calm				
Symbolic Racism × Anger	0.26** (0.12)			
Constant	0.27*** (0.04)			
Controls	No	No	No	No
Observations	182	182	142	40
R^2	0.46			

Table 3.5 Symbolic racism, anger, and opposition to race-conscious policies I

Standard errors in parentheses. *$p < 0.10$; **$p < 0.05$; ***$p < 0.01$ (two-tailed)

individual is angry and 0 otherwise, and *Symbolic Racism* is a continuous variable that measures an individual's level of racial resentment on a 0–1 scale. Our results are shown in the first column of Table 3.5.

a. Use the results from the standard interactive model specification shown in the first column of Table 3.5 to identify as much information as you can about the coefficients, standard errors, and R^2 that we'd obtain if we used the alternative interactive specification in which we break *Anger* up into two dichotomous variables, *Anger* and *Calm*, that are then interacted with *Symbolic Racism*. Put this information in the second column of Table 3.5.

b. Use the results from the standard interactive specification shown in the first column of Table 3.5 to identify as much information as

you can about the results we'd obtain if we were to split our
sample and estimate an additive model on a sub-sample where
Anger $= 0$ (Calm) and a sub-sample where *Anger* $= 1$ (Angry). Put
this information in the third and fourth columns of Table 3.5.

c. When Banks and Valentino (2012) estimated their model, they
included a variety of control variables. Imagine that the results in
the first column of Table 3.5 are based on a model that includes
control variables. What information can we identify about the
coefficients, standard errors, and R^2 if we were to now split our
sample and estimate additive models on the sub-sample where
Anger $= 0$ and the sub-sample where *Anger* $= 1$? Why is this
the case?

5. In Chapter 5, we replicate an analysis by Block, Golder and Golder
(2023) that examines how gender and race affects support for the
Republican Party in the United States. Given the complex nature of
gender, race, and partisan politics in the United States, we might
expect that the level of support for the Republican Party would vary
depending on whether an individual is a White female, Black female,
White male, or Black male. Using data from the 2016 American
National Election Study (2019), we estimate the following model,

$$Like\ Republican\ Party = \gamma_0 + \gamma_1\ White\ Female + \gamma_2\ Black\ Male$$
$$+ \gamma_3\ Black\ Female + \gamma_4\ Age + \epsilon, \qquad (3.50)$$

in which *Like Republican Party* is a 0–10 index where 0 indicates
that an individual strongly dislikes the Republican Party and 10
indicates that an individual strongly likes the Republican Party,
White Female is a dichotomous variable that equals 1 if an individual
is a White female and 0 otherwise, *Black Male* is a dichotomous
variable that equals 1 if an individual is a Black male and 0
otherwise, *Black Female* is a dichotomous variable that equals 1 if an
individual is a Black female and 0 otherwise, and *Age* is a continuous
variable that indicates an individual's age in years. Our results are
shown in the first column of Table 3.6.

a. Is the model shown in Eq. 3.50 an interaction model? How do
you know?

b. How do we interpret the coefficients on *White Female, Black
Male,* and *Black Female*?

c. What is the "interaction effect" of gender and race on support for
the Republican Party?

d. The model specification shown in Eq. 3.50 treats White men as
the reference category. Suppose that we now re-estimate our
model but treat Black women as the reference category. Use the

Table 3.6 Gender, race, and support for the Republican Party in the United States

Dependent Variable: *Like Republican Party, 0–10*

	Alternative Specification I	Alternative Specification II	Standard Specification
White Female	−0.04 (0.12)		
Black Male	−1.50*** (0.27)		
Black Female	−2.57*** (0.22)		
White Male			
Female			
Black			
Female×Black			
Age	0.02*** (0.003)		
Constant	4.47*** (0.18)		
Observations	2,858	2,858	2,858
R^2	0.07		

Standard errors in parentheses. $^*p < 0.10$; $^{**}p < 0.05$; $^{***}p < 0.01$ (two-tailed)

results from the model specification shown in the first column of Table 3.6 to identify as much information as you can about the coefficients, standard errors, and R^2 that we'd obtain from re-estimating our model in this way. Put this information in the second column of Table 3.6.

e. Suppose that we now estimate the following interactive model,

$$Like\ Republican\ Party = \beta_0 + \beta_1 Female + \beta_2 Black$$
$$+ \beta_3 Female \times Black + \epsilon, \qquad (3.51)$$

where *Female* is a dichotomous variable that equals 1 if an individual is female and 0 if an individual is male, and *Black* is a dichotomous variable that equals 1 if an individual is Black and 0 if an individual is White. Use the results from the model specification shown in the first column of Table 3.6 to identify as much information as you can about the coefficients, standard errors, and R^2 that we'd obtain from estimating our model in this way. Put this information in the third column of Table 3.6.

4 Interpreting Quantities of Interest
Effects, Predicted Values, and Measures of Uncertainty

So, you've developed a theory positing interaction, and you've carefully derived several hypotheses. You've also specified an appropriate model that can test these hypotheses and obtained some results. What's next? In this chapter, we'll take a more in depth look at how to calculate and interpret the key quantities of interest from an interaction model. As before, we'll continue to focus our attention on linear-interactive models that posit interaction between two independent variables X and Z on a continuous dependent variable Y.

We begin the chapter by discussing how to interpret the effects of both continuous and discrete independent variables in the context of interaction models. Our main piece of advice is to think about the effects of the variables in terms of derivatives and differences rather than coefficients. This can help to avoid some of the most common inferential errors that people make when interpreting the results from an interaction model. To evaluate the hypotheses that come from a theory positing interaction, we must determine the statistical and substantive significance of our results. In the second part of the chapter, we show how to calculate appropriate measures of uncertainty for the effects of our independent variables and the predicted values associated with different scenarios. While calculating measures of uncertainty isn't necessarily difficult to do, it does typically require that we look beyond the information contained in our immediate regression output or in a traditional table of results. After briefly reminding ourselves of the key quantities of interest that are necessary to fully evaluate a theory positing interaction, the third part of the chapter examines several sets of prototypical results that we might obtain when estimating an interaction model in order to assess the extent to which we'd feel comfortable claiming support for the underlying theory. Our discussion here suggests that we should avoid establishing hard and fast rules about exactly what combinations of evidence regarding the key quantities of interest constitute support for a theory positing interaction. That said, finding evidence of an "interaction effect" should be treated as necessary for concluding that such a theory is supported.

4.1 THINK ABOUT EFFECTS IN TERMS OF DERIVATIVES AND DIFFERENCES

While we've briefly discussed how to interpret the results from a linear-interactive model in the previous chapters, it's worth looking in more detail at how to interpret the effects of continuous and discrete independent variables in the context of these types of models. In a standard linear-additive model, the individual coefficients tell us the direction and magnitude of the effects of our independent variables on the dependent variable. However, this is *not* the case in a linear-interactive model. To understand how the value of a dependent variable responds to a change in the value of an independent variable, we need to think in terms of *derivatives* and *differences*. The reason why we can look at the individual coefficients in a linear-additive model to see the effects of our variables is that the coefficients *are* derivatives or differences. This isn't the case for the individual coefficients in a linear-interactive model. Scholars who are interested in evaluating the effects of their variables in an interaction model must look beyond the individual coefficients and calculate the appropriate derivatives or differences.

Consider the following linear-additive model with two independent variables X and Z:

$$Y = \delta_0 + \delta_1 X + \delta_2 Z + \upsilon. \tag{4.1}$$

We'll assume for illustrative purposes that X is continuous, and Z is discrete. Conceptually, there are two distinct ways to think about the effect of a continuous variable like X on Y. First, we can think in terms of derivatives and differentiate Y with respect to X,

$$\frac{\partial Y}{\partial X} = \delta_1. \tag{4.2}$$

As you can see, the coefficient on X is a derivative, and it gives us what's called the *marginal effect* of X on Y. A derivative gets at the effect of a continuous independent variable on a dependent variable because it represents a slope or a rate of change. In effect, the derivative tells us the slope relationship between Y and X or, equivalently, the rate at which Y changes when X changes. Technically, the derivative tells us how Y changes with an infinitesimally small, or "marginal," increase in X divided by the size of the increase in X.[1] However, since our model is linear in X, the slope relationship between Y and X is constant, and so the derivative, or marginal effect, can also be interpreted as the effect of a *one-unit* increase in X on Y. In other words, the coefficient on X tells us that a one-unit increase in X leads to a δ_1 increase in Y.

Second, we can think of the effect of X on Y in terms of differences. Thinking in this way doesn't require knowledge of calculus and simply involves looking at how the predicted value of the dependent variable $\hat{Y}$ responds to a given change in X while holding everything else constant. The difference in the predicted value of Y when we change X from some "baseline" value X_b to some "counterfactual" value X_c is

$$\hat{Y}_{X_c} - \hat{Y}_{X_b} = (\delta_0 + \delta_1 X_c + \delta_2 Z) - (\delta_0 + \delta_1 X_b + \gamma_2 Z)$$
$$= \delta_1 (X_c - X_b). \tag{4.3}$$

In the case where we increase X by one unit from its baseline value to its counterfactual value, $X_c = X_b + 1$, our difference simplifies to

$$\hat{Y}_{X_c} - \hat{Y}_{X_b} = \delta_1 (X_b + 1 - X_b) = \delta_1. \tag{4.4}$$

As you can see, the coefficient on X is a difference and tells us that a one-unit increase in X leads to a δ_1 increase in Y. This confirms our earlier claim that the marginal effect of X on Y is exactly the same as the effect of a one-unit increase in X on Y. This equivalence isn't a coincidence, and these two quantities will always be the same in linear-additive models. They'll also be the same in linear-interactive models in which the interaction is itself linear.[2]

Conceptually, there's only one way to think about the effect of a discrete variable like Z on Y. The concept of a derivative doesn't make sense when a variable is discrete as we can't have an infinitesimally small or "marginal" change in a discrete variable. This means that when we have a discrete variable like Z, we need to think, at least conceptually, in terms of differences to evaluate its effect. The difference in the predicted value of Y when we change Z by some discrete amount from some "baseline" value Z_b to some "counterfactual" value Z_c is

---

[1] To be even more precise, it tells us the slope relationship (or rate of change) between Y and X *at a particular value of* X. For linear models, though, the slope relationship between Y and X is the same at all values of X, and so we can essentially ignore this point. As we'll see in Chapter 7, though, things are different in non-linear models as the slope relationship between Y and X depends (among other things) on the value of X.

[2] As we'll see in Chapter 7, these quantities are not the same in non-linear models. This is one of the reasons why it's important to be aware of the conceptual distinction between derivatives and differences.

$$\hat{Y}_{Z_c} - \hat{Y}_{Z_b} = (\delta_0 + \delta_1 X + \gamma_2 Z_c) - (\delta_0 + \delta_1 X + \delta_2 Z_b)$$
$$= \delta_2 (Z_c - Z_b). \tag{4.5}$$

In the case where we increase Z by one unit from its baseline value to its counterfactual value, $Z_c = Z_b + 1$, our difference simplifies to

$$\hat{Y}_{Z_c} - \hat{Y}_{Z_b} = \delta_2 (Z_b + 1 - Z_b) = \delta_2. \tag{4.6}$$

As you can see, the coefficient on Z is a difference and tells us that a one-unit increase in Z is associated with a δ_2 increase in Y.

Calculating differences can be rather tedious. As we saw earlier, calculating the effect of a one-unit difference in some variable produces the exact same quantity in a linear-additive setting as calculating the marginal effect of that variable. What this means in practice is that if we know how to differentiate, we can also find the effect of a one-unit increase in a discrete variable like Z by treating it "as if" it were continuous and calculating a derivative,

$$\frac{\partial Y}{\partial Z} = \delta_2. \tag{4.7}$$

This is much simpler. We just have to remember that Z isn't actually continuous, and so we can only talk about having "unit," and not "marginal," increases in Z.

The main point from our discussion so far is that the individual coefficients in a linear-additive model are derivatives or differences, and they therefore tell us the effects of our independent variables. Things are different, though, in a linear-interactive model. This is because the individual coefficients associated with the variables involved in an interaction are not derivatives or differences. What this means in practice is that if we want to know the effect of the variables in an interaction model, we have to look beyond the individual coefficients.

Consider the following linear-interactive model specification:

$$Y = \beta_0 + \beta_1 X + \beta_2 Z + \beta_3 XZ + \beta_4 W + \epsilon, \tag{4.8}$$

where X is a continuous variable, Z is a discrete variable, and W is some continuous or discrete control variable. Recall that we refer to X and Z as constitutive terms as when they're multiplied together they "constitute" the interaction term XZ. Since X is continuous, we can think of its effect on Y in terms of a derivative,

$$\frac{\partial Y}{\partial X} = \beta_1 + \beta_3 Z. \tag{4.9}$$

We can also think of its effect in terms of a difference where we increase X from some baseline value X_b to some counterfactual value X_c,

$$
\begin{aligned}
\hat{Y}_{X_c} - \hat{Y}_{X_b} &= (\beta_0 + \beta_1 X_c + \beta_2 Z + \beta_3 X_c Z + \beta_4 W) \\
&\quad - (\beta_0 + \beta_1 X_b + \beta_2 Z + \beta_3 X_b Z + \beta_4 W) \\
&= \beta_1 X_c - \beta_1 X_b + \beta_3 X_c Z - \beta_3 X_b Z \\
&= (X_c - X_b)(\beta_1 + \beta_3 Z).
\end{aligned}
\tag{4.10}
$$

If we have a one-unit increase in X such that $X_c = X_b + 1$, then this difference simplifies to

$$
\hat{Y}_{X_c} - \hat{Y}_{X_b} = (X_b + 1 - X_b)(\beta_1 + \beta_3 Z) = \beta_1 + \beta_3 Z.
\tag{4.11}
$$

Whether we think in terms of a derivative or a difference, we see that a one-unit increase in X leads to a $\beta_1 + \beta_3 Z$ increase in Y. We can immediately see from this that, unlike in the linear-additive world, the coefficient on X doesn't tell us the effect of X on Y. Indeed, we can see that so long as $\beta_3 \neq 0$, there's no single magnitude for the effect of X on Y. The magnitude of the effect of X depends on the value of Z. When $Z = 0$, the effect of X on Y is β_1. When $Z = 1$, the effect of X is $\beta_1 + \beta_3$. When $Z = 2$, the effect of X is $\beta_1 + 2\beta_3$. When $Z = 3$, the effect of X is $\beta_1 + 3\beta_3$. And so on. This is why we sometimes refer to the effect of X on Y in an interaction model as the *conditional* effect of X; the effect of X on Y is *conditional* on the value of Z.

Since Z is discrete, we can find the effect of Z by calculating a difference where we increase Z by some discrete amount from a baseline value Z_b to some counterfactual value Z_c,

$$
\begin{aligned}
\hat{Y}_{Z_c} - \hat{Y}_{Z_b} &= (\beta_0 + \beta_1 X + \beta_2 Z_c + \beta_3 X Z_c + \beta_4 W) \\
&\quad - (\beta_0 + \beta_1 X + \beta_2 Z_b + \beta_3 X Z_b + \beta_4 W) \\
&= \beta_2 Z_c - \beta_2 Z_b + \beta_3 X Z_c - \beta_3 X Z_b \\
&= (Z_c - Z_b)(\beta_2 + \beta_3 X).
\end{aligned}
\tag{4.12}
$$

If we have a one-unit increase in Z, such that $Z_c = Z_b + 1$, then this difference simplifies to

$$
\hat{Y}_{Z_c} - \hat{Y}_{Z_b} = (Z_b + 1 - Z_b)(\beta_2 + \beta_3 X) = \beta_2 + \beta_3 X.
\tag{4.13}
$$

This tells us that a one-unit increase in Z leads to a $\beta_2 + \beta_3 X$ increase in Y. As we noted earlier, in linear-interactive models in which the interaction is itself linear, we can also obtain the effect of a one-unit increase in Z by treating Z "as if" it were continuous and calculating a derivative,

$$
\frac{\partial Y}{\partial Z} = \beta_2 + \beta_3 X.
\tag{4.14}
$$

Whichever method we use to calculate this quantity, we can immediately see that the coefficient on Z doesn't tell us the effect of Z on Y. So long as $\beta_3 \neq 0$, there's no single magnitude for the effect of Z on Y. The magnitude of the effect of Z depends on the value of X. When $X = 0$, the effect of Z on Y is β_2. When $X = 1$, the effect of Z is $\beta_2 + \beta_3$. When $X = 2$, the effect of Z is $\beta_2 + 2\beta_3$. When $X = 3$, the effect of Z is $\beta_2 + 3\beta_3$. And so on. As with X, we sometimes refer to the effect of Z on Y in an interaction model as the *conditional* effect of Z; the effect of Z on Y is *conditional* on the value of X.

Unlike X and Z, the control variable W in Eq. 4.8 is entered in an additive manner and, as a result, we *can*, in this case, just look at the coefficient on W to see the effect of W on Y. Whether we're thinking in terms of a derivative or a difference, a one-unit increase in W is associated with a β_4 increase in Y.

> ⚠ **Important:** Scholars should not think in terms of individual coefficients when interpreting results from an interaction model. This is because the coefficients do not tell us the effect of the variables involved in an interaction on the dependent variable. Instead, scholars should think in terms of derivatives and differences.

The key point to take away from all of this is that, unlike in the linear-additive world, we can't just look at the individual coefficients to see the effects of the variables involved in an interaction. Instead, we have to calculate the appropriate derivative or difference. This is why we strongly recommend that scholars stop thinking in terms of coefficients when interpreting the results from an interaction model and start thinking in terms of derivatives and differences.[3]

As should be clear, this doesn't mean that the individual coefficients from a linear-interaction model are meaningless. As we've seen, the coefficient on X, β_1, tells us the effect of X on Y when Z is zero, and the coefficient on Z, β_2, tells us the effect of Z on Y when X is zero. In other words, the coefficients on the constitutive terms *can* be interpreted and

---

[3] Later in the book, we'll look at more theoretically complex forms of conditionality than what we've seen so far. This will require us to make changes to either the left- and/or right-hand side of our interaction models. For example, we'll look at situations where our dependent variable is no longer continuous or we need to add additional interaction terms. No matter what we do to the model specification, though, we always think in terms of derivatives and differences when interpreting the results from our models. Doing this helps us to avoid making inferential errors. We recommend that you always do the same when confronted with theoretical scenarios and interaction models that aren't explicitly examined in this book.

may be substantively interesting depending on whether it makes theoretical sense to think about the effect of X or Z on Y when the other variable is zero. The critical thing to recognize is that, so long as $\beta_3 \neq 0$, the individual coefficients on the constitutive terms never tell us the unconditional effect of X or Z on Y. In our experience, many scholars make this mistake and slip into interpreting the coefficients on constitutive terms as if they capture the unconditional effect of these variables. Because the terminology used to discuss constitutive terms varies across, and even within, various disciplines, we'd like to reiterate that claiming that the coefficients on the constitutive terms represent the "unconditional," "main," "independent," or "average" effect is simply wrong, since it implies that these coefficients can be interpreted in the same way as they would in a linear-additive model.[4] Just as we've come to recognize that the coefficients in logit and probit models can't be interpreted as unconditional effects, we need to recognize that the coefficients on constitutive terms in interaction models can't be interpreted in this way either.

> ⚠ **Important:** Do not interpret the coefficients on the constitutive terms X and Z in a linear-interactive model with an interaction term XZ as if they indicate an unconditional, average, independent, or main effect. They do not. These coefficients indicate only the effect of X or Z on Y when the value of the other constitutive term is zero.

As we saw in Chapter 2, we *can* look at the individual coefficient on the interaction term, β_3, to see the "interaction effect"; that is, how Z modifies the effect of X on Y and how X modifies the effect of Z on Y. This is because the coefficient on the interaction term is a (second) derivative or difference. One way to find how Z modifies the marginal effect of X on Y involves differentiating the marginal effect of X on Y with respect to Z,

$$\frac{\partial\left(\frac{\partial Y}{\partial X}\right)}{\partial Z} = \frac{\partial\left(\beta_1 + \beta_3 Z\right)}{\partial X} = \beta_3. \tag{4.15}$$

---

[4] In our opinion, a particularly problematic practice involves referring to the coefficient on a constitutive term such as X as the "main effect" of X and the coefficient on the interaction term as the "interactive effect" of X. First, there's nothing "main" about the coefficient on the constitutive term, and the unfortunate nomenclature wrongly encourages scholars to think that the coefficient on the constitutive term represents the unconditional, independent, or average effect of the variable. Second, X doesn't have two conceptually distinct effects on Y. There's only ever one conditional effect of X on Y, $\beta_1 + \beta_3 Z$, whose magnitude varies with Z, and it doesn't make conceptual or theoretical sense to break it up into separate "main" and "interactive" components.

This tells us that a one-unit increase in Z increases the one-unit effect of X on Y by β_3. One way to find how X modifies the effect of Z on Y involves differentiating the "marginal effect" of Z on Y with respect to X,

$$\frac{\partial\left(\frac{\partial Y}{\partial Z}\right)}{\partial X} = \frac{\partial\left(\beta_2 + \beta_3 X\right)}{\partial X} = \beta_3. \tag{4.16}$$

This tells us that a one-unit increase in X increases the one-unit effect of Z on Y by β_3. The fact that these two modifying effects – the modifying effect of Z on the effect of X on Y and the modifying effect of X on the effect of Z on Y – are the same shouldn't come as a surprise as this was the basis for our discussion in Chapter 2 regarding the inherent symmetry of interactions. The main point here is that we can look at the individual coefficient on the interaction term in a linear-interactive model to determine if there's an interaction effect. This is because the coefficient on the interaction term is a (second) derivative or difference.

4.2 CALCULATE APPROPRIATE MEASURES OF UNCERTAINTY

One thing we've studiously ignored to this point in the book is a discussion of statistical significance and measures of uncertainty in the context of interaction models. Such a discussion was relevant neither for our examination of theories positing interaction nor our examination of how to appropriately specify interaction models. However, now that we're talking about how to interpret the results from interaction models, it's time to address this important topic.

4.2.1 Effects and Measures of Uncertainty

As we've seen, the individual coefficients in a linear-additive model tell us the estimated effects of the independent variables on a dependent variable. Specifically, they tell us how the value of the dependent variable changes in response to a one-unit increase in the independent variables. The coefficient standard errors provide us with a measure of uncertainty for these estimated effects. We can use the standard errors to test claims about the effects of the independent variables. For example, we might use the standard errors to conduct a t-test to determine whether some independent variable has a statistically significant non-zero effect on the dependent variable. Or we might use the standard errors to construct a confidence interval around the estimated effect of an independent variable.

Things are a little different in the linear-interactive world. As we've already seen, the individual coefficients in a linear-interactive model don't tell us the effects of the independent variables on a dependent variable.

It follows, therefore, that the standard errors associated with the individual coefficients don't provide us with a measure of uncertainty for the effects of our independent variables either. Consider again the following linear-interactive model:

$$Y = \beta_0 + \beta_1 X + \beta_2 Z + \beta_3 XZ + \beta_4 W + \epsilon. \tag{4.17}$$

To simplify our discussion, we'll assume that X, Z, and W are now all continuous variables.[5] As we saw in the last section, the marginal effect of X on Y is

$$\frac{\partial Y}{\partial X} = \beta_1 + \beta_3 Z. \tag{4.18}$$

This tells us that a one-unit increase in X is associated with a $\beta_1 + \beta_3 Z$ increase in Y. As Eq. 4.18 indicates, there's no single magnitude for the effect of X on Y; the magnitude depends on the value of Z. It follows that we can't answer the question of whether X has a statistically significant effect on Y in some unconditional sense. What we can determine, though, is whether X has a statistically significant effect on Y for a particular value of Z (or for particular values of Z). If we evaluate whether the effect of X on Y is statistically significant across a range of values of Z, we might find that the effect of X is always statistically significant, that the effect of X is never statistically significant, or that the effect of X is statistically significant for some values in the range of Z but not for others.

To evaluate if and when X has a statistically significant effect on Y, we need an appropriate measure of uncertainty for the marginal effect of X on Y shown in Eq. 4.18.[6] The estimated variance of this effect is calculated as

$$\text{var}\left(\frac{\partial Y}{\partial X}\right) = \text{var}\,(\beta_1 + \beta_3 Z) = \text{var}\,(\beta_1) + Z^2 \text{var}\,(\beta_3) + 2Z\text{cov}\,(\beta_1, \beta_3), \tag{4.19}$$

and the estimated standard error is calculated as the square root of this quantity,

$$\text{se}\left(\frac{\partial Y}{\partial X}\right) = \text{se}\,(\beta_1 + \beta_3 Z) = \sqrt{\text{var}\,(\beta_1) + Z^2 \text{var}\,(\beta_3) + 2Z\text{cov}\,(\beta_1, \beta_3)}. \tag{4.20}$$

---

[5] Recall that finding the effect of a variable by calculating a derivative or a one-unit difference produces the exact same mathematical quantity in a linear-interactive model in which the interaction is linear. This means that our upcoming discussion of how to calculate measures of uncertainty for the effects of the variables in a model that includes an interaction term XZ applies equally well irrespective of whether our variables are continuous or discrete.

[6] Calculating an appropriate measure of uncertainty for a quantity of interest typically starts with finding the estimated variance for that quantity. To help with this process, we present some of the basic properties of variances in Appendix A. These properties help to explain where the estimated variances for various quantities of interest that we report in the main text come from.

Two things are worth noting here.[7] The first is that the estimated variance and standard error for the effect of X depend on the value of Z. This means that there's a different estimated variance and standard error for each possible magnitude of the effect of X on Y depending on the value of Z. Thus, just as we can talk about the conditional effect of X on Y, we can also talk about the conditional variance or standard error for the effect of X on Y. When $Z = 0$, the marginal effect of X is β_1, and its estimated standard error is $\sqrt{\text{var}(\beta_1)}$ or, more simply, $\text{se}(\beta_1)$. In other words, the estimated standard error associated with the coefficient on X in Eq. 4.17 is just the standard error for the effect of X when Z is zero; it's not the standard error for the effect of X in some unconditional or average sense. When $Z = 1$, the marginal effect of X is $\beta_1 + \beta_3$, and its estimated standard error is $\sqrt{\text{var}(\beta_1) + \text{var}(\beta_3) + 2\text{cov}(\beta_1, \beta_3)}$. When $Z = 2$, the marginal effect of X is $\beta_1 + 2\beta_3$, and its estimated standard error is $\sqrt{\text{var}(\beta_1) + 4\text{var}(\beta_3) + 4\text{cov}(\beta_1, \beta_3)}$. When $Z = 3$, the marginal effect of X is $\beta_1 + 3\beta_3$, and its estimated standard error is $\sqrt{\text{var}(\beta_1) + 9\text{var}(\beta_3) + 6\text{cov}(\beta_1, \beta_3)}$. And so on.

> ⚠ **Important:** The standard errors associated with the coefficients on the constitutive terms X and Z are just the standard errors for the effects of X and Z on Y when the value of the other constitutive term is zero. They do not indicate the standard errors for the effect of X and Z on Y in some unconditional or average sense.

The size of the estimated standard error associated with the effect of X on Y clearly varies with the value of Z. As we demonstrate in Appendix C, we can find the value of Z at which the estimated standard error is smallest by taking the derivative of the standard error shown in Eq. 4.20, setting this equal to zero, and solving for Z,

$$Z^* = -\frac{\text{cov}(\beta_1, \beta_3)}{\text{var}(\beta_3)}. \qquad (4.21)$$

This value isn't necessarily the sample mean of Z as some have claimed (Kam and Franzese, 2007, 65). The standard error associated with the effect of X on Y increases in magnitude (at an increasing rate) as the value of Z becomes smaller or larger than Z^*. It increases at an increasing rate due to the appearance of the Z^2 term in the expression for the estimated standard error in Eq. 4.20.

---

[7] The estimated effects and associated variances from various commonly-used linear-interactive model specifications are shown in Tables B.1 and B.2 in Appendix B.

The second thing to note is that not all of the information necessary to calculate the standard errors for the different possible magnitudes of the effect of X on Y is contained in a traditional table of results. While it's possible to calculate the marginal effect of X for any value of Z from a typical results table using a little algebra, it's not possible to do the same for the standard errors. We can immediately read off of a table of results the standard error for the marginal effect of X when Z is 0 as this is just the standard error associated with the estimated coefficient on X. However, we can't use a typical table of results to identify the standard error for the marginal effect of X on Y at any other value of Z. This is because the quantity $\text{cov}(\beta_1, \beta_3)$, which we need to calculate these standard errors, is almost never reported. It follows that we must almost always look beyond the information contained in our immediate regression output or a traditional table of results if we want to fully evaluate the hypotheses that come from a theory positing interaction.

The quantity $\text{cov}(\beta_1, \beta_3)$ refers to the estimated covariance between β_1 and β_3 and has to do with how the two estimated coefficients are jointly related. The covariance is positive when higher values of one coefficient are associated with higher values of the other coefficient, and it's negative when higher values of one coefficient are associated with lower values of the other coefficient. In practice, $\text{cov}(\beta_1, \beta_3)$ is often negative because the corresponding variables X and XZ tend to be positively correlated (Kam and Franzese, 2007, 65). The covariance between any pair of estimated coefficients can be found in the variance-covariance matrix for the coefficients that's calculated whenever we estimate a regression model. The variance-covariance matrix is a square matrix that contains all of the variances and covariances associated with the estimated coefficients from a regression model. The diagonal elements of the matrix, from the top left to bottom right, contain the variances of the coefficients. If we were to take the square root of these elements, we'd obtain the coefficient standard errors that are reported in our statistical output and in most tables of results. The off-diagonal elements contain the covariances between all of the pairs of coefficients. The variance-covariance matrix for the linear-interactive model shown in Eq. 4.17 would look like this:

$$V(\beta) = \begin{pmatrix} \text{var}(\beta_1) & \text{cov}(\beta_1,\beta_2) & \text{cov}(\beta_1,\beta_3) & \text{cov}(\beta_1,\beta_4) & \text{cov}(\beta_1,\beta_0) \\ \text{cov}(\beta_2,\beta_1) & \text{var}(\beta_2) & \text{cov}(\beta_2,\beta_3) & \text{cov}(\beta_2,\beta_4) & \text{cov}(\beta_2,\beta_0) \\ \text{cov}(\beta_3,\beta_1) & \text{cov}(\beta_3,\beta_2) & \text{var}(\beta_3) & \text{cov}(\beta_3,\beta_4) & \text{cov}(\beta_3,\beta_0) \\ \text{cov}(\beta_4,\beta_1) & \text{cov}(\beta_4,\beta_2) & \text{cov}(\beta_4,\beta_3) & \text{var}(\beta_4) & \text{cov}(\beta_4,\beta_0) \\ \text{cov}(\beta_0,\beta_1) & \text{cov}(\beta_0,\beta_2) & \text{cov}(\beta_0,\beta_3) & \text{cov}(\beta_0,\beta_4) & \text{var}(\beta_0) \end{pmatrix}.$$

$$(4.22)$$

The variances are shown in gray and the covariances are shown in black. The variance-covariance matrix is symmetric in that the covariance between any two coefficients is the same irrespective of the the order in which the coefficients are listed. For example, $\mathrm{cov}(\beta_1,\beta_3)$ is the same as $\mathrm{cov}(\beta_3,\beta_1)$. This means that the covariance between each pair of coefficients appears twice in the full variance-covariance matrix. Given its symmetric nature, all of the information in the variance-covariance matrix can be found in the "bottom triangle"

$$
V(\beta) = \begin{pmatrix}
\mathrm{var}(\beta_1) & & & & \\
\mathrm{cov}(\beta_2,\beta_1) & \mathrm{var}(\beta_2) & & & \\
\mathrm{cov}(\beta_3,\beta_1) & \mathrm{cov}(\beta_3,\beta_2) & \mathrm{var}(\beta_3) & & \\
\mathrm{cov}(\beta_4,\beta_1) & \mathrm{cov}(\beta_4,\beta_2) & \mathrm{cov}(\beta_4,\beta_3) & \mathrm{var}(\beta_4) & \\
\mathrm{cov}(\beta_0,\beta_1) & \mathrm{cov}(\beta_0,\beta_2) & \mathrm{cov}(\beta_0,\beta_3) & \mathrm{cov}(\beta_0,\beta_4) & \mathrm{var}(\beta_0)
\end{pmatrix}
\tag{4.23}
$$

or the "top triangle"

$$
V(\beta) = \begin{pmatrix}
\mathrm{var}(\beta_1) & \mathrm{cov}(\beta_1,\beta_2) & \mathrm{cov}(\beta_1,\beta_3) & \mathrm{cov}(\beta_1,\beta_4) & \mathrm{cov}(\beta_1,\beta_0) \\
& \mathrm{var}(\beta_2) & \mathrm{cov}(\beta_2,\beta_3) & \mathrm{cov}(\beta_2,\beta_4) & \mathrm{cov}(\beta_2,\beta_0) \\
& & \mathrm{var}(\beta_3) & \mathrm{cov}(\beta_3,\beta_4) & \mathrm{cov}(\beta_3,\beta_0) \\
& & & \mathrm{var}(\beta_4) & \mathrm{cov}(\beta_4,\beta_0) \\
& & & & \mathrm{var}(\beta_0)
\end{pmatrix}
\tag{4.24}
$$

of the full variance-covariance matrix.

While most statistical packages don't automatically report the variance-covariance matrix for the coefficients on the computer screen when we estimate a regression model, it's almost always stored "in the background" and can be easily accessed. We can use the information about $\mathrm{cov}(\beta_1,\beta_3)$ in the variance-covariance matrix to calculate the standard error for the marginal effect of X on Y at any value of Z.

Once we have the standard error for the marginal effect of X on Y at a given value of Z, we can employ a standard t-test to evaluate claims about this effect. For example, we might want to know whether the effect of X on Y at a given value of Z is significantly different from zero. The test statistic in this case is

$$
t = \frac{(\beta_1 + \beta_3 Z) - 0}{\mathrm{se}\left(\frac{\partial Y}{\partial X}\right)} = \frac{\beta_1 + \beta_3 Z}{\sqrt{\mathrm{var}(\beta_1) + Z^2 \mathrm{var}(\beta_3) + 2Z\mathrm{cov}(\beta_1,\beta_3)}}.
\tag{4.25}
$$

This test statistic follows a t-distribution with $n - k$ degrees of freedom, where n is the number of observations, and k is the number of independent variables including the constant. As with any t-test, we can use the appropriate t-distribution to evaluate the statistical significance of the effect of X on Y at a given value of Z by either choosing a particular α level of significance or by calculating a p-value.

We can also use the standard error to calculate a confidence interval for the effect of X on Y at a given value of Z. For example, a two-sided confidence interval for the effect of X on Y with confidence coefficient $1 - \alpha$ is

$$\beta_1 + \beta_3 Z \pm t_{n-k,\alpha/2} \times \text{se}\left(\frac{\partial Y}{\partial X}\right)$$

or

$$\beta_1 + \beta_3 Z \pm t_{n-k,\alpha/2} \times \sqrt{\text{var}(\beta_1) + Z^2 \text{var}(\beta_3) + 2Z\text{cov}(\beta_1, \beta_3)}, \quad (4.26)$$

where $t_{n-k,\alpha/2}$ refers to the critical value from a t-distribution table for a particular α level of significance and $n - k$ degrees of freedom.

⚠ **Important:** Consider the following linear-interactive model,

$$Y = \beta_0 + \beta_1 X + \beta_2 Z + \beta_3 XZ + \beta_4 W + \epsilon.$$

The marginal effect of X on Y is

$$\frac{\partial Y}{\partial X} = \beta_1 + \beta_3 Z,$$

and its estimated standard error is

$$\text{se}\left(\frac{\partial Y}{\partial X}\right) = \sqrt{\text{var}(\beta_1) + Z^2 \text{var}(\beta_3) + 2Z\text{cov}(\beta_1, \beta_3)}.$$

The marginal effect of Z on Y is

$$\frac{\partial Y}{\partial Z} = \beta_2 + \beta_3 X,$$

and its estimated standard error is

$$\text{se}\left(\frac{\partial Y}{\partial Z}\right) = \sqrt{\text{var}(\beta_2) + X^2 \text{var}(\beta_3) + 2X\text{cov}(\beta_2, \beta_3)}.$$

The marginal effect of W on Y is

$$\frac{\partial Y}{\partial W} = \beta_4,$$

and its estimated standard error is

$$\text{se}\left(\frac{\partial Y}{\partial W}\right) = \sqrt{\text{var}(\beta_4)} = \text{se}(\beta_4).$$

So far we've focused on calculating measures of uncertainty for the marginal effect of X in the linear-interactive model shown in Eq. 4.17. However, we can calculate equivalent measures of uncertainty for the marginal effect of Z,

$$\frac{\partial Y}{\partial Z} = \beta_2 + \beta_3 X. \tag{4.27}$$

The estimated variance of the marginal effect of Z is

$$\text{var}\left(\frac{\partial Y}{\partial Z}\right) = \text{var}\left(\beta_2 + \beta_3 X\right) = \text{var}\left(\beta_2\right) + X^2 \text{var}\left(\beta_3\right) + 2X\text{cov}\left(\beta_2, \beta_3\right),$$
$$\tag{4.28}$$

and the estimated standard error is

$$\text{se}\left(\frac{\partial Y}{\partial Z}\right) = \text{se}\left(\beta_2 + \beta_3 X\right) = \sqrt{\text{var}\left(\beta_2\right) + X^2 \text{var}\left(\beta_3\right) + 2X\text{cov}\left(\beta_2, \beta_3\right)}. \tag{4.29}$$

A two-sided confidence interval for the marginal effect of Z on Y with confidence coefficient $1 - \alpha$ is

$$\beta_2 + \beta_3 X \pm t_{n-k,\alpha/2} \times \sqrt{\text{var}\left(\beta_2\right) + X^2 \text{var}\left(\beta_3\right) + 2X\text{cov}\left(\beta_2, \beta_3\right)}. \tag{4.30}$$

While it might seem somewhat tedious to calculate the marginal effects of the variables involved in an interaction and their associated measures of uncertainty by hand, most statistical packages make our lives easier by doing the necessary calculations for us with a simple post-estimation command.

The measures of uncertainty for variables that are included in a purely additive manner in a linear-interactive specification are calculated in the usual way. For example, the marginal effect of W in Eq. 4.17 is β_4. The estimated variance for this effect is $\text{var}(\beta_4)$, and the estimated standard error is $\text{se}(\beta_4)$. A two-sided confidence interval for the marginal effect of W is $\beta_4 \pm t_{n-k,\alpha/2} \times \text{se}(\beta_4)$.

4.2.2 Predicted Values and Measures of Uncertainty

In addition to calculating estimated effects and their measures of uncertainty, we may also be interested in calculating the predicted value of Y in particular scenarios. Suppose we're interested in the mean predicted value of Y from Eq. 4.17 when our independent variables have the following values: $X = X_b$, $Z = Z_b$, and $W = W_b$.[8] In this case, the predicted value of Y in this "baseline" scenario is

---

[8] Scholars may be interested in either the mean predicted value of Y in a given scenario or the predicted value of Y for an individual observation in a given scenario. While the point estimates for the "mean prediction" and the "individual prediction" are identical, the variances and standard errors differ slightly (Gujarati, 2003, 940–942). In the text, we focus on the *mean* predicted value of Y and its associated measure of uncertainty.

$$\hat{Y}_b = \beta_0 + \beta_1 X_b + \beta_2 Z_b + \beta_3 X_b Z_b + \beta_4 W_b. \tag{4.31}$$

The predicted value of Y is a sum of random variables (coefficient estimates) multiplied by constants (the values for our independent variables). As the formula for calculating the variance of a linear combination of random variables in Appendix A indicates, the estimated variance of this sum is equal to the sum of the variances multiplied by their associated constants squared plus two times all of the covariances multiplied by the product of their constant cofactors (Kam and Franzese, 2007, 80–81),

$$\begin{aligned}
\operatorname{var}\left(\hat{Y}_b\right) = {} & \operatorname{var}\left(\beta_0 + \beta_1 X_b + \beta_2 Z_b + \beta_3 X_b Z_b + \beta_4 W_b\right) \\
= {} & \operatorname{var}(\beta_0) + X_b^2 \operatorname{var}(\beta_1) + Z_b^2 \operatorname{var}(\beta_2) + (X_b Z_b)^2 \operatorname{var}(\beta_3) + W_b^2 \operatorname{var}(\beta_4) \\
& + 2 X_b \operatorname{cov}(\beta_0, \beta_1) + 2 Z_b \operatorname{cov}(\beta_0, \beta_2) + 2 X_b Z_b \operatorname{cov}(\beta_0, \beta_3) + 2 W_b \operatorname{cov}(\beta_0, \beta_4) \\
& + 2 X_b Z_b \operatorname{cov}(\beta_1, \beta_2) + 2 X_b (X_b Z_b) \operatorname{cov}(\beta_1, \beta_3) + 2 X_b W_b \operatorname{cov}(\beta_1, \beta_4) \\
& + 2 Z_b (X_b Z_b) \operatorname{cov}(\beta_2, \beta_3) + 2 Z_b W_b \operatorname{cov}(\beta_2, \beta_4) \\
& + 2 W_b (X_b Z_b) \operatorname{cov}(\beta_3, \beta_4) .
\end{aligned} \tag{4.32}$$

As you can see, the estimated variance associated with the mean predicted value is rather messy when written in scalar notation. However, it can be written more succinctly in matrix notation,

$$\operatorname{var}\left(\hat{Y}_b\right) = M_b V(\beta) M_b', \tag{4.33}$$

where M_b is a row vector

$$M_b = \begin{pmatrix} X_b & Z_b & X_b Z_b & W_b & 1 \end{pmatrix} \tag{4.34}$$

that contains the values for the independent variables and the constant for the baseline scenario in which we're interested, $V(\beta)$ is the estimated variance-covariance matrix for the coefficients shown in Eq. 4.22, and M_b' is a column vector equal to the transpose of M_b,

$$M_b' = \begin{pmatrix} X_b \\ Z_b \\ X_b Z_b \\ W_b \\ 1 \end{pmatrix}. \tag{4.35}$$

Taking the square root of the variance gives us the estimated standard error for the predicted value of Y in the baseline scenario, $\operatorname{se}\left(\hat{Y}_b\right)$. We can use this standard error to calculate an appropriate confidence interval,

$$\hat{Y}_b \pm t_{n-k, \alpha/2} \times \operatorname{se}\left(\hat{Y}_b\right). \tag{4.36}$$

While it might seem taxing to calculate a predicted value and its corresponding standard error by hand, most statistical packages facilitate things by doing the necessary calculations for us with a simple post-estimation command.

As we saw in the previous section, we can think about the effects of our independent variables in terms of derivatives or differences. Thinking in terms of differences involves looking at how the predicted value of Y responds to a given change in one of our independent variables. For example, we might be interested in thinking about the effect of X in terms of a difference where we increase X from some baseline value X_b to some counterfactual value X_c while the other variables are all held at their baseline values.[9] This difference is calculated as

$$\hat{Y}_{X_c} - \hat{Y}_{X_b} = (\beta_0 + \beta_1 X_c + \beta_2 Z + \beta_3 X_c Z + \beta_4 W)$$
$$- (\beta_0 + \beta_1 X_b + \beta_2 Z + \beta_3 X_b Z + \beta_4 W)$$
$$= \beta_1 X_c - \beta_1 X_b + \beta_3 X_c Z - \beta_3 X_b Z$$
$$= (X_c - X_b)(\beta_1 + \beta_3 Z). \tag{4.37}$$

The estimated variance of this difference is

$$\text{var}\left(\hat{Y}_{X_c} - \hat{Y}_{X_b}\right) = \text{var}\left[(X_c - X_b)(\beta_1 + \beta_3 Z)\right]$$
$$= (X_c - X_b)^2 \,\text{var}\,(\beta_1 + \beta_3 Z)$$
$$= (X_c - X_b)^2 \left[\text{var}\,(\beta_1) + Z^2 \text{var}\,(\beta_3) + 2Z\text{cov}\,(\beta_1, \beta_3)\right]. \tag{4.38}$$

If we think in terms of a one-unit increase in X such that $X_c = X_b + 1$, then the difference shown in Eq. 4.37 simplifies to

$$\hat{Y}_{X_b+1} - \hat{Y}_{X_b} = (X_b + 1 - X_b)(\beta_1 + \beta_3 Z) = \beta_1 + \beta_3 Z, \tag{4.39}$$

and the estimated variance in Eq. 4.38 simplifies to

$$\text{var}\left(\hat{Y}_{X_b+1} - \hat{Y}_{X_b}\right) = (X_b + 1 - X_b)^2 \left[\text{var}\,(\beta_1) + Z^2 \text{var}\,(\beta_3) + 2Z\text{cov}\,(\beta_1, \beta_3)\right]$$
$$= \text{var}\,(\beta_1) + Z^2 \text{var}\,(\beta_3) + 2Z\text{cov}\,(\beta_1, \beta_3). \tag{4.40}$$

As expected, the difference in the mean predicted value of Y associated with increasing X by one unit shown in Eq. 4.39 is exactly the same as the marginal effect of X on Y shown in Eq. 4.11. Similarly, the variance associated with the effect of a one-unit change in X on the mean predicted value of Y shown in Eq. 4.40 is exactly the same as the variance associated with the marginal effect of X on Y shown in Eq. 4.19. To reiterate, thinking in terms of derivatives and marginal effects or one-unit differences and

---

[9] To simplify the presentation, we remove the subscript b from the variables Z and W. The important point to remember is that the values of Z and W aren't changing as we move from our baseline scenario in which $X = X_b$ to the counterfactual scenario in which $X = X_c$.

changes in predicted values is effectively equivalent in linear-interactive models in which the interaction is linear.

4.2.3 Substantive Significance

This section has focused on how to calculate appropriate measures of uncertainty to go with our estimated effects and predicted values. These measures of uncertainty help us to talk about whether an effect is significant or not. We want to take this opportunity to briefly flag that there's obviously a difference between statistical and substantive significance. *Statistical significance* refers to whether an effect is significantly different from some critical value θ at some specified α level of significance. Unless we explicitly state otherwise, "statistical significance" in this book implies that an effect is significantly different from 0 at a specified α level of significance. *Substantive significance* refers to whether an effect is large enough to be deemed of nontrivial magnitude. The minimum magnitude required for substantive significance is almost always subjective. Given this distinction between different types of significance, it's possible for an effect to be both statistically and substantively significant, statistically significant but substantively insignificant, substantively significant but statistically insignificant, and neither statistically nor substantively significant (Achen, 1982, 46–51).

While identifying whether an effect is statistically significant is important, we're ultimately more interested in knowing whether an effect has reliably been shown to be substantively significant. As a result, we encourage scholars to consider the substantive, as well as the statistical, significance of the results obtained from an interaction model. This means not immediately assuming that an effect is substantively meaningless just because it's statistically insignificant or that an effect is substantively meaningful just because it's statistically significant.

In our opinion, there's no single correct way of establishing substantive significance. That said, at least three pieces of information are typically needed to evaluate the substantive significance of an effect. The first concerns the appropriate size of the change in an independent variable we should be considering when evaluating the substantive significance of its effect.[10] Any change we're considering should be small enough to be

---

[10] Note that we don't have to think of the "change" in the independent variable as a within-individual or within-unit change. For many variables, conceptualizing the change in this way makes sense. For example, it's straightforward to think about changing someone's level of education and examining how the value of our dependent variable changes. Thinking in this way is less reasonable, though, when we have an independent variable capturing, say, an individual's gender and race. While gender reassignment is possible, and race can be fluid (Saperstein and Penner,

realistic in the sense that it's likely or possible to happen in the real world but large enough to be perceptible to the reader. In effect, this has to do with whether the counterfactual under consideration is both plausible and meaningful. The second concerns the estimated magnitude of the change in the dependent variable associated with the "appropriate" change in our independent variable. The third concerns the distribution of the dependent variable prior to any change in the value of our independent variable. Knowing, for example, the mean level of the dependent variable for a given context provides information in a sense about the "typical" baseline scenario and can therefore act as a useful metric for evaluating whether the estimated magnitude of an effect is substantively meaningful. All three pieces of information are important when evaluating the substantive significance of statistical results. We discuss issues of substantive significance further in the upcoming applications in the next chapter.

4.3 KEY QUANTITIES OF INTEREST AND SOME PROTOTYPICAL RESULTS

In the next chapter, we'll use three substantive applications to take a practical look at how to calculate, interpret, and present the key quantities of interest from a linear-interactive model. Before turning to these applications, though, we first remind ourselves of the key quantities of interest that are necessary to fully evaluate a theory positing interaction and briefly discuss some prototypical results that we might get from estimating an interaction model.

In Chapter 2, we noted that most theories positing an interaction between X and Z are typically strong enough to produce five key predictions:

1. P_{XZ}: The interaction effect between X and Z is [positive, negative].
2. $P_{X|Z_{min}}$: The marginal effect of X on Y is [positive, negative, zero] when Z is at its *lowest* value.
3. $P_{X|Z_{max}}$: The marginal effect of X on Y is [positive, negative, zero] when Z is at its *highest* value.
4. $P_{Z|X_{min}}$: The marginal effect of Z on Y is [positive, negative, zero] when X is at its *lowest* value.
5. $P_{Z|X_{max}}$: The marginal effect of Z on Y is [positive, negative, zero] when X is at its *highest* value.

2012; Davenport, 2018, 2020), gender and racial categories tend to be relatively fixed. In such a situation, we can conceptualize the "change" in the independent variable as a comparison between individuals who differ in terms of their gender or race. This suggests that some care should be taken when talking about the "effects" of these types of variables (Zuberi, 2000).

These predictions speak to the direction of the interaction between X and Z, the direction of the marginal effect of X on Y, and the direction of the marginal effect of Z on Y. As we discussed, these predictions can usually be incorporated into two hypotheses, one about how the marginal effect of X on Y varies with Z and one about how the marginal effect of Z on Y varies with X. We can test all five of these key predictions using the following generic interaction model:

$$Y = \beta_0 + \beta_1 X + \beta_2 Z + \beta_3 XZ + \epsilon. \tag{4.41}$$

The evidence in favor of a theory positing interaction is strongest when there's clear support for each of our five key predictions. But should we require that *all* of the predictions be corroborated before we claim any empirical support for a theory and reject a theory if any of the predictions are falsified? This is probably an unrealistically strong standard for empirical evidence, and it would be a mistake to treat all situations in which at least one of these predictions is falsified as equivalent. Although a firm knowledge that one of the five predictions has been falsified would be sufficient logical grounds for concluding that the underlying theory is false, it's important to recognize that statistical tests can't tell us with certainty whether any of the predictions are false; all they offer is information about the risks of a false inference if one rejects the null hypothesis that a quantity of interest equals a particular value, such as 0. For this reason, it's inappropriate to establish "hard and fast" rules about what combinations of evidence regarding the five key predictions constitute support for one's underlying theory (Berry, Golder and Milton, 2012, 659). Ultimately, the decision as to whether the empirical evidence corroborates or falsifies a theory always relies on the skill, expertise, and subjective evaluation of the analyst.[11]

That said, finding empirical evidence for prediction P_{XZ} is largely necessary for concluding that a theory positing interaction is supported. A defining characteristic of any such theory is that there's an interaction. All theories positing an interaction between X and Z imply that the effect of X depends on the value of Z and that the effect of Z depends on the value of X. As we saw in Chapter 2, how the value of Z modifies the effect of X is, due to the symmetry of interaction, identical to how the value of X modifies

[11] It's also worth remembering that our ability to draw inferences about a theory from the results of any empirical test also depends on things such as whether the model is correctly specified and whether we have any measurement error. In addition, we note that in practice our decision is rarely ever one between believing a theory and nothing. Instead, our decision is typically between believing some theory X or some other theory Z. In this sense, it often takes some other theory, rather than just empirical evidence, to kill a theory.

the effect of Z. These modifying effects are known as the interaction effect and are calculated as

$$\frac{\partial \left(\frac{\partial Y}{\partial X} \right)}{\partial Z} = \frac{\partial \left(\frac{\partial Y}{\partial Z} \right)}{\partial X} = \beta_3. \tag{4.42}$$

Prediction P_{XZ} has to do with whether this interaction effect, β_3, is negative or positive. We need only conduct a standard t-test on the interaction term coefficient β_3 to test the null hypothesis of no conditionality or no interaction. If we are unable to reject the null hypothesis that $\beta_3 = 0$ (and we determined that the magnitude of β_3 was substantively trivial), then this would be strong evidence that our underlying theory is wrong. Similarly, if we find that β_3 has the "wrong" sign, then this would also be strong evidence that our theory is wrong. In both of these cases, there would be little point going on to further evaluate the empirical support for the underlying theory positing interaction by testing the other four predictions. Of course, this doesn't mean that we should conclude that a theory positing interaction is correct simply because we find support for prediction P_{XZ}. As we saw in Chapter 2, a predicted interaction effect is always consistent with a wide variety of ways in which X and Z interact to determine Y, some of which may be inconsistent with the underlying theory. The point here is simply that when determining the level of support for a theory positing interaction, we should always evaluate prediction P_{XZ} first and then move on to evaluating the other predictions only if prediction P_{XZ} is corroborated.[12] The fact that the interaction effect can always be evaluated directly from the regression output in a standard interaction model is why many scholars prefer this particular type of interactive specification to the alternative specifications discussed in Chapter 3 that can be employed when one or both of the modifying variables are discrete.

⚠ **Important:** Finding empirical evidence for prediction P_{XZ}, which relates to the sign of the interaction effect, should be considered a necessary (but not sufficient) condition for concluding that a theory positing interaction is supported.

While there may be no hard and fast rules about what combinations of evidence regarding the five key predictions constitute support for one's

---

[12] We *can*, of course, go on to examine how the effect of X on Y varies with Z and how the effect of Z on Y varies with X even if there's no support for prediction P_{XZ}. While this may be interesting, we're engaging, at this point, in a purely empirical exercise as the underlying theory positing interaction from which our predictions were derived has already been falsified.

underlying theory, we can examine several prototypical sets of results we might get when estimating a linear-interactive model similar to the one shown in Eq. 4.41 and, for each, assess the extent to which we'd be willing to claim support for the underlying theory given the empirical evidence presented.[13]

To ground the discussion, let's assume that we seek to evaluate a theory positing interaction from which we can derive the following two hypotheses:

- $H_{X|Z}$: The marginal effect of X on Y is positive at all values of Z; this effect is strongest when Z is at its lowest value and declines in magnitude as Z increases.
- $H_{Z|X}$: The marginal effect of Z on Y is positive when X is at its lowest level. This effect declines in magnitude as X increases; at some value of X, Z has no effect on Y. As X rises further, the effect of Z becomes negative and strengthens in magnitude as X increases.

Taken together, these two hypotheses include all five of the key predictions listed above. Hypothesis $H_{X|Z}$ implies that the marginal effect of X is positive at both the lowest and highest values of Z and so makes predictions $P_{X|Z_{min}}$ and $P_{X|Z_{max}}$. Hypothesis $H_{Z|X}$ implies that the marginal effect of Z is positive when X is at its lowest value and negative when X is at its highest value, thereby offering predictions $P_{Z|X_{min}}$ and $P_{Z|X_{max}}$. There's no need to state a separate hypothesis that Z has a negative effect on the relationship between Y and X and that X has a negative effect on the relationship between Y and Z because these predictions about the interaction effect – P_{XZ} – are implicit in *both* $H_{X|Z}$ and $H_{Z|X}$.

As we'll discuss in more detail in the next chapter, scholars can often greatly increase their ability to convey substantively meaningful information from interaction models by using their parameter estimates to construct a *marginal effect plot* for an independent variable. As we've seen, a marginal effect plot is a graph that shows how the marginal effect of a variable, say X, varies with the value of another variable, say Z.[14]

---

[13] Our upcoming examination of different sets of prototypical results closely follows the discussion found in Berry, Golder and Milton (2012, 659–662).

[14] When the independent variable of interest is discrete, the graph shows how the effect of a one (or some other) unit increase in X on Y varies with the value of Z. While such a graph doesn't technically depict a *marginal* effect, it's still commonly referred to as a marginal effect plot. Moreover, as we saw earlier in the chapter, the effect of a one-unit increase in a discrete variable is mathematically equivalent to treating the variable as continuous and calculating a marginal effect in a linear-interactive model in which the interaction is linear. For these reasons, we'll refer to these graphs as marginal effect plots even when they show how the effect of a one (or some) unit change in a discrete variable varies with the value of another independent variable.

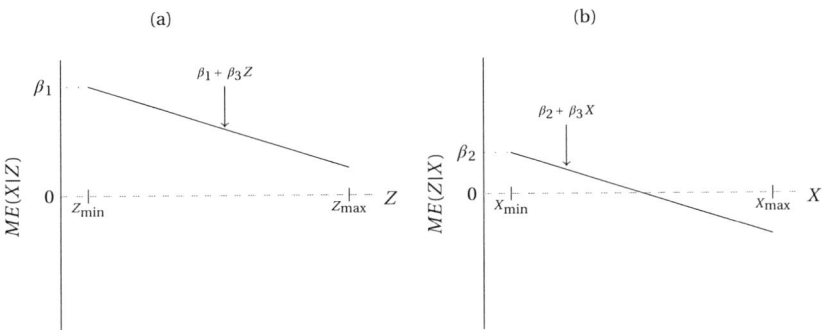

Figure 4.1 (a)–(b) Marginal effect plots consistent with hypotheses $H_{X|Z}$ and $H_{Z|X}$

Note: Panel (a) shows a predicted marginal effect plot consistent with hypothesis $H_{X|Z}$, while panel (b) shows a predicted marginal effect plot consistent with hypothesis $H_{Z|X}$.

Before actually estimating an interaction model, we encourage scholars to always think about what their hypotheses and the predictions they contain mean for what these marginal effect plots should look like. In Figure 4.1, we present *predicted* marginal effect plots for X and Z that are consistent with hypotheses $H_{X|Z}$ and $H_{Z|X}$. The marginal effect plot in Figure 4.1a graphically captures the three predictions contained in $H_{X|Z}$: (1) $ME(X|Z = Z_{min}) > 0$, (2) $ME(X|Z = z_{max}) > 0$, and (3) $\beta_3 < 0$, while the marginal effect plot in Figure 4.1b graphically captures the three predictions contained in $H_{Z|X}$: (1) $ME(Z|X = x_{min}) > 0$, (2) $ME(Z|X = x_{max}) < 0$, and (3) $\beta_3 < 0$. With these *predicted* marginal effect plots in hand, we can evaluate how closely they are approximated by the *estimated* marginal effect plots we produce from our interaction model.

A strong test of our underlying theory would obviously require that we use our estimated model coefficients and standard errors to evaluate all five of the predictions contained in $H_{X|Z}$ and $H_{Z|X}$. For illustrative purposes here, though, we focus on only hypothesis $H_{X|Z}$ and the three predictions that it contains. In Figure 4.2, we plot the marginal effect of X on Y across the observed range of Z based on six different prototypical sets of results that we might get when using the linear-interaction model outlined in Eq. 4.41 to test the theory underlying hypothesis $H_{X|Z}$. Each of the plots in Figure 4.2 should be compared with the predicted marginal effect plot for X in Figure 4.1a. We've identified the values of Z at which the marginal effect of X on Y is considered significant by making the horizontal axis in each plot bold at these values.[15] Under each plot, we also indicate whether

---

[15] We adopt this "bold axis" convention here because it's useful for portraying and discussing hypothetical results about "generic" X, Z, and Y variables. As will

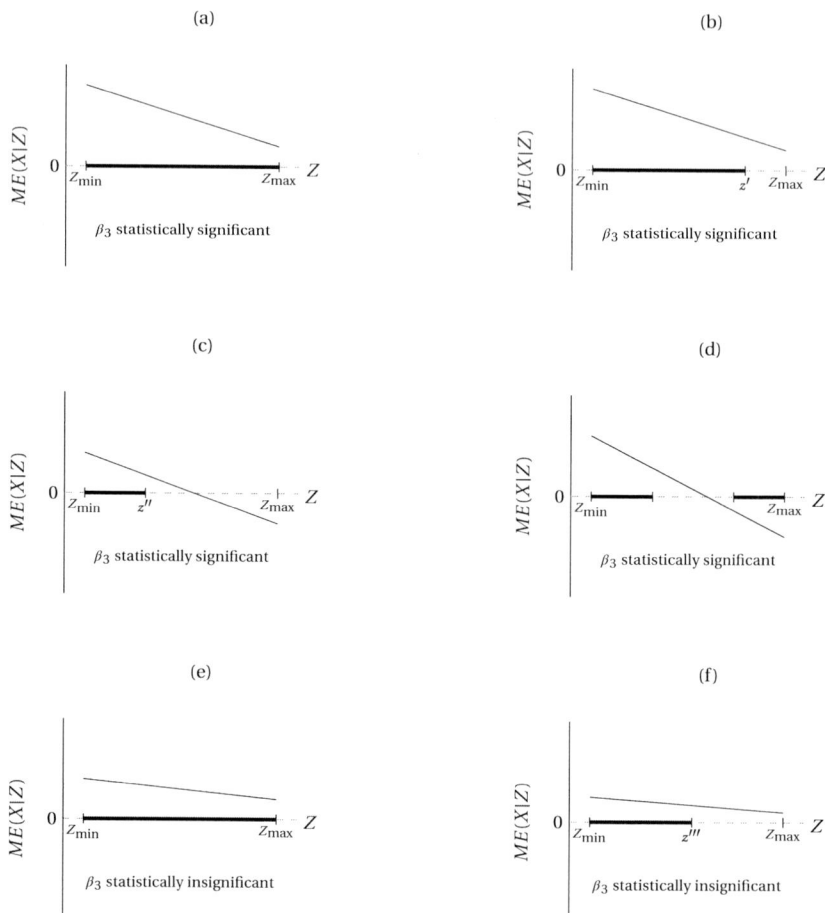

Figure 4.2 (a)–(f) Plots of the marginal effect of X on Y across the observed range of Z reflecting several prototypical sets of empirical results

Note: The horizontal axis is bold for all values of Z at which the marginal effect of X on Y is significant. Z_{min} and Z_{max} indicate the lowest and highest observed values of Z.

the coefficient on the interaction term, β_3, is significant. This information is critical for determining whether there's empirical evidence of interaction between X and Z – that is, for testing prediction P_{XZ}.

How comfortable would we be claiming support for hypothesis $H_{X|Z}$ if the parameter estimates from our interaction model led to a marginal effect plot like one of those shown in Figure 4.2? In panel (a), the marginal

become clear in the next chapter, we don't recommend that scholars adopt this convention when reporting actual results.

effect of X is positive and significant across the observed range of Z, and the coefficient on the interaction term β_3 is negative and significant. This plot provides unambiguously strong evidence for hypothesis $H_{X|Z}$ as each of its three predictions receive strong empirical support. The marginal effect plot in panel (b) is similar except that the marginal effect of X is no longer significant when Z is at its highest value. Note, though, that hypothesis $H_{X|Z}$ predicts that the marginal effect of X on Y declines in magnitude as Z increases. This leaves open the possibility that the positive effect of X on Y will be weak by the time that Z gets large. It's for this reason that we're not too concerned by the fact that $ME(X|Z = z_{max})$ fails to be significant. In this situation, we're comfortable concluding that there's still strong support for hypothesis $H_{X|Z}$ even though the value for $ME(X|Z = z_{max})$ isn't significant.[16]

The marginal effect plot shown in panel (c) represents a more ambiguous case. As before, the significant coefficient on the interaction term provides clear evidence in support of prediction P_{XZ} and its claim of interaction between X and Z. The difference in this plot compared to the one in panel (b) is that the range of values for Z for which the marginal effect of X is positive and significant is now much smaller and the point estimate for $ME(X|Z = z_{max})$ is negative rather than positive. In this particular case, we're not too concerned that the point estimate for $ME(X|Z = z_{max})$ has the "wrong" sign because it's insignificant. How supportive these results are of hypothesis $H_{X|Z}$ hinges on the percentage of observations that have values of Z at which the marginal effect of X is positive and significant; that is, for which $Z < z''$ in panel (c). The higher this percentage, the more we'd be willing to accept the empirical evidence as supportive. Of course, exactly how high the percentage needs to be to justify a claim of support is subjective. As a consequence, we recommend that researchers report the percentage of observations that fall within the region of significance. Indeed, it would be helpful if researchers provided a frequency distribution for the variable plotted on the horizontal axis so that readers can assess for themselves the relative density of observations across the full range of Z. We illustrate how such a frequency distribution might be incorporated into a marginal effect plot in our upcoming substantive applications in the next chapter.

When it comes to evaluating theories positing interaction, researchers shouldn't get into a "counting game" in which their conclusion is based solely on the number of predictions for which there's statistical support.

---

[16] In an ideal world, the theory underlying hypothesis $H_{X|Z}$ would be strong enough to generate a prediction about whether the effect of X on Y should remain strong or decline to near zero when Z reaches its maximum. However, we recognize that most theories in the social sciences are rarely capable of making such a fine distinction.

Note that the marginal effect plot in panel (d) provides support for two of the three predictions contained in $H_{X|Z}$, namely $ME(X|Z = Z_{\min}) > 0$ and $\beta_3 < 0$. The fact that the coefficient on the interaction term, β_3, is significant provides strong empirical evidence of interaction between X and Z. Significantly, though, the plot in panel (d) indicates that this interaction takes an appreciably different form from that predicted by hypothesis $H_{X|Z}$. While X has the predicted positive effect on Y when Z is low, X has a significant *negative* effect on Y when Z gets large. Researchers shouldn't downplay this inconsistency with hypothesis $H_{X|Z}$ by claiming that two out of three of their predictions are supported. Evidence that an increase in X leads to a significant reduction in Y when Z is high, rather than the predicted increase, should raise serious concerns about the theory underpinning hypothesis $H_{X|Z}$.

The marginal effect plot shown in panel (e) illustrates a more extreme case in which claiming support for hypothesis $H_{X|Z}$ based on two of the three predictions receiving statistical support would be unwarranted. In this case, X has a positive and significant effect on Y across the observed range of Z, thereby indicating support for predictions $P_{X|Z_{\min}}$ and $P_{X|Z_{\max}}$. The problem comes with the coefficient on the interaction term, β_3. Although this coefficient is negative, as predicted, it lacks significance, and the nearly flat marginal effect line is drawn this way to imply that the magnitude of β_3 is substantively trivial. In effect, there's no evidence of a meaningful interaction between X and Z. Indeed, this sort of plot – with a marginal effect line that's sloped slightly upward or downward – is exactly what we'd expect to find if we were to estimate a linear-interactive model when each of X and Z has a strong positive effect on Y but their effects are additive rather than interactive. In sum, the evidence presented in panel (e) seriously challenges the theory predicting that X and Z interact in influencing Y. Our reasoning here is in line with our earlier point that a significant coefficient with the predicted sign on the interaction term should be considered necessary for concluding that a theory positing interaction is supported.

The marginal effect plot shown in panel (f) is based on a final set of prototypical results. The coefficient on the interaction term is negative but insignificant, and the marginal effect line is again drawn as almost flat to imply that the magnitude of β_3 is substantively trivial. The fact that the marginal effect of X changes from significant when $Z < z'''$ to insignificant when $Z > z'''$ might seem to suggest that there's interaction between X and Z. Indeed, Brambor, Clark and Golder (2006, 74) imply precisely this when they claim that a situation in which the effect of X on Y is significant for some values of Z but not for others might be interpreted as a sign of meaningful interaction even when the coefficient on the interaction term is insignificant. As we've seen, though, this is incorrect. Without a significant coefficient on the interaction term, there's no evidence of meaningful interaction. The nearly-flat line in panel (f) represents a case in which the marginal effect of X has a t-ratio barely above the threshold for significance when Z is low and a t-ratio barely below the threshold when Z is high. If we were to capitalize on the fact that $ME(X|Z)$ changes from significant to insignificant as Z surpasses z''' to claim evidence of interaction, we'd be placing too much reliance on an arbitrarily chosen level of significance. If this level were set slightly higher, $ME(X|Z)$ would be significant over the entire range for Z. If the level were set slightly lower, $ME(X|Z)$ would not be significant at any value of Z. The more relevant information is that the coefficient on the interaction term, β_3, isn't significant and has a small magnitude. This indicates that the marginal effect of X varies only trivially with Z, and on this basis we should reject the theory positing interaction that underpins $H_{X|Z}$.

In this chapter, we've looked in more detail at how to calculate and interpret key quantities of interest from a linear-interactive model. In addition, we've discussed how to calculate appropriate measures of uncertainty to accompany our quantities of interest. In the next chapter, we take a more practical look at how to interpret and present the results from an interaction model. We do so by walking through three substantive applications that differ in terms of whether the variables involved in an interaction are continuous and/or discrete.

4.4 EXERCISES

1. In the exercises at the end of Chapter 3, we introduced a 2012 study by Banks and Valentino (2012) that looks at how opposition by Whites toward race-conscious policies such as affirmative action depends on the interaction between an individual's level of symbolic racism (or racial resentment) and their level of anger. Using their data, we estimated the following linear-interactive model specification:

$$Oppose\ Race\ Policies = \beta_0 + \beta_1 Symbolic\ Racism + \beta_2 Anger$$
$$+ \beta_3 Symbolic\ Racism \times Anger + \epsilon, \quad (4.43)$$

where *Oppose Race Policies* is a continuous variable on a 0–1 scale that captures an individual's level of opposition to race-conscious policies, *Anger* is a dichotomous variable that equals 1 when an individual is angry and 0 otherwise, and *Symbolic Racism* is a continuous variable that measures an individual's level of racial resentment on a 0–1 scale. The results from our analysis are shown in

Table 4.1 Symbolic racism, anger, and opposition to race-conscious policies II

Dependent Variable: *Oppose Race Policies*

	Model 1
Symbolic Racism	0.52126***
	( )
Anger	−0.11196
	( )
Symbolic Racism × *Anger*	0.25792**
	( )
Constant	0.26545***
	( )
Observations	182
R^2	0.46

Standard errors in parentheses. *$p < 0.10$; **$p < 0.05$; ***$p < 0.01$ (two-tailed)

Table 4.1, and the estimated variance-covariance matrix for the coefficients is,

$$V(\beta) = \begin{pmatrix} 0.00298512 & 0.00199709 & -0.00298512 & -0.00199709 \\ 0.00199709 & 0.00915827 & -0.01125366 & -0.00155169 \\ -0.00298512 & -0.01125366 & 0.01550988 & 0.00199709 \\ -0.00199709 & -0.00155169 & 0.00199709 & 0.00155169 \end{pmatrix}.$$

(4.44)

a. You'll notice that we haven't reported the standard errors for any of the coefficients in Table 4.1. Use the information in the variance-covariance matrix to identify the "missing" standard errors and put them in the appropriate empty parentheses in Table 4.1.

b. Looking at the variance-covariance matrix, what's $\text{cov}(\beta_1, \beta_3)$? What's $\text{cov}(\beta_3, \beta_1)$? What's $\text{cov}(\beta_2, \beta_3)$? And what's $\text{cov}(\beta_3, \beta_2)$?

c. What's the marginal effect of *Symbolic Racism* on *Oppose Race Policies* when someone isn't angry? What are the variance, standard error, and t-statistic associated with this marginal effect? Use a t-distribution table to determine the p-value associated with this marginal effect assuming a two-tailed hypothesis. What's the two-tailed 95% confidence interval for this marginal effect?

d. What's the marginal effect of *Symbolic Racism* on *Oppose Race Policies* when someone's angry? What are the variance, standard error, and t-statistic associated with this marginal effect? Use a t-distribution table to determine the p-value associated with this marginal effect assuming a two-tailed hypothesis. What's the two-tailed 95% confidence interval for this marginal effect?

e. What's the effect of being angry on *Oppose Race Policies* when an individual's level of racial resentment is 0? What are the variance, standard error, and t-statistic associated with this effect? Use a t-distribution table to determine the p-value associated with this effect? What are the two-tailed 95% confidence interval for this effect?

f. What's the effect of being angry on *Oppose Race Policies* when an individual's level of racial resentment is 0.5? What are the variance, standard error, and t-statistic associated with this effect? Use a t-distribution table to determine the p-value associated with this effect assuming a two-tailed hypothesis. What's the two-tailed 95% confidence interval for this effect?

g. What's the effect of being angry on *Oppose Race Policies* when an individual's level of racial resentment is 1? What are the variance, standard error, and t-statistic associated with this effect? Use a t-distribution table to determine the p-value associated with this marginal effect assuming a two-tailed hypothesis. What's the two-tailed 95% confidence interval for this marginal effect?

h. What's the predicted value of *Oppose Race Policies* when an individual is calm and has a level of racial resentment equal to 0.5. What's the two-tailed 95% confidence interval for this predicted value?

i. What's the predicted value of *Oppose Race Policies* when an individual is angry and has a level of racial resentment equal to 0.5. What's the two-tailed 95% confidence interval for this predicted value?

j. What's the difference in the predicted value of *Oppose Race Policies* between an individual who has a level of racial resentment equal to 0.5 and is angry and an equivalent individual who's calm. What's the two-tailed 95% confidence interval for this difference?

5 Three Substantive Applications

Interpretation and Presentation

In this chapter, we use what we've learned over the last three chapters to walk through three substantive applications. The substantive applications differ in terms of whether the variables that interact are continuous and/or discrete. The first application looks at how race and gender affect attitudes toward political parties in the United States. The second application examines how voter concerns with ideology and descriptive representation influence support for political candidates in the United States. And the third application investigates why there are more women legislators in some countries than others. One of the primary goals of this chapter is to illustrate different ways to constructively present the results from a linear-interactive model that specifies a possible interaction between two variables X and Z on a continuous dependent variable Y.

5.1 WHEN X AND Z ARE BOTH DISCRETE: GENDER, RACE, AND SUPPORT FOR THE REPUBLICAN PARTY

How do race and gender affect attitudes toward political parties in the United States? In this substantive application, which is based on a replication of Block, Golder and Golder (2023), we examine how an individual's race (Black/White) and gender (female/male) affected how much they liked the Republican Party at the time of the 2016 presidential election in the United States. In terms of race, Blacks are expected to exhibit less support for the Republican Party than Whites. This is because the Republican Party espouses a policy position that is conservative on the issues of civil rights and race. These issues were particularly salient during the 2016 election campaign due to the heightened racial tensions following the presidency of Barack Obama and the racialized rhetoric and support for individuals and messages associated with white supremacy from the Republican candidate Donald Trump (Huber, 2016; Newman, Shah and Collingwood, 2018; Swain, 2018). In terms of gender, women are expected to exhibit less support for the Republican Party than men. This is because the Republican Party holds a conservative position on a host of policy issues related to things like healthcare, same sex marriages, restrictions on firearms, and government activism, where women have historically held a more progressive position than men. The sexist language used by Donald Trump during the campaign is likely to have reinforced the partisan gender gap that has seen women consistently favor the Democratic Party over the Republican Party since the early 1980s (Bock, Byrd-Craven and Burkley, 2017; Frasure-Yokley, 2018; Schaffner, MacWilliams and Nteta, 2018; Cassese and Barnes, 2019).

Rather than assume that gender and race have separate and additive effects on how much an individual likes the Republican Party, Block, Golder and Golder (2023) argue that there are reasons to think that they interact to determine Republican Party support. Scholars of intersectionality, for example, have long argued that it's not always appropriate to treat groups, such as women, men, Blacks, and Whites, as homogenous and that we need to recognize that social identities such as race and gender "interact to form qualitatively different meanings and experiences" (Crenshaw, 1989; McCall, 2005; Weldon, 2006; Hancock, 2007, 2016; Warner, 2007, 454). In our particular application, women and men may exhibit different levels of support for the Republican Party depending on their race, and Blacks and Whites may exhibit different levels of support depending on their gender. There are several potential reasons for this. One is that Black women have frequently been stigmatized and framed in particularly negative terms – "welfare queen" or "crack-mother" – by political elite discourse such as that coming from the Republican Party (Jordan-Zachary, 2003; Hancock, 2004; Gillespie and Brown, 2019). A second reason is that Black men often hold more conservative attitudes relative to Black women than White men do relative to White women (Dawson, 2001; Lewis, 2013; Rigueur, 2014; Smith, 2018). Finally, the relative absence of Black men due to phenomena like mass incarceration means that Black women tend to play a more politically active role in the community compared to Black men than White women do compared to White men (Weaver, 2010; Nellis, 2016; Anderson, 2018; Subramanian, Riley and Mai, 2018). All of this suggests that Black women will exhibit a particularly negative reaction to the Republican Party relative to both White women and Black men. In sum, there are reasons to believe that gender and race interact to determine Republican Party support.

The theory presented here implies the following two hypotheses:

Female Hypothesis: Women will always like the Republican Party less than men. This negative effect is larger among Blacks than Whites.

Black Hypothesis: Blacks will always like the Republican Party less than Whites. This negative effect is larger among women than men.

Together, these two hypotheses contain all five of the key predictions that we recommend for a theory positing interaction like the one presented here. The *Female Hypothesis* predicts that the effect of being female is negative for both Whites and Blacks, thereby providing predictions $P_{Female|Black_{min}}$ and $P_{Female|Black_{max}}$. The *Black Hypothesis* predicts that the the effect of being Black is negative for both men and women, thereby offering predictions $P_{Black|Female_{min}}$ and $P_{Black|Female_{max}}$. Both hypotheses imply that there's a negative interaction between being female and being Black, $P_{Female \times Black}$, because the negative effect of being female is predicted to be stronger among Blacks than Whites and because the negative effect of being Black is expected to be stronger among women than men.

We use data from the 2019 version of the American National Election Studies 2016 Time Series Study to test our hypotheses (American National Election Studies, 2019). The dependent variable, *Like Republican Party*, is based on a survey question in which respondents are asked to indicate how much they like the Republican Party on a 0–10 scale, where 0 indicates they strongly dislike the Republican Party, and 10 indicates they strongly like it.[1] *Like Republican Party* has a mean of 4.95 and a standard deviation of 3.03. In terms of our key independent variables, *Female* is a dichotomous variable that equals 1 if an individual self-identifies as female and 0 otherwise, *Black* is a dichotomous variable that equals 1 if an individual self-identifies as Black and 0 otherwise, and *Female × Black* is an interaction term created by multiplying together the constitutive terms *Female* and *Black*.[2] As a control variable, we include the respondent's *Age* in years.[3] We treat our dependent variable as continuous and estimate an ordinary least squares regression with the following "standard" interactive model specification:

$$Like\ Republican\ Party = \beta_0 + \beta_1 Female + \beta_2 Black + \beta_3 Female \times Black$$
$$+ \beta_4 Age + \epsilon. \tag{5.1}$$

Past experience has told us that it's often helpful to explicitly present one's interactive model specification to the reader as we've done here.

---

[1] This particular question formed part of the post-election survey where pre-election respondents were re-interviewed between November 9, 2016 and January 8, 2017.

[2] Given the purposes of our analysis here, we omit those respondents who don't self-identify as female or male, as well as those who don't self-identify as Black or White.

[3] Our specification is almost certainly underspecified in terms of control variables. However, our goal here isn't to estimate the best possible model for evaluating the effects of race and gender on support for the Republican Party but rather to show how to correctly interpret and present the results from an interaction model with two discrete modifying variables. Our results and inferences are robust to also controlling for an individual's education and income.

While some readers can immediately "see" what regression model we plan to estimate when we discuss our research design, others get a handle on it only when they see the exact specification presented to them.[4] One advantage of presenting the precise model specification is that it facilitates a discussion of our predictions for the empirical analysis. On this point, we strongly recommend that researchers estimating an interaction model always state their predictions for the key model parameters and other quantities of interest. Such a practice forces us to carefully think through exactly how our verbal hypotheses translate into predictions about specific model parameters. It also makes it easier for the reader who wishes to evaluate the level of empirical support for a proposed theory to know exactly what to look for as they transition from the theoretical and research design sections of a study to the section discussing the actual results.[5]

⚠ **Important:** When estimating an interaction model, it is often helpful to explicitly state one's predictions for the key model parameters and other quantities of interest.

Our two hypotheses speak to the effect of gender and race on Republican Party support. The effect of being female as opposed to male can be calculated as

$$\frac{\partial Like\ Republican\ Party}{\partial Female} = \beta_1 + \beta_3 Black. \tag{5.2}$$

From this, we see that β_1 indicates the effect of being female among Whites (when $Black = 0$) and that $\beta_1 + \beta_3$ indicates the effect of being female among Blacks (when $Black = 1$). According to our *Female Hypothesis*, women should always exhibit less support than men for the Republican Party, but this negative effect should be larger for Blacks than Whites. It follows, therefore, that β_1 and $\beta_1 + \beta_3$ should both be negative. Since the negative effect of being female should be larger among Blacks, $\beta_1 + \beta_3 < \beta_1$, it follows that β_3 should also be negative.

---

[4] We're not necessarily recommending that scholars present an interactive specification that includes *all* of the variables. For example, a researcher might simply include a variable called *Controls* to note that there are a variety of control variables in the model. However, researchers should include those variables that are relevant for testing their main theoretical hypotheses.

[5] Where space is limited, researchers might consider moving the presentation of the model specification and their discussion of the predicted quantities of interest to an appendix.

The effect of being Black as opposed to White can be calculated as

$$\frac{\partial Like\ Republican\ Party}{\partial Black} = \beta_2 + \beta_3 Female. \tag{5.3}$$

From this, we see that β_2 indicates the effect of being Black among men (when $Female = 0$) and that $\beta_2 + \beta_3$ indicates the effect of being Black among women (when $Female = 1$). According to our *Black Hypothesis*, Blacks should always exhibit less support than Whites for the Republican Party, but this negative effect should be larger for women than men. It follows, therefore, that β_2 and $\beta_2 + \beta_3$ should both be negative. Since the negative effect of being Black should be larger among women, $\beta_2 + \beta_3 < \beta_2$, it follows again that β_3 should be negative. As expected, both hypotheses make the same prediction about the sign of the coefficient on the interaction term. This is necessarily the case due to the symmetry of interactions. If we ever specify hypotheses or make predictions where the modifying effects of the variables involved in an interaction have different signs, we immediately know that we've done something wrong.

The results from the standard interaction model shown in Eq. 5.1 are presented in the first column of Table 5.1. We start by discussing how to interpret these results. The coefficient on *Female* is negative but is substantively small (-0.04) and statistically insignificant. This tells us that being female as opposed to male has no significant effect on how much someone likes the Republican Party among Whites ($Black = 0$). Put differently, there's no significant difference in how much White women and White men like the Republican Party. The coefficient on *Black* is negative and statistically significant. This tells us that being Black as opposed to White has a significant negative effect on support for the Republican Party among men ($Female = 0$). More specifically, Black men like the Republican Party about 1.5 units less than White men. The magnitude of this effect is substantively meaningful given that the dependent variable *Like Republican Party* has a mean of 5.25 among White men. The important thing to remember here is that the coefficients on *Female* and *Black* don't indicate the effect of gender and race in an unconditional or average sense; instead, they only indicate the effect of being female *among Whites* and the effect of being Black *among men*. The coefficient on the interaction term *Female×Black* is negative and statistically significant. This provides the important evidence of conditionality and indicates both that being Black has a negative effect on how much being female affects support for the Republican Party and that being female has a negative effect on how much being Black affects support for the Republican Party. We'll discuss the substantive significance of the interaction effect shortly. The coefficient on *Age* is positive and statistically significant, indicating that

Table 5.1 Gender, race, and support for the Republican Party in the United States at the 2016 presidential elections

Dependent Variable: *Like Republican Party, 0–10*

	Standard Interactive Specification	Alternative Interactive Specification I	Alternative Interactive Specification II
Female	−0.04 (0.12)		
Black	−1.50*** (0.27)		
Female×Black	−1.03*** (0.35)		
White Female		−0.04 (0.12)	2.53*** (0.22)
Black Male		−1.50*** (0.27)	1.07*** (0.33)
Black Female		−2.57*** (0.22)	
White Male			2.57*** (0.22)
Age	0.02*** (0.003)	0.02*** (0.003)	0.02*** (0.003)
Constant	4.47*** (0.18)	4.47*** (0.18)	1.90*** (0.25)
Observations	2,858	2,858	2,858
R^2	0.07	0.07	0.07

Standard errors in parentheses. $^*p < 0.10$; $^{**}p < 0.05$; $^{***}p < 0.01$ (two-tailed)

Note: The two alternative interactive specifications differ in terms of the group that acts as the omitted, and hence reference, category. In Alternative Interactive Specification I, White men act as the reference category, and in Alternative Interactive Specification II, Black women act as the reference category. All three models shown in Table 5.1 are equivalent and produce the exact same quantities of interest.

support for the Republican Party increases by 0.02 units for each year of an individual's life.

To fully evaluate our theory, we need to examine all five of the key predictions in the *Female Hypothesis* and the *Black Hypothesis*. The results in the first column of Table 5.1 provide the information necessary to evaluate three of these predictions: (1) the interaction effect, (2) the effect of *Female* when *Black* is at is lowest value of 0, and (3) the effect of *Black* when *Female* is at its lowest value of 0. We've already discussed these effects in the previous paragraph. What we can't see directly from the regression output is the effect of *Female* when *Black* is at its highest value of 1 and the effect of *Black* when *Female* is at its highest value of 1. To evaluate these particular effects, we need to make additional calculations.[6]

Based on Eq. 5.2, we see that the effect of being female as opposed to male when an individual is Black is $\beta_1 + \beta_3$ or $-0.04 + (-1.03) = -1.07$. This tells us that Black women like the Republican Party 1.07 units less than Black men. The magnitude of this effect is substantively meaningful given that the dependent variable *Like Republican Party* has a mean of 3.61 among Black men. Is this effect also statistically significant? Substituting our variable names into Eq. 4.20, we see that the standard error associated with the effect of being female is

$$\text{se}\left(Female\right) = \text{se}\left(\beta_1 + \beta_3 Black\right)$$
$$= \sqrt{\text{var}\left(\beta_1\right) + \left(Black\right)^2 \text{var}\left(\beta_3\right) + 2 \times Black \times \text{cov}\left(\beta_1, \beta_3\right)}.$$
$$(5.4)$$

We want to know the standard error associated with the effect of being female among Blacks, and so we set $Black = 1$, and Eq. 5.4 simplifies to

$$\text{se}\left(Female|Black = 1\right) = \text{se}\left(\beta_1 + \beta_3\right) = \sqrt{\text{var}\left(\beta_1\right) + \text{var}\left(\beta_3\right) + 2 \times \text{cov}\left(\beta_1, \beta_3\right)}.$$
$$(5.5)$$

To calculate this, we must gather some information from the estimated variance-covariance matrix for the coefficients,

$$V(\beta) = \begin{pmatrix} 0.013665 & 0.007221 & -0.013652 & -0.000004 & -0.007031 \\ 0.007221 & 0.075446 & -0.075144 & 0.000069 & -0.010774 \\ -0.013652 & -0.075144 & 0.123598 & -0.000026 & -0.008573 \\ -0.000004 & 0.000069 & -0.000026 & 0.000010 & -0.000505 \\ -0.007031 & -0.010774 & 0.008573 & -0.000505 & 0.032923 \end{pmatrix}.$$
$$(5.6)$$

---

[6] In what follows, we show how to manually do the required calculations. As we've noted previously, most statistical packages provide a variety of simple post-estimation commands that can do the necessary calculations for us.

The two variances that we need are shown in bold, while the covariance we need is shown in gray.[7] Substituting the highlighted values into Eq. 5.5, we now have

$$\text{se}\left(Female|Black = 1\right) = \sqrt{\mathbf{0.013665} + \mathbf{0.123598} + 2 \times -0.013652}$$
$$= \sqrt{0.109959} = 0.3316. \tag{5.7}$$

Combining our information, we see that the effect of being female among Blacks is -1.07 with a standard error of 0.3316. The t-statistic for an effect of this size is

$$t = \frac{-1.07}{0.3316} = -3.23. \tag{5.8}$$

This test statistic follows a t-distribution with $n - k$ degrees of freedom, where in our case $n = 2,858$ and $k = 5$. Consulting a t-distribution table and assuming a two-tailed hypothesis, we find that the p-value associated with our test statistic and degrees of freedom is $p = 0.001$. Following the formula shown in Eq. 4.26, the two-tailed 95% confidence interval for the effect of being female among Blacks is calculated as

$$-1.07 \pm t_{2,853,0.025} \times 0.3316$$
$$-1.07 \pm 1.96 \times 0.3316$$
$$-1.07 \pm 0.65. \tag{5.9}$$

In other words, the effect of being female among Blacks is -1.07, and the confidence interval is $(-1.72, -0.42)$. Whether we look at the p-value or the confidence interval, we conclude that the negative effect of being female on support for the Republican Party among Blacks is statistically significant. Put differently, Black women like the Republican Party significantly less than Black men.

We can make similar calculations to evaluate the effect of being Black as opposed to White among females. Based on Eq. 5.3, we see that this effect is $\beta_2 + \beta_3$ or $-1.50 + (-1.03) = -2.53$. The magnitude of this effect is substantively meaningful given that the dependent variable *Like Republican Party* has a mean of 5.19 among White women. The standard error for the effect of being Black is

$$\text{se}\left(Black\right) = \text{se}\left(\beta_2 + \beta_3 Female\right)$$
$$= \sqrt{\text{var}\left(\beta_2\right) + \left(Female\right)^2 \text{var}\left(\beta_3\right) + 2 \times Female \times \text{cov}\left(\beta_2, \beta_3\right)}. \tag{5.10}$$

---

[7] As a reminder, the variance-covariance matrix is symmetric, and so the covariance we need appears twice, once in the first column as $\text{cov}(\beta_3, \beta_1)$ and once in the first row as $\text{cov}(\beta_1, \beta_3)$.

Setting *Female* $= 1$ and obtaining the relevant information from the variance-covariance matrix in Eq. 5.6, we have

$$\text{se}\left(Black|Female = 1\right) = \sqrt{\text{var}\left(\beta_2\right) + \text{var}\left(\beta_3\right) + 2 \times \text{cov}\left(\beta_2, \beta_3\right)}$$
$$= \sqrt{0.075446 + 0.123598 + 2 \times -0.075144}$$
$$= \sqrt{0.049} = 0.221. \tag{5.11}$$

Putting our information together, we see that the effect of being Black among females is -2.53 with a standard error of 0.221. The *t*-statistic for this effect is

$$t = \frac{-2.53}{0.221} = -11.45. \tag{5.12}$$

The *p*-value associated with this test statistic is exceedingly small. The two-tailed 95% confidence interval for the effect of being Black among females is

$$-2.53 \pm 1.96 \times 0.221$$
$$-2.53 \pm 0.43. \tag{5.13}$$

In other words, the effect of being Black among females is -2.53 $(-2.96,$ $-2.10)$. Whether we look at the *p*-value or the confidence interval, we conclude that the negative effect of being Black among females on support for the Republican Party is statistically significant. Put differently, Black women like the Republican Party significantly less than White women.

We now have all of the quantities of interest we need to evaluate our five key predictions. How should we present this information? In the case that we have here, where both modifying variables are dichotomous, perhaps the easiest and most efficient way to present the quantities of interest is directly in the text. With respect to gender, the effect of being female is -0.04 $(-0.27, 0.19)$ among Whites and -1.07 $(-1.72, -0.42)$ among Blacks. In other words, there's no significant difference between White women and White men when it comes to liking the Republican Party, but Black women like the Republican Party significantly less than Black men. Put differently, gender doesn't seem to matter among Whites when it comes to liking the Republican Party. With respect to race, the effect of being Black is -1.50 $(-2.04, -0.96)$ among men and -2.53 $(-2.96,$ $-2.10)$ among women. In other words, Black men like the Republican Party significantly less than White men, and Black women like the Republican Party significantly less than White women. Put differently, race always matters when it comes to liking the Republican Party. The interaction effect is -1.03 (-1.72, -0.34).[8] This indicates that race and gender interact negatively to determine support for the Republican Party. Specifically, it indicates that the magnitude of the negative effect of being female on liking the Republican Party is 1.03 units larger for Blacks than Whites and that the magnitude of the negative effect of being Black is 1.03 units larger for women than men. The magnitude of the interaction effect is substantively meaningful. One way to evaluate this is to compare the effects that we've calculated for gender among Blacks and Whites and the effects for race among women and men. When we do this, we see that the negative impact of gender, or being female, is about $\frac{-1.07}{-0.04} = 25.36$ times larger for Blacks than Whites and that the negative effect of race, or being Black, is about $\frac{-2.53}{-1.50} = 1.69$ times or 69% larger among women than men.

In terms of our theory, four of our five key predictions receive unambiguous support. The only prediction that doesn't receive complete support is the one that White women will like the Republican Party less than White men. The estimated effect of being female for Whites is negative, as predicted, but it's substantively small and statistically insignificant. It turns out that this particular result is consistent with several recent studies suggesting that gender may not play a significant role in determining support for the Republican Party among Whites (Huddy, Cassese and Lizotte, 2012; Dittmar, 2016; Cassese, 2017; Junn, 2017; Williams, 2017; Cassese and Barnes, 2019; Junn and Masuoka, 2020). Overall, the results support the idea that gender and race interact to determine how much individuals like the Republican Party and strengthen the assertion by scholars of intersectionality that we should pay more attention to groups that exist at the intersections of traditional gender and race categories.

There are alternative ways to present the key quantities of interest when both modifying variables are dichotomous. One possibility is to present them graphically rather than in the text. For example, we could use a "marginal effect" plot to graph (1) the interaction effect for gender and race, (2) the effects of being female among Whites and Blacks, and (3) the effects of being Black among men and women. Such a marginal effect plot is shown in Figure 5.1. Figure 5.1 shows all of the information necessary to evaluate the five key predictions that are jointly contained in the *Female Hypothesis* and the *Black Hypothesis*. Each of the five "effects" is shown as a small circle along with their corresponding two-tailed 95% confidence intervals. The dashed vertical gray line is included to aid the reader in determining whether the effects are significantly different from

---

[8] You may be wondering where we got the confidence intervals for the effect of being female among Whites, the effect of being Black among men, and the interaction effect. Although we didn't report these confidence intervals in Table 5.1, they were provided in the regression output for the coefficients on *Female*, *Black*, and *Female×Black*.

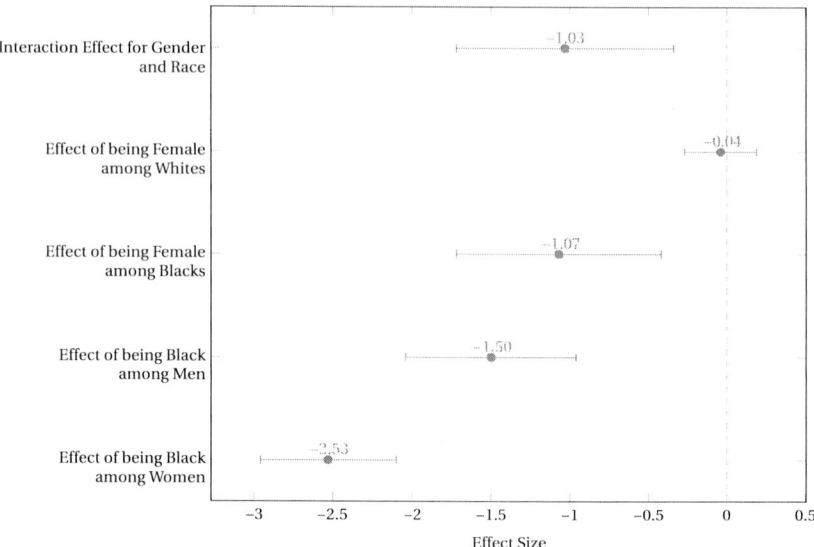

Figure 5.1 The conditional effects of gender and race on support for the Republican Party in the 2016 US presidential elections

Note: The plot shows the interaction effect for gender and race, as well as the effects of being female among Whites and Blacks and the effects of being Black among men and women on how much individuals report liking the Republican Party in the 2016 US presidential elections. The dependent variable, *Like Republican Party*, is measured on a 0–10 scale. The effects are based on the results from the standard interaction model in the first column of Table 5.1.

zero. Whenever the confidence interval contains the vertical line, we can't reject the possibility that the effect is zero. This is the case for the effect of being female among Whites but not for any of the other effects. We recommend that scholars report the numerical value for the estimated effect sizes in the plot so as to maintain the type of precision that one would get from simply reporting the effects in the text. Depending on the circumstances, researchers may also wish to report the numerical values for the lower and upper bounds of the confidence intervals.

An alternative approach to simply presenting the "effects" of the interacting variables is to show both predicted values and differences. To a large extent, we can talk of having four categories of observations when we have two dichotomous variables that interact. In our case, where the interacting variables are *Female* and *Black*, the four categories correspond to White men $(0,0)$, White women $(1,0)$, Black men $(0,1)$, and Black women $(1,1)$. We might be interested in calculating the predicted level of support for the Republican Party exhibited by each of these categories of people in a given scenario. For example, we might want to know how much, say, a 40 year old White male, White female, Black male, and Black female likes the Republican Party based on the regression results shown in the first column of Table 5.1. Following the logic presented in panel (b) of Figure 3.2 in Chapter 3, we see that the predicted level of support for the Republican Party is $\beta_0 + \beta_4 \times 40 = 4.47 + 0.02 \times 40 = 5.08$ for a 40 year old White male, it's $\beta_0 + \beta_1 + \beta_4 \times 40 = 4.47 - 0.04 + 0.02 \times 40 = 5.04$ for a 40 year old White female, it's $\beta_0 + \beta_2 + \beta_4 \times 40 = 4.47 - 1.50 + 0.02 \times 40 = 3.58$ for a 40 year old Black male, and it's $\beta_0 + \beta_1 + \beta_2 + \beta_3 + \beta_4 \times 40 = 4.47 - 0.04 - 1.50 - 1.03 + 0.02 \times 40 = 2.51$ for a 40 year old Black female.[9]

In the previous chapter in Section 4.2.2, we saw how to calculate variances for predicted values. Rather than use the scalar notation shown in Eq. 4.32, which works but is rather messy, we're going to use the matrix notation that we saw in Eq. 4.33 to calculate our variances. Recall that the variance of a predicted value for some baseline scenario b is

$$\operatorname{var}\left(\hat{Y}_b\right) = M_b V(\beta) M_b', \tag{5.14}$$

where M_b is a row vector that contains the values for the independent variables and the constant in our scenario of interest, $V(\beta)$ is the estimated variance-covariance matrix for the coefficients, and M_b' is the transpose of M_b. Let's start with the scenario in which we have a 40 year old White male. In this scenario, $M_b = \begin{pmatrix} 0 & 0 & 0 & 40 & 1 \end{pmatrix}$, indicating that $Female = 0$, $Black = 0$, $Female \times Black = 0$, $Age = 40$, and the $Constant = 1$. The variance for the predicted value in this case is

$$\operatorname{var}\left(\hat{Y}_b\right) = \begin{pmatrix} 0 & 0 & 0 & 40 & 1 \end{pmatrix} \begin{pmatrix} 0.013665 & 0.007221 & -0.013652 & -0.000004 & -0.007031 \\ 0.007221 & 0.075446 & -0.075144 & 0.000069 & -0.010774 \\ -0.013652 & -0.075144 & 0.123598 & -0.000026 & -0.008573 \\ -0.000004 & 0.000069 & -0.000026 & 0.000010 & -0.000505 \\ -0.007031 & -0.010774 & 0.008573 & -0.000505 & 0.032923 \end{pmatrix} \begin{pmatrix} 0 \\ 0 \\ 0 \\ 40 \\ 1 \end{pmatrix}$$

$$= 0.0084. \tag{5.15}$$

Taking the square root of this variance, we find that the standard error for the predicted level of support for the Republican Party exhibited by a 40 year old White male is $\sqrt{0.0084} = 0.09$. We can now use the standard error to calculate a two-tailed 95% confidence interval,

$$5.08 \pm 1.96 \times 0.09$$

$$5.08 \pm 0.1764. \tag{5.16}$$

---

[9] If you look carefully, you'll notice that these sums appear to be slightly off. This is because we've rounded the coefficients to two decimal places in the text but completed the sums using the full coefficients reported in the regression output. Similar issues arise elsewhere in the book.

Finally, we can say that the predicted level of support for the Republican Party for a 40 year old White male is 5.08 (4.90, 5.26).

Rather than calculate the variances for our four predicted values separately, we can actually calculate them all at once. We do this by turning the row vector M_b into a matrix where each row indicates the values for the independent variables and constant in one of our four scenarios. This also means that its transpose, M'_b, also becomes a matrix. We now have

$$
\text{var}\left(\hat{Y}_b\right)
$$

$$
= \begin{pmatrix} 0 & 0 & 0 & 40 & 1 \\ 1 & 0 & 0 & 40 & 1 \\ 0 & 1 & 0 & 40 & 1 \\ 1 & 1 & 1 & 40 & 1 \end{pmatrix} \begin{pmatrix} 0.013665 & 0.007221 & -0.013652 & -0.000004 & -0.007031 \\ 0.007221 & 0.075446 & -0.075144 & 0.000069 & -0.010774 \\ -0.013652 & -0.075144 & 0.123598 & -0.000026 & -0.008573 \\ -0.000004 & 0.000069 & -0.000026 & 0.000010 & -0.000505 \\ -0.007031 & -0.010774 & 0.008573 & -0.000505 & 0.032923 \end{pmatrix} \begin{pmatrix} 0 & 1 & 0 & 1 \\ 0 & 0 & 1 & 1 \\ 0 & 0 & 0 & 1 \\ 40 & 40 & 40 & 40 \\ 1 & 1 & 1 & 1 \end{pmatrix}
$$

$$
= \begin{pmatrix} 0.0084 & 0.0077 & 0.0679 & 0.0426 \end{pmatrix}, \tag{5.17}
$$

where each element in the resulting row vector corresponds to the variance for the predicted value in one of our four scenarios. Taking the square roots of these elements gives us standard errors, which we can use to calculate two-tailed 95% confidence intervals for our four predicted values. We show the mean predicted values and their associated confidence intervals for a 40 year old White male, White female, Black male, and Black female in the top-left shaded square in Figure 5.2.

With this setup we can think of the effects of gender and race on support for the Republican Party in terms of differences in predicted values. For example, we might want to calculate the effect of changing the gender of our forty year old individual from male to female. This is equivalent to calculating the differences in the predicted values as we move from the top row to the bottom row of Figure 5.2. Building on the insights we saw in Chapter 4 in Eq. 4.39, the difference in predicted values when we change an individual's gender by one unit from male to female is

$$
\hat{Y}_{\text{Female}} - \hat{Y}_{\text{Male}} = \beta_1 + \beta_3 Black. \tag{5.18}
$$

As you'll have realized, we've already calculated this effect for both Black and White individuals along with the associated standard errors and confidence intervals. For example, the difference in predicted values as we change our 40 year old individual from male to female is -0.04 (-0.27, 0.19) if they're White and -1.07 (-1.72, -0.42) if they're Black. These differences are reported in the first two cells of the bottom row in Figure 5.2.

We can also calculate the effect of changing the race of our 40 year old from White to Black. This is equivalent to calculating the differences in predicted values as we move from the left column to the right column of Figure 5.2. Again building on the insights we saw earlier in Eq. 4.39, the

Race

	White	Black	Difference
Male	5.08 (4.90, 5.26)	3.58 (3.07, 4.09)	−1.50 (−2.04, −0.96)
Female	5.04 (4.86, 5.21)	2.51 (2.10, 2.91)	−2.53 (−2.96, −2.10)
Difference	−0.04 (−0.27, 0.19)	−1.07 (−1.72, −0.42)	−1.03 (−1.72, −0.34)

(Gender is labeled on the left, spanning Male / Female / Difference rows.)

Figure 5.2 Predicted values and the conditional effects of gender and race on support for the Republican Party in the 2016 US presidential elections

Note: The top-left shaded square shows how the predicted level of support for the Republican Party varies depending on whether a forty year old individual is a White male, White female, Black male, or a Black female. The dependent variable, *Like Republican Party*, is measured on a 0–10 scale. The effect of gender (male ⟶ female) for Whites and Blacks is shown in the bottom row, while the effect of race (White ⟶ Black) for males and females is shown in the right column. The interaction effect is shown in gray in the bottom right cell. The information is based on the results from the standard interaction model in the first column of Table 5.1.

difference in predicted values when we change an individual's race by one unit from White to Black is

$$\hat{Y}_{\text{Black}} - \hat{Y}_{\text{White}} = \beta_2 + \beta_3 Female. \tag{5.19}$$

As before, we've already calculated this effect for both male and female individuals along with the associated standard errors and confidence intervals. The difference in predicted values as we change our 40 year old from White to Black is −1.50 (−2.04, −0.96) if they're male and −2.53 (−2.96, −2.10) if they're female. These differences appear in the top two cells of the right column in Figure 5.2.

The difference in the two differences in the bottom row of Figure 5.2 indicates how the effect of gender varies depending on an individual's race,

$$\text{Difference}_{\text{Gender|Black}} - \text{Difference}_{\text{Gender|White}} = -1.07 - (-0.04) = -1.03. \quad (5.20)$$

As you'll no doubt realize, this is the "interaction effect," and we already know the associated standard error and confidence interval for this. Similarly, the difference in the two differences in the right column of Figure 5.2 indicates how the effect of race varies depending on an individual's gender,

$$\text{Difference}_{\text{Race|Female}} - \text{Difference}_{\text{Race|Male}} = -2.53 - (-1.50) = -1.03. \quad (5.21)$$

Due to the symmetry of interactions, this is also the interaction effect. The interaction effect is shown in gray in the lower right cell in Figure 5.2.

The value of Figure 5.2 as opposed to Figure 5.1 is that we can present predicted values in addition to the "effects" that are necessary to evaluate the five key predictions contained in most theories positing interaction between two dichotomous modifying variables. This provides the reader with potentially useful extra information with which to evaluate the substantive importance of the results. In particular, the predicted values, by providing information about the level of the dependent variable in particular scenarios, can act as metrics that allow the reader to better evaluate the substantive significance of the effect sizes.

> ⚠ **Important:** The key quantities of interest from an interaction model in which both modifying variables are dichotomous can be usefully presented in the text, in a marginal effect plot, or in a table that combines predicted values and differences.

Before moving on to our second substantive application, we first examine the alternative interactive specification that scholars can use when both modifying variables are discrete. This alternative specification was discussed in Section 3.4 of Chapter 3. In our particular application, both of our modifying variables are dichotomous. *Female* indicates whether an individual is female (1) or male (0), while *Black* indicates whether they're Black (1) or White (0). Building on Eq. 3.32, the alternative interactive specification for our setup can be written as

$$\textit{Like Republican Party} = \gamma_0 + \gamma_1 \textit{Female}_1 \times \textit{Black}_0 + \gamma_2 \textit{Female}_0 \times \textit{Black}_1$$
$$+ \gamma_3 \textit{Female}_1 \times \textit{Black}_1 + \gamma_4 \textit{Age} + \epsilon,$$
$$(5.22)$$

where $Female_0$ is a dichotomous variable that equals 1 when $Female = 0$ and 0 otherwise, $Female_1$ is a dichotomous variable that equals 1 when $Female = 1$ and 0 otherwise, $Black_0$ is a dichotomous variable that equals 1 when $Black = 0$ and 0 otherwise, and $Black_1$ is a dichotomous variable that equals 1 when $Black = 1$ and 0 otherwise. This can be written more succinctly as

$$Like\ Republican\ Party = \gamma_0 + \gamma_1 White\ Female + \gamma_2 Black\ Male$$
$$+ \gamma_3 Black\ Female + \gamma_4 Age + \epsilon, \qquad (5.23)$$

where $White\ Female$ is a dichotomous variable that equals 1 if an individual is a White female and 0 otherwise, $Black\ Male$ is a dichotomous variable that equals 1 if an individual is a Black male and 0 otherwise, $Black\ Female$ is a dichotomous variable that equals 1 if an individual is a Black female and 0 otherwise, and $White\ Male$, or equivalently $Female_0 \times Black_0$, is the omitted identity category.[10] We show the more complicated-looking specification in Eq. 5.22 to remind you that the model in Eq. 5.23 is an interaction model even though it looks just like an additive model that includes a series of dichotomous variables. As Eq. 5.22 indicates, variables like $White\ Female$, $Black\ Male$, $Black\ Female$, and $White\ Male$ are all dichotomous *interaction terms*. As we saw in Chapter 3, the alternative specification in Eq. 5.23 is exactly equivalent to the standard interactive specification in Eq. 5.1 in that they both produce the same quantities of interest.

The results from the alternative interactive specification are presented in the second column of Table 5.1. We start by discussing how to interpret these results. The first thing to note is that the omitted identity category relates to White males. As a result, White males become the baseline or reference category against which the other identity categories are compared. As an example, the coefficient on $White\ Female$ indicates the effect of being a White female as opposed to a White male. Recognizing this, the results in the second column of Table 5.1 tell us that White females like the Republican Party 0.04 units less than White males but that this difference isn't statistically significant. They also indicate that Black males like the Republican Party 1.50 units less than White males and that Black females like the Republican Party 2.57 units less than White males.

As we saw in Eq. 3.38 in Chapter 3, the coefficients on $White\ Female$ and $Black\ Male$ from the alternative specification shown in the second column of Table 5.1 are identical to the coefficients on $Female$ and $Black$ from the standard interactive specification shown in the first column. This is

---

[10] Recall that we can't include dichotomous variables for all four of our identity categories as this will result in perfect multicollinearity.

because the coefficients on *White Female* and *Female* both tell us the effect of changing an individual's gender from male to female among Whites and because the coefficients on *Black Male* and *Black* both tell us the effect of changing an individuals race from White to Black among males. As Eq. 3.38 also indicates, the coefficient on *Black Female* in the alternative interactive specification doesn't have a direct equivalent in the standard interactive specification. Instead it's equal to the sum of the coefficients on *Female*, *Black*, and *Female×Black* in the standard interactive specification. In other words, $\gamma_3 = \beta_1 + \beta_2 + \beta_3$ or $-2.57 = -0.04 - 1.50 - 1.03$.

Just as we had to look beyond the individual coefficients in the standard interactive specification to calculate the effect of being female among Blacks and the effect of being Black among females, we have to do the same in the alternative specification. The effect of being female among Blacks in the alternative specification is the coefficient on *Black Female* minus the coefficient on *Black Male*, or $\gamma_3 - \gamma_2 = -2.57 - (-1.50) = -1.07$. The effect of being Black among females in the alternative specification is the coefficient on *Black Female* minus the coefficient on *White Female*, or $\gamma_3 - \gamma_1 = -2.57 - (-0.04) = -2.53$. As expected, these effects are identical to those we obtained earlier when we calculated the same quantities using the results from the standard interactive specification.

But what about the associated measures of uncertainty for these effects? Leveraging information about the basic properties of variances in Appendix A, the estimated standard error for the effect of being female among Blacks from the alternative interactive specification is

$$\text{se}\left(Female|Black = 1\right) = \text{se}\left(\gamma_3 - \gamma_2\right) = \sqrt{\text{var}\left(\gamma_2\right) + \text{var}\left(\gamma_3\right) - 2\text{cov}\left(\gamma_2, \gamma_3\right)}, \tag{5.24}$$

and the estimated standard error for the effect of being black among females is

$$\text{se}\left(Black|Female = 1\right) = \text{se}\left(\gamma_3 - \gamma_1\right) = \sqrt{\text{var}\left(\gamma_1\right) + \text{var}\left(\gamma_3\right) - 2\text{cov}\left(\gamma_1, \gamma_3\right)}. \tag{5.25}$$

To calculate these quantities, we must look at the estimated variance-covariance matrix for the coefficients from the alternative interactive specification,

$$V(\beta) = \begin{pmatrix} 0.013665 & 0.007221 & 0.007234 & -0.000004 & -0.007031 \\ 0.007221 & 0.075446 & 0.007523 & 0.000069 & -0.010774 \\ 0.007234 & 0.007523 & 0.049558 & 0.000039 & -0.009233 \\ -0.000004 & 0.000069 & 0.000039 & 0.000010 & -0.000505 \\ -0.007031 & -0.010774 & -0.009233 & -0.000505 & 0.032923 \end{pmatrix}. \tag{5.26}$$

From this, we see that the estimated standard error for the effect of being female among Blacks is

$$\text{se}\left(\textit{Female}|\textit{Black} = 1\right) = \sqrt{0.075446 + 0.049558 - 2 \times 0.007523}$$
$$= \sqrt{0.109959} = 0.3316. \tag{5.27}$$

You'll notice that this is exactly the same as the standard error shown earlier in Eq. 5.7 where we calculated the same quantity using the results from the standard interactive specification. The estimated standard error for the effect of being Black among females is

$$\text{se}\left(\textit{Black}|\textit{Female} = 1\right) = \sqrt{0.013665 + 0.049558 - 2 \times 0.007234}$$
$$= \sqrt{0.049} = 0.221. \tag{5.28}$$

This is also exactly the same as the standard error shown earlier in Eq. 5.11 that was calculated using the results from the standard interactive specification.

One of the drawbacks of using the alternative interactive specification rather than the standard interactive specification is that we can't see the "interaction effect" directly from the regression output. While the coefficient on the interaction term $\textit{Female} \times \textit{Black}$ in the standard interactive specification tells us the interaction effect and therefore indicates whether there's any evidence of conditionality, there's no equivalent coefficient in the alternative interactive specification. Recall from our discussion in Chapter 3 that the coefficient on $\textit{Black Female}$, γ_3, tells us only the effect of jointly changing the race and gender of a White male and not the interaction effect of race and gender. This means that we need to actually calculate the interaction effect with the alternative interactive specification. This is very important to remember. The coefficients on the included variables, such as $\textit{White Female}$, $\textit{Black Male}$, and $\textit{Black Female}$, may all be different and statistically significant, but this doesn't necessarily mean that we have evidence of interaction or conditionality. Without explicitly calculating the interaction effect, it's not possible to know whether the results from the alternative specification are consistent with a world in which the two variables X and Z have an interactive or additive effect on Y. In terms of our particular application, we can't tell from simply looking at the results from the alternative specification in the second (or third) column of Table 5.1 whether gender and race have an interactive or additive effect on liking the Republican Party. As Eq. 3.38 shows in Chapter 3, we calculate the interaction effect from the results in the alternative interactive specification as $\gamma_3 - \gamma_1 - \gamma_2 = -2.57 - (-0.04) - (-1.50) = -1.03$. As expected, this is identical to the coefficient on the interaction term in the standard interactive specification. We can calculate the variance of the interaction effect as

$$\begin{aligned}
\text{var}\,(\gamma_3 - \gamma_1 - \gamma_2) &= \text{var}\,(\gamma_1) + \text{var}\,(\gamma_2) + \text{var}\,(\gamma_3) \\
&\quad + 2\text{cov}\,(\gamma_1, \gamma_2) - 2\text{cov}\,(\gamma_1, \gamma_3) - 2\text{cov}\,(\gamma_2, \gamma_3) \\
&= 0.013665 + 0.075446 + 0.049558 \\
&\quad + 0.007221 - 0.007234 - 0.007523 \\
&= 0.124.
\end{aligned}$$

(5.29)

Taking the square root tells us that the standard error for the interaction effect is $\sqrt{0.124} = 0.35$. This is exactly the same as the standard error associated with the coefficient on the interaction term in the first column of Table 5.1.

In addition to being used to calculate "effects," the results from the alternative interactive specification can also be used to calculate predicted values. As an example, the predicted level of support for the Republican Party is $\gamma_0 + \gamma_4 \times 40 = 4.47 + 0.02 \times 40 = 5.08$ for a forty year old White male, it's $\gamma_0 + \gamma_2 + \gamma_4 \times 40 = 4.47 - 1.50 + 0.02 \times 40 = 3.58$ for a forty year old Black male, it's $\gamma_0 + \gamma_1 + \gamma_4 \times 40 = 4.47 - 0.04 + 0.02 \times 40 = 5.04$ for a forty year old White female, and it's $\gamma_0 + \gamma_3 + \gamma_4 \times 40 = 4.47 - 2.57 + 0.02 \times 40 = 2.51$ for a 40 year old Black female. As expected, these predicted values are identical to those we calculated earlier using the results from the standard interactive specification. Although we don't show it here, the estimated measures of uncertainty associated with these predicted values are also identical to the ones we calculated earlier.

The bottom line is that the estimated effects, predicted values, and measures of uncertainty calculated using the results from the alternative interactive specification shown in Eq. 5.23 are identical to these same quantities of interest calculated using the results from the standard interactive specification shown in Eq. 5.1. This is because these two models, while they look different, are exactly equivalent to one another. This means that either model specification can be used to create something like the marginal effect plot shown in Figure 5.1 or the table showing predicted values and differences in Figure 5.2.

Whether we use the standard interactive or alternative interactive specification, we have to look beyond the individual coefficients directly reported in the regression output if we wish to fully evaluate all five of the key predictions that can be derived from a theory positing interaction between two discrete modifying variables. Each specification, though, differs in how easy it is to see or calculate particular quantities of interest. This implies that we can usefully switch between estimating these different specifications when we want to "see" particular quantities. For example, we might estimate the alternative specification so that we can easily see whether some particular category, such as Black women, differs in a significant way from some other category, such as White men, but then switch to the standard interactive specification to more easily see if there's a significant interaction effect.

Finally, does it matter what category of observations we omit in the alternative model specification? The answer is "no," in the sense that we always calculate the same quantities of interest no matter which category of observations is omitted. The omitted category acts as the baseline or reference category against which the included categories are compared, and this naturally affects how the coefficients should be interpreted. In the alternative specification shown in the second column of Table 5.1, the omitted category is White males, and so the coefficients tell us the effect of being a White female, a Black male, or a Black female as opposed to a White male. In the third column of Table 5.1, we present the results from a different alternative specification in which the omitted category is Black females. The coefficients in this column tell us the effect of being a White female, Black male, or White male as opposed to a Black female. The sets of coefficients differ across these two models because they make different comparisons. The important thing to note, though, is that when we use the coefficients from the two models to make the *same* comparison, they produce identical results.

Consider the coefficient on *Black Female* in the first alternative specification. This coefficient is −2.57 and tells us the effect of being a Black female instead of a White male. Now consider the coefficient on *White Male* in the second alternative specification. This coefficient is 2.57 and tells us the effect of being a White male instead of a Black female. These two coefficients capture the same comparison but from opposite directions (Black female vs. White male as opposed to White male vs. Black female). This is why the two coefficients are identical except for the fact that they have the opposite sign. Now consider the coefficients on *White Female* and *Black Male* in the second alternative specification. These coefficients are 2.53 and 1.07, respectively, and tell us the effect of being a White female instead of a Black female and the effect of being a Black male instead of a Black female. These specific comparisons aren't directly made by the regression output from the first alternative specification. As we saw earlier, though, we can use the results from the first alternative specification to make these comparisons. When we did so, we found that the effect of being a Black female instead of a White female, or equivalently, the effect of being Black among females, was −2.53 and that the effect of being a Black female instead of a Black male, or equivalently, the effect of being female among Blacks, was −1.07. These quantities are identical to the coefficients on *White Female* and *Black Male* in the second alternative specification except that they have the opposite sign due to the fact that the comparisons are conducted from the opposite direction. This suggests a potentially useful strategy for the researcher. When estimating an alternative interactive specification, we can immediately make comparisons between the included categories and the omitted category. However, we can't necessarily immediately make comparisons between the included categories to see if they're significantly different from each other. As we saw earlier, this is something that we have to explicitly evaluate by testing whether certain coefficients or predicted values are significantly different. However, we now see that an alternative strategy is to simply reestimate the alternative specification and carefully choose the omitted category to make the desired comparison. For example, if we want to easily see whether White females are significantly different from Black females, then we'd estimate an alternative specification in which Black females (or White females) are the omitted category. If we then want to determine whether Black females are significantly different from Black males, we can reestimate our model with Black males (or Black females) as the omitted category. The main point here, though, is that the choice of the omitted category doesn't affect our inferences or the quantities of interest that we calculate.

5.2 WHEN X IS CONTINUOUS AND Z IS DISCRETE: IDEOLOGY, RACE, AND SUPPORT FOR BARACK OBAMA

How does ideology and race affect support for political candidates in the United States? In this substantive application, based on a partial replication of Block and Golder (2023), we examine how an individual's ideological position and their race (Black/White) affected their support for Barack Obama at the time of the 2012 presidential election in the United States. The 2012 presidential election was between the Democratic incumbent Barack Obama and the Republican challenger Mitt Romney. In terms of ideology, spatial theories of politics predict that individuals will exhibit more support for candidates who share their ideological position than those who don't (Downs, 1957; Black, 1958). In effect, voters are expected to "punish" candidates who are ideologically distant from them. In terms of race, theories of descriptive representation predict that Blacks will exhibit more support for Obama than Whites because of their shared racial background (Pitkin, 1967; Phillips, 2014; Parker, 2016; Tillery, 2019). Descriptive representation calls for political representatives who resemble their constituents and is thought to be particularly important in situations where there are high levels of mistrust between groups or a history of discrimination (Phillips, 1998; Mansbridge, 1999). Descriptive representation is often considered valuable in its own right but also because individuals who share similar descriptive characteristics are likely to have shared experiences and developed a sense of linked fate that generates a common set of perspectives and substantive interests (Dawson, 1995; Young, 2002).

There are reasons to think that ideology and race won't have separate and additive effects on support for Obama but will instead have an interactive effect. Recall that the 2012 presidential election saw a head-to-head contest between a Black candidate and a White candidate. In these circumstances, Black voters are likely to weigh both spatial policy considerations and descriptive representation considerations when evaluating the candidates. Given the low level of descriptive representation for Blacks in the United States, Black voters are expected to trade off policy congruence for higher levels of descriptive representation. This implies that ideological incongruence is likely to matter less for Blacks than Whites. Put differently, Blacks are less likely to "punish" Obama for not sharing their ideological preferences than Whites because of their shared descriptive characteristics. This leads to the *Ideological Incongruence Hypothesis*.

Ideological Incongruence Hypothesis: An increase in the ideological distance between individuals and Obama leads to less support for Obama. This negative effect of ideological incongruence will be smaller and may even disappear for Blacks as opposed to Whites.

As we know by now, all interactions are symmetric. This means that if race modifies the effect of ideological incongruence on support for Obama, then it logically follows that ideological incongruence must also modify the effect of race on support for Obama. This, in itself, doesn't tell us whether race will have a positive, negative, or zero effect on support for Obama. Note, though, that our underlying theory clearly speaks to the effect of race on support for Obama, and so we should make this explicit. Specifically, our theory predicts that concerns with descriptive representation will lead Blacks to always exhibit more support for Obama than Whites. This positive gap in support for Obama between Blacks and Whites is expected to grow with ideological incongruence because Blacks will "punish" Obama less for his ideological incongruence than Whites. This leads to the *Black Hypothesis*.

Black Hypothesis: Blacks always exhibit greater support for Obama than Whites. This positive effect grows with the ideological distance between individuals and Obama.

Together, these two hypotheses contain all five of the key predictions that we recommend for a theory positing interaction like the one presented here. By predicting that the effect of increased ideological incongruence will lower support for Obama among Whites and will either lower support or have no effect among Blacks, the *Ideological Incongruence Hypothesis* provides predictions $P_{Ideological\ Incongruence|Black_{min}}$ and $P_{Ideological\ Incongruence|Black_{max}}$. By predicting that being Black always increases support for Obama, the *Black Hypothesis* offers predictions $P_{Black|Ideological\ Incongruence_{min}}$ and $P_{Black|Ideological\ Incongruence_{max}}$. Both hypotheses imply that there's a positive interaction between ideological incongruence and being Black, $P_{Ideological\ Incongruence \times Black}$, because the negative effect of ideological incongruence is expected to be smaller for Blacks than Whites and because the positive effect of being Black is expected to grow with ideological incongruence.

We test our hypotheses using data from the 2016 release of the American National Election Studies 2012 Time Series Study (American National Election Studies, 2014). Our dependent variable, *Feeling Thermometer Obama*, is based on a "feeling thermometer" survey question in which respondents are asked to indicate how they feel about Barack Obama on a 0–100 scale, where 0 is a very negative (very cold) feeling, 50 is neutral or no feeling at all (neither warm nor cold), and 100 is a very positive (very warm) feeling. *Feeling Thermometer Obama* has a mean of 56.31 and a standard deviation of 35.24. In terms of our key independent variables, *Ideological Incongruence* is measured on a 0–6 scale and captures the absolute ideological distance between the respondent's self-placement on the left-right dimension and the respondent's placement of Obama on the same dimension.[11] *Black* is a dichotomous variable that equals 1 if an individual self-identifies as Black and 0 if they self-identify as White.[12] *Ideological Incongruence×Black* is an interaction term created by multiplying together the constitutive terms *Ideological Incongruence* and *Black*. In terms of our analysis, we treat the dependent variable as continuous and estimate an ordinary least squares regression with the following "standard" interactive model specification:

$$Feeling\ Thermometer\ Obama = \beta_0 + \beta_1 Ideological\ Incongruence + \beta_2 Black$$
$$+ \beta_3 Ideological\ Incongruence \times Black + \epsilon.$$

$$(5.30)$$

---

[11] Respondents are asked to place themselves and Obama on a 1–7 scale, where 1 indicates they're extremely liberal, 4 indicates they're moderate, and 7 indicates they're extremely conservative. *Ideological Incongruence* is the absolute distance between these two placements. While the concept of ideological incongruence is continuous, its operationalization as *Ideological Incongruence* technically leads to a discrete variable with values 0, 1, 2, 3, 4, 5, and 6. For the purposes of this application, we treat *Ideological Incongruence* "as if" it were continuous.

[12] Given the purposes of our analysis here, we omit those respondents who don't self-identify as Black or White.

Initially, we don't include any control variables. Later we add control variables for the respondent's age, gender, education, income, and partisan identification.

Our two hypotheses speak to the effect of ideological incongruence and race on support for Obama. The marginal effect of ideological incongruence is

$$\frac{\partial Feeling\ Thermometer\ Obama}{\partial Ideological\ Incongruence} = \beta_1 + \beta_3 Black. \tag{5.31}$$

We see that β_1 indicates the effect of ideological incongruence among Whites (when $Black = 0$) and that $\beta_1 + \beta_3$ indicates the effect of ideological incongruence among Blacks (when $Black = 1$). According to our *Ideological Incongruence Hypothesis*, an increase in the ideological distance between a respondent and Obama should always result in lower support for Obama among Whites. As a result, β_1 should be negative. We expect the negative effect of ideological incongruence to be smaller for Blacks, and so β_3 should be positive. Our theory is unclear as to whether the effect of ideological incongruence remains negative or disappears for Blacks. Under no circumstances, though, should ideological incongruence have a positive effect on support for Obama among Blacks. As a result, we predict that $\beta_1 + \beta_3 \leq 0$.

The effect of being Black as opposed to White can be calculated as

$$\frac{\partial Feeling\ Thermometer\ Obama}{\partial Black} = \beta_2 + \beta_3 Ideological\ Incongruence.$$
$$\tag{5.32}$$

We see that β_2 indicates the effect of being Black when there's no ideological distance between the respondent and Obama (when *Ideological Incongruence = 0*). According to our *Black Hypothesis*, Blacks should always exhibit more support for Obama than Whites. As a result, β_2 should be positive and $\beta_2 + \beta_3 Ideological\ Incongruence$ should be positive for all observed values of *Ideological Incongruence*. We expect the positive effect of being Black to grow with ideological incongruence, and so β_3 should be positive. As expected, both of our hypotheses make the same prediction about the sign of the coefficient on the interaction term.

As an overview, Table 5.2 lists each of our five key predictions along with their associated derivative and quantity of interest. Each key prediction speaks to a particular effect. These effects are calculated with a specific derivative. Our quantities of interest refer to the predicted signs of these derivatives.

The results from the interactive specification shown in Eq. 5.30 are presented in the first column of Table 5.3. We start by discussing how to

Table 5.2 Key predictions, derivatives, and quantities of interest

Key Prediction	Derivative	Quantity of Interest
$P_{Ideological\ Distance \times Black}$	$\partial \left(\dfrac{\partial Feeling\ Thermometer\ Obama}{\partial Ideological\ Incongruence} \right) \Big/ \partial Black = \partial \left(\dfrac{\partial Feeling\ Thermometer\ Obama}{\partial Black} \right) \Big/ \partial Ideological\ Incongruence$	$\beta_3 > 0$
$P_{Ideological\ Incongruence|Black_{min}}$	$\dfrac{\partial Feeling\ Thermometer\ Obama}{\partial Ideological\ Incongruence} \Big|_{Black=0}$	$\beta_1 < 0$
$P_{Ideological\ Incongruence|Black_{max}}$	$\dfrac{\partial Feeling\ Thermometer\ Obama}{\partial Ideological\ Incongruence} \Big|_{Black=1}$	$\beta_1 + \beta_3 \leq 0$
$P_{Black|Ideological\ Incongruence_{min}}$	$\dfrac{\partial Feeling\ Thermometer\ Obama}{\partial Black} \Big|_{Ideological\ Incongruence=0}$	$\beta_2 > 0$
$P_{Black|Ideological\ Incongruence_{max}}$	$\dfrac{\partial Feeling\ Thermometer\ Obama}{\partial Black} \Big|_{Ideological\ Incongruence=6}$	$\beta_2 + 6\beta_3 > 0$

Table 5.3 Ideology, race, and support for Barack Obama in the 2012 US presidential elections

Dependent Variable: *Feeling Thermometer Obama, 0–100*

	Standard Interactive Specification	Alternative Interactive Specification	Split Sample	
			Whites Only	Blacks Only
Ideological Incongruence	−13.55***	−13.55***	−13.55***	−3.39***
	(0.21)	(1.23)	(0.22)	(0.44)
Black	12.51***	12.51***		
	(1.23)	(1.23)		
Ideological Incongruence×White		−13.55***		
		(0.21)		
Ideological Incongruence×Black	10.17***	−3.39***		
	(0.64)	(0.60)		
Constant	80.38***	80.38***	80.38***	92.89***
	(0.64)	(0.64)	(0.67)	(0.78)
Observations	4,060	4,060	3,274	786
R^2	0.63	0.63	0.54	0.07

Standard errors in parentheses. $* p < 0.10$; $** p < 0.05$; $*** p < 0.01$ (two-tailed)

151

interpret these results. As predicted, the coefficient on *Ideological Incongruence* is negative and statistically significant. This tells us that support for Obama among Whites (*Black* = 0) declines as his ideological position becomes more incongruent with theirs. Specifically, a one-unit increase in the ideological distance between a White respondent and Obama leads to a 13.55 unit reduction in support for Obama. The magnitude of this effect is substantively meaningful given that the dependent variable *Feeling Thermometer Obama* has a mean of 46.68 among Whites. As predicted, the coefficient on *Black* is positive and statistically significant. This indicates that Blacks who are perfectly congruent with Obama's ideological position (*Ideological Incongruence* = 0) support Obama more than similarly ideologically-inclined Whites. Specifically, Blacks who are ideologically congruent like Obama 12.51 units more than Whites. A case can be made that this effect is substantively meaningful given that the dependent variable *Feeling Thermometer Obama* has a mean of 74.95 among Whites who are ideologically congruent with Obama. The important thing to remember here is that the coefficients on *Ideological Incongruence* and *Black* don't indicate the effect of ideological incongruence and being Black in an unconditional or average sense; instead, they indicate only the effect of ideological incongruence *among Whites* and the effect of being Black among those who are *perfectly congruent with Obama's ideological position.*

As predicted, the coefficient on the interaction term is positive and statistically significant. This provides the important evidence of conditionality and indicates both that race modifies the effect of ideological incongruence on support for Obama and that ideological incongruence modifies the effect of race. The magnitude of the interaction effect is substantively meaningful. One way to evaluate this is to examine how the impact of ideological incongruence depends on someone's race. Here we look at the impact of a one unit increase in ideological incongruence among Whites and Blacks. A one unit increase in ideological incongruence represents a plausible and meaningful change for both groups given that *Ideological Incongruence* has a mean of 2.52 and a standard deviation of 1.83 among Whites and a mean of 1.20 and a standard deviation of 1.28 among Blacks. The effect of increasing ideological incongruence by one unit is associated with a $\beta_1 = -13.55$ unit change in support for Obama among Whites and a $\beta_1 + 1 \times \beta_3 = -13.55 + 1 \times 10.17 = -3.39$ unit change among Blacks. In other words, the negative impact of a one unit increase in ideological incongruence on support for Obama is $\frac{-13.55}{-3.39} = 4.00$ times larger among Whites than Blacks. This indicates that Blacks "punish" Obama about 75% less than Whites when it comes to ideological incongruence and is consistent with our claim that Blacks are willing to trade off ideological considerations for descriptive representation.

We can also evaluate the substantive significance of the interaction effect by looking at how the effect of race depends on the level of ideological incongruence. Here we look at how the effect of being Black changes depending on whether the value of *Ideological Incongruence* is 1 or 2. This counterfactual comparison is plausible and meaningful given that ideological incongruence has a mean of 1.2 and a standard deviation of 1.28 among Blacks. The effect of being Black is associated with a $\beta_2 + 1 \times \beta_3 = 12.51 + 1 \times 10.17 = 22.68$ unit increase in support for Obama when someone is one unit away from Obama's ideological position and a $\beta_2 + 2 \times \beta_3 = 12.51 + 2 \times 10.17 = 32.84$ unit increase when someone is two units away from Obama's position. In other words, the effect of race is 1.45 times, or 45%, larger when *Ideological Incongruence* is 2 instead of 1. This result is consistent with our claim that the difference in support for Obama between Blacks and Whites increases with ideological incongruence.

To fully evaluate our theory, we need to examine all five of the key predictions jointly contained in the *Ideological Incongruence Hypothesis* and the *Black Hypothesis*. The results shown in the first column of Table 5.3 provide the information necessary to evaluate three of these predictions: (1) the interaction effect, (2) the effect of *Ideological Incongruence* when *Black* is at its lowest value of 0, and (3) the effect of *Black* when *Ideological Incongruence* is at its lowest value of 0.[13] We've already discussed these effects. What we can't see directly from the regression output is the effect of *Ideological Incongruence* when *Black* is at its highest value of 1 and the effect of *Black* when *Ideological Incongruence* is at its highest value of 6 (or indeed any value other than 0). To evaluate these particular effects, we need to make additional calculations.

Based on Eq. 5.31, we see that the effect of ideological incongruence among Blacks is $\beta_1 + \beta_3$ or $-13.55 + 10.17 = -3.39$.[14] This tells us that a one-unit increase in ideological distance between a Black respondent and Obama leads to a 3.39 unit reduction in support for Obama. What about the standard error? Substituting our variable names into Eq. 4.20, we see that the standard error associated with the effect of ideological incongruence is

---

[13] It isn't always the case that the regression output provides us with so much of the information needed to evaluate our key predictions. In both this substantive application and the previous one, the lowest value for both of our interacting variables is zero, and so the coefficients on the constitutive terms speak directly to some of our key predictions. In many cases, the lowest value for one or more of our interacting variables will be non-zero. In these cases, the regression output may speak directly to only one of our five key predictions – the interaction effect prediction P_{XZ} – and we'll have to do additional calculations to fully evaluate our theory.

[14] As a reminder, some of the sums that appear in the text may appear to be slightly off. This is because we round the coefficients to two decimal places in the text but complete the sums using the full coefficients reported in the regression output.

$$se\,(Ideological\ Incongruence)$$
$$= se\,(\beta_1 + \beta_3 Black)$$
$$= \sqrt{var\,(\beta_1) + (Black)^2\,var\,(\beta_3) + 2 \times Black \times cov\,(\beta_1, \beta_3).} \quad (5.33)$$

We want to know the standard error associated with the effect of ideological incongruence among Blacks, and so we set $Black = 1$, and Eq. 5.33 simplifies to

$$se\,(Ideological\ Incongruence|Black = 1) = se\,(\beta_1 + \beta_3)$$
$$= \sqrt{var\,(\beta_1) + var\,(\beta_3) + 2 \times cov\,(\beta_1, \beta_3).}$$
$$(5.34)$$

To calculate this, we must gather some information from the estimated variance-covariance matrix for the coefficients,

$$V(\beta) = \begin{pmatrix} 0.042303 & 0.106545 & -0.042303 & -0.106545 \\ 0.106545 & 1.514758 & -0.539033 & -0.409408 \\ -0.042303 & -0.539033 & 0.403552 & 0.106545 \\ -0.106545 & -0.409408 & 0.106545 & 0.409408 \end{pmatrix}.$$
$$(5.35)$$

Substituting the appropriate values from this into Eq. 5.34, we have

$$se\,(Ideological\ Incongruence|Black = 1)$$
$$= \sqrt{0.042303 + 0.403552 + 2 \times -0.042303}$$
$$= \sqrt{0.36125} = 0.60. \quad (5.36)$$

We now see that the effect of ideological incongruence on support for Obama among Blacks is -3.39 with a standard error of 0.60. The associated t-statistic is

$$t = \frac{-3.39}{0.60} = -5.63. \quad (5.37)$$

Consulting a t-distribution table and assuming a two-tailed hypothesis, we find that the p-value associated with a t-statistic of -5.63 and degrees of freedom equal to $n - k = 4,060 - 4 = 4,056$ is very, very small. Following the formula shown in Eq. 4.26, the two-tailed 95% confidence interval for the effect of ideological incongruence among Blacks is

$$-3.39 \pm t_{4,056,0.025} \times 0.60$$
$$-3.39 \pm 1.96 \times 0.60$$
$$-3.39 \pm 1.176. \quad (5.38)$$

In other words, the effect of ideological incongruence among Blacks is -3.39, and the confidence interval is $(-4.56, -2.21)$. Whether we look at the *p*-value or the confidence interval, we conclude that ideological incongruence has a statistically significant negative effect on how Blacks evaluate Obama.

Although the negative effect of ideological incongruence among Blacks is statistically significant, it's not clear that it's substantively significant. The dependent variable *Feeling Thermometer Obama* has a mean of 89.39 among Blacks. Given the very high average level of support for Obama among Blacks, it's perhaps not unreasonable to think that a 3.39 unit reduction in support for Obama is quite small. We realize that the 3.39 unit reduction considered here is just the effect of a *one-unit* increase in ideological incongruence and that the reduction in support for Obama would be larger if we considered a larger change in ideological incongruence. It's important to recognize, though, that there's very little observed variation in the ideological distance to Obama among Blacks. To be specific, the standard deviation for ideological incongruence among Blacks is just 1.28. This means that changes in ideological incongruence among Blacks that are substantially larger than one are not especially likely. Our discussion here highlights the importance of considering the mean level of the dependent variable in a typical baseline scenario and choosing an appropriately sized change for an independent variable when evaluating substantive significance.

We now have all of the quantities of interest we need to evaluate our *Ideological Incongruence Hypothesis*. Given that the modifying variable, *Black*, is dichotomous, we could choose to present these quantities directly in the text. The effect of a one-unit change in the ideological distance between a respondent and Obama leads to a 13.15 (13.55, 13.96) unit reduction in support for Obama among Whites and a 3.39 (2.21, 4.56) unit reduction among Blacks. This difference in the effect of ideological distance across racial groups is statistically significant, as indicated by the the interaction effect of 10.17 (8.92, 11.41).

An alternative way to present the results is to do so graphically using a marginal effect plot. The marginal effect plot in Figure 5.3 shows all of the information necessary to evaluate the three predictions contained in the *Ideological Incongruence Hypothesis*. Each of the three effects is shown as a small circle along with their two-tailed 95% confidence intervals. The dashed vertical gray line is included to aid the reader in determining whether the effects are significantly different from zero. Whenever a confidence interval contains the vertical line, we can't reject the possibility that the effect is zero. All three predictions contained in the *Ideological Incongruence Hypothesis* receive support from the plot shown in Figure 5.3.

While marginal effect plots can be very useful for conveying information about the size of estimated effects and their statistical significance, we

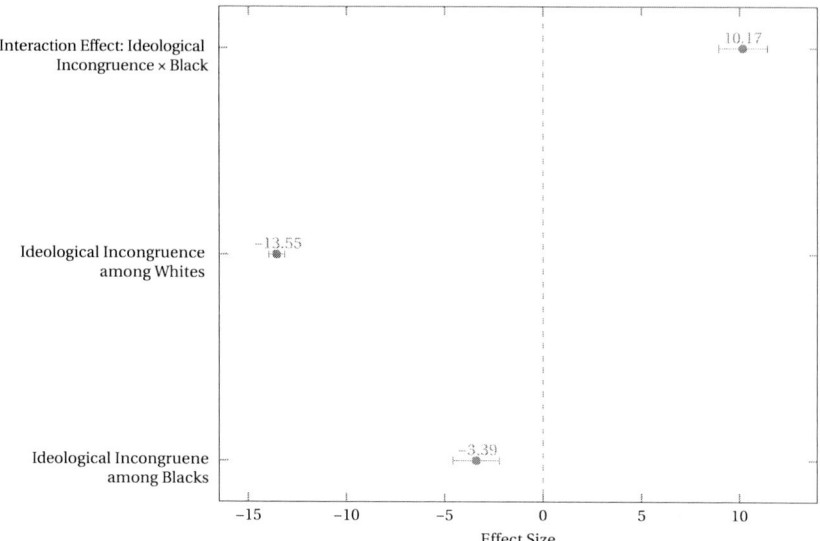

Figure 5.3 The conditional effect of ideological incongruence on support for Barack Obama in the 2012 US presidential elections

Note: The plot shows the interaction effect for ideological incongruence and race, as well as the effects of a one-unit increase in ideological incongruence between a respondent and Obama on support for Obama for Whites and Blacks. The dependent variable, *Feeling Thermometer Obama*, is measured on a 0–100 scale. The effects are based on the results from the standard interaction model in the first column of Table 5.3.

note that they don't typically convey all of the information necessary to evaluate substantive significance. For example, is a magnitude of -3.39 for the effect of a one-unit change in ideological incongruence among Blacks substantively meaningful? The information contained in Figure 5.3 is not sufficient to answer this question. As we discussed earlier, it's open to debate as to whether an effect of this size should be considered substantively large given the already very high level of support for Obama among Blacks in a typical baseline scenario. The main point here is that we encourage researchers to always combine their marginal effect plots with a discussion of the substantive significance of the reported effects in the text.

We now turn to evaluating our *Black Hypothesis*. Recall that we only know directly from the regression output the effect of being Black when *Ideological Incongruence* is at its lowest value of 0. The theoretical and observed range of *Ideological Incongruence* is 0–6. What's the effect of being Black at other values of *Ideological Distance*? Suppose we're interested in the effect of being Black when *Ideological Incongruence* is at its highest value of 6? Based on Eq. 5.32, we see that this effect is $\beta_2 + \beta_3 \times 6$

or $12.51 + 10.17 \times 6 = 73.51$. The standard error for the effect of being Black is

$$
\begin{aligned}
&\text{se}\,(Black) \\
&= \text{se}\,(\beta_2 + \beta_3 Ideological\ Incongruence) \\
&= \sqrt{\text{var}\,(\beta_2) + (Ideological\ Incongruence)^2\,\text{var}\,(\beta_3) + 2 \times Ideological\ Incongruence \times \text{cov}\,(\beta_2, \beta_3)}.
\end{aligned}
\tag{5.39}
$$

Setting *Ideological Incongruence* $= 6$ and obtaining the relevant information from the variance-covariance matrix in Eq. 5.35, we have

$$
\begin{aligned}
&\text{se}\,(Black|Ideological\ Incongruence = 6) \\
&= \sqrt{\text{var}\,(\beta_2) + 36 \times \text{var}\,(\beta_3) + 12 \times \text{cov}\,(\beta_2, \beta_3)} \\
&= \sqrt{1.514758 + 36 \times 0.403552 + 12 \times -0.539033} \\
&= \sqrt{9.57} = 3.09.
\end{aligned}
\tag{5.40}
$$

This tells us that the effect of being Black instead of White when someone is six units away from Obama in the ideological space is 73.51 with a standard error of 3.09. The associated t-statistic is

$$
t = \frac{73.51}{3.09} = 23.76.
\tag{5.41}
$$

The p-value associated with this test statistic is exceedingly small, and the two-tailed 95% confidence interval is

$$
73.51 \pm 1.96 \times 3.09
$$
$$
73.51 \pm 6.07.
\tag{5.42}
$$

In other words, the effect of being Black when *Ideological Incongruence* $= 6$ is 73.51 (67.44, 79.58). Whether we look at the p-value or the confidence interval, we conclude that this effect is statistically significant.

In Chapter 2, we noted that our call to make predictions about the marginal effect of a modifying variable when the other modifying variable is at its lowest and highest values wasn't meant to imply that we should necessarily focus our greatest attention on the *estimated* marginal effects at these extreme values. Instead, we emphasized the predictions at the "extremes" because of the fact that if one assumes linearity, as we've been doing, such predictions automatically imply predictions at values between the extremes. So far, we've found that the effect of *Black* is positive and statistically significant when *Ideological Incongruence* is at its lowest and highest values. We've also seen from the positive and significant coefficient on the interaction term that the positive effect of *Black* grows significantly in magnitude with higher values of *Ideological Incongruence*. These results are exactly in line with our predictions. However, it might be informative to know the effect of *Black* at other, possibly more common or substantively interesting, values of *Ideological Incongruence*.

One thing we could do is calculate the effect of *Black* at particular values across the observed range of *Ideological Incongruence*, say, 0, 1, 2, 3, 4, 5, and 6. We already know the effect of *Black* when *Ideological Incongruence* is 0 and 6. We can easily employ the same approach we just used to calculate the effect of *Black* at the other desired values of *Ideological Incongruence*. This information can be presented in a table. As Table 5.4 indicates, the magnitude of the effect of *Black* is 12.51 (the coefficient on the constitutive term *Black*) when *Ideological Incongruence* = 0 and increases by 10.17 (the coefficient on the interaction term *Ideological Incongruence × Black*) for each unit increase in *Ideological Incongruence*. As expected, the sizes of the confidence intervals around the reported effects vary across the different values of *Ideological Incongruence*. Specifically, the magnitudes of the confidence intervals vary from a low of 3.59 when *Ideological Incongruence* is 1 to a high of 12.13 when *Ideological Incongruence* is 6.

When presenting marginal effects in a table, we encourage researchers to also report information about the value of the dependent variable in a corresponding typical baseline scenario. This allows the reader to better evaluate the substantive significance of the effect sizes. Along these lines, Table 5.4 reports the mean level of support for Obama among Whites at each level of ideological incongruence directly next to the corresponding effect of *Black* to make this evaluation as easy as possible.

Another thing that we strongly recommend when the modifying variable is continuous is that scholars provide the reader with information about its frequency distribution. There are at least two reasons for this. The first reason is that this information speaks to the plausibility of the counterfactual under consideration and hence the substantive significance of the effects being reported. As Table 5.4 indicates, only 3.2% of the respondents hold an ideological position that is 6 units away from Obama. As a result, we're only likely to see an effect of being Black as large as 73.51 among a relatively small share of the population. Put differently, the gap in support for Obama between Blacks and Whites is only rarely going to be as large as 73.51 units. The second reason is that providing information about the frequency distribution of the modifying variable helps us to see whether the reported effects are estimated "far from the available data" (King and Zeng, 2006, 131). This speaks to the issue of "common support" and the extent to which the reported effects are based purely on the extrapolation or interpolation of the assumed linear functional form for the interaction to a location where there's no, or only sparse, data (Hainmueller, Mummolo and Zu, 2019, 165).

Table 5.4 The conditional effect of *Black* on support for Obama at different values of *Ideological Incongruence*				
Ideological Incongruence	Black Respondents	White Respondents	Effect of *Black*	Mean Support Among Whites
0	289 (7.11%)	508 (12.51%)	12.51*** (10.10, 14.92)	74.95
1	241 (5.93%)	707 (17.41%)	22.68*** (20.88, 24.47)	70.64
2	134 (3.30%)	531 (13.07%)	32.84*** (30.91, 34.78)	58.36
3	82 (2.02%)	439 (10.81%)	43.01*** (40.30, 45.72)	35.79
4	22 (0.54%)	407 (10.02%)	53.18*** (49.43, 56.93)	24.61
5	10 (0.25%)	560 (13.79)	63.34*** (58.46, 68.23)	10.41
6	8 (0.20%)	122 (3.00%)	73.51*** (67.44, 79.58)	7.02
	786 (19.36%)	3,274 (80.64%)		

95% confidence intervals in parentheses. $*p < 0.10$; $**p < 0.05$; $***p < 0.01$ (two-tailed)

Note: The table reports the effect of *Black* at different values of *Ideological Incongruence*. The dependent variable, *Feeling Thermometer Obama*, is measured on a 0–100 scale. The effects are based on the results from the standard interaction model shown in the first column of Table 5.3. The second and third columns report the number and percentage of Black and White respondents at each value of *Ideological Incongruence*. The last column indicates the mean level of support for Obama among Whites.

To illustrate this last point, consider our current substantive application. The effects reported in Table 5.4 involve attempted counterfactuals where we compare the support for Obama exhibited by a Black respondent with that exhibited by an otherwise similar White respondent at particular values of ideological incongruence with Obama. To estimate the effect of a counterfactual, it's helpful to have "common support." Common support requires that there's a sufficient number of respondents at the chosen level of *Ideological Incongruence* and that these respondents actually vary in terms of their race. If, for example, there are only White respondents at the chosen level of *Ideological Incongruence*, we'd lack common support, and since there are no Black respondents, any estimated effect for the counterfactual will be driven entirely by extrapolation or interpolation based on the model's assumption of a linear interaction. Estimated effects that lack common support can be problematic because they tend to be model dependent and fragile in the sense that small changes to the data or assumed functional form can result in very different answers (King and Zeng, 2006). Of course, this is only an issue if the assumption of a linear interaction is incorrect. In other words, the estimated effects will remain consistent and unbiased if the model is correctly specified, even if there are regions of the modifying variable that lack common support.[15] When the variable whose effect we're evaluating in the counterfactual is dichotomous, as is the case here with *Black*, we recommend that researchers report the frequency distribution of the modifying variable separately for those observations where the value of the dichotomous variable is zero and those observations where it's one (Hainmueller, Mummolo and Zu, 2019). This way the reader gets a sense of both the frequency of observations at particular values of the modifying variable and the extent to which these observations vary in terms of the counterfactual effect we're estimating. This is why we report the number and percentage of both Black and White respondents at each value of *Ideological Incongruence* in Table 5.4.

As an alternative, we can graphically present much of the information in Table 5.4 in the form of a marginal effect plot like the one shown in Figure 5.4.[16] The effects of being Black for respondents who differ in their ideological distance from Obama are shown as small circles along with their two-tailed 95% confidence intervals. The magnitude of the effects and the confidence intervals can be evaluated using the vertical axis on the left. The dashed horizontal gray line is included to aid the reader in determining whether the effects are significantly different from zero. To further help the reader to evaluate the substantive importance of these effects and evaluate any issues with common support, the plot includes two histograms in the background that show the percentage of Black (dark gray) and White (light gray) respondents at different levels of ideological distance from Obama.

---

[15] For strategies to evaluate whether the data are consistent with a linear interaction assumption, see Section 7.1.2 in Chapter 7 and Hainmueller, Mummolo and Zu (2019).

[16] Unlike Table 5.4, the marginal effect plot shown in Figure 5.4 doesn't include information about the mean level of support for Obama among Whites at the different levels of ideological incongruence. As a result, it doesn't convey information about the values of the dependent variable in the "typical" baseline scenarios. The omission of this information makes it harder to evaluate the substantive importance of the effects being reported.

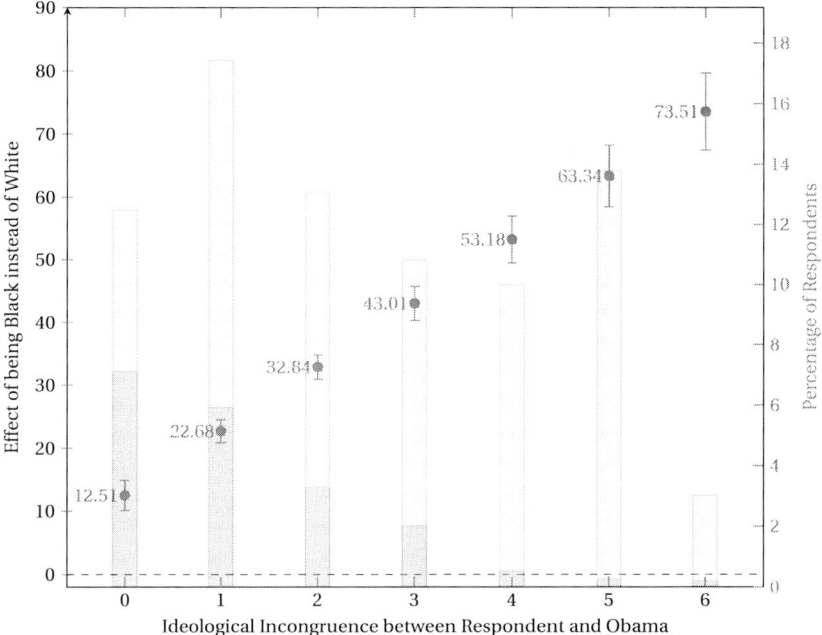

Figure 5.4 The conditional effect of *Black* on *Feeling Thermometer Obama* across various values of *Ideological Incongruence*

Note: The plot shows the effects of being Black instead of White on support for Obama in the 2012 US presidential elections for respondents who vary in their ideological distance to Obama. The dependent variable, *Feeling Thermometer Obama*, is measured on a 0–100 scale. The effects are based on the results from the standard interaction model in the first column of Table 5.3. The two histograms show the percentage of Black (dark gray) and White (light gray) respondents at different levels of ideological distance from Obama. The vertical axis for the histograms is shown on the right-hand side of the plot.

The vertical axis for the histograms is shown in gray on the right-hand side of the plot.

An advantage of using a marginal effect plot rather than a table to show the effect of a variable like *Black* when the modifying variable is continuous is that we can essentially plot the effect of *Black* for as many values of the modifying variable as we want. In Figure 5.4, we calculated the effect of *Black* as we increased the value of *Ideological Incongruence* in increments of 1 from its lowest observed value of 0 to its highest observed value of 6. We could repeat this process, though, using smaller increments in *Ideological Incongruence* such as 0.1 or 0.01 or 0.001. We could then join up the estimated effects and the respective upper and lower bounds of the confidence intervals to create a marginal effect plot that shows the effect of *Black* for "all" values of *Ideological Incongruence*.[17] Such a marginal effect plot is shown in Figure 5.5. The upward-sloping solid line indicates how the effect of being Black on support for Obama varies with the ideological distance between a respondent and Obama. The two dashed lines on either side represent the upper and lower bounds of a two-tailed 95% confidence interval.

Marginal effect plots such as the one shown in Figure 5.5 are very informative and increasingly common in empirical studies that employ interaction models. As a result, it's worth spending a few moments discussing their core features. Among other things, a good understanding of marginal effect plots can be helpful in determining whether they've been constructed correctly. First, the vertical intercept of the marginal effect line – the value of the marginal effect when the modifying variable is zero – should be equal to the coefficient on the constitutive term for the variable whose effect we're calculating. In other words, the vertical intercept in Figure 5.5 should be equal to the estimated coefficient on *Black* (12.51) shown in the first column of Table 5.3. Second, the slope of the marginal effect line – the interaction effect – should be equal to the estimated coefficient on the interaction term. In other words, the slope of the marginal effect line in Figure 5.5 should be equal to the coefficient on *Ideological Incongruence* × *Black* (10.17) in the first column of Table 5.3. It should be clear from this how we can use the information in the regression output to eyeball whether a marginal effect line has been constructed correctly.

A third core feature of a marginal effect plot is that the confidence intervals around the marginal effect line should have an hour-glass shape. If they don't, you know that something's wrong.[18] Recall that the estimated standard error associated with the effect of some variable like *Black* in an interaction model varies with the value of the other modifying variable. This implies that there's a value of the other modifying variable at which the standard error is smallest. We'll call this MV^*. The standard error increases

---

[17] Recall that the size of the estimated standard error, and hence the size of the confidence interval, for the effect of a variable like *Black* varies in a non-linear way with the value of the modifying variable. As a result, the smaller the incremental increase we use in our calculations, the smoother the curves for the upper and lower bounds of the confidence interval will look. The size of the incremental increase doesn't matter for plotting the actual effect of *Black* across the different values of *Ideological Incongruence* as this relationship is linear.

[18] In some cases, like the marginal effect plot in Figure 5.5, the curves of the upper and lower bounds of the confidence interval are very shallow. Nonetheless, the upper and lower bounds will never be linear.

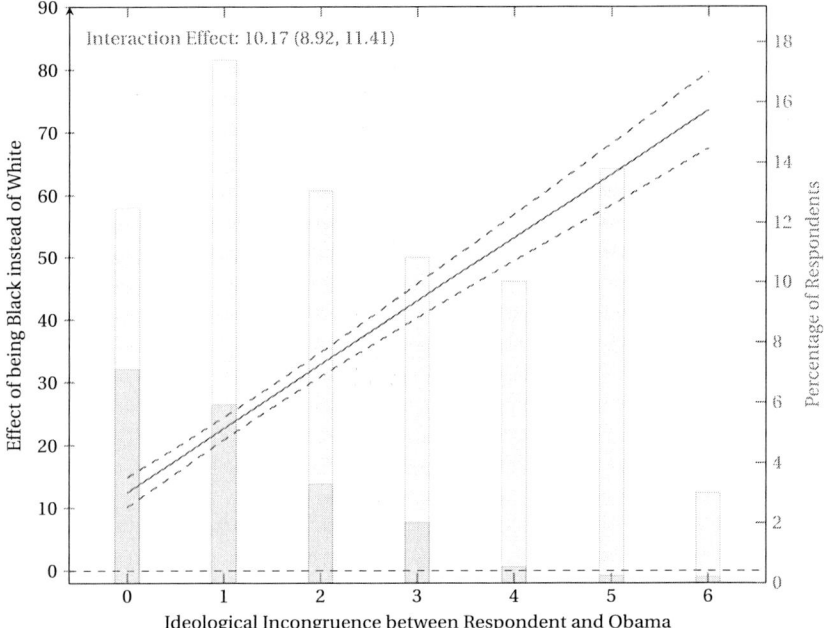

Figure 5.5 The conditional effect of *Black* on *Feeling Thermometer Obama* across the observed range of *Ideological Incongruence*

Note: The solid line shows the effect of being Black on support for Barack Obama for respondents who vary in their ideological distance to Obama. The two dashed lines represent the upper and lower bounds of a two-tailed 95% confidence interval. The interaction effect and its confidence interval is shown in the top left. The dependent variable, *Feeling Thermometer Obama*, is measured on a 0–100 scale. The two histograms show the percentage of Black (dark gray) and White (light gray) respondents at different levels of ideological distance from Obama. The vertical axis for the histogram is shown on the right. The plot is based on the results from the standard interaction model shown in the first column of Table 5.3.

(at an increasing rate) as the value of the modifying variable becomes smaller or larger than MV^*. It increases at an increasing rate due to the appearance of the squared term in the formula for a conditional standard error. In our particular case, the squared term refers to (*Ideological Incongruence*)^2 in Eq. 5.39. Given how the size of the standard error varies with the modifying variable, the confidence intervals in a marginal effect plot will always have an hour-glass shape and will be narrowest when the value of the modifying variable on the horizontal axis is equal to MV^*. Following Eq. 4.21, we see that the standard error for the effect of *Black* is smallest, and the confidence interval in Figure 5.5 is therefore narrowest, when *Ideological Incongruence* is

$$Ideological\ Incongruence^* = -\left[\frac{\text{cov}\left(\beta_2, \beta_3\right)}{\text{var}\left(\beta_3\right)}\right] = -\left[\frac{-0.539033}{0.403552}\right] = 1.34.$$

$$(5.43)$$

The size of the confidence interval around the effect of *Black* grows at an increasing rate as the value of *Ideological Incongruence* becomes smaller or larger than 1.34. It's worth noting that *Ideological Incongruence** is not, as some have claimed, the mean value of *Ideological Incongruence*, which in our sample is 2.26.

We have several recommendations to maximize the information contained in a marginal effect plot. First, plot the marginal effect across the observed range of the modifying variable. Plotting marginal effects over a wider range than this risks misleading readers by portraying out-of-sample inferences, whereas plotting marginal effects over a narrower range ignores information that might be relevant for evaluating hypotheses (Berry, Golder and Milton, 2012, 668).

Second, present a confidence interval with the marginal effect line. This is important if we want to evaluate the estimated uncertainty associated with a given marginal effect or if we wish to determine the values of the modifying variable at which the marginal effect is statistically significant. We can further aid the reader in this regard by providing a horizontal (or vertical) "zero line" depicting when a marginal effect is zero. Such a line is useful as it helps to indicate when the marginal effect is statistically significant. The marginal effect is statistically significant whenever the upper and lower bounds of the confidence interval are both on the same side of the zero line. And it's statistically insignificant whenever the zero line lies between the upper and lower bounds of the confidence interval.[19]

Third, report the interaction effect, along with its *t*-ratio or confidence interval, somewhere in the plot. At present, scholars rarely if ever report the interaction effect in their marginal effect plots. However, this is a crucial piece of information when it comes to evaluating the empirical support for a hypothesis derived from a theory positing interaction. The curved confidence intervals in a marginal effect plot tell us whether the effect of some variable is significantly different from zero at different values of the modifying variable. What they don't tell us is whether the marginal effects associated with the different values of the modifying variable are

---

[19] When we discussed prototypical results in the previous chapter, we used a "bold axis" convention to indicate the values of the modifying variable at which the marginal effect is statistically significant. While this was a useful strategy given our discussion of "generic" results, we noted at the time that we didn't recommend the adoption of this convention for reporting actual research results. It should be clear that the use of confidence intervals provides readers with significantly more information than they'd have if the bold axis convention were adopted.

significantly different from each other. In other words, there's no way of knowing whether the slope of the marginal effect line is significantly different from zero. Put differently, one can't tell from simply looking at a typical marginal effect plot whether there's any evidence of conditionality. We can obviously see if the marginal effect line slopes up or down. But this, in itself, isn't evidence of a statistically significant interaction. Only by adding the interaction effect and some measure of its uncertainty will a marginal effect plot contain the necessary information to evaluate all three of the predictions that are jointly contained in a hypothesis like our *Black Hypothesis*, $P_{Ideological\ Incongruence \times Black}$, $P_{Black|Ideological\ Incongruence_{min}}$, and $P_{Black|Ideological\ Incongruence_{max}}$.[20]

Fourth, incorporate information about the distribution of the modifying variable (Berry, Golder and Milton, 2012, 668). We've recommended plotting the marginal effect of the variable we're interested in across the observed range of the modifying variable. However, not all values of the modifying variable shown on the horizontal axis are equally important. If, for example, the minimum and maximum observed values for the modifying variable are relative outliers in the sample, then it seems reasonable to think that the estimated marginal effects at the extremes are less relevant for assessing a hypothesis than the marginal effects closer to the center of the distribution where the observations are presumably more concentrated. Of course, there may be ranges of values closer to the center of the distribution of the modifying variable on the horizontal axis at which there are also few observations or where common support is lacking. In these situations, it's important to remember that the validity of any inferences we might make about the marginal effect at such values rests on the linearity assumption of the model being correct.

One strategy we recommend is to superimpose the marginal effect plot over a "frequency" distribution for the modifying variable so that readers have a sense of the relative density of observations at different locations. When the variable whose effect we're evaluating in a counterfactual is dichotomous, we can usefully provide even more information by showing separate histograms for those observations where the value of the dichotomous variable is zero and those observations where it's one. As an example, we employed separate histograms for Whites and Blacks in Figure 5.5 showing the percentage of respondents at different values of

---

[20] We realize that the interaction effect is reported in the table of results in the form of the coefficient on the interaction term. However, we strongly believe that figures (and tables) should be stand-alone, in the sense that readers can evaluate the information they contain without having to read any of the text around them. If incorporating the interaction effect into the plot itself makes it look messy, one can simply add it to the note that accompanies the plot.

the modifying variable *Ideological Incongruence*. We placed the histograms in the background and made them fairly light so that readers could concentrate on the marginal effect line and confidence interval and only "see" the frequency distributions if they wanted. When using a histogram, we should choose the number and width of the bins to facilitate the reader's ability to interpret the frequency distribution. An appealing strategy in some contexts is to combine a histogram with a "rug plot" (Berry, Golder and Milton, 2012, 666). A rug plot uses short perpendicular markers along the horizontal axis to indicate the specific values of the observations for the modifying variable in the sample. While the histogram provides readers with a general overview of the frequency distribution and a quick sense of the percentage of observations that fall into various regions, a rug plot can be helpful because it provides details about the values of individual observations. The value of incorporating a rug plot often declines as sample size increases because the individual perpendicular markers along the horizontal axis blend together and become indistinguishable.

Fifth, provide a detailed note beneath the marginal effect plot. As usual, the note should include information on exactly what's shown in the marginal effect plot. It can also be useful, however, to include information about the dependent variable, such as the scale on which it's measured and its distribution. This can help the reader determine if and when the magnitude of the marginal effect is substantively significant. In cases where we've estimated several slightly different model specifications, the note should also indicate precisely which model specification was used as the "source" for the marginal effect plot. In general, the note that accompanies the marginal effect plot should be sufficiently detailed that the plot can stand alone in the sense that it can be evaluated by a reader without them having to consult the surrounding text.

As we've seen, the plot in Figure 5.3 provides the necessary information to evaluate all three of the predictions jointly contained in our *Ideological Incongruence Hypothesis*, and the plot in Figure 5.5 provides the necessary information to evaluate all three of the predictions jointly contained in our *Black Hypothesis*. One appealing strategy that we can adopt is to combine these two plots in the same figure. Together they provide the necessary information to evaluate all five of the key predictions that can be derived from our theory. Figure 5.6 shows what such a figure might look like.

⚠ **Important:** Researchers can maximize the information contained in a marginal effect plot by implementing the following recommendations:

(a) The Conditional Effect of *Ideological Incongruence*

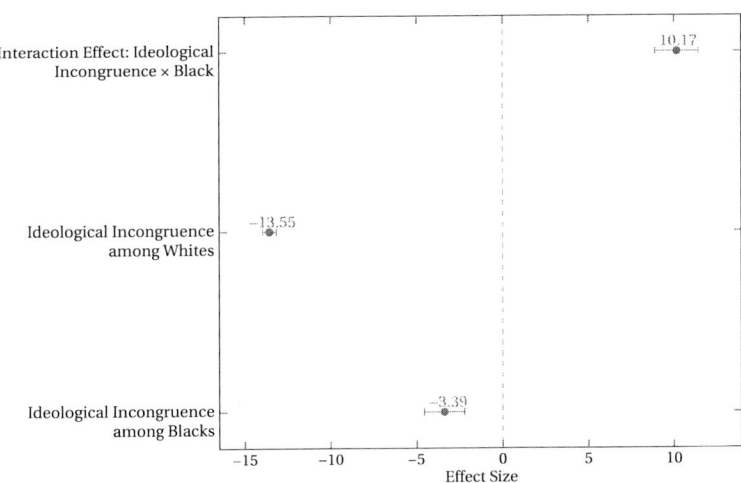

(b) The Conditional Effect of *Black*

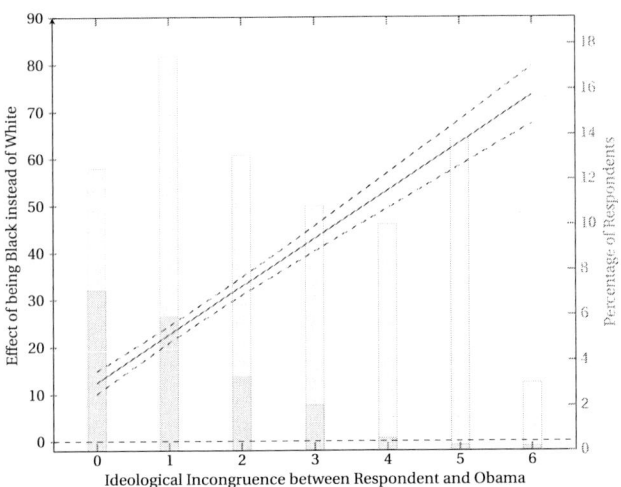

Figure 5.6 The conditional effects of *Ideological Incongruence* and *Black* on support for Obama

Note: The plot in panel (a) shows the effects of a one-unit increase in ideological distance between a respondent and Obama for Whites and Blacks along with 95% confidence intervals. It also shows the interaction effect between ideological incongruence and race. The plot in panel (b) shows the effect of being Black for individuals who vary in their ideological distance to Obama along with 95% confidence intervals. The histograms show the percentage of Whites (light gray) and Blacks (dark gray) at different levels of ideological incongruence. The dependent variable, *Feeling Thermometer Obama*, is measured on a 0–100 scale. The two plots are based on the results from the standard interaction model in the first column of Table 5.3.

- Plot the marginal effect across the observed range of the modifying variable.
- Present a confidence interval to go with the marginal effect.
- Report the interaction effect along with a measure of its uncertainty.
- Incorporate information about the frequency distribution of the modifying variable.
- Provide a detailed note so that the marginal effect plot can more easily be read as a stand-alone figure.

Exactly how these recommendations should be implemented will vary depending on the substantive application.

Before moving on to our third substantive application, we first examine the alternative interactive specification that scholars can use when one of their modifying variables is discrete. This alternative specification was discussed in Section 3.3. In our particular application, the modifying variable *Black* is dichotomous. Building on Eq. 3.12, the alternative interactive specification for our setup can be written as

$$Feeling\ Thermometer\ Obama = \gamma_0 + \gamma_1 Black$$
$$+ \gamma_2 Ideological\ Incongruence \times White$$
$$+ \gamma_3 Ideological\ Incongruence \times Black + \epsilon,$$
$$(5.44)$$

where *Black* and *Ideological Incongruence* are calculated as before, and *White* is equal to 1−*Black* and is therefore a dichotomous variable that equals 1 if an individual self-identifies as White and 0 if they self-identify as Black. As we saw in Section 3.3, the alternative specification in Eq. 5.44 is exactly equivalent to the standard interactive specification in Eq. 5.30 in that they both produce the same quantities of interest.

The results from the alternative specification are presented in the second column of Table 5.3. As previously mentioned, the advantage of the alternative specification is that it's possible to immediately see directly from the regression output the effect of the continuous modifying variable *Ideological Incongruence* for *both* values of the dichotomous modifying variable *Black*. The coefficient on *Ideological Incongruence* × *White* tells us that a one-unit increase in the ideological distance between a *White* respondent and Obama leads to a statistically significant 13.55 unit decrease in support for Obama. This coefficient and its associated standard error are identical to the coefficient and standard error associated with *Ideological Incongruence* in the standard interactive specification shown in the first column. This is because they represent the same quantity of interest. The coefficient on *Ideological Incongruence* × *Black* tells us that a one-unit increase in the ideological distance between a *Black* respondent and Obama leads to a statistically significant 3.39 unit decrease in support for Obama. You'll recall that this quantity of interest was not directly available in the regression output from the standard interactive specification; instead, we had to calculate it.

The primary drawback of the alternative interactive specification compared to the standard interactive one is that it's not always possible to see whether there's a statistically significant interaction effect directly from the regression output. In our case, we'd like to know whether the effect of ideological incongruence is *significantly* different for Blacks as opposed to Whites. The two estimated coefficients $\gamma_2 = -13.55$ $(-13.96, -13.55)$ and $\gamma_3 = -3.39$ $(-4.56, -2.21)$ are different, and both are significantly different from zero, but are they significantly different from each other? In this particular case, we can determine directly from the regression output that the effect of ideological incongruence is *significantly* different for Blacks and Whites because the confidence intervals for γ_2 and γ_3 don't overlap. In effect, the regression output in the second column of Table 5.3 tells us that the interaction effect is $\gamma_3 - \gamma_2 = -3.39 - (-13.55) = 10.17$ and that it's statistically significant. Of course, if we wanted to know the actual standard error or confidence interval for the interaction effect, we'd have to calculate $se(\gamma_3 - \gamma_2)$. Had the two confidence intervals for γ_2 and γ_3 overlapped in our application, we'd have had to calculate $se(\gamma_3 - \gamma_2)$ to know whether the interaction effect was statistically significant. In this particular scenario, it would perhaps be easier to recognize the equivalence between the alternative and standard interaction models and simply reestimate the model using the standard interactive specification and look at the coefficient and standard error on the interaction term.

So far, we've focused on how we can use the results from the alternative interactive specification to evaluate the *Ideological Incongruence Hypothesis*. But what about the *Black Hypothesis*? As we noted in Section 3.3, we need to be a little careful when calculating the effect of the dichotomous modifying variable in the alternative interactive specification. When looking at the specification in Eq. 5.44, it looks like the modifying variable *Black* appears twice, once as the constitutive term *Black* and once as part of the interaction term *Ideological Distance* × *Black*. On this basis, we might think that the effect of *Black* can be calculated as

$$\frac{\partial Feeling\ Thermometer\ Obama}{\partial Black} = \gamma_1 + \gamma_3 Ideological\ Incongruence.$$
(5.45)

However, this is incorrect. The problem arises in not recognizing that the variable *Black* appears for a third time in Eq. 5.44 through the presence of *White* in the interaction term *Ideological Incongruence × White*. Recall that *White* $= 1 - Black$, and so we can rewrite the alternative specification in Eq. 5.44 as

$$
\begin{aligned}
Feeling\ &Thermometer\ Obama \\
&= \gamma_0 + \gamma_1 Black + \gamma_2 Ideological\ Incongruence \times (1 - Black) \\
&\quad + \gamma_3 Ideological\ Incongruence \times Black + \epsilon
\end{aligned}
$$

$$
\begin{aligned}
&= \gamma_0 + \gamma_1 Black + \gamma_2 Ideological\ Incongruence \\
&\quad - \gamma_2 Ideological\ Incongruence \times Black \\
&\quad + \gamma_3 Ideological\ Incongruence \times Black + \epsilon.
\end{aligned}
\tag{5.46}
$$

Once we recognize this, we see that the effect of *Black* is actually calculated as

$$
\frac{\partial Feeling\ Thermometer\ Obama}{\partial Black} = \gamma_1 + (\gamma_3 - \gamma_2)\,Ideological\ Incongruence.
\tag{5.47}
$$

As expected, this is exactly the same effect for *Black* that we estimated from the standard interactive model in Eq. 5.32, $\beta_1 + \beta_3$ *Ideological Incongruence*, given that $\gamma_1 = \beta_1$ and $\gamma_3 - \gamma_2 = \beta_3$. With this in mind, the coefficients on *Black* in the standard and alternative interactive specifications shown in Table 5.3 are identical as they both capture the same quantity of interest. Specifically, the coefficients on *Black* tell us that Black respondents who are ideologically congruent with Obama support Obama 12.51 units more than ideologically similar White respondents. We can use Eq. 5.47 to calculate the effects of being Black for those respondents who vary in their ideological distance to Obama. These effects will be identical to those that we calculated earlier from the standard interactive specification and presented in Table 5.4 and Figure 5.5.

As we can see, the alternative interactive specification is equivalent to the standard interactive specification when one of our modifying variables is dichotomous, and, as a result, the exact same quantities of interest can be calculated from both specifications. In this sense, the choice of which specification to use is largely one of personal choice.

As we noted in Section 3.3.1, scholars sometimes adopt a split-sample strategy when one of their modifying variables is discrete. In our particular application, the split-sample strategy involves estimating an additive model that includes *Ideological Incongruence* on sub-samples that include only White or Black respondents. The results from the split-sample strategy are shown in the third and fourth columns of Table 5.3. The coefficient on *Ideological Incongruence* in the Whites-only model indicates that a one-unit increase in ideological distance between a White respondent and Obama leads to a 13.55 unit reduction in support for Obama, while the same coefficient in the Blacks-only model indicates that a one-unit increase in ideological distance between a Black respondent and Obama leads to a 3.39 unit reduction in support for Obama. These effects are exactly the same as those we obtained based on the results from the standard and alternative interactive models. While the point estimates for the effects of ideological incongruence are identical, it isn't the case that the split-sample strategy is exactly equivalent to estimating a standard or alternative interaction model. This is because the standard errors produced by the split-sample strategy are different due to the smaller sample sizes and the fact that the observed variation of the independent variable *Ideological Incongruence* and the residual variation in the dependent variable differ across the two sub-samples. Measures of model fit from the split-sample strategy, such as the R^2, are also different. For example, the R^2 is an identical 0.63 in the two "pooled" interactive models but 0.54 in the Whites-only model and a much smaller 0.07 in the Blacks-only model.

One of the main drawbacks of the split-sample strategy, as with the alternative interactive specification, is that we can't determine directly from the regression output whether the marginal effect of *Ideological Incongruence* among Whites is significantly different from its effect among Blacks. To do this, we'd need to test whether the coefficients on *Ideological Incongruence* in the two sub-sample models are equal to each other. The fact that these two coefficients come from different models means that this test, while possible, isn't as straightforward as testing whether $\gamma_2 = \gamma_3$ in the alternative interactive specification. Again, perhaps the easiest strategy to determine whether the two effects are significantly different is to estimate the standard interaction model and see if the coefficient on the interaction term is statistically significant. A second drawback is that the split-sample strategy makes it harder to test the *Black Hypothesis*. To calculate the effect of *Black*, for example, we must compare sums of coefficients across the two split-sample models. Based on the discussion in Chapter 3, the effect of *Black* is calculated as $(92.89 - 80.38) + (-3.39 + 13.55) \times Ideological\ Incongruence = 12.51 + 10.17 \times Ideological\ Incongruence$. While calculating a measure of uncertainty for this effect is possible, it's not as straightforward as it would be if we were estimating one of the pooled interaction models. The alternative interactive specification offers all the benefits of the split-sample strategy, in particular the ability to see the marginal effect of *Ideological Incongruence* among Whites and Blacks directly from the regression output, without some of the drawbacks. As a result, we see no reason to ever prefer the split-sample strategy to using the alternative interactive specification.

As a final point, we remind you that the split-sample strategy will only produce identical point estimates for the effects of the variables involved in an interaction to the pooled interactive models when there are no control variables. As we noted in Section 3.3, the standard and alternative interactive models remain equivalent with the inclusion of controls. However, this isn't the case with the split-sample strategy. While the split-sample strategy produces the same coefficients but slightly different standard errors when there are no control variables, it produces different coefficients and standard errors when control variables are included. This is because the split-sample strategy allows the effects of *all* of the variables, including the control variables, to vary across the sub-samples, whereas the standard and alternative interactive specifications only allow the effects of those variables that interact with the dichotomous modifying variable to vary. Since the effects of the control variables are allowed to vary across the sub-samples in the split-sample strategy, we get different estimates for the coefficients on the variables involved in the interaction. One can, of course, make the pooled interactive models equivalent to the split-sample strategy, at least with respect to the coefficients, by adding interactions between each of the control variables and the dichotomous modifying variable.

To illustrate this, we now add some control variables to our earlier models. Specifically, we control for a respondent's age, gender, education, income, and partisan identification. The new results from the two pooled interactive models and the split-sample strategy are shown in Table 5.5. Given our purposes, we explicitly report only those coefficients for the control variables dealing with a respondent's partisan identification. Respondents self-identify as Democrat, Republican, or Independent. *Republican* is a dichotomous variable that equals 1 if a respondent is a Republican and 0 otherwise, while *Democrat* is a dichotomous variable that equals 1 if a respondent is a Democrat and 0 otherwise. The omitted category is *Independents*. As a result, the coefficients on *Democrat* and *Republican* indicate the effect of being a Democrat or a Republican as opposed to an Independent on support for Obama. As we'd expect, the positive coefficients on *Democrat* across all of the models indicate that Democrats exhibit more support for the Democrat Obama than Independents. Similarly, the negative coefficient on *Republican* across all of the models indicate that Republicans exhibit less support for the Democrat Obama than Independents.

From the alternative interactive specification, we can immediately see that the marginal effect of *Ideological Incongruence* is -9.96 among Whites and -1.95 among Blacks. From the standard interactive specification, we see that the marginal effect of *Ideological Incongruence* is also -9.96 among Whites and $-9.96 + 8.01 = -1.95$ among Blacks.

Table 5.5 Ideology, race, and support for Barack Obama in the 2012 US presidential elections (with controls)

Dependent Variable: Feeling Thermometer Obama, 0–100			Split Sample	
	Standard Interactive Specification	Alternative Interactive Specification	Whites Only	Blacks Only
Ideological Incongruence	−9.96***		−9.83***	−2.33***
	(0.26)		(0.28)	(0.44)
Black	9.07***	9.07***		
	(1.24)	(1.24)		
Ideological Incongruence×White		−9.96***		
		(0.26)		
Ideological Incongruence×Black	8.01***	−1.95***		
	(0.64)	(0.60)		
Democrat	16.91***	16.91***	17.88***	11.63***
	(0.86)	(0.86)	(1.01)	(1.44)
Republican	−9.56***	−9.56***	−9.63***	−7.97***
	(0.95)	(0.95)	(1.03)	(3.06)
Constant	67.94***	67.94***	66.63***	80.66***
	(1.89)	(1.89)	(2.26)	(2.81)
Other Controls	Yes	Yes	Yes	Yes
Observations	3,635	3,635	2,948	687
R^2	0.69	0.69	0.62	0.24

Standard errors in parentheses. *$p < 0.10$; **$p < 0.05$; ***$p < 0.01$ (two-tailed)

Note: "Other Controls" include variables that capture an individual's age, gender, education, and income.

(a) The Conditional Effect of *Ideological Incongruence*

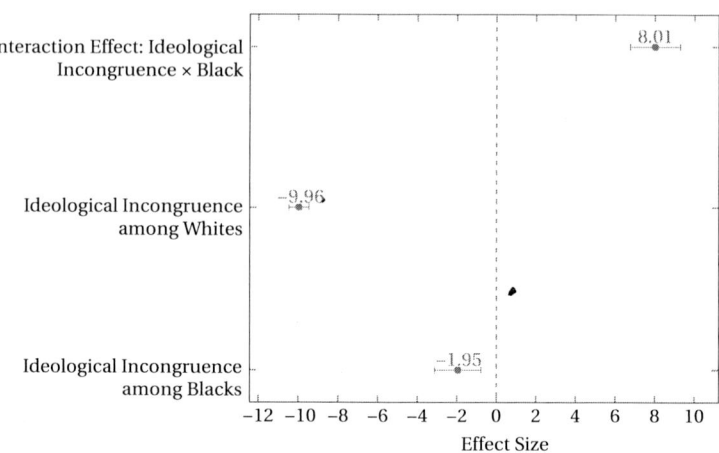

(b) The Conditional Effect of *Black*

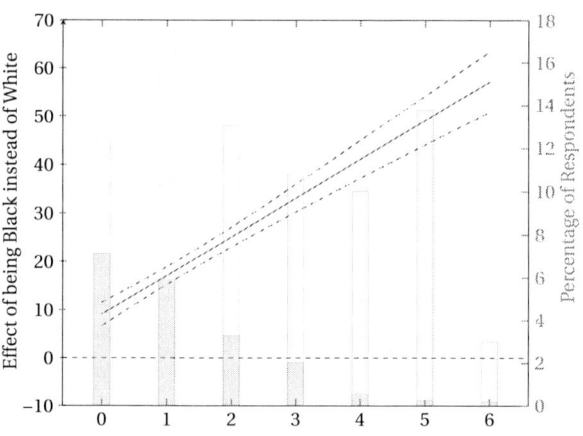

Figure 5.7 The conditional effects of *Ideological Distance* and *Black* on support for Obama (with controls)

Note: The plot in panel (a) shows the effects of a one-unit increase in ideological distance between a respondent and Obama for Whites and Blacks along with 95% confidence intervals. It also shows the interaction effect between ideological incongruence and race. The plot in panel (b) shows the effect of being Black for individuals who vary in their ideological distance to Obama along with 95% confidence intervals. The histograms show the percentage of Whites (light gray) and Blacks (dark gray) at different levels of ideological incongruence. The dependent variable, *Feeling Thermometer Obama*, is measured on a 0–100 scale. The two plots are based on the results from the standard interaction model shown in the first column of Table 5.5.

As expected, the two pooled interactive models produce identical quantities of interest. The key point to take away from Table 5.5 is that the split-sample strategy produces different quantities of interest. According to the split-sample strategy, the marginal effect of *Ideological Incongruence* is −9.83 among Whites and −2.33 among Blacks. The reason for the difference is because the split-sample strategy allows the effects of the control variables to vary across the two sub-samples, whereas the two pooled interactive models don't. The split-sample strategy, for example, estimates two different coefficients for both *Democrat* and *Republican*, one for White respondents and one for Black respondents. In contrast, the two pooled interactive models each estimate only one coefficient each for *Democrat* and *Republican*.

A researcher who's thinking of using the split-sample strategy must determine whether it makes theoretical sense to allow the effects of *all* of the variables in a model to vary with the dichotomous modifying variable – that is, to vary across the two sub-samples. It's worth noting, though, that even if we conclude that a theory calls for a model in which all of the variables are allowed to vary with the dichotomous modifying variable, it's still not clear that we should adopt the split-sample strategy. As we've seen, the alternative interactive specification can do everything that the split-sample strategy can do and more without some of its drawbacks.

For those who are interested, Figure 5.7 reproduces the plots shown in Figure 5.6 using the results from the standard interactive specification with control variables in Table 5.5. The plots, and hence the inferences we'd draw, are remarkably similar to those we made earlier based on the standard interactive model without control variables.

5.3 WHEN X AND Z ARE BOTH CONTINUOUS: DEMAND AND SUPPLY EFFECTS ON WOMEN'S LEGISLATIVE REPRESENTATION

Why are there more women legislators in some countries than others? In this application, which is based on a replication of Dhima (2022), we examine how demand-side and supply-side factors influence the level of women's legislative representation around the world. In 2018, the percentage of women in national legislatures ranged from a low of 0% in countries like Papua New Guinea and Vanuatu to a high of 61.3% in Rwanda. On average, women comprised just 21.7% of the representatives in lower house legislatures globally (Inter-Parliamentary Union, 2019). Existing theory explains women's legislative representation in terms of demand-side and supply-side factors (Norris and Lovenduski, 1993; Inglehart and Norris, 2003; Matland, 2005; Paxton, Kunovich and Hughes, 2007).

Demand-side factors have to do with the preferences that people have for women legislators. When people want women legislators, the "demand" for women's representation is high. Demand for women legislators is low when individuals hold traditional attitudes regarding gender roles. *Supply-side factors* determine the size of the pool of women with the experience and willingness to effectively compete for political office. When women have the resources and ambition to run for office, the "supply" for women's representation is high. The supply of qualified women is high when there are high levels of female education and when women participate at high rates in the labor market.

To date, scholars interested in explaining women's representation have typically focused on either demand-side factors or supply-side factors. The few scholars who've included both demand-side and supply-side factors in their analyses have done so only in an additive manner (Yoon, 2004; Tripp and Kang, 2008; Stockemer, 2011). As Dhima (2022) notes, though, there's an inherent conditionality built into the supply and demand theory of women's representation. The demand for female representatives should have little effect on women's representation when there's no supply of qualified women candidates. As the pool of qualified women candidates expands, though, voter demand for female legislators should increasingly be reflected in actual women's representation. Likewise, the supply of qualified women candidates should have little effect on women's representation when there's no demand for female representatives. As voter demand for female legislators grows, though, increases in the supply of qualified women candidates should be more accurately reflected in higher levels of women's representation. Put differently, we should only expect high levels of women's legislative representation when both supply and demand are sufficiently high. This argument leads to the *Demand Hypothesis* and the *Supply Hypothesis*.

Demand Hypothesis: An increase in the demand for female representatives has little effect on women's legislative representation when the supply of qualified female candidates is low. However, it has an increasingly large positive effect as the supply of qualified female candidates expands.

Supply Hypothesis: An increase in the supply of qualified female candidates has little effect on women's legislative representation when the demand for female representatives is low. However, it has an increasingly large positive effect as the demand for female representatives grows.

Together, these two hypotheses contain all five of the predictions that we recommend for a theory positing interaction like the one presented here. By predicting that increased demand for female representatives will only increase women's representation when the supply of qualified female candidates is sufficiently high, the *Demand Hypothesis* provides predictions $P_{Demand|Supply_{min}}$ and $P_{Demand|Supply_{max}}$. By predicting that the increased supply of qualified female candidates will only increase women's representation when the demand for female representatives is sufficiently high, the *Supply Hypothesis* offers predictions $P_{Supply|Demand_{min}}$ and $P_{Supply|Demand_{max}}$. Both hypotheses imply that there's an interaction between demand-side and supply-side factors, $P_{Demand \times Supply}$, because the magnitudes of their effects are conditional on each other's value.

Our dependent variable *Women's Representation* captures the percentage of female representatives in lower house legislatures around the world from 1990 to 2018. The data come from Paxton, Kunovich and Hughes (2007) for 1990–2003 and the Inter-Parliamentary Union (2019) for 2004–2018. *Women's Representation* has a mean of 14.8 and a standard deviation of 10.9. To capture the demand for women's representation, we use the pooled 1981–2016 World and European Values Surveys (EVS, 2015; WVS, 2015). *Demand* captures the percentage of respondents in a country who disagree or strongly disagree with the claim that men make better political leaders than women. Higher values indicate less support for traditional gender roles and greater demand for women's political representation.[21] The *Demand* variable, which is based on information from 203 surveys in 92 different countries, has a mean of 52.01 and varies from a low of 7.98% in Egypt in 2008 to a high of 91.82% in Sweden in 2006. To capture the supply for women's representation, we turn to the 2018 World Development Indicators from the World Bank (2017). *Supply* captures the ratio of female to male labor force participation rates.[22] The idea is that the supply of qualified female political candidates increases as more women leave the home and participate in the labor market. The *Supply* variable has a mean of 67.88 and varies from a low of 8.61 in the Republic of Yemen in 2017 to a high of 111.01 in Mozambique in 2016.[23] *Demand* × *Supply* is an interaction term created by multiplying together the constitutive terms *Demand* and *Supply*. In the upcoming analysis, we also take account of whether a country has an effective gender quota and

---

[21] Similar results are obtained if demand is operationalized in terms of the percentage of respondents who disagree with the claim that men have more right to a job than women when jobs are scarce (Valdini, 2012).

[22] Similar results are obtained if supply is operationalized in terms of women's enrollment in tertiary education.

[23] To be clear, when *Supply* is 100, women participate in the labor market at the same rate as men. Values less than 100 indicate that women participate in the labor market less than men, and values more than 100 indicate that women participate in the labor market more than men.

its level of democracy.[24] We treat the dependent variable as continuous and estimate an ordinary least squares regression with the following interactive specification:

$$\begin{aligned} \textit{Women's Representation} = {} & \beta_0 + \beta_1 \textit{Demand} + \beta_2 \textit{Supply} \\ & + \beta_3 \textit{Demand} \times \textit{Supply} \\ & + \beta_4 \textit{Effective Gender Quota} \\ & + \beta_5 \textit{Democracy} + \epsilon. \end{aligned} \tag{5.48}$$

The marginal effect of *Demand* is

$$\frac{\partial \textit{Women's Representation}}{\partial \textit{Demand}} = \beta_1 + \beta_3 \textit{Supply}. \tag{5.49}$$

According to the *Demand Hypothesis*, an increase in demand for female representatives should have little effect on women's representation when there's no supply of qualified female candidates but should have an increasingly large positive effect as the supply of qualified female candidates increases. As a result, β_1 should be close to 0, β_3 should be positive, and $\beta_1 + \beta_3 \textit{Supply}_{max}$ should be positive.

The marginal effect of *Supply* is

$$\frac{\partial \textit{Women's Representation}}{\partial \textit{Supply}} = \beta_2 + \beta_3 \textit{Demand}. \tag{5.50}$$

According to the *Supply Hypothesis*, an increase in the supply of qualified female candidates should have little effect on women's representation when there's no demand for women legislators but should have an increasingly large positive effect as the demand for women legislators increases. As a result, β_2 should be close to 0, β_3 should be positive, and $\beta_2 + \beta_3 \textit{Demand}_{max}$ should be positive.

The results from the interactive specification shown in Eq. 5.48 are presented in the first column of Table 5.6. The coefficient on *Demand* is negative and statistically significant. This tells us that a one-unit increase in *Demand* has a significant negative effect on women's representation when *Supply* = 0. Not too much should be read into this coefficient, though, as *Supply* is never 0 in the real world; as we noted earlier, the lowest observed value of *Supply* is 8.61. The coefficient on *Supply* is also negative and statistically significant. This tells us that a one-unit increase in *Supply* has a significant negative effect on women's representation when

[24] A gender quota is considered effective if there's a 10% de facto threshold for either candidate or reserved seat quotas; there must also be strong sanctions for noncompliance with the quota or strong placement mandates on party lists (Hughes et al., 2017, 2019). The democracy variable runs from -10 to $+10$, with higher numbers indicating higher levels of democracy (Marshall, Gurr and Jaggers, 2019).

Table 5.6 Demand, supply, and women's legislative representation

Dependent Variable: *Women's Representation, 0 − 100*

	Standard Interactive Specification		Centered Interactive Specification
Demand	−0.28**	$Demand_c$	0.23***
	(0.13)		(0.04)
Supply	−0.12*	$Supply_c$	0.27***
	(0.07)		(0.04)
Demand × Supply	0.008***	$Demand_c × Supply_c$	0.008***
	(0.002)		(0.002)
Effective Gender Quota	9.26***		9.26***
	(1.57)		(1.57)
Democracy	−0.23*		−0.23*
	(0.13)		(0.13)
Constant	11.57**		15.32***
	(4.90)		(1.02)
Observations	186		186
R^2	0.52		0.52

Standard errors in parentheses. $^*p < 0.10$; $^{**}p < 0.05$; $^{***}p < 0.01$ (two-tailed)

Demand $= 0$. Again, though, not too much should be read into this as *Demand* is also never 0 in the real world; the lowest observed value of *Demand* is 7.98. These particular results remind us that while the coefficients on constitutive terms in interaction models, in this case *Demand* and *Supply*, can be interpreted and have a specific meaning, they're not necessarily of substantive interest. Much depends on whether it makes substantive sense to examine the effect of a modifying variable when the value of the other is 0. As predicted, the coefficient on the interaction term is positive and statistically significant. This provides the important evidence of conditionality and indicates, as Dhima (2022) argues, that demand and supply interact in determining the level of women's representation. In terms of the control variables, we see that countries with an effective gender quota have 9.26 percentage points more women legislators than those that don't have such a quota. There's also some evidence that more democratic countries have lower levels of women's representation.

⚠ **Important:** While the coefficients on constitutive terms can be interpreted and have a specific meaning, they're not necessarily of substantive interest. Much will depend on whether it makes substantive sense to examine the effect of a modifying variable when the value of the other is zero.

Given that our modifying variables are both continuous, it's useful to construct marginal effect plots for both *Demand* and *Supply*. Based on the results in Table 5.6 and Eq. 5.49, we see that the estimated marginal effect of *Demand* is $-0.28 + 0.008 \times Supply$. Since the observed range of *Supply* is 8.61 to 111.01, the observed marginal effect of *Demand* ranges from $-0.28 + 0.008 \times 8.61 = -0.21$ to $-0.28 + 0.008 \times 111.01 = 0.56$. The standard error associated with the marginal effect of *Demand* is

$$\text{se}\left(Demand\right) = \text{se}\left(\beta_1 + \beta_3 Supply\right)$$
$$= \sqrt{\text{var}\left(\beta_1\right) + \left(Supply\right)^2 \text{var}\left(\beta_3\right) + 2 \times Supply \times \text{cov}\left(\beta_1, \beta_3\right)}. \quad (5.51)$$

We can calculate this for any value of *Supply* using the information contained in the estimated variance-covariance matrix for the coefficients,

$$V(\beta) = \begin{pmatrix} 0.016526 & 0.007235 & -0.000207 & 0.003814 & -0.005884 & -0.557778 \\ 0.007235 & 0.005190 & -0.000107 & 0.012601 & -0.000977 & -0.328715 \\ -0.000207 & -0.000107 & 0.000003 & -0.000163 & 0.000042 & 0.007364 \\ 0.003814 & 0.012601 & -0.000163 & 2.454234 & 0.014583 & -0.931239 \\ -0.005884 & -0.000977 & 0.000042 & 0.014583 & 0.016942 & 0.124258 \\ -0.557778 & -0.328715 & 0.007364 & -0.931239 & 0.124258 & 23.99028 \end{pmatrix}.$$

$$(5.52)$$

With this information, and following the strategy we used in the previous substantive application, we constructed the marginal effect plot for *Demand* across the observed range of *Supply* shown in the top panel of Figure 5.8. As predicted, the effect of demand for women legislators on the level of women's legislative representation varies with the supply of qualified female candidates.

We see, for example, that there's a range of values for *Supply* where the effect of *Demand* is negative and statistically insignificant, a range of values where it's positive and statistically insignificant, and a range where it's positive and statistically significant. We can be more precise about these ranges. *Demand* has a negative and statistically insignificant effect when *Supply* is less than $\frac{-\beta_1}{\beta_3} = \frac{0.28}{0.008} = 36.94$. This is the point at which the marginal effect line, $\beta_1 + \beta_3 Supply$, crosses the horizontal zero line. We know that the effect of *Demand* is statistically insignificant in this range because the horizontal zero line falls between the upper and lower bounds

(a) The Conditional Effect of *Demand*

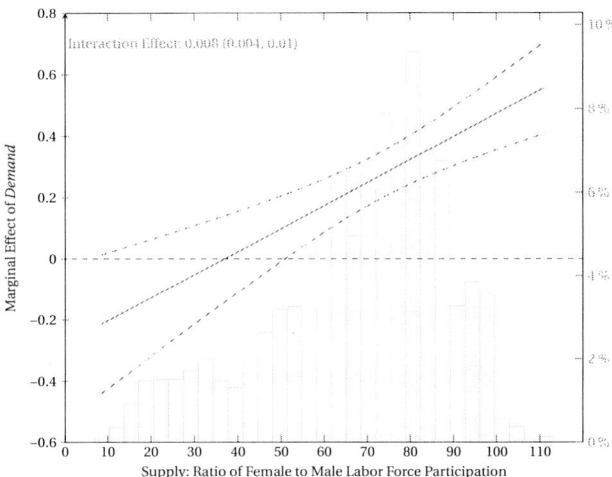

(b) The Conditional Effect of *Supply*

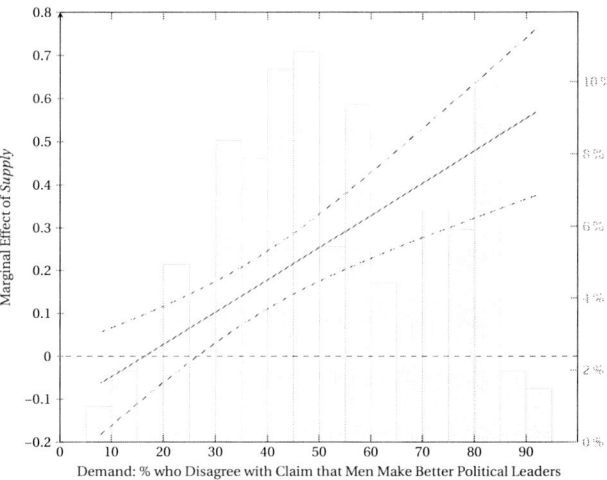

Figure 5.8 The conditional effects of *Demand* and *Supply* on the level of women's legislative representation

Note: The plot in panel (a) shows the effect of a one-unit increase in *Demand* on women's legislative representation across the observed range of *Supply*. It also shows the interaction effect between *Demand* and *Supply*. The plot in panel (b) shows the effect of a one-unit increase in *Supply* on women's legislative representation across the observed range of *Demand*. The curved dashed lines indicate two-tailed 95% confidence intervals. The histograms show the percentage of observations at different values of each modifying variable. The two plots are based on the results from the standard interaction model shown in the first column of Table 5.6.

of the confidence interval. Roughly 10.9% of the sample observations fall in this region. *Demand* has a positive and statistically insignificant effect when $36.94 < Supply < 50.88$. The right cut-point for this particular range is found by identifying where the lower bound of the confidence interval crosses the horizontal zero line. One way to do this is to simply eyeball this point on the marginal effect plot. Another is to examine the numerical values that were used to construct the lower bound of the confidence in the marginal effect plot. The most precise method is to use a little algebra to calculate the value of *Supply* at which the lower bound of the confidence interval,

$$\beta_1 + \beta_3 Supply - t_{n-k,\alpha/2} \times \sqrt{\text{var}\,(\beta_1) + (Supply)^2 \text{var}\,(\beta_3) + 2 \times Supply \times \text{cov}\,(\beta_1, \beta_3)},$$

(5.53)

equals 0. We show how to do this in Appendix D. Roughly 8.8% of the sample observations fall in this second region. *Demand* has a positive and statistically significant effect when $Supply > 50.88$. We know that the effect of *Demand* is statistically significant in this range because the upper and lower bounds of the confidence interval are both on the same side (above) of the horizontal zero line. Roughly 80.4% of the sample observations fall in this third region. Overall, the marginal effect plot in panel (a) is consistent with the *Demand Hypothesis's* prediction that demand for women legislators has a positive and increasing effect on women's legislative representation but only once the supply of qualified women candidates is sufficiently high. The fact that the effect of *Demand* is negative when the level of *Supply* is very low could be read as inconsistent with the theory. However, the negative effect of *Demand* is never statistically significant for observed values of *Supply* and relatively few observations in our sample have such low values of *Supply*.[25]

What about the substantive significance of these results? Increasing *Demand* by 20 percentage points, which is slightly less than one standard deviation (20.71), when *Supply* is at its mean value (67.88) increases women's legislative representation by $(-0.28 + 0.008 \times 67.88) \times 20 = 4.65$ (3.10, 6.20) percentage points; two-tailed 95% confidence intervals are shown in parentheses. This effect size is equivalent to a 21.4% increase in

---

[25] The negative effect of *Demand* at very low levels of *Supply* may also be an artifact of having employed a model specification that assumes a *linear* interaction effect. To some extent, our theory of women's legislative representation predicts a threshold effect in which *Demand* has a positive effect only once the value of *Supply* is sufficiently high. Theories like this that posit some kind of threshold effect imply that the relationship between the marginal effect of a variable like *Demand* and the other modifying variable, in this case *Supply*, is only piecewise linear; it's not linear over the whole range of the other modifying variable. For one way of modeling this sort of piecewise linear relationship, see Berry, Golder and Milton (2012, 669–671). Piecewise linear regression models are also briefly discussed in Chapter 7.

the 2018 mean level of women's legislative representation (21.7) around the world. Increasing *Demand* by the same amount when *Supply* is at its maximum observed value (111.01) increases women's legislative representation by $(-0.28 + 0.008 \times 111.01) \times 20 = 11.14$ (8.17, 14.12) percentage points. This effect size is equivalent to a 51.3% increase in the 2018 mean level of women's legislative representation. These effect sizes are substantively meaningful. The modifying effect of *Supply* is also substantively large. As we've seen, the effect of a 20 percentage point increase in *Demand* is 239.4% larger when *Supply* is at its maximum as opposed to its mean. This difference in effect size is equivalent to 29.9% of the 2018 mean level of women's legislative representation. As predicted, the positive interaction effect indicates that the demand for female representatives is more accurately translated into actual female representatives as the supply of qualified female candidates increases. For example, each percentage point increase in demand leads to a 0.23 percentage point increase in women's representation when supply is at its mean observed value and a 0.56 percentage point increase when supply is at its maximum observed value.

Based on the results in Table 5.6 and Eq. 5.50, the estimated marginal effect of *Supply* is $-0.12 + 0.008 \times Demand$. Since the observed range of *Demand* is 7.98 to 91.82, the observed marginal effect of *Supply* ranges from a low of $-0.12 + 0.008 \times 7.98 = -0.06$ to a high of $-0.12 + 0.008 \times 91.82 = 0.57$. The standard error associated with the marginal effect of *Supply* is

$$
\begin{aligned}
\text{se}\,(Supply) &= \text{se}\,(\beta_2 + \beta_3 Demand) \\
&= \sqrt{\text{var}\,(\beta_2) + (Demand)^2\,\text{var}\,(\beta_3) + 2 \times Demand \times \text{cov}\,(\beta_2, \beta_3)}.
\end{aligned}
\tag{5.54}
$$

We can calculate this for any value of *Demand* using the information contained in the estimated variance-covariance matrix for the coefficients shown in Eq. 5.52. Using this information, we constructed the marginal effect plot for *Supply* across the observed range of *Demand* shown in the bottom panel of Figure 5.8. As predicted, the effect of increasing the supply of qualified female candidates on the level of women's legislative representation varies with the demand for women legislators. As the marginal effect plot indicates, *Supply* has a negative but statistically insignificant effect when *Demand* < 16.36 (3.4%), it has a positive but statistically insignificant effect when 16.35 < *Demand* < 26.52 (9.4%), and it has a positive and statistically significant effect when *Demand* > 26.52 (87.2%). Parentheses indicate the percentage of observations that fall within each of these ranges. These results are consistent with the *Supply Hypothesis*, which predicts that the supply of qualified female candidates will have a positive and increasing effect on women's legislative representation but only once the demand for women legislators is sufficiently large. The fact that the effect of *Supply* is negative when the level of *Demand* is very low could be read as inconsistent with our theory. Again, though, the negative effect of *Supply* is never statistically significant for observed values of *Demand*, and almost no observations have such low values of *Demand*.

In terms of substantive significance, increasing *Supply* by 20 points, which is slightly less than one standard deviation (20.51), when *Demand* is at its mean (52.01) increases women's legislative representation by $(-0.12 + 0.008 \times 52.01) \times 20 = 5.36$ (3.72, 7.01) percentage points. This effect size is equivalent to a 24.7% increase in the 2018 mean level of women's legislative representation around the world. Increasing *Supply* by the same amount when *Demand* is at its maximum observed value (91.82) increases women's legislative representation by $(-0.12 + 0.008 \times 91.82) \times 20 = 11.35$ (7.50, 15.20) percentage points. This effect size is equivalent to a 52.3% increase in the 2018 mean level of women's legislative representation. These effect sizes are substantively large. The modifying effect of *Demand* is also substantively large. As we've seen, the effect of a 20 point increase in *Supply* is 212% larger when *Demand* is at its maximum as opposed to its mean. This difference in effect size is equivalent to 27.6% of the 2018 mean level of women's legislative representation. As predicted, the positive interaction effect indicates that the supply of qualified female candidates is more accurately translated into actual female representatives as the demand for female legislators increases. For example, a one point increase in supply leads to a 0.27 percentage point increase in women's representation when demand is at its mean and a 0.57 percentage point increase when it's at its maximum observed value.

Together, the combined marginal effect plots shown in Figure 5.8 provide all of the necessary information to fully evaluate the predictions jointly contained in the *Demand Hypothesis* and the *Supply Hypothesis*. When both modifying variables are continuous, as is the case here, scholars sometimes choose to present their results using a three-dimensional (3-D) plot of their dependent variable against the two modifying variables. While such 3-D plots are often pretty and can help us to visualize how the two modifying variables affect the dependent variable, they don't typically convey the necessary information to evaluate the hypotheses that come from a theory positing interaction. In this respect, these types of 3-D plots tend to place more emphasis on style than substance.

To examine this issue further, we present a 3-D predicted value plot for our substantive application in Figure 5.9. The "surface" indicates the predicted level of women's representation for different possible combinations

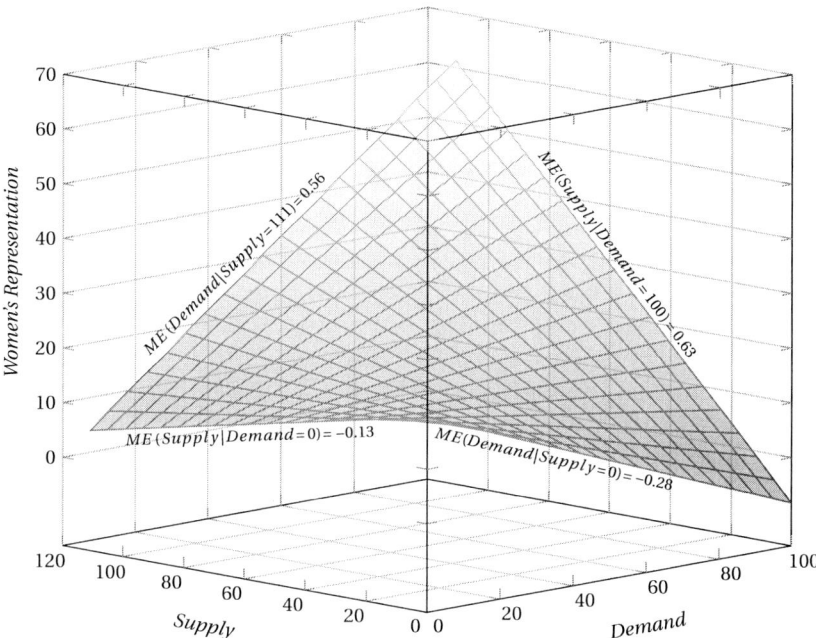

Figure 5.9 Visualizing the relationship between *Supply*, *Demand*, and *Women's Representation*

Note: The surface shows how the predicted value of *Women's Representation* varies across *Supply* and *Demand* and is based on the standard interaction model shown in the first column of Table 5.6; we assume that we have a democracy (*Democracy* = 6) with an effective gender quota. *Women's Representation* captures the percentage of female legislative representatives. *Supply* captures the ratio of female to male labor force participation rates. *Demand* captures the percentage of respondents who disagree or strongly disagree with the claim that men make better political leaders than women.

of values for *Supply* and *Demand*. Darker shades indicate lower levels of women's representation, while lighter shades indicate higher levels. We've calculated the predicted values assuming that we're dealing with a democracy (*Democracy* = 6) that has an effective gender quota. As an example, the front corner of the surface indicates the predicted value, $11.57 + 9.26 - 0.23 \times 6 = 19.46$, for a democracy with an effective gender quota where the values of *Demand* and *Supply* are both 0. In contrast, the rear corner of the surface indicates the predicted value, $11.57 - 0.28 \times 100 - 0.12 \times 111 + 0.008 \times 100 \times 111 + 9.26 - 0.23 \times 6 = 61.51$, for the same country but where *Demand* is 100 and *Supply* is 111.

The slopes of each edge of the "surface" are shown in small text. The slope going left to right on the front right edge, -0.28, tells us the marginal effect of *Demand* when *Supply* = 0 and is equivalent to the coefficient on

Demand in the first column of Table 5.6. The slope going left to right on the back left edge, $-0.28 + 0.008 \times 111 = 0.56$, tells us the marginal effect of *Demand* when *Supply* $= 111$. The left-to-right slope of the surface between the front right edge and the back left edge indicates the marginal effect of *Demand* at intermediate values of *Supply*. The changing left-to-right slope of the surface as you move from the front right edge to the back left edge is what's shown in the marginal effect plot for *Demand* in the top panel of Figure 5.8.[26] The slope going right to left on the front left edge, -0.13, tells us the marginal effect of *Supply* when *Demand* $= 0$ and is equivalent to the coefficient on *Supply* in the first column of Table 5.6. The slope going right to left on the back right edge, $-0.13 + 0.008 \times 100 = 0.63$, tells us the marginal effect of *Supply* when *Demand* $= 100$. The right-to-left slope of the surface between the front left edge and the back right edge indicates the marginal effect of *Supply* at intermediate values of *Demand*. The changing right-to-left slope of the surface as you move from the front left edge to the back right edge is what's shown in the marginal effect plot for *Supply* in the bottom panel of Figure 5.8.[27] In line with our theory, the 3-D plot indicates that women's representation is at its highest when both demand and supply are high.

There are several issues with 3-D plots like this one. One is that they don't easily allow us to show the location of the underlying data for the predicted surface. As noted previously, information about the frequency distributions for the modifying variables is helpful for addressing the substantive importance of our estimated effects and for evaluating "common support" issues. In the marginal effect plots shown in Figure 5.8, we easily conveyed information about the location of the underlying data by incorporating histograms showing the frequency distributions for the modifying variable on the horizontal axis. Doing something similar with a 3-D plot is more challenging. One possibility is to incorporate a two-dimensional scatterplot that shows the "coordinate" values of the modifying variables for the sample observations on the floor or ceiling of the "cube" housing the 3-D plot. The value of this particular strategy, as with the strategy of including a rug plot below a marginal effect plot, declines with the sample size as the markers for the individual observations will blend together and become indistinguishable. Another

---

[26] This isn't entirely accurate as the top panel in Figure 5.8 plots the marginal effect of *Demand* only over the observed range of *Supply*, 8.61–111.01. In contrast, the surface of the 3-D plot shows how the marginal effect of *Demand* changes as *Supply* increases from 0 to 111.

[27] Again, this isn't quite accurate as the bottom panel in Figure 5.8 plots the marginal effect of *Supply* only over the observed range of *Demand*, 7.98–91.82. In contrast, the surface of the 3-D plot shows how the marginal effect of *Supply* changes as *Demand* increases from 0 to 100.

strategy is to use some kind of contour plot to display the joint density of the two modifying variables on the floor of the "cube" housing the 3-D plot. Both of these strategies, though, can make the 3-D plot look rather busy.

A more significant issue with 3-D plots like this is that they emphasize predicted values rather than effects or differences. Our theoretical hypotheses, however, nearly always make claims about how changes in the value of the independent variables lead to *changes* or *differences* in the predicted value of the dependent variable. To put this in context, the 3-D plot in Figure 5.9 makes it relatively easy to see when the predicted level of women's legislative representation is high or low. It's not so easy, though, to determine the precise value of the slope of the surface – and hence the effect of *Demand* and *Supply* – for different values of the relevant modifying variable. Indeed, it's not always easy to see when the slope of the surface is positive or negative. The fact that we've indicated the slopes of each edge of the surface in small text helps to some extent, but it's still difficult to determine the precise slopes at intermediate values of *Demand* and *Supply*. For example, we suspect that many of you can't identify, say, the value of *Supply* at which the marginal effect of *Demand* becomes positive simply by looking at the surface in Figure 5.9. Contrast this with the marginal effect plots in Figure 5.8, where we can easily see not only if and when the effects of *Demand* and *Supply* are negative or positive but also the magnitude of these effects.

Even more importantly, 3-D predicted value plots typically don't convey appropriate measures of uncertainty. We might think to shade the surface of the plot so that we can easily see those areas where the predicted values are significantly different from 0. Remember, though, that our hypotheses usually don't make claims about the predicted values of our dependent variable. Instead, they make claims about the *effects* of our modifying variables. Put differently, shading those areas of the surface where the predicted values are significantly different from 0 doesn't help us to evaluate the empirical support for our hypotheses. We could instead shade those areas where the slopes, and hence the effects of our modifying variables, are statistically significant. However, this probably requires producing *two* 3-D plots – one in which we shade the surface where the slopes for the effect of *Demand* are statistically significant and one in which we shade the surface where the slopes for the effect of *Supply* are statistically significant. While it's perhaps possible to convey some information about the direction and statistical significance of the effects of the variables in a 3-D predicted value plot, the same information is presented in a much more transparent and accessible way in the marginal effect plots shown in Figure 5.8. Indeed, the fact that the plots in Figure

5.8 present confidence intervals for the marginal effects provides readers with much more nuanced information about the effects of the modifying variables than the more discrete significant vs. not-significant shading that we would have to use in a 3-D plot.

> ⚠ **Important:** 3-D predicted value plots are good for visualizing the overall relationship between the modifying variables and a dependent variable. However, they offer little leverage when it comes to testing the implications of a theory positing interaction.

The bottom line is that 3-D predicted value plots are good for visualizing the overall relationship between the modifying variables and a dependent variable. They can also be helpful, as we saw in Chapter 3, when we wish to distinguish between different types of conditional relationships. However, they offer little leverage when it comes to testing the implications of a theory positing interaction. In practice, they provide almost none of the information that we need to evaluate the empirical support for our hypotheses. As a result, we strongly recommend that researchers use marginal effect plots rather than 3-D predicted value plots when testing conditional hypotheses. Marginal effect plots, so long as they also indicate the interaction effect, provide all of the information necessary to evaluate the five key predictions that can be derived from most theories positing interaction between two modifying variables.

Before moving on to the next chapter, we recognize that some scholars choose to center their modifying variables before estimating an interaction model. In our particular application, this would mean estimating the following "centered" interactive model:

$$Women's\ Representation = \delta_0 + \delta_1 Demand_c + \delta_2 Supply_c$$
$$+ \delta_3 Demand_c \times Supply_c$$
$$+ \delta Controls + \epsilon_c, \tag{5.55}$$

where the sample means have been subtracted from *Demand* and *Supply*, such that $Demand_c = Demand - \overline{Demand}$ and $Supply_c = Supply - \overline{Supply}$.[28] Recall from earlier that $\overline{Demand} = 52.01$ and $\overline{Supply} = 67.88$. The results from the centered model in Eq. 5.55 are shown in the last column of Table 5.6.

---

[28] It's important to recognize that the interaction term between X and Z in a centered interactive model is $X_c Z_c$, where $X_c = X - \overline{X}$ and $Z_c = Z - \overline{Z}$, and not $(XZ)_c$, where $(XZ)_c = XZ - \overline{XZ}$. This is because $X_c Z_c \neq (XZ)_c$.

As discussed in Chapter 3, this centered model is exactly equivalent to the standard interactive model in Eq. 5.48. The coefficients on the interaction term and control variables are identical across the two models. While the coefficients on the constitutive terms and the constant are different, this is only because they capture different quantities of interest. Whenever the two models are used to calculate the same quantities of interest, they produce identical values.

The coefficient on the constitutive term *Demand* in the standard interactive model tells us that a one-unit increase in *Demand* is associated with a 0.28 percentage point reduction in women's legislative representation when *Supply* is 0. In contrast, the coefficient on the constitutive term $Demand_c$ in the centered model tells us that a one-unit increase in *Demand* is associated with a 0.23 percentage point increase in women's legislative representation when *Supply* is at its mean value of 67.88. If we look at the top panel of Figure 5.8, we see that the results from the standard interactive specification also show that the marginal effect of *Demand* is 0.23 when *Supply* is 67.88. Similarly, we can use the results from the centered model to calculate the marginal effect of *Demand* when *Supply* is 0 instead of 0.23, $\delta_1 - \delta_3 \overline{Supply} = 0.23 - 0.008 \times 67.88 = -0.28$. As expected, this is identical to the coefficient on *Demand* in the standard interactive model.

The coefficient on the constitutive term *Supply* in the standard interactive model tells us that a one-unit increase in *Supply* is associated with a 0.12 percentage point reduction in women's legislative representation when *Demand* is 0. In contrast, the coefficient on the constitutive term $Supply_c$ in the centered model tells us that a one-unit increase in *Supply* is associated with a 0.27 percentage point increase in women's legislative representation when *Demand* is at its mean value of 52.01. If we look at the bottom panel of Figure 5.8, we see that the results from the standard interactive specification also show that the marginal effect of *Supply* is 0.27 when *Demand* is 52.01. Similarly, we can use the results from the centered model to calculate the marginal effect of *Supply* when *Demand* is 0 instead of 52.01, $\delta_2 - \delta_3 \overline{Demand} = 0.27 - 0.008 \times 52.01 = -0.12$. As expected, this is identical to the coefficient on *Supply* in the standard interactive model.[29] If we used the results from the centered interaction model to produce marginal effect plots for *Demand* and *Supply*, they would be identical to the plots shown in Figure 5.8, which are based on the results from the uncentered interaction model.

[29] The constants across the two models differ. However, as we showed in Eq. 3.10, the constant in the standard interactive model, 11.57, is equivalent to
$$\delta_0 - \delta_1 \overline{Demand} - \delta_2 \overline{Supply} + \delta_3 \overline{Demand} \times \overline{Supply} =$$
$$15.32 - 0.23 \times 52.01 - 0.27 \times 67.88 + 0.008 \times 52.01 \times 67.88 = 11.57.$$

In our experience, readers are much less familiar with centered interactive models than the more typical uncentered ones and are therefore less sure about how to interpret the results. Given this, and the fact that centered interactive models produce identical quantities of interest, it's not clear to us why anyone would choose to estimate a centered interactive model rather than a standard uncentered one. The reasons that scholars typically give for doing so are either flawed or unpersuasive. The most common reason given rests on the claim that centering the modifying variables reduces issues with multicollinearity. As we demonstrated in Chapter 3, though, this claim is simply false. Centering "alters nothing important statistically and nothing at all substantively" (Kam and Franzese, 2007). Some scholars also argue that centering the modifying variables makes the coefficients on the constitutive terms more substantively meaningful. It's certainly true that centering changes the interpretation of the coefficients on the constitutive terms. But whether the marginal effect of, say, X on Y is more substantively meaningful when the modifying variable $Z = \overline{Z}$ than it is when $Z = 0$ will depend on the frequency distribution of Z and the particular application. Some claim that the marginal effect of X at the mean value of Z is substantively important because this is the value of Z at which the effect of X is most likely to be statistically significant. As noted in the previous chapter, though, this belief is also false. The standard error associated with the marginal effect of X on Y is minimized when the value of Z is $Z^* = -\frac{\text{cov}(\beta_1, \beta_3)}{\text{var}(\beta_3)}$ and not $\overline{Z}$. Notably, the marginal effect of X at the mean value of Z doesn't speak to any of the three key predictions that are typically contained in a conditional hypothesis about the effect of X on Y. Moreover, the precise quantity of interest captured by the coefficients on the constitutive terms is never too important anyway as researchers always have to move beyond the information contained in the regression output to fully evaluate the empirical support for a theory positing interaction.

In this chapter, we've applied what we've learned so far in this book to three substantive cases that differ in terms of whether the variables involved in an interaction are discrete and/or continuous. We focused in particular on how best to present the results from interaction models. Among other things, we saw that the judicious use of marginal effect plots is especially valuable when evaluating the empirical support for theories positing interaction.

The first part of this book has addressed the use of linear-interactive models that allow for an interaction between two variables X and Z on a continuous dependent variable Y. In the second part of the book, we retain our focus on interaction models where the dependent variable is continuous. However, we shift our attention to situations in which our theory implies more complex forms of conditionality. In the next chapter, we look at the case where our theory predicts that the effect of some independent variable X on Y depends on the value of more than one modifying variable.

5.4 EXERCISES

1. Figure 5.8 presents marginal effect plots for *Demand* and *Supply* from our third substantive application.
   a. The confidence intervals in both marginal effect plots aren't linear; instead, they have an hour-glass shape. This isn't something peculiar to this particular application but is something that's common to all marginal effect plots in which the interaction effect is linear. Explain why this is the case.
   b. Using the information you are provided with in the application, determine the value of *Supply* at which the confidence interval for the marginal effect of *Demand* in the top panel is narrowest. Is it the mean value of *Supply*?
   c. Determine the value of *Demand* at which the confidence interval for the marginal effect of *Supply* in the bottom panel is narrowest. Is it the mean value of *Demand*?
   d. Use the information in Appendix D to calculate the precise values of *Demand* at which the upper and lower bounds of the confidence interval for the marginal effect of *Supply* equal 0.

2. In this exercise, we return once again to the 2012 study by Banks and Valentino (2012) looking at how opposition by Whites toward race-conscious policies such as affirmative action depends on the interaction between an individual's level of symbolic racism (or racial resentment) and their level of anger. Recall from exercises in previous chapters that we used data from Banks and Valentino to estimate the following linear-interactive specification:

$$Oppose\ Race\ Policies = \beta_0 + \beta_1 Symbolic\ Racism + \beta_2 Anger$$
$$+ \beta_3 Symbolic\ Racism \times Anger + \epsilon, \quad (5.56)$$

where *Oppose Race Policies* is a continuous variable on a 0–1 scale that captures an individual's level of opposition to race-conscious policies, *Anger* is a dichotomous variable that equals 1 when an individual is angry and 0 otherwise, and *Symbolic Racism* is a continuous variable that measures an individual's level of racial resentment on a 0–1 scale. The results from our analysis are shown

Table 5.7 Symbolic racism, anger, and opposition to race-conscious policies III

Dependent Variable: Oppose Race Policies

	Model 1
Symbolic Racism	0.52***
	(0.05)
Anger	−0.11
	(0.10)
Symbolic Racism × Anger	0.26**
	(0.12)
Constant	0.27***
	(0.04)
Observations	182
R^2	0.46

Standard errors in parentheses. $*p < 0.10$; $**p < 0.05$; $***p < 0.01$ (two-tailed)

in Table 5.7, and the estimated variance-covariance matrix for the coefficients is shown in Eq. 5.57.

$$V(\beta) = \begin{pmatrix} 0.00298512 & 0.00199709 & -0.00298512 & -0.00199709 \\ 0.00199709 & 0.00915827 & -0.01125366 & -0.00155169 \\ -0.00298512 & -0.01125366 & 0.01550988 & 0.00199709 \\ -0.00199709 & -0.00155169 & 0.00199709 & 0.00155169 \end{pmatrix}.$$
(5.57)

a. Use the information that's available to you to sketch a marginal effect plot for *Symbolic Racism*. If you completed Exercise 1 at the end of Chapter 4, you'll find some of the information you calculated there helpful.

b. At what value of *Symbolic Racism* is the standard error associated with the effect of *Anger* minimized?

c. Use the information that's available to you to sketch a marginal effect plot for *Anger*. If you completed Exercise 1 at the end of Chapter 4, you'll find that some of the information you calculated there can be used to make your plot more accurate.

d. What other information would you like to know in order to make your marginal effect plot for *Anger* more informative?

PART II

More Complex Forms of Conditionality

6 When We Have More Than One Modifying Variable

So far we've focused on theories that posit an interaction between two independent variables X and Z on a continuous dependent variable Y. We've discussed how to derive hypotheses, how to correctly specify an interaction model, and how to effectively calculate, interpret, and present key quantities of interest. But what if our theory isn't as simple as those we've examined so far? In this chapter, we begin to look at some more theoretically complex forms of conditionality. Specifically, we turn our attention to theories that imply that the effect of an independent variable such as X depends on the value of more than one other modifying variable. While the effect of X on Y could depend on the value of any number of other variables, we'll focus our attention here on those situations in which it depends in some way on the value of two other variables Z and W.[1] To a large extent, the skills and knowledge we've developed in the first part of the book transfer directly over to this new scenario.

There are two distinct theoretical cases to consider. The first case occurs when the modifying effects of Z and W on the effect of X are *independent* or *unconditional*. In this case, how Z modifies the effect of X on Y doesn't depend on the value of W, and how W modifies the effect of X on Y doesn't depend on the value of Z. The basic structure of this case is graphically illustrated in panel (a) of Figure 6.1. The second case occurs when the modifying effects of Z and W on the effect of X are *dependent* or *conditional*. In this case, how Z modifies the effect of X on Y depends on the value of W, and how W modifies the effect of X on Y depends on the value of Z. The basic structure of this case is graphically illustrated in panel (b) of Figure 6.1. The distinction between the two cases will become clearer as we examine each of them in more detail.

---

[1] All of the points we make generalize easily to cases in which X interacts with more than two other variables.

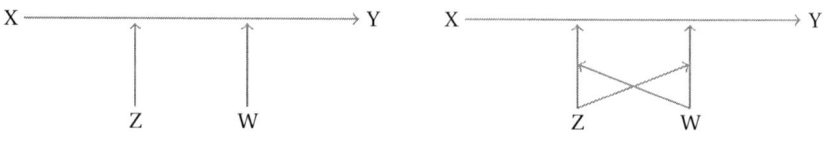

(a) Independent Modifying Effects   (b) Dependent Modifying Effects

Figure 6.1 (a)–(b) Two modifying variables: Independent and dependent modifying effects

> ⚠ **Important:** There are two possible theoretical cases when we think
> that the effect of X on Y depends on the values of two other independent
> variables Z and W:
>
> 1. *The modifying effects of Z and W are independent or
>   unconditional.* In this case, how Z modifies the effect of X on Y
>   does not depend on the value of W, and how W modifies the effect
>   of X on Y does not depend on the value of Z.
> 2. *The modifying effects of Z and W are dependent or conditional.* In
>   this case, how Z modifies the effect of X on Y depends on the value
>   of W, and how W modifies the effect of X on Y depends on the
>   value of Z.

6.1 WHEN THE MODIFYING EFFECTS OF Z AND W ARE INDEPENDENT

We begin with theory. In Chapter 2, we noted that scholars often focus
on the predicted sign of the interaction effect between their independent
variables when stating hypotheses and evaluating the empirical support for
their conditional claims. In the current context, this would mean focusing
on how the effect of X on Y varies with the values of Z and W. A scholar
might, for example, claim that the effect of X on Y increases with Z and
decreases with W. As we indicated in Section 2.4, though, observing the
predicted signs for these interaction effects, while necessary, isn't sufficient
to corroborate a conditional claim about the effect of X on Y. This is
because any set of observed interactions is always consistent with a wide
variety of ways in which X interacts with Z and W to determine Y, some of
which may be inconsistent with the underlying theory. Importantly, simply
knowing the signs of the interaction effects says nothing about whether X
has a positive, negative, or zero effect on Y for any value of Z or W. Only
proposing and testing predictions about the signs of the interaction effect

between X and Z and between X and W constitutes an extremely weak, and often substantively uninformative, test of a conditional claim about the effect of X.

When X interacted with only Z, we recommended that scholars always supplement a prediction about the sign of the interaction effect between X and Z with a prediction about the sign of the marginal effect of X on Y when Z is at its lowest and highest values. Doing so significantly narrows the range of conditional relationships that are consistent with one's underlying theory, thereby strengthening any empirical test. Things are more complicated in the case where X interacts with two variables Z and W. This is because the sign of the marginal effect of X depends *jointly* on the values of Z and W. In these circumstances, we recommend that scholars supplement their predictions about the signs of the interaction effect between X and Z and between X and W with predictions about the sign of the marginal effect of X at different combinations of values for Z and W. In particular, scholars should make predictions about the signs of the marginal effect of X when Z and W are at each possible combination of their minimum and maximum values.[2] Altogether, we encourage scholars to make six key predictions about the marginal effect of X and how this varies with the values of Z and W: P_{XZ}, P_{XW}, $P_{X|Z_{min}, W_{min}}$, $P_{X|Z_{min}, W_{max}}$, $P_{X|Z_{max}, W_{min}}$, and $P_{X|Z_{max}, W_{max}}$.[3] Different combinations of signs for these six predictions correspond to different possible claims about the conditional effect of X on Y. Only by deriving and testing all six of these predictions is it possible to know whether the data support our particular claim about the conditional effect of X on Y or some other claim.[4]

⚠ **Important:** Theories positing that Z and W *independently* modify the effect of X on Y are typically strong enough to generate six key predictions about the marginal effect of X and how this varies with Z and W. These predictions speak to the direction of the interaction between

---

[2] As before, our recommendation to make predictions about the marginal effect of X when Z and W are at their minimum and maximum values doesn't mean that analysts should necessarily focus their greatest attention on the *estimated* marginal effect of X at these "extreme" values. Instead, we emphasize predictions at the extremes because they automatically imply predictions at values between the extremes.

[3] Recall from Chapter 2 that P_{XZ} and P_{XW} refer to predictions about the signs of the interaction effects between X and Z and between X and W. $P_{X|Z_{min}, W_{min}}$, $P_{X|Z_{min}, W_{max}}$, $P_{X|Z_{max}, W_{min}}$, and $P_{X|Z_{max}, W_{max}}$ refer to predictions about the signs of the marginal effect of X for the different possible combinations of the minimum and maximum values of Z and W.

[4] In some theoretical contexts, it *may* be possible to also derive a prediction about whether the modifying effect of Z is larger, smaller, or the same size as the modifying effect of W.

> X and Z, the direction of the interaction between X and W, and the direction of the marginal effect of X at different combinations of values for Z and W.

As a hypothetical example, let's assume we have a theory predicting that an increase in X always has a positive effect on Y but that the magnitude of this positive effect is expected to increase with higher values of Z and decrease with higher values of W. In the language of Chapter 2, Z "reinforces" the positive effect of X on Y, while W "mitigates" the positive effect of X on Y. We can capture these predictions in a single hypothesis about the marginal effect of X on Y and how this changes with Z and W,

$H_{X|Z,W}$: The effect of an increase in X on Y is always positive. The magnitude of this positive effect increases with values of Z but decreases with values of W.

As recommended, hypothesis $H_{X|Z,W}$ contains a prediction about the sign of the marginal effect of X on Y for all possible combinations of values for Z and W, as well as predictions about how the effect of X changes with Z and W. Other than not also making a prediction about the relative size of the modifying effects of Z and W, our hypothesis provides as complete a description of the expected effect of X as we could offer without providing specific magnitudes for the marginal effect of X at particular combinations of values for Z and W.

At this point, it's worth recalling that all interactions are symmetric. Thus, our claim that the marginal effect of X varies with Z and W necessarily implies that the marginal effects of Z and W each vary with X. Simply by stating hypothesis $H_{X|Z,W}$, and therefore without any additional theorizing, we automatically introduce some predictions about the effects of Z and W. To be specific, hypothesis $H_{X|Z,W}$ implies that the effect of Z on Y increases with X and that the effect of W decreases with X. As things stand, we don't have a prediction as to whether the marginal effects of W and Z are positive, negative, or zero for any value of X. As we've indicated previously, though, further examination of the underlying theory that led to a hypothesis like $H_{X|Z,W}$ can often lead to predictions about the signs for the marginal effects of W and Z. This additional theorizing is valuable as it further narrows the range of estimated conditional relationships that are consistent with the underlying theory, thereby increasing the strength of any empirical test. This is why we encourage scholars, *whenever possible*, to supplement a hypothesis about the marginal effect of X and how this varies with Z and W with two additional hypotheses about the marginal effects of Z and W and how they vary with X. In addition to confirming the predictions about the interaction effects between X and Z and between X

and W, these hypotheses should also make predictions about the marginal effects of Z and W when X is at its minimum and maximum values: $P_{Z|X_{\min}}$, $P_{Z|X_{\max}}$, $P_{W|X_{\min}}$, and $P_{Z|X_{\max}}$. Adding these four new predictions to the six predictions contained in a hypothesis about the marginal effect of X, we see that theories positing that Z and W independently modify the effect of X on Y are often strong enough to make ten key predictions in total. Different signs for these key predictions (positive, negative, zero) lead to thousands of possible ways in which X, Z, and W interact to affect some outcome of interest. Only by making all ten of these predictions can we know whether the data support our particular theory of interaction between X, Z, and W as opposed to one of the other thousands of possible interactive relationships.

For our purposes here, we'll assume that the underlying theory that produced hypothesis $H_{X|Z,W}$ also produces the following hypotheses about the marginal effects of Z and W:

$H_{Z|X}$: The effect of an increase in Z on Y is negative when X is at its lowest value and positive when X is at its highest value.

$H_{W|X}$: The effect of an increase in W on Y is positive when X is at its lowest value and zero when X is at its highest value.

From this we see that X "transforms" the effect of Z on Y from negative to positive and "inhibits" the negative effect of W on Y.

> ⚠ **Important:** Theories positing that Z and W *independently* modify the effect of X on Y are typically strong enough to generate ten key predictions in total. These predictions speak to the direction of the marginal effect of X and how this varies with Z and W, the direction of the marginal effect of Z and how this varies with X, and the direction of the marginal effect of W and how this varies with X.

We can test all three of our hypotheses with the following interaction model:

$$Y = \beta_0 + \beta_1 X + \beta_2 Z + \beta_3 W + \beta_4 XZ + \beta_5 XW + \epsilon. \tag{6.1}$$

We'll assume that the variables X, Z, and W are all continuous unless otherwise stated. The multiplicative interaction terms XZ and XW are included to capture the proposed conditional relationships between X and Z and between X and W. As always with a "standard" setup like this, we should include all of the constitutive terms X, Z, and W. To see why this interaction model is appropriate for testing our hypotheses, it's helpful to consider the marginal effects of our variables.

We start with the marginal effect of X on Y,

$$\frac{\partial Y}{\partial X} = \beta_1 + \beta_4 Z + \beta_5 W. \tag{6.2}$$

So long as $\beta_4 \neq 0$ and $\beta_5 \neq 0$, the marginal effect of X depends on the value of Z *and* on the value of W. The effect of a one-unit increase in X on Y is β_1 when Z and W are *both* 0. In other words, the coefficient on the constitutive term X in Eq. 6.1 doesn't tell us the average, unconditional, independent, or main effect of X on Y; instead, it tells us the marginal effect of X only when Z and W are both 0. To see how the marginal effect of X changes with the values of Z and W, note that the marginal effect of X on Y is β_1 when $Z = W = 0$, it's $\beta_1 + \beta_4$ when $Z = 1$ and $W = 0$, it's $\beta_1 + \beta_4 + \beta_5$ when $Z = 1$ and $W = 1$, it's $\beta_1 + 2\beta_4 + \beta_5$ when $Z = 2$ and $W = 1$, it's $\beta_1 + 2\beta_4 + 2\beta_5$ when $Z = 2$ and $W = 2$, and so on.

The modifying effect of Z on the marginal effect of X is given by the coefficient on the interaction term XZ,

$$\frac{\partial \left(\frac{\partial Y}{\partial X} \right)}{\partial Z} = \frac{\partial \left(\beta_1 + \beta_4 Z + \beta_5 W \right)}{\partial Z} = \beta_4. \tag{6.3}$$

In other words, each one-unit increase in Z increases the marginal effect of X by β_4. From this, we see that there's a linear relationship between the marginal effect of X and Z. A simple t-test on β_4 can be used to test whether the effect of X on Y depends on Z.

> ⚠ **Important:** Do not interpret the coefficient on the constitutive term X in a linear-interactive model with interaction terms XZ and XW as if it indicates the unconditional, average, independent, or main effect of X. It does not. This coefficient indicates the marginal effect of X on Y only when the values of Z and W are both 0.

Similarly, the modifying effect of W on the marginal effect of X is given by the coefficient on the interaction term XW,

$$\frac{\partial \left(\frac{\partial Y}{\partial X} \right)}{\partial W} = \frac{\partial \left(\beta_1 + \beta_4 Z + \beta_5 W \right)}{\partial W} = \beta_5. \tag{6.4}$$

This tells us that each one-unit increase in W increases the marginal effect of X by β_5. Thus, there's also a linear relationship between the marginal effect of X and W. A t-test on β_5 can be used to test whether the effect of X depends on W.

As we can see, we now have two separate "interaction effects," one capturing the modifying relationship between X and Z and one capturing the modifying relationship between X and W. As the marginal effect for X in Eq. 6.2 indicates, the modifying effects of Z and W are additive and therefore independent from one another. Put differently, and as Eq. 6.3 and Eq. 6.4 explicitly show, the modifying effect of Z on the marginal effect of X doesn't depend on the value of W, and the modifying effect of W on the marginal effect of X doesn't depend on the value of Z. It's in this sense that the modifying effects of Z and W are said to be *unconditional*. This independence between the modifying effects, which is posited in hypothesis $H_{X|Z,W}$, is assumed in the interaction model shown in Eq. 6.1.[5]

According to hypothesis $H_{X|Z,W}$, the marginal effect of X should always be positive. As a result, $\beta_1 + \beta_4 Z + \beta_5 W$ should be positive for all observed combinations of values for Z and W. It follows that β_1 should be positive. The magnitude of X's positive effect is expected to increase with Z and decrease with W. Thus, β_4 should be positive and β_5 should be negative. It's worth noting that our hypothesis actually makes a stronger prediction regarding the modifying effect of W, β_5. The precise nature of this prediction depends on the minimum value of Z and the maximum value of W. Suppose that the minimum value of Z is 0 and that the maximum value of W is positive. According to our hypothesis, the marginal effect of X is always positive, the modifying effect of Z is positive, and the modifying effect of W is negative. It follows that the marginal effect of X must be positive when Z is at its minimum value of 0 and W is at its maximum value, $\beta_1 + \beta_5 W_{\max} > 0$. Solving for β_5 and combining with our prediction that β_5 is negative, we have $-\frac{\beta_1}{W_{\max}} < \beta_5 < 0$. In effect, hypothesis $H_{X|Z,W}$ implies a constraint on exactly how large the negative modifying effect of W can be. We encourage scholars to look out for these "hidden" predictions in their conditional claims.

When X interacted with only one modifying variable Z, we could fully visualize the marginal effect of X with a 2-D marginal effect plot for X. As we'll see, this isn't always possible when X interacts with two modifying variables Z and W. Given that the modifying effects of Z and W on the marginal effect of X are independent, we might think that we could visualize the marginal effect of X by providing a 2-D marginal effect plot for X across the observed range of Z and another 2-D marginal effect

---

[5] As we'll see, it's possible, if one wishes, to test, rather than assert, the independence of the modifying effects using the interaction model discussed in the next section. We can see the merits of both strategies. If it's a theoretical prediction that the two modifying effects are independent and additive, why not test it? At the same time, we rarely feel it necessary to test for all of the possible interactions (and higher-order interactions) between the independent variables when our theory predicts that they should have only an additive and independent effect on the outcome of interest. We leave it to the analyst to justify their choice of model specification here.

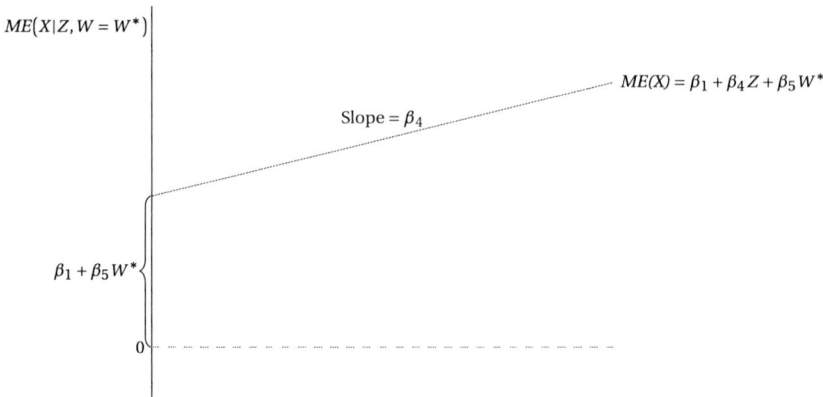

Figure 6.2 A 2-D marginal effect plot for X when $W = W^*$

Note: Figure 6.2 shows the marginal effect of X across the observed range of Z when $W = W^*$.

plot for X across the observed range of W. However, such a strategy is problematic.

Consider a 2-D marginal effect plot for X across the observed range of Z. The slope of the marginal effect line with respect to Z is determined by the coefficient on the interaction term XZ, β_4. Significantly, the intercept of the line, $\beta_1 + \beta_5 W$, depends on the value of W. In other words, to draw a marginal effect line for X across the observed range of Z, we must choose a particular value of W, W^*. In Figure 6.2, we show what such a 2-D marginal effect plot might look like. This particular marginal effect plot is consistent with the predictions in hypothesis $H_{X|Z, W}$ in that the marginal effect of X is always positive, and its magnitude increases with Z. The important point here is that the marginal effect line in Figure 6.2 is only one of infinitely many possible marginal effect lines for X across the observed range of Z. If we were to choose a different value of W, the slope of the marginal effect line would remain the same but the intercept would differ. This is important because a change in the intercept can alter the sign of the marginal effect of X. We may find, for example, that there are some values of W and Z at which the marginal effect of X isn't positive, thereby contradicting hypothesis $H_{X|Z, W}$.[6] A similar logic applies to constructing and evaluating a 2-D marginal effect plot for X across the observed range of W.

---

[6] Given that $H_{X|Z, W}$ predicts that the marginal effect of X is *always* positive and that the modifying effect of W is negative, it would be especially informative to produce a marginal effect plot for X when W is at its maximum value to see if the marginal effect of X remains positive.

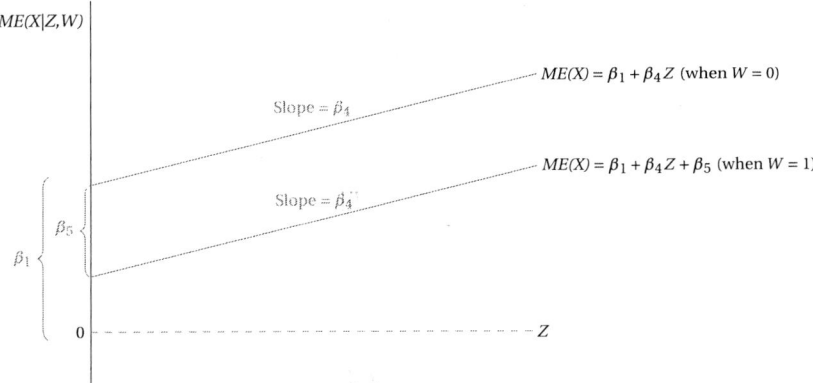

Figure 6.3 A 2-D marginal effect plot for X consistent with hypothesis $H_{X|Z,\,W}$ when Z is continuous and W is dichotomous

Note: Figure 6.3 shows the marginal effect of X across the observed range of Z when $W = 0$ and when $W = 1$.

We could, of course, construct a 2-D marginal effect plot showing marginal effect lines for X across the observed range of Z (or W) for a few different substantively relevant values of W (or Z). This particular strategy is quite compelling when one of the modifying variables is discrete. This is because the resulting 2-D marginal effect plot allows us to *fully* visualize the marginal effect of X on Y. In Figure 6.3, we show what a 2-D marginal effect plot for X that's consistent with hypothesis $H_{X|Z,\,W}$ might look like if Z is continuous and W is dichotomous (0/1). The key is that there are now two marginal effect lines for X across the observed range of Z, one for when $W = 0$ and one for when $W = 1$. Both marginal effect lines are above the gray dashed zero line for all values of Z, indicating that the marginal effect of X is positive for all possible combined values of Z and W. And both marginal effect lines have a positive slope, indicating that the marginal effect of X always increases with Z. The slopes of the two marginal effect lines are identical as the modifying effect of Z is assumed to be independent of W. The marginal effect line for the case when W is 1 is lower than the marginal effect line for the case when W is 0, indicating that the modifying effect of W is negative. We could construct a similar 2-D marginal effect plot for X across the observed range of Z if W were continuous rather than discrete. However, such a plot would not allow us to fully visualize the marginal effect of X because it would include marginal effect lines for X for only some selected values of W.

Previously, we encouraged scholars in their *estimated* marginal effect plots to include a confidence interval with their marginal effect lines and to incorporate information about the distribution of the modifying variables. Doing this raises some issues when X interacts with two modifying variables. The practical issue that arises when incorporating confidence intervals with each of the marginal effect lines in a plot like the one shown in Figure 6.3 is that the plot can become messy if the different sets of confidence intervals overlap. One way to avoid this is to "break up" the marginal effect plot for X into a *series* of marginal effect plots where each plot shows the marginal effect for X across the observed range of Z for a specific value of W. Incorporating information about the frequency distributions of the modifying variables into a marginal effect plot for X when Z and W are both continuous is unlikely to be informative or even possible. This is because there are likely to be few actual observations that have the specific *combinations* of values for Z and W depicted by the various points along each marginal effect line. Things are better when one of the modifying variables, say W, is discrete. If W is dichotomous, for example, we could overlay a marginal effect plot like the one shown in Figure 6.3 with two histograms showing the percentage of observations where $W = 0$ and $W = 1$ at different values of Z. This is similar to the strategy we adopted in Figure 5.4 when we were looking at how race affected the support for Obama. If we were to present the two marginal effect lines for X shown in Figure 6.3 in separate plots, we could easily incorporate a histogram in each showing the frequency distribution for Z when $W = 0$ or $W = 1$.

Constructing a 2-D marginal effect plot that allows us to fully visualize the marginal effect of X is also possible if both of the modifying variables are discrete. In Figure 6.4, we show what such a plot might look like if Z and W are both dichotomous (0/1). Again, we've drawn the plot to be consistent with hypothesis $H_{X|Z,W}$. The plot shows the marginal effect of X for each of the four possible combinations of values for Z and W, as well as the interaction effects between X and Z and between X and W. The positive interaction effect for XZ toward the bottom of the plot indicates that the modifying effect of Z is positive, while the negative interaction effect for XW indicates that the modifying effect of W is negative. That the modifying effect of Z is positive and the modifying effect of W is negative explains why the marginal effect of X is largest when $Z = 1$ and $W = 0$ and is smallest when $Z = 0$ and $W = 1$. As it's currently written, hypothesis $H_{X|Z,W}$ doesn't tell us whether the positive modifying effect of Z, β_4, is larger in absolute terms than the negative modifying effect of W, β_5. All we know is that the modifying effect of Z should be positive and that the magnitude of the negative modifying effect of W is constrained such that $-\frac{\beta_1}{W_{max}} < -\beta_1 < \beta_5 < 0$. It follows that we can't predict whether the marginal effect of X is larger when $Z = 1$ and $W = 1$ than when $Z = 0$ and $W = 0$. In Figure

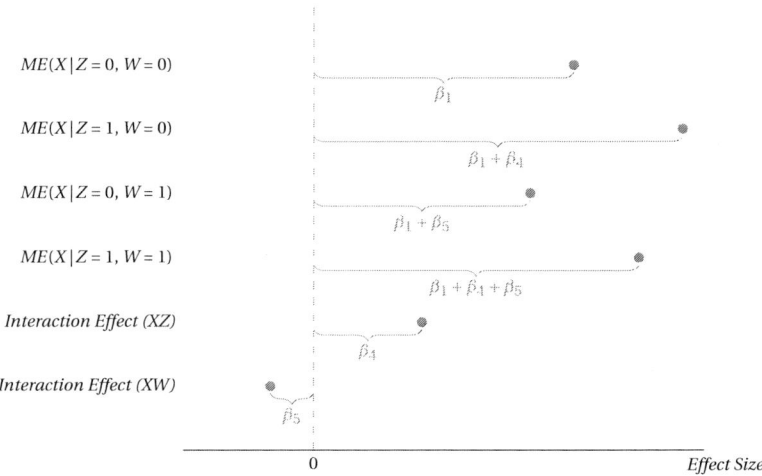

Figure 6.4 A 2-D marginal effect plot for X consistent with hypothesis $H_{X|Z, W}$ when Z and W are both dichotomous

6.4, we've assumed that the positive modifying effect of Z is larger in absolute terms than the negative modifying effect of W. This leads to the prediction shown in Figure 6.4 that $ME(X|Z = 1, W = 1) > ME(X|Z = 0, W = 0)$. We encourage scholars in situations like this, where Z and W are both dichotomous, to make clear predictions, if they can, about the relative magnitudes of the marginal effect of X for all four possible combinations of values for Z and W. This will involve making a prediction about whether the modifying effect of Z is larger or smaller than the modifying effect of W.

Researchers who wish to visualize the marginal effect of X may choose to use a tabular format rather than a marginal effect plot. Such an approach is particularly appealing when both of the modifying variables are discrete. In Figure 6.5, we show what this might look like if Z and W are both dichotomous $(0/1)$. The marginal effect of X for each possible combination of values for Z and W is shown in the four cells. The modifying effect of Z is shown in bold and indicates how the marginal effect of X changes when we move from the left column ($Z = 0$) to the right column ($Z = 1$). The modifying effect of W is shown in gray and indicates how the marginal effect of X changes when we move from the bottom row ($W = 0$) to the top row ($W = 1$). According to hypothesis $H_{X|Z, W}$, the marginal effect of X reported in each cell should be positive. The magnitude of these positive effects should grow as we increase the value of Z (move from the left column to the right column) and decrease as we increase the value of W (move from the bottom row to the top row). We could present the marginal effect of X in a similar tabular format if one or both of the modifying

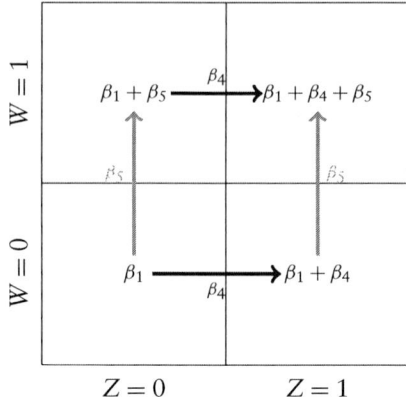

Figure 6.5 The marginal effect of X when Z and W are both dichotomous

Note: The modifying effect of Z is shown in bold and indicates how the marginal effect of X changes when we move from the left column to the right column. The modifying effect of W is shown in gray and indicates how the marginal effect of X changes when we move from the bottom row to the top row.

variables were continuous. However, doing so would not allow us to fully visualize the marginal effect of X because the cells, no matter how many there are, would show marginal effects for X for only some selected values of Z and/or W.

As we've seen, it's not possible to fully visualize the marginal effect of X with a 2-D marginal effect plot or a table if both modifying variables are continuous. However, it is possible to do so if we use a 3-D marginal effect plot. A 3-D marginal effect plot consistent with hypothesis $H_{X|Z, W}$ is shown in Figure 6.6.[7] We've assumed arbitrarily that the values of Z and W both run between 0 and 10, and that $\beta_1 = 4$, $\beta_4 = 0.4$, and $\beta_5 = -0.2$. The shaded mesh surface indicates how the marginal effect of X on Y varies with Z and W. Darker colors indicate smaller values for the marginal effect of X. The solid colored "zero surface" is equivalent to the "zero line" used in a 2-D marginal effect plot to help indicate if and when the marginal effect of X is 0. The shaded surface in Figure 6.6 is always above the zero surface, indicating that the marginal effect of X is positive for all possible combinations of values for Z and W. The surface slopes upward to the right, indicating that the marginal effect of X increases with Z. And

[7] Note that 3-D *marginal effect* plots like this one are different from the 3-D *predicted value* plots we've seen in previous chapters. Whereas 3-D predicted value plots allow us to visualize how the predicted value of a dependent variable Y varies with the values of two modifying variables X and Z, 3-D marginal effect plots like the one in Figure 6.6 helps us to visualize how the marginal effect of X on Y varies with the values of two modifying variables Z and W.

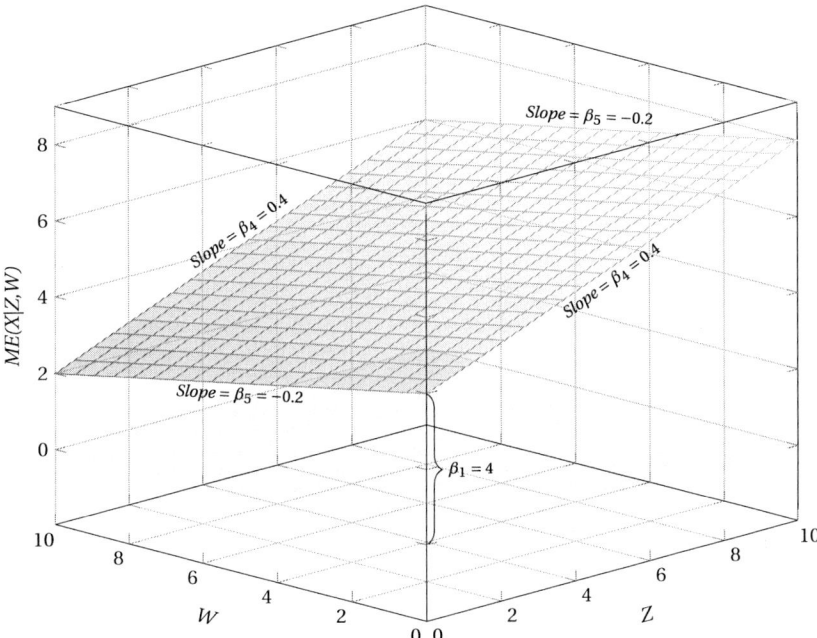

Figure 6.6 A 3-D plot of the marginal effect of X on Y consistent with hypothesis $H_{X|Z,W}$ (assuming $\beta_1 = 4$, $\beta_4 = 0.4$, and $\beta_2 = -0.2$)

Note: The shaded mesh surface shows how the marginal effect of X on Y varies with Z and W. Darker colors indicate smaller values for the marginal effect of X. The solid colored "zero surface" is equivalent to the "zero line" we used in a 2-D marginal effect plot to indicate if and when the marginal effect of X is 0.

it slopes downward to the left, indicating that the marginal effect of X decreases with W. The surface, while sloped, is perfectly flat, indicating that the positive modifying effect of Z (the slope upward to the right) doesn't change with the value of W and that the negative modifying effect of W (the slope downward to the left) doesn't change with the value of Z. The shaded surface is flat because we've assumed, in line with our theory, that the modifying effects of Z and W are independent of each other. The darker colors in the bottom left corner of the surface show that the positive effect of X on Y is smallest when Z is at its lowest value and W is at its highest value. The lighter colors in the top right corner show that the positive effect of X is largest when Z is at its highest value and W is at its lowest value.[8]

---

[8] Our theory doesn't make a specific prediction about whether the positive effect of X will be greater in the back corner where W and Z are both at their maximum value than in the front corner where W and Z are both at their minimum values. This will depend on whether the predicted positive modifying effect of Z is greater than the

When constructing an *estimated* 3-D marginal effect plot, we can color code or shade the surface to indicate the combinations of values for Z and W at which the marginal effect of X is statistically significant and use the axis label values to evaluate the magnitude of the marginal effect of X. Although it can sometimes make the plot look busy, it's possible to incorporate information about the frequency distributions for the modifying variables. One way to do this is by including a two-dimensional scatterplot that shows the "coordinate" values of the modifying variables for the sample observations on the floor or ceiling of the "cube" housing the 3-D plot. This particular strategy becomes less valuable as the sample size increases because the markers for the individual observations will tend to blend together. Another strategy is to use some kind of contour plot to display the joint density of Z and W on the floor of the "cube" housing the 3-D plot.

In order to fully evaluate hypothesis $H_{X|Z, W}$ and its conditional claim about the effect of X on Y,

$$\frac{\partial Y}{\partial X} = \beta_1 + \beta_4 Z + \beta_5 W, \tag{6.5}$$

we must calculate an appropriate measure of uncertainty for the marginal effect of X. Using the basic properties of variances shown in Appendix A, we calculate the variance of the marginal effect of X as

$$\begin{aligned}
\text{var}\left(\frac{\partial Y}{\partial X}\right) &= \text{var}\left(\beta_1 + \beta_4 Z + \beta_5 W\right) \\
&= \text{var}(\beta_1) + Z^2 \text{var}(\beta_4) + W^2 \text{var}(\beta_5) \\
&\quad + 2Z\text{cov}(\beta_1, \beta_4) + 2W\text{cov}(\beta_1, \beta_5) + 2ZW\text{cov}(\beta_4, \beta_5).
\end{aligned} \tag{6.6}$$

Like the marginal effect of X on Y, the variance depends jointly on the values of Z and W. This means that each marginal effect of X and associated variance that we calculate is for a specific *combination* of values for Z and W. This is important to understand when making claims about when the marginal effect of X is or isn't statistically significant. As usual, we obtain the standard error for the marginal effect of X by taking the square root of the variance,

predicted negative modifying effect of W and the scales on which Z and W are measured. In the marginal effect plot shown in Figure 6.6, we've assumed that the scales for Z and W are the same and that the magnitude of Z's positive modifying effect is greater than that of W's negative modifying effect. As a result, the positive effect of X is depicted to be greater when Z and W are both at their maximum values than when they're both at their minimum values.

$$\text{se}\left(\frac{\partial Y}{\partial X}\right) = \text{se}\,(\beta_1 + \beta_4 Z + \beta_5 W)$$

$$= \Big\{ \text{var}\,(\beta_1) + Z^2 \text{var}\,(\beta_4) + W^2 \text{var}\,(\beta_5)$$

$$+ 2Z\text{cov}\,(\beta_1, \beta_4) + 2W\text{cov}\,(\beta_1, \beta_5) + 2ZW\text{cov}\,(\beta_4, \beta_5) \Big\}^{\frac{1}{2}}.$$

$$(6.7)$$

We can use the standard error to conduct t-tests in the normal way to evaluate claims we might have about the marginal effect of X.[9] We can also use it to construct confidence intervals for the marginal effect of X on Y at given combinations of values for Z and W. For example, a two-sided confidence interval for the effect of X on Y with confidence coefficient $1 - \alpha$ is

$$\beta_1 + \beta_4 Z + \beta_5 W \pm t_{n-k,\alpha/2} \times \text{se}\left(\frac{\partial Y}{\partial X}\right), \qquad (6.8)$$

where $t_{n-k,\alpha/2}$ refers to the critical value from a t-distribution table for a particular α level of significance and $n - k$ degrees of freedom.

Having dealt with the marginal effect of X on Y, we now turn briefly to the marginal effect of Z,

$$\frac{\partial Y}{\partial Z} = \beta_2 + \beta_4 X. \qquad (6.9)$$

So long as $\beta_4 \neq 0$, then the marginal effect of Z depends on the value of X. The effect of a one-unit increase in Z on Y is β_2 when $X = 0$. In other words, the coefficient on the constitutive term Z in Eq. 6.1 doesn't tell us the average, unconditional, independent, or main effect of Z on Y; instead, it tells us the effect of Z only when X is 0. We see that the modifying effect of X on the marginal effect of Z is given by the coefficient on the interaction term XZ, β_4, which, as expected, is identical to the modifying effect of Z on the marginal effect of X. In other words, each one-unit increase in X increases the marginal effect of Z on Y by β_4. The modifying effect of X on the marginal effect of Z doesn't depend on the value of any other variable. This is because the interaction model specified in Eq. 6.1 assumes that the modifying effects of Z, W, and X are all independent.

According to hypothesis $H_{Z|X}$, the marginal effect of Z should be negative when X is at its lowest value and positive when X is at its highest value. For the purposes of our example, we'll assume again that X, Z, and W are all continuous variables whose values run between 0 and 10. Given

---

[9] For those who are unfamiliar with the mathematical notation used in Eq. 6.7, we note that taking the square root of a quantity such as X is the same as raising that quantity to the power of one half; that is, $\sqrt{X} = X^{1/2}$.

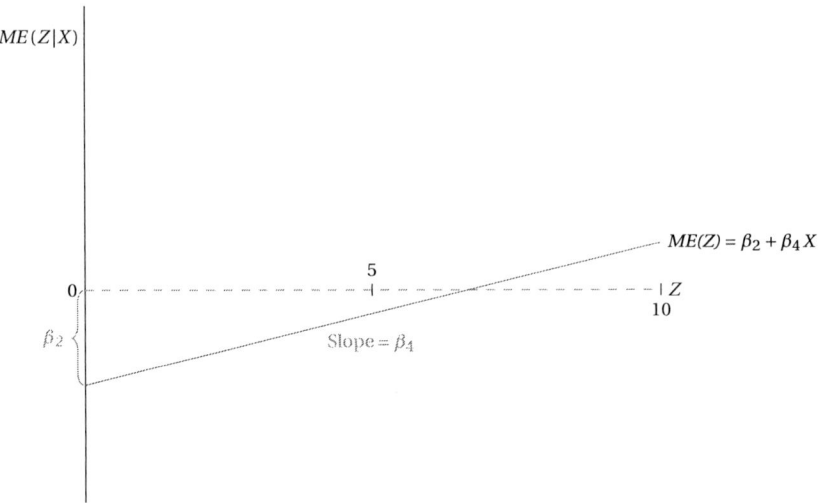

Figure 6.7 A marginal effect plot for Z consistent with hypothesis $H_{Z|X}$

this, hypothesis $H_{Z|X}$ predicts that β_2 should be negative and that $\beta_2 + \beta_4 X$ will become positive once X is sufficiently large. Our theory isn't precise enough to specify exactly when X is sufficiently large for the marginal effect of Z to be positive. Since the marginal effect of Z is expected to switch from negative to positive at some value of X, β_4 is expected to be positive. Given that Z interacts with only X, we can easily visualize the marginal effect of Z with a 2-D marginal effect plot. A plot consistent with hypothesis $H_{Z|X}$ is shown in Figure 6.7.

The variance of the marginal effect of Z on Y is

$$\text{var}\left(\frac{\partial Y}{\partial Z}\right) = \text{var}(\beta_2 + \beta_4 X) = \text{var}(\beta_2) + X^2 \text{var}(\beta_4) + 2X\text{cov}(\beta_2, \beta_4),$$

$$(6.10)$$

and the standard error is

$$\text{se}\left(\frac{\partial Y}{\partial Z}\right) = \text{se}(\beta_2 + \beta_4 X) = \sqrt{\text{var}(\beta_2) + X^2 \text{var}(\beta_4) + 2X\text{cov}(\beta_2, \beta_4)}.$$

$$(6.11)$$

A two-sided confidence interval for the marginal effect of Z with confidence coefficient $1 - \alpha$ is

$$\beta_2 + \beta_4 X \pm t_{n-k, \alpha/2} \times \sqrt{\text{var}(\beta_2) + X^2 \text{var}(\beta_4) + 2X\text{cov}(\beta_2, \beta_4)}.$$

$$(6.12)$$

Finally, we turn to the marginal effect of W on Y,

$$\frac{\partial Y}{\partial W} = \beta_3 + \beta_5 X. \tag{6.13}$$

So long as $\beta_5 \neq 0$, then the marginal effect of W depends on the value of X. The effect of a one-unit increase in W on Y is β_3 when $X = 0$. In other words, the coefficient on the constitutive term W in Eq. 6.1 tells us the effect of W only when X is 0. The modifying effect of X on the marginal effect of W is given by the coefficient on the interaction term XW, β_5, which, as expected, is identical to the modifying effect of W on the marginal effect of X. In other words, each one-unit increase in X increases the marginal effect of W by β_5. The modifying effect of X on the marginal effect of W doesn't depend on the value of any other variable. Again, this is because the interaction model shown in Eq. 6.1 assumes that the modifying effects of Z, W, and X are all independent.

According to hypothesis $H_{W|X}$, the marginal effect of W should be positive when X is at its lowest value and 0 when X is at its highest value. As a result, β_3 should be positive and $\beta_3 + \beta_5 X$ should be 0 when X is at its maximum observed value of 10. Since the marginal effect of W declines with higher values of X, β_5 should be negative. Given that W interacts with only X, we can visualize the marginal effect of W with a 2-D marginal effect plot for W. A plot consistent with hypothesis $H_{W|X}$ is shown in Figure 6.8.

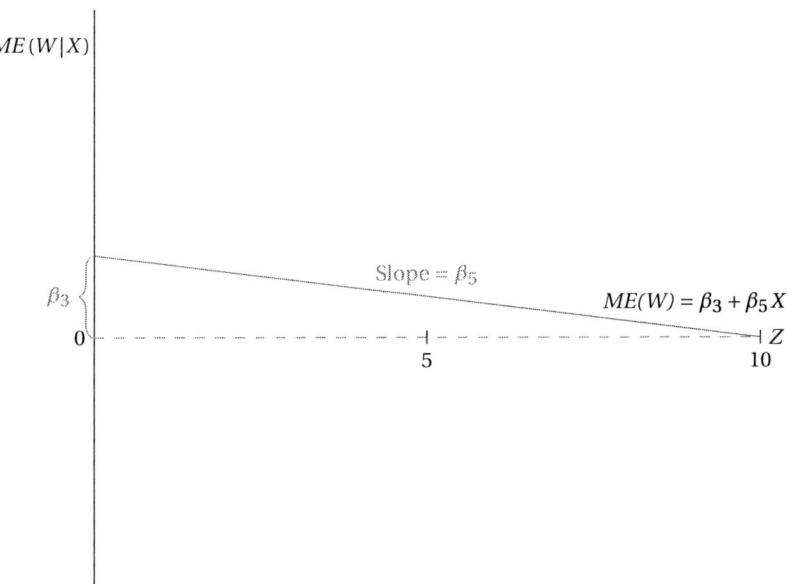

Figure 6.8 A marginal effect plot for W consistent with hypothesis $H_{W|X}$

The variance of the marginal effect of W on Y is

$$\text{var}\left(\frac{\partial Y}{\partial W}\right) = \text{var}\left(\beta_3 + \beta_5 X\right) = \text{var}\left(\beta_3\right) + X^2 \text{var}\left(\beta_5\right) + 2X\text{cov}\left(\beta_3, \beta_5\right),$$

(6.14)

and the standard error is

$$\text{se}\left(\frac{\partial Y}{\partial W}\right) = \text{se}\left(\beta_3 + \beta_5 X\right) = \sqrt{\text{var}\left(\beta_3\right) + X^2 \text{var}\left(\beta_5\right) + 2X\text{cov}\left(\beta_3, \beta_5\right)}.$$

(6.15)

A two-sided confidence interval for the marginal effect of W with confidence coefficient $1 - \alpha$ is

$$\beta_3 + \beta_5 X \pm t_{n-k,\alpha/2} \times \sqrt{\text{var}\left(\beta_3\right) + X^2 \text{var}\left(\beta_5\right) + 2X\text{cov}\left(\beta_3, \beta_5\right)}.$$

(6.16)

⚠ **Important:** Consider the following linear-interactive model:

$$Y = \beta_0 + \beta_1 X + \beta_2 Z + \beta_3 W + \beta_4 XZ + \beta_5 XW + \epsilon.$$

The marginal effect of X on Y is

$$\frac{\partial Y}{\partial X} = \beta_1 + \beta_4 Z + \beta_5 W,$$

and its estimated standard error is

$$\text{se}\left(\frac{\partial Y}{\partial X}\right) = \left\{\text{var}\left(\beta_1\right) + Z^2 \text{var}\left(\beta_4\right) + W^2 \text{var}\left(\beta_5\right)\right.$$
$$\left. + 2Z\text{cov}\left(\beta_1, \beta_4\right) + 2W\text{cov}\left(\beta_1, \beta_5\right) + 2ZW\text{cov}\left(\beta_4, \beta_5\right)\right\}^{\frac{1}{2}}.$$

The marginal effect of Z on Y is

$$\frac{\partial Y}{\partial Z} = \beta_2 + \beta_4 X,$$

and its estimated standard error is

$$\text{se}\left(\frac{\partial Y}{\partial Z}\right) = \sqrt{\text{var}\left(\beta_2\right) + X^2 \text{var}\left(\beta_4\right) + 2X\text{cov}\left(\beta_2, \beta_4\right)}.$$

The marginal effect of W on Y is

$$\frac{\partial Y}{\partial W} = \beta_3 + \beta_5 X,$$

and its estimated standard error is

$$\text{se}\left(\frac{\partial Y}{\partial W}\right) = \sqrt{\text{var}\left(\beta_3\right) + X^2 \text{var}\left(\beta_5\right) + 2X\text{cov}\left(\beta_3, \beta_5\right)}.$$

6.1.1 Substantive Application: Gender, Education, Age, and Support for Feminism

In this substantive application, we look at how gender interacts with educational attainment and age to determine support for feminism in the United States.

In terms of gender, women are expected to exhibit more support for feminism than men. This is because women, given the nature of gender discrimination in many domains, are expected to benefit more than men from attempts to guarantee equal rights and opportunities between the sexes (Davis and Robinson, 1991; Bolzendahl and Myers, 2004). Educational attainment and, in particular, obtaining a university education is expected to be associated with stronger support for feminism. One reason for this is that exposure to new ideas and ways of thinking that challenge traditional gender norms grows as people move through the education system, leading many to develop a sense of feminist consciousness (Davis and Robinson, 1991; Rhodebeck, 1996). Educational attainment may also matter because it increases the likelihood that women participate in the labor force, which in turn is expected to increase their support for feminism because, among other things, they are then exposed to more non-traditional experiences and they get to see types of discrimination that are typically hidden from homemakers (Huber and Spitze, 1983; Klein, 1984; Plutzer, 1988; Rhodebeck, 1996; Pampel, 2011). Thus, we expect educational attainment to have both a direct and indirect positive effect on support for feminism.

There are reasons to think that gender and educational attainment have an interactive effect on support for feminism. In particular, the positive effect of educational attainment on support for feminism is likely to be greater (and may only exist) among women. From an interest-based perspective, any positive effect of exposure to new ideas that challenge traditional gender norms should be larger for women because they're likely to perceive themselves to be the primary beneficiaries of increased equality between the sexes. From an exposure-based perspective, we'd expect men to have less educational exposure than women to ideas and ways of thinking that challenge traditional gender relations due to their self-selection into different types of classes. Men, for example, are, for a variety of reasons, much less likely than women to self-select into classes dealing with women, gender, and sexuality issues (Alilunas, 2011). Due to the symmetry of interactions, our claim that the positive effect of educational attainment is larger for women than men necessarily means that the expected positive effect of being female on support for feminism should grow with educational attainment. In other words, the positive gender gap between women and men when it comes to support for feminism should be greater, all else equal, among the highly educated than among the less educated.

There are reasons to think that age will also interact with gender to affect support for feminism.[10] We might expect that older individuals will simply hold more conservative attitudes and that, as a result, support for feminism will decline with age.[11] Indeed, several studies have found that support for various "feminist" policies tends to decline with age (Plutzer, 1988; Banaszak and Plutzer, 1993; Bolzendahl and Myers, 2004). However, it's perhaps important here to make a conceptual distinction between *feminist identity*, which speaks to a sense of psychological closeness or group association with feminism, and *feminist opinion*, which speaks to the expression of particular policy preferences associated with the goals of feminism (Rhodebeck, 1996). While much of the existing literature addresses feminist opinion, our upcoming analysis focuses instead on feminist identity.

When it comes to women, we expect age to be negatively associated with support for feminism. One reason for this is that feminism, both as a scholarly and activist pursuit, often focuses on the experiences of young and, to some extent, middle-aged women rather than older women (Calasanti, Slevin and King, 2006, 14). This is reflected in survey data, which shows that older women are much less likely to think that the feminist movement is "empowering" or "focused on changes that they want" than younger women and are twice as likely to think that feminism is "outdated" (Cai and Clement, 2016).

When it comes to men, we expect age to be positively associated with support for feminism. From an interest-based perspective, this is because the magnitude of any male privilege arguably shrinks with old age. Power, and with it male privilege, is often tied to economic activity. While women tend to be more economically dependent on men and the state in their younger years, men become increasingly dependent on the state in

[10] One of the central goals of intersectionality research is to examine how structures of power, or axes of inequality, interact to create and perpetuate inequalities between different groups (Weldon, 2006; Cho, Crenshaw and McCall, 2013; Tomlinson, 2013). The vast majority of existing research focuses on interactions between the "holy trinity" (Davis and Zarkov, 2017, 319) or "triptych" (Beckwith and Baldez, 2007) of gender, race, and class, and increasingly sexuality. In contrast, relatively little attention has been paid to how age acts as a social organizing principle that overlaps with other axes of power such as gender (Calasanti and Slevin, 2006). On the whole, "feminists have given little thought to how age might influence the ways that women and men might do gender" (Calasanti, Slevin and King, 2006, 26).

[11] The cross-sectional nature of the data we employ in the upcoming empirical analysis means that we can't technically distinguish between *age effects* and *cohort effects*. Given the pedagogical purposes of this application, we put this issue to one side in what follows.

their old age. Although men continue to enjoy many privileges in their old age, this reliance on the state and redistributive policies often means they "end up in a position regarded as unmanly" (Calasanti and Slevin, 2001; Calasanti, Slevin and King, 2006, 23). Evidence also suggests that men enjoy smaller social and support networks than women, something that becomes increasingly important in old age (Ajrouch, Blandon and Antonucci, 2005). Men also tend to take on more care-giving duties in old age than when they were younger (Kramer and Thompson, 2002), with some studies suggesting that spousal caregivers in old age exhibit few significant gender differences with respect to things like time spent providing care (Thompson, 2000). It may also be the case that the increased wealth that often comes with age means that it's less costly for older men to give up their male privilege than it is for younger men. We suspect that these types of changes related to the ageing process, many of which challenge traditional gender norms and stereotypes of masculinity, may lead to increased feminist consciousness and support for feminism among men.

Due to the symmetry of interactions, it follows from our discussion about how the effect of age may vary with gender that the expected positive effect of being female on support for feminism should decrease with age. In other words, the positive gender gap between women and men when it comes to support for feminism should be greater, all else equal, among the young than the old.

Our theoretical discussion leads to the following three hypotheses:

Female Hypothesis: Women always exhibit more support for feminism than men. The size of this positive gender gap increases with educational attainment but decreases with age.

Education Hypothesis: Increased educational attainment has a positive effect on support for feminism among women. The magnitude of this positive effect is lower (and may be zero) among men.

Age Hypothesis: Increased age has a negative effect on support for feminism among women but a positive effect among men.

Together, these hypotheses contain all ten of the key predictions that we recommend for a theory positing interaction like the one presented here. The *Female Hypothesis* contains a prediction about the sign of the effect of being female for all possible value combinations of educational attainment and age, as well as a prediction about how the effect of being female varies with educational attainment and age. The *Education Hypothesis* contains a prediction about the sign of the effect of educational attainment for all values of gender and a prediction about how the effect of educational attainment varies with gender. Finally, the *Age Hypothesis* contains a prediction about the sign of the effect of age for all values of gender and a prediction about how the effect of age varies with gender.

We use data from the 2019 release of the American National Election Studies 2016 Time Series Study to test our hypotheses (American National Election Studies, 2019). The dependent variable, *Feminism*, is designed to capture feminist consciousness or feminist identity and is based on a "feeling thermometer" survey question in which respondents are asked to indicate how they feel about feminists on a 0–100 scale, where 0 is a very negative (very cold) feeling, 50 is neutral or no feeling at all (neither warm nor cold), and 100 is a very positive (very warm) feeling (Rhodebeck, 1996, 390). *Feminism* has a mean of 56.06 and a standard deviation of 26.00. In terms of our key independent variables, *Female* is a dichotomous variable that equals 1 if an individual self-identifies as female and 0 if they self-identify as male. *College Educated* is a dichotomous variable that equals 1 if an individual has completed a Bachelor's degree or higher and 0 otherwise. We focus on college education because we believe that university is the place in the education system where individuals are most likely to be exposed to ideas and ways of thinking that challenge traditional gender norms and because a college education is often key to accessing important sectors of the labor market. *Age* is a continuous variable that equals an individual's age in years. *Female×College Educated* and *Female×Age* are interaction terms created by multiplying the constitutive term *Female* with either *College Educated* or *Age*. We include control variables for an individual's race, their level of religiosity, and their left-right ideological position.[12] In terms of race, we use a series of dichotomous variables for White (omitted reference category), Black, Asian, Native American, Hispanic, and other. In terms of religiosity, we include a dichotomous variable indicating whether religion is an important part of a respondent's life (1) or not (0). For ideological position, we use a seven-point liberal-conservative scale that runs from one to seven, where higher numbers indicate a more conservative position.[13]

[12] Our upcoming inferences regarding the interactive effects of gender, educational attainment, and age on support for feminism are qualitatively similar if we also control for an individual's income, as well as their marital and employment status. None of these additional controls have a statistically significant effect on *Feminism*.

[13] While early research suggested that support for feminism develops differently for men and women (Klein, 1984), later research has found few significant gender differences in the development of feminist consciousness (Bolzendahl and Myers, 2004). As part of our analysis, we have chosen to examine whether the effects of educational attainment and age vary be gender. We could easily extend our analysis to examine other possible gender differences in the effects of any number of variables by including additional interactions between the chosen control variables and *Female*. While we could also examine other possible gender differences by splitting our sample into sub-samples of female and male respondents, we refer the reader back to our discussion in Chapter 3 where we suggest that this type of strategy is

In terms of our analysis, we treat the dependent variable as continuous and estimate an ordinary least squares regression with the following "standard" interactive model specification:

$$
\begin{aligned}
Feminism = \beta_0 + &\beta_1 Female + \beta_2 College\ Educated + \beta_3 Age \\
&+ \beta_4 Female \times College\ Educated \\
&+ \beta_5 Female \times Age \\
&+ \beta Controls + \epsilon.
\end{aligned}
\tag{6.17}
$$

Our three hypotheses speak to the effect of being female, educational attainment, and age on support for feminism. The effect of being female as opposed to male, or equivalently the gender gap, can be calculated as

$$
\frac{\partial Feminism}{\partial Female} = \beta_1 + \beta_4 College\ Educated + \beta_5 Age.
\tag{6.18}
$$

According to the *Female Hypothesis*, women should always exhibit more support for feminism than men, and, as a result, the quantity shown in Eq. 6.18 should be positive for all possible combinations of observed values for *College Educated* and *Age*. This positive gender gap between women and men when it comes to support for feminism is expected to increase with a college education and decrease with age. It follows, therefore, that β_4 should be positive and that β_5 should be negative. Note that β_1 tells us the effect of being female only for those with no college education *and* no age. Since there's obviously no respondent in our sample with these combined characteristics, this coefficient is not substantively interesting. Nonetheless, our theory does make a specific prediction about the sign of this coefficient. Recall that the effect of being female is expected to be positive when *College Educated* is at its minimum value of 0 and *Age* is at its maximum value, which in our sample is 90. This means that $\beta_1 + \beta_5 \times 90 > 0$ and, thus, that $\beta_1 > -90 \times \beta_5$. Since our theory predicts that β_5 is negative, it follows that β_1 should be positive. From this, we also see that there's a constraint on β_5 beyond our earlier prediction that it should be negative. Specifically, we see that $\frac{-\beta_1}{90} < \beta_5 < 0$.

The effect of being college educated can be calculated as

$$
\frac{\partial Feminism}{\partial College\ Educated} = \beta_2 + \beta_4 Female.
\tag{6.19}
$$

weakly dominated by a pooled interactive approach. For those who are interested, our upcoming inferences regarding the effects of gender, educational attainment, and age on support for feminism are robust to allowing the effects of all of our control variables to vary with gender. The only control variable, including income, marital status, and employment status, that varies significantly with gender is *Native American*.

We see that β_2 indicates the effect of a college education on support for feminism among men (when *Female* = 0) and that $\beta_2 + \beta_4$ indicates the effect of a college education among women (when *Female* = 1). According to the *Education Hypothesis*, obtaining a college education should increase support for feminism among women, meaning that $\beta_2 + \beta_4$ should be positive. The magnitude of this positive effect is expected to be lower and may be zero among men. It follows that β_4 should be negative and that β_2 should be non-negative.

The marginal effect of age is

$$\frac{\partial Feminism}{\partial Age} = \beta_3 + \beta_5 Female. \qquad (6.20)$$

We see that β_3 indicates the effect of an additional year in age on feminism among men (when *Female* = 0) and that $\beta_3 + \beta_5$ indicates the effect of an additional year in age among women (when *Female* = 1). According to the *Age Hypothesis*, support for feminism should increase with age among men and decrease with age among women. This means that β_3 should be positive and that β_5 and $\beta_3 + \beta_5$ should both be negative.

The results from the interactive specification shown in Eq. 6.17 are presented in Table 6.1. We start by discussing what they tell us about the *Female Hypothesis*. As predicted, the coefficient on *Female* is positive and statistically significant. Recall, though, that this coefficient is not substantively interesting as it tells us that the effect of being female as opposed to male on support for feminism is positive only among those with no college education *and* zero age, a scenario that doesn't occur in the real world. The coefficient on *Female×College Educated* is positive and statistically significant. This provides the important evidence of conditionality between gender and educational attainment and indicates, in line with our predictions, that obtaining a college education increases the effect of being female on support for feminism. Put differently, the gender gap between women and men when it comes to supporting feminism is 3.46 units higher among those with a college education than those without a college education. The coefficient on *Female×Age* is negative and statistically significant. This provides the equally important evidence of conditionality between gender and age and indicates, as predicted, that the effect of being female declines with age. To be specific, the gender gap between women and men when it comes to supporting feminism decreases by 0.18 units for each year of age.

Although we've found empirical support for the interaction effects predicted by our theory, we haven't yet ascertained if and when the effect of being female on support for feminism is positive. In Figure 6.9, we show the marginal effect of being female across the observed

Table 6.1 Gender, educational attainment, age, and support for feminism I	
Dependent Variable: Feminism, 0–100	
	Model 1
Female	13.98***
	(2.67)
Education	2.41**
	(1.21)
Age	0.22***
	(0.04)
Female×Education	3.46**
	(1.69)
Female×Age	−0.18***
	(0.05)
Controls	
Black	5.59***
	(1.72)
Asian	4.92**
	(2.49)
Native American	3.96
	(6.36)
Hispanic	3.73**
	(1.51)
Other Race	4.53**
	(2.26)
Religion Important	−2.67***
	(0.97)
Left-Right Ideology	−8.05***
	(0.29)
Constant	74.80***
	(2.18)
Observations	2,710
R^2	0.31

Standard errors in parentheses. $*p < 0.10$; $**p < 0.05$; $***p < 0.01$ (two-tailed)

(a) The Conditional Effect of being Female *among the Non-College Educated*

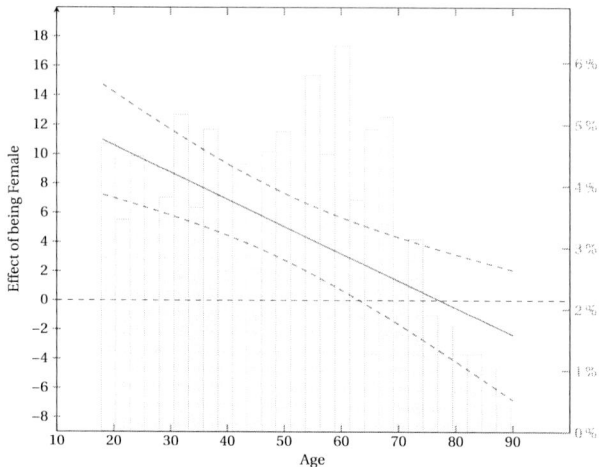

(b) The Conditional Effect of being Female *among the College Educated*

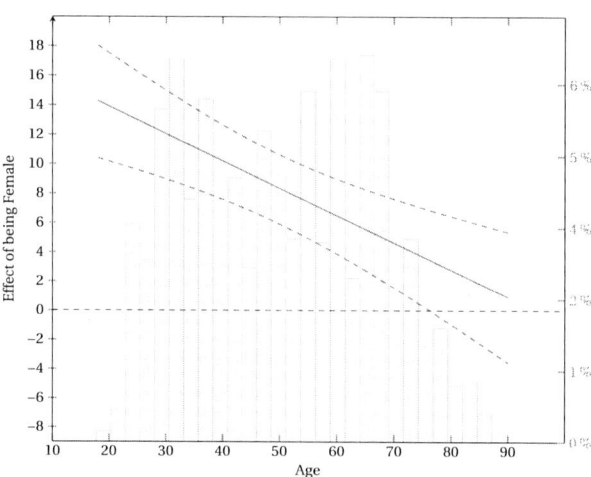

Interaction Effects: *Female×Age*: –0.18 (–0.27, –0.08); *Female×Education*: 3.46 (0.14, 6.78)

Figure 6.9 The conditional effects of gender on support for feminism across age and educational attainment

Note: The two plots show the effect of being female across the observed range of respondent age for those without a college education (a) and for those with a college education (b). The curved dashed lines indicate two-tailed 95% confidence intervals. The histograms show the percentage of respondents across the observed range of *Age* for the two education groups. The two plots are based on the results from the standard interaction model shown in the first column of Table 6.1.

range of age for those without a college education (top panel) and for those with a college education (bottom panel). The marginal effect lines in the two plots have the same negative slope. The negative slope of -0.18 visually shows how the effect of being female, or equivalently the gender gap in support for feminism, declines with age. The slopes in the two plots are identical because we've assumed that the modifying effects of educational attainment and age on the effect of being female are independent of each other. The marginal effect line in the bottom plot is 3.46 units higher than the one in the top plot, illustrating how the effect of being female increases with educational attainment. We report the interaction effects and their two-tailed 95% confidence intervals at the bottom to remind the reader that the negative slope in the two plots and the difference in the heights of the two marginal effect lines are statistically significant.

What do the two plots tell us about if and when the effect of being female on support for feminism is positive? Consider first the scenario where individuals don't have a college education. The marginal effect plot in panel (a) indicates that the effect of being female is positive and statistically significant when *Age* is less than 63.7 (1,926, 76.01%), it's positive and statistically insignificant when *Age* is between 63.6 and 78.5 (478, 18.86%), and it's negative and statistically insignificant when *Age* is greater than 78.4 (130, 5.13%). Figures in parentheses here indicate the number and percentage of non-college educated respondents who fall within each of these regions. Now consider the scenario where individuals do have a college education. The marginal effect plot in panel (b) indicates that the effect of being female is positive and statistically significant when *Age* is less than 77.7 (1,505, 95.19%) and is positive and statistically insignificant when *Age* is greater than 77.6 (130, 4.81%). Figures in parentheses indicate the number and percentage of college educated respondents who fall within each of these regions. In sum, 96.8% of our respondents have an age where the effect of being female on support for feminism is positive, and 83.4% have an age where the effect is positive and statistically significant. There's no evidence that being female ever has a statistically significant negative effect on support for feminism. Given that our theory predicts that the effect of being female declines with age, we're not too concerned that there's a very small percentage of older age respondents where the effect is either positive and statistically insignificant or negative and statistically insignificant. Taken together, the results provide strong support for our *Female Hypothesis*.

What about the substantive significance of these results? The effect of being female rather than male for a 25 year old college graduate is $13.98 + 3.46 - 0.18 \times 25 = 12.98$ [9.58, 16.38]; two-tailed 95% confidence intervals are shown in brackets. In other words, among 25 year old college graduates, women exhibit 12.98 units more support for feminism than men. This gender gap is substantively large and represents a 23.9% increase from the average level of support for feminism exhibited by male college graduates in their twenties (54.2). The effect of being female for a 25 year old with no college education is $13.98 - 0.18 \times 25 = 9.52$ [6.26, 12.78]. The magnitude of this effect is also substantively large and represents a 20.5% increase from the average level of support for feminism exhibited by men without a college education in their twenties (46.4). As we'd expect from the coefficient on the interaction term *Female × College Educated*, the modifying effect of a college education across our hypothetical scenarios (and, indeed, any comparison of hypothetical scenarios) is $12.98 - 9.52 = 3.46$. Given that *Feeling Thermometer Feminism* has a mean of 56.06 and a standard deviation of 26.0, the magnitude of the modifying effect of educational attainment is substantively modest.

To examine the substantive significance of the modifying effect of age, we can change the age of the individual in our hypothetical scenarios. The effect of being female rather than male for a 55 year old college graduate is $13.98 + 3.46 - 0.18 \times 55 = 7.63$ [5.13, 10.13]. The magnitude of this effect is 41.2% smaller than it was when the same hypothetical individual was 25 years old. The effect of being female rather than male for a 55 year old with no college education is $13.98 - 0.18 \times 55 = 4.17$ [1.84, 6.50]. The magnitude of this effect is 56.2% smaller than it was when the same hypothetical individual was 25 years old. These differences across age are substantively meaningful. Indeed, the marginal effect plots in Figure 6.9 clearly indicate that the magnitude of the gender gap when it comes to support for feminism, which is substantively large when people are young, shrinks significantly with age and eventually disappears.

We can present much of this information regarding the substantive significance of the gender gap in support for feminism in a table like the one shown in Figure 6.10. The shaded square shows the marginal effect of being female for our four hypothetical individuals who differ in terms of their age and educational attainment. Positive values indicate that support for feminism is higher if the individuals are female rather than male. The modifying effect of changing the age of our hypothetical individuals from 25 to 55 is shown in the right "Difference" column, while the modifying effect of changing their educational attainment is shown in the bottom "Difference" row. To have a metric for evaluating whether the reported gender gaps are substantively meaningful, we show the mean level of support for feminism among males in each of the four "baseline scenarios" in small gray font in the bottom right corner of each cell. For example, the 54.2 written in gray in the bottom corner of the top left cell indicates the

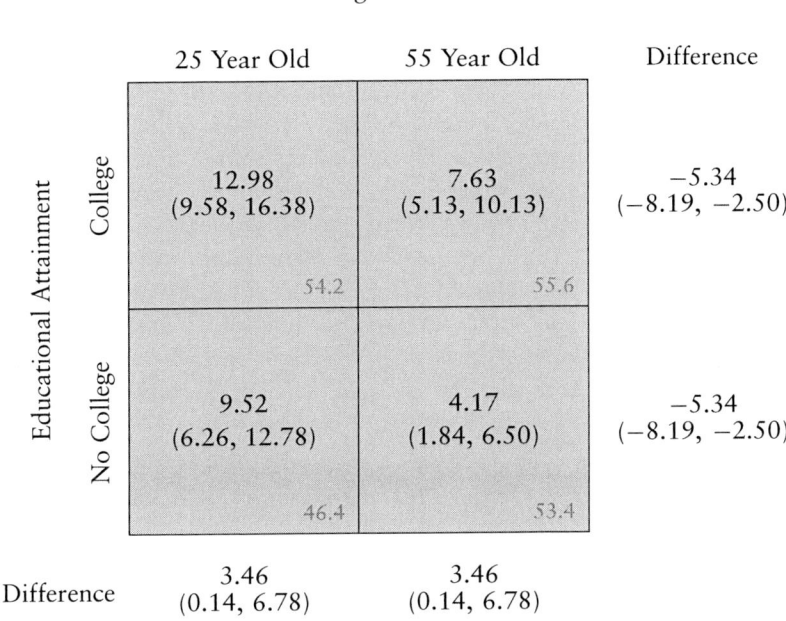

Figure 6.10 Differences in support for feminism between women and men

Note: The shaded square shows the effect of being female on support for feminism along with 95% confidence intervals for four hypothetical individuals who differ in their age and educational attainment. Positive values indicate that support for feminism is higher if the individuals are female rather than male. The modifying effect of changing the age of our hypothetical individuals is shown in the right "Difference" column, while the modifying effect of changing their level of educational attainment is shown in the bottom "Difference" row. The small gray numbers in the bottom right corner of each cell indicate the mean level of support for feminism among males in each of the four "baseline scenarios" used to calculate the effect of being female. The dependent variable, *Feminism*, is measured on a 0–100 scale.

mean level of support for feminism among college-educated males in their twenties.[14]

What about the effect of educational attainment? As predicted, the results in Table 6.1 show that the effect of being college educated for men is positive and statistically significant, 2.41 [0.03, 4.79]. This means that college-educated men exhibit 2.41 units more support for feminism than

---

[14] We realize, in this case, that the baseline scenario technically refers only to a 25 *year old* college educated male. However, there are few respondents for a given age like this, and so we report the means for males (with and without a college education) in their twenties or in their fifties.

similarly situated men who don't have a college education. Given that the mean level of support for feminism among men with no college education is 49.99, we might consider an effect of this size to be substantively small. As predicted, the effect of being college educated for women is also positive and statistically significant, $2.41 + 3.46 = 5.87$ [3.51, 8.23]. This means that college-educated women exhibit 5.87 units more support for feminism than similarly situated women who don't have a college education. The effect of a college education is clearly much larger (2.43 times larger) for women than men. Given that the mean level of support for feminism among women with no college education is 56.04, it's arguable that an effect size of 5.87 could be considered substantively meaningful. This suggests that while a college education has a statistically significant positive effect on support for feminism for both men and women, it has a substantively meaningful effect only for women. This is entirely consistent with our *Education Hypothesis*.

What about the effect of age? As predicted, the results in Table 6.1 show that the effect of an additional year in age for men is positive and statistically significant, 0.22 [0.15, 0.29]. This means that a 55 year old man will exhibit $0.22 \times 30 = 6.67$ units more support for feminism than a similarly situated 25 year old man. Given that the mean level of support for feminism among men in their twenties is 49.02, an effect of this size is likely to be considered substantively meaningful. In contrast, the effect of an additional year in age for women is both substantively small and statistically insignificant, 0.04 [−0.02, 0.11]. This particular result runs counter to our claim that older women will exhibit less support for feminism than similarly situated younger women. These results are consistent with the idea that the age-related reduction in the gender gap with respect to support for feminism that we saw in Figure 6.9 is driven primarily by men becoming more supportive of feminism as they age rather than women becoming less supportive.[15]

Finally, in terms of the control variables, we find that Blacks, Asians, and Hispanics all exhibit significantly higher levels of support for feminism than Whites. Those on the right of the political spectrum and those who consider religion to be an important part of their lives exhibit less support for feminism.

---

[15] Our results here are partially consistent with research by Rhodebeck (1996) showing that age has a positive and statistically significant effect on feminist identity for men but not women. They're also partially consistent with much earlier research by Bernard (1974) suggesting that the gender gap with respect to feminism decreases (and even reverses) with age.

6.2 WHEN THE MODIFYING EFFECTS OF Z AND W ARE DEPENDENT

In the previous section, we examined conditional claims from a theory positing that the modifying effects of Z and W on the marginal effect of X are *independent* or *unconditional*. We now turn our attention to conditional claims from a theory positing that the modifying effects of Z and W are *dependent* or *conditional*. These types of claims predict that the modifying effect of Z on the marginal effect of X depends on the value of W and that the modifying effect of W on the marginal effect of X depends on the value of Z. In this sense, we might say that these types of conditional claims predict that the relationship between X, Z, and W is "fully interactive."

What do theories that produce these types of conditional claims look like? To provide some substantive context up front, we start by presenting a couple of very brief examples of theories in which the impact of three independent variables X, Z, and W on some dependent variable Y is expected to be "fully interactive." Our first example has to do with the connection between elections and macroeconomic policy (Clark, 2003). The core assumption is that incumbent governments want to use fiscal and monetary policy before an election to stimulate the economy in order to improve their chances of reelection. Governments, though, are constrained in their ability to use these policy tools by the institutional environment in which they act. The extent to which the institutional environment constrains the use of fiscal and monetary policy depends, in particular, on three factors: (1) the degree of central bank independence, (2) the extent of capital mobility, and (3) whether the exchange rate regime is fixed or flexible. The degree of central bank independence is important because it determines whether fiscal and monetary policy both remain in the hands of the government or just fiscal policy. The extent of capital mobility and the exchange rate regime are important because, according to the Mundell–Fleming model of macroeconomics (Fleming, 1962; Mundell, 1963), they jointly determine the relative effectiveness of fiscal and monetary policy. According to this theoretical story, the different possible combinations, or "interactions," of these three factors determine the extent to which we'll see monetary or fiscal expansions at election time.

Our second example comes from the literature on intersectionality theory. Perhaps the core goal of intersectionality theory is to highlight how structures of power interact to create and perpetuate inequalities between different groups (Crenshaw, 1989, 1991; Weldon, 2006; Cho, Crenshaw and McCall, 2013; Tomlinson, 2013). In one of our substantive applications in the previous chapter, we looked at how the interaction of two particular axes of inequality related to gender and race influenced support for the Republican Party at the time of the 2016 US presidential elections (Block, Golder and Golder, 2023). However, there's no need to limit our focus to examining the effects of just these two categories of difference. Much of the intersectionality literature, for example, focuses on how the interaction of gender, race, *and* class affects some outcome of interest. The idea here would be that, say, the interactive impact of gender and race on the outcome of interest will depend on the social class to which the individual belongs. Similar to the case with our political economy example, the intersectional perspective adopted here implies that the different possible combinations of values for gender, race, and class are associated with different outcomes.

To better see exactly what's going on in these types of theoretical stories, consider Figure 6.11, which helps to visualize a "fully interactive" relationship between X, Z, and W. To aid our discussion, we've assumed that X, Z, and W are all dichotomous variables and that there are no control variables. To provide some substance, we can think that X captures someone's gender, where 1 indicates a female and 0 indicates a male, that Z captures their race, where 1 indicates Black and 0 indicates White, and that W indicates their class, where 1 indicates "higher" class and 0 indicates "lower" class.[16] The different possible combinations of values for our three dichotomous variables X, Z, and W define $2^3 = 8$ types of distinct observations or categories of individual. Each of the shaded cells in Figure 6.11 corresponds to one of these eight categories and indicates the mean level of the dependent variable Y for that category. We've adopted the convention $\bar{Y}_{x,z,w}$ to indicate the mean level of Y when $X = x$, $Z = z$, and $W = w$. This means, for example, that $\bar{Y}_{1,0,0}$ indicates the mean level of Y when $X = 1$ (Female), $Z = 0$ (White), and $W = 0$ (lower class); that is, the mean level of Y for a lower class White female. We've seen these types of shaded 2×2 "squares" before when we used one to illustrate the interactive impact of gender and race on support for the Republican Party in Figure 5.2. Then, we needed only one of these squares to capture the interactive relationship between gender and race. Here, though, we need two to capture the interactive relationship between gender and race, one for the case when $W = 0$ (lower class) and one for the case where $W = 1$ (higher class). In other words, the scenario we're examining here allows the interactive relationship between gender and race to vary with the value

---

[16] Note that we don't intend to apply any normative judgement when referring to "higher" and "lower" classes. Our labels are simply designed to recognize that most societies are stratified into a hierarchical arrangement of social classes that are determined by a whole host of things such as wealth, income, educational attainment, occupation, social network, and "status" (Cohen et al., 2017; Lindh and McCall, 2020).

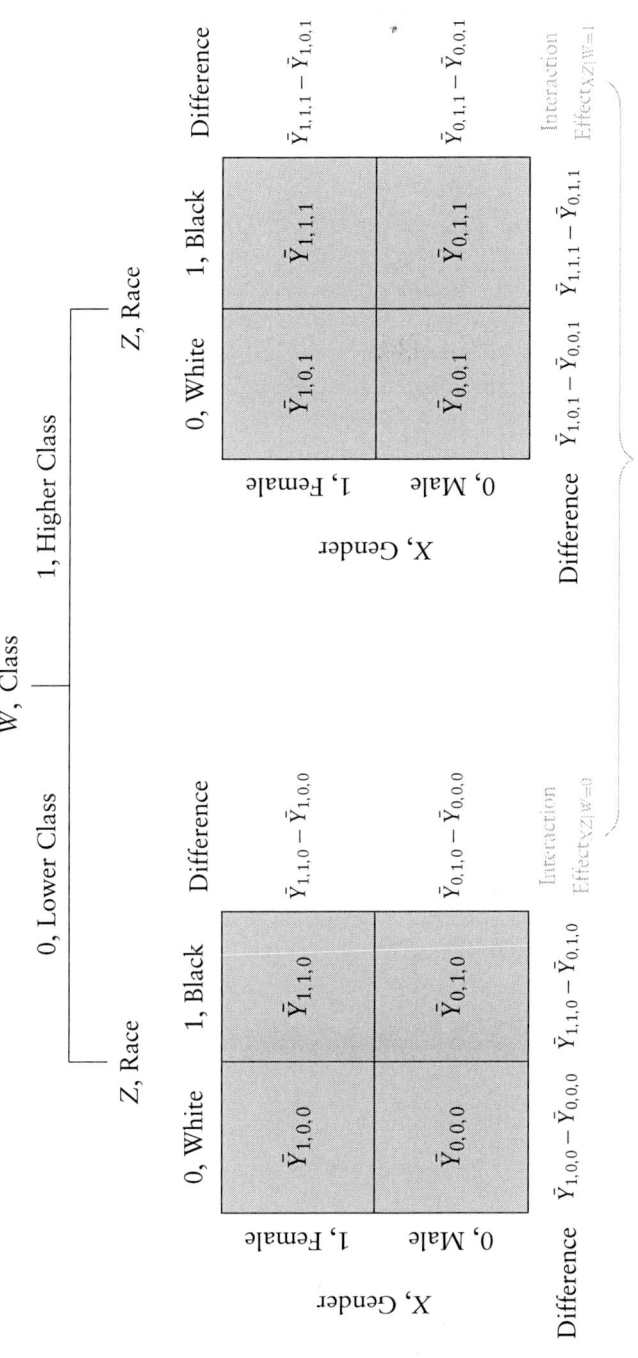

Figure 6.11 Visualizing a fully interactive relationship between X, Z, and W

Note: The cells in the two shaded squares show the predicted value of Y for all eight possible combinations of values for the dichotomous variables X, Z, and W. The subscripts associated with each predicted value of Y indicate the values (High, 1 or Low, 0) for X, Z, and W. The effect of X under the different scenarios is shown in the bottom row, while the effect of Z under the different scenarios is shown in the right columns. The interaction effect between X and Z for each possible value of W is shown in gray in the bottom right cell. The modifying effect of W on the interaction effect between X and Z is shown in gray below the bracket at the bottom.

of a third variable W (*Class*). As we'll see, it also allows the interactive relationship between gender and class to vary with race and the interactive relationship between class and race to vary with gender. Indeed, the way that each of these interactive relationships vary with a third variable will be identical due to the symmetry of interactions.

To see this, let's delve a little deeper. For each 2×2 square, we can identify the effect of X (gender) or Z (race) in various scenarios by calculating differences in our category means. The "Difference" row at the bottom of the left square indicates the effect of being female instead of male among Whites (left cell) and among Blacks (right cell) for the lower class. Put differently, it indicates the differences between women and men among lower-class Whites and among lower-class Blacks. The "Difference" row at the bottom of the right square indicates the same quantities but for the higher class. The "Difference" column to the right of the left square indicates the effect of being Black instead of White among women (top cell) and among men (bottom cell) for the lower class. In other words, it indicates the differences between Blacks and Whites among lower-class women and among lower-class men. The "Difference" column to the right of the right square indicates the same quantities but for the higher class. We can immediately see how this type of setup allows the effect of X (gender) to vary with both Z (race) and W (class) and the effect of Z (race) to vary with both X (gender) and W (class).

For each 2×2 square, we can also identify the "interaction effect" between X (gender) and Z (race) among the lower class (left square) and the higher class (right square). The interaction effects reported in gray toward the bottom right of each square are just differences of differences and tell us how the effect of X (gender) varies by Z (race) *and* how the effect of Z (race) varies by X (gender) for a particular value of W (class). For example, the interaction effect toward the bottom right of the left square tells us both how Z (race) modifies the effect of X (gender) when $W = 0$ (lower class),

$$\text{Interaction Effect}_{XZ|W=0} = \left(\bar{Y}_{1,1,0} - \bar{Y}_{0,1,0}\right) - \left(\bar{Y}_{1,0,0} - \bar{Y}_{0,0,0}\right)$$
$$= \bar{Y}_{1,1,0} - \bar{Y}_{0,1,0} - \bar{Y}_{1,0,0} + \bar{Y}_{0,0,0}, \quad (6.21)$$

and how X (gender) modifies the effect of Z (race) when $W = 0$ (lower class),

$$\text{Interaction Effect}_{XZ|W=0} = \left(\bar{Y}_{1,1,0} - \bar{Y}_{1,0,0}\right) - \left(\bar{Y}_{0,1,0} - \bar{Y}_{0,0,0}\right)$$
$$= \bar{Y}_{1,1,0} - \bar{Y}_{0,1,0} - \bar{Y}_{1,0,0} + \bar{Y}_{0,0,0}. \quad (6.22)$$

The interaction effect towards the bottom right of the right square tells us the same quantities among the higher class.

We can determine whether the modifying effect of Z (race) on the effect of X (gender), or equivalently the modifying effect of X (gender) on the effect of Z (race), is "dependent" or "conditional" on W (class) by seeing whether the interaction effects associated with each of the 2×2 squares, *Interaction Effect*$_{XZ|W=0}$ and *Interaction Effect*$_{XZ|W=1}$, are different. The difference in these interaction effects tells us both how W (class) modifies the modifying effect of Z (race) on the effect of X (gender) and how W (class) modifies the modifying effect of X (gender) on the effect of Z (race). Since it represents a modifying effect on a modifying effect, we can think of this difference as a *second-order* interaction effect,

$$
\begin{aligned}
\textit{Interaction Effect}_{XZW} \\
= \textit{Interaction Effect}_{XZ|W=1} - \textit{Interaction Effect}_{XZ|W=0} \\
= \left(\bar{Y}_{1,1,1} - \bar{Y}_{0,1,1} - \bar{Y}_{1,0,1} + \bar{Y}_{0,0,1} \right) - \left(\bar{Y}_{1,1,0} - \bar{Y}_{0,1,0} - \bar{Y}_{1,0,0} + \bar{Y}_{0,0,0} \right) \\
= \bar{Y}_{1,1,1} - \bar{Y}_{0,1,1} - \bar{Y}_{1,0,1} + \bar{Y}_{0,0,1} - \bar{Y}_{1,1,0} + \bar{Y}_{0,1,0} + \bar{Y}_{1,0,0} - \bar{Y}_{0,0,0}.
\end{aligned}
\tag{6.23}
$$

It's this quantity that indicates whether our empirical results are consistent with a theoretical claim that there's a "fully interactive" relationship between X (gender), Z (race), and W (class). That W modifies both the modifying effect of Z on the effect of X and the modifying effect of X on the effect of Z in exactly the same way arises due to the inherent symmetry of interactions that continues to work at this higher level.

Although it's slightly less easy to see, we can glean more information from Figure 6.11 about the interactive impact of X, Z, and W on Y. Just as we calculated the effect of X (gender) and Z (race) for various scenarios by calculating differences in our category means, we can do the same for the effect of W (class). Rather than calculating differences between cells *within* the same 2×2 square, this requires calculating differences between similarly-situated cells *across* the 2×2 squares. For example, we can calculate the effect of being higher class rather than lower class for White women by taking the difference between the value in the top left cell in the right 2×2 square and the value in the top left cell in the left 2×2 square, $\bar{Y}_{1,0,1} - \bar{Y}_{1,0,0}$. The effect of being higher class rather than lower class for White men is the difference between the value in the bottom left cell in the 2×2 square on the right and the value in the bottom left cell in the 2×2 square on the left, $\bar{Y}_{0,0,1} - \bar{Y}_{0,0,0}$. The difference between these two differences tells us how X (gender) modifies the effect of W (class) when $Z = 0$ (White),

$$
\begin{aligned}
\textit{Interaction Effect}_{XW|Z=0} &= \left(\bar{Y}_{1,0,1} - \bar{Y}_{1,0,0} \right) - \left(\bar{Y}_{0,0,1} - \bar{Y}_{0,0,0} \right) \\
&= \bar{Y}_{1,0,1} - \bar{Y}_{1,0,0} - \bar{Y}_{0,0,1} + \bar{Y}_{0,0,0}.
\end{aligned}
\tag{6.24}
$$

The effect of being higher class rather than lower class for Black women is the difference between the value in the top right cell in the right square and the value in the top right cell in the left square, $\bar{Y}_{1,1,1} - \bar{Y}_{1,1,0}$. The effect of being higher class rather than lower class for Black men is the difference between the value in the bottom right cell in the square on the right and the value in the bottom right cell in the square on the left, $\bar{Y}_{0,1,1} - \bar{Y}_{0,1,0}$. The difference between these two differences tells us how X (gender) modifies the effect on W (class) when $Z = 1$ (Black),

$$\textit{Interaction Effect}_{XW|Z=1} = \left(\bar{Y}_{1,1,1} - \bar{Y}_{1,1,0}\right) - \left(\bar{Y}_{0,1,1} - \bar{Y}_{0,1,0}\right)$$
$$= \bar{Y}_{1,1,1} - \bar{Y}_{1,1,0} - \bar{Y}_{0,1,1} + \bar{Y}_{0,1,0}. \quad (6.25)$$

The difference between these two interaction effects tells us how Z (race) modifies the modifying effect of X (gender) on the effect of W (class), and is therefore a second-order "interaction effect,"

$$\textit{Interaction Effect}_{XZW}$$
$$= \textit{Interaction Effect}_{XW|Z=1} - \textit{Interaction Effect}_{XW|Z=0}$$
$$= \left(\bar{Y}_{1,1,1} - \bar{Y}_{1,1,0} - \bar{Y}_{0,1,1} + \bar{Y}_{0,1,0}\right) - \left(\bar{Y}_{1,0,1} - \bar{Y}_{1,0,0} - \bar{Y}_{0,0,1} + \bar{Y}_{0,0,0}\right)$$
$$= \bar{Y}_{1,1,1} - \bar{Y}_{1,1,0} - \bar{Y}_{0,1,1} + \bar{Y}_{0,1,0} - \bar{Y}_{1,0,1} + \bar{Y}_{1,0,0} + \bar{Y}_{0,0,1} - \bar{Y}_{0,0,0}$$
$$= \bar{Y}_{1,1,1} - \bar{Y}_{0,1,1} - \bar{Y}_{1,0,1} + \bar{Y}_{0,0,1} - \bar{Y}_{1,1,0} + \bar{Y}_{0,1,0} + \bar{Y}_{1,0,0} - \bar{Y}_{0,0,0}. \quad (6.26)$$

As expected, the inherent symmetry of interactions means that this interaction effect is identical to the second-order interaction effect shown in Eq. 6.23.

⚠ **Important:** Theories positing a fully interactive relationship between X, Z, and W are typically strong enough to generate nine key predictions about the marginal effect of X and how this varies with Z and W. These predictions speak to the direction of the second-order interaction between X, Z, and W, the direction of the interaction between X and Z at different values of W, the direction of the interaction between X and W at different values of Z, and the direction of the marginal effect of X at different combinations of values for Z and W.

It's perhaps useful at this point to take stock of the information we've been able to glean from looking at Figure 6.11. This is because it throws light on the possible predictions we can make from a "fully interactive" theory involving X, Z, and W. Let's start by looking at what a hypothesis about the effect of X (gender) on some dependent variable Y might say. Our discussion has shown us that we're able to identify (1) the effect of

X (gender) for all possible combinations of values for Z (race) and W (class), (2) the interaction effect between X (gender) and Z (race) for all possible values of W (class), (3) the interaction effect between X (gender) and W (class) for all possible values of Z (race), and (4) the second-order interaction effect between X (gender), Z (race), and W (class). In the specific case shown in Figure 6.11, where X, Z, and W are all dichotomous, this amounts to nine distinct effects. Different combinations of signs for these nine predictions correspond to different possible claims about the conditional effect of X (gender) on Y. Only by deriving and testing all nine of these predictions is it possible to know whether the data support our particular claim about the conditional effect of X (gender) on Y or some other claim. In the case where X, Z, and W aren't all dichotomous, scholars should still make nine predictions: P_{XZW}, $P_{XZ|W_{min}}$, $P_{XZ|W_{max}}$, $P_{XW|Z_{min}}$, $P_{XW|Z_{max}}$, $P_{X|Z_{min},W_{min}}$, $P_{X|Z_{min},W_{max}}$, $P_{X|Z_{max},W_{min}}$, and $P_{X|Z_{max},W_{max}}$. All nine of these predictions can easily be captured in a single hypothesis about the marginal effect of X and how it varies with Z and W.

Equivalent predictions can be made for the effects of Z and W on Y. In line with our previous advice, we encourage scholars, *where their theory allows*, to make predictions about the signs of these same effects when making hypotheses about the effects of Z and W on Y. Note that due to the symmetry of interactions, some of the predictions regarding interaction effects will be common across the claims about the effects of X, Z, and W. As a result, the total number of effects about which we can make predictions from the interactive research design in Figure 6.11 is 19 and not $3 \times 9 = 27$.[17] The ten additional predictions not contained in a hypothesis about the marginal effect of X are $P_{WZ|X_{min}}$, $P_{WZ|X_{max}}$, $P_{Z|X_{min},W_{min}}$, $P_{Z|X_{min},W_{max}}$, $P_{Z|X_{max},W_{min}}$, $P_{Z|X_{max},W_{max}}$, $P_{W|X_{min},Z_{max}}$, $P_{W|X_{min},Z_{max}}$, $P_{W|X_{max},Z_{min}}$, and $P_{W|X_{max},Z_{max}}$. Different signs for our nineteen key predictions (positive, negative, zero) lead to thousands of theoretically possible ways in which X, Z, and W interact to affect some outcome of interest. Only by making all nineteen of these key predictions can scholars know whether the data support their particular theory of interaction between X, Z, and W as opposed to one of the other thousands of possible interactive relationships.

⚠ **Important:** Theories positing a fully interactive relationship between X, Z, and W are typically strong enough to generate nineteen key predictions. These predictions speak to the direction of the marginal

---

[17] Note that we can actually make more than nineteen predictions if our theory is strong enough to also make predictions about the relative sizes of the modifying effects of X, Z, and W.

effect of X and how this varies with Z and W, the direction of the marginal effect of Z and how this varies with X and W, and the direction of the marginal effect of W and how this varies with X and Z.

As a hypothetical example, let's return to earlier in the chapter where we assumed that we had a theory predicting that an increase in X always has a positive effect on Y but that the magnitude of this positive effect is expected to increase with higher values of Z and decrease with higher values of W. As it stands, our theory predicts that the modifying effects of Z and W on the effect of X are unconditional or independent. Suppose now, though, that our theory instead predicts that the modifying effects of Z and W are conditional or dependent. Specifically, we'll assume that our theory predicts that the modifying effect of Z is always positive but declines in magnitude with higher values of W and that the modifying effect of W is positive when Z is low but negative when Z is high. We can capture these predictions in a single hypothesis about the marginal effect of X on Y and how this changes with Z and W:

$H_{X|Z,W}$: The effect of an increase in X on Y is always positive. The magnitude of X's positive effect increases with Z but less so when W is high. The magnitude of X's positive effect increases with W when Z is low but decreases with W when Z is high.

As recommended, hypothesis $H_{X|Z,W}$ contains all nine of the predictions we recommend for a hypothesis about the conditional effect of X from a theory positing a fully interactive relationship between X, Z, and W. It provides predictions about the sign of the marginal effect of X on Y for all possible value combinations of Z and W, predictions about the sign of the modifying effect of Z at the minimum and maximum values of W, predictions about the sign of the modifying effect of W at the minimum and maximum values of Z, and a prediction about the second-order interaction effect indicating both how the modifying effect of Z varies with W and how the modifying effect of W varies with Z. Other than not also making a prediction about the relative size of the modifying effects of Z and W, our hypothesis provides as complete a description of the expected marginal effect of X on Y as we could offer without providing specific magnitudes for the marginal effect at particular combinations of values for Z and W.

Our claim that the marginal effect of X on Y varies with Z and W necessarily implies that the marginal effects of Z and W on Y each vary with X. Simply by stating hypothesis $H_{X|Z,W}$, and therefore without any additional theorizing, we automatically introduce some predictions about the effects of Z and W. To be specific, hypothesis $H_{X|Z,W}$ implies that the effect of Z on Y always increases with X but less so when W is high and that the effect of W on Y increases with X when Z is low but less so when Z is high. As things stand, we don't have predictions as to whether the marginal effect of Z is positive, negative, or zero for any combination of values for X and W, we don't have predictions as to whether the marginal effect of W is positive, negative, or zero for any combination of values for X and W, and we don't have predictions about the sign of the modifying effect of W on the marginal effect of Z for any value of X or equivalently the sign of the modifying effect of Z on the marginal effect of W for any value of X. As we've indicated previously, though, further examination of the underlying theory that led to a hypothesis like $H_{X|Z,W}$ can often allow us to make these missing predictions. This additional theorizing is valuable as it further narrows the range of conditional relationships that are consistent with the underlying theory, thereby increasing the strength of any empirical test. In this regard, let's suppose that the underlying theory that produced hypothesis $H_{X|Z,W}$ also produces the following hypotheses about the marginal effects of Z and W:

$H_{Z|X,W}$: The effect of an increase in Z on Y is always negative. The magnitude of Z's negative effect decreases with X but less so when W is high. The magnitude of Z's negative effect increases with W and increasingly so with higher values of X.

$H_{W|X,Z}$: The effect of an increase in W is zero when X and Z are both at their minimum value, positive when X is at its maximum value and Z is at its minimum value, negative when X is at its minimum value and Z is at its maximum value, and negative when X and Z are both at their maximum values. The modifying effect of X is positive when Z is at its minimum value but negative when Z is at its maximum value. The modifying effect of Z is always negative and increasingly so with higher values of X.

These hypotheses are clearly more complicated than the ones we've seen previously. However, they are no more complicated than necessary to convey all of the predictions needed to distinguish the hypothesized interactive relationship between X, Z, and W from all of the possible alternative interactive relationships we might find in the data.

We can test all three of our hypotheses with the following interaction model:

$$Y = \beta_0 + \beta_1 X + \beta_2 Z + \beta_3 W + \beta_4 XZ + \beta_5 XW + \beta_6 ZW + \beta_7 XZW + \epsilon. \tag{6.27}$$

For this example, we'll assume that the variables X, Z, and W are all continuous unless otherwise stated. As always with a "standard" setup like this, we should include all of the constitutive elements X, Z, W, XZ, XW, and ZW of the various interaction terms. To see why this interaction model is appropriate for testing our hypotheses, it's helpful to consider the marginal effects of our variables.

We start with the marginal effect of X on Y,

$$\frac{\partial Y}{\partial X} = \beta_1 + \beta_4 Z + \beta_5 W + \beta_7 Z W. \tag{6.28}$$

From this, we see that the marginal effect of X is allowed to depend on the values of Z and W. Like when we treated the modifying effects of Z and W as independent or unconditional, the effect of a one-unit increase in X on Y is β_1 only when Z and W are both 0. In other words, the coefficient on X in Eq. 6.27 doesn't tell us the average or unconditional effect of X; instead, it tells us the marginal effect of X only when Z and W are both 0. Unlike when we were assuming that that modifying effects of Z and W were independent, though, we see that the marginal effect of X depends not only on the values of Z and W but also on the value of ZW. As we'll see, it's this that allows the modifying effects of Z and W to be *dependent* or *conditional*. To get a sense of how the marginal effect of X changes with the values of Z and W, note that the marginal effect of X is β_1 when $Z = W = 0$, it's $\beta_1 + \beta_4$ when $Z = 1$ and $W = 0$, it's $\beta_1 + \beta_4 + \beta_5 + \beta_7$ when $Z = 1$ and $W = 1$, its $\beta_1 + 2\beta_4 + \beta_5 + 2\beta_7$ when $Z = 2$ and $W = 1$, it's $\beta_1 + 2\beta_4 + 2\beta_5 + 4\beta_7$ when $Z = 2$ and $W = 2$, and so on.

The modifying effect of Z on the marginal effect of X is

$$\frac{\partial \left(\frac{\partial Y}{\partial X} \right)}{\partial Z} = \frac{\partial (\beta_1 + \beta_4 Z + \beta_5 W + \beta_7 Z W)}{\partial Z} = \beta_4 + \beta_7 W. \tag{6.29}$$

This is the "interaction effect" between X and Z and tells us how the marginal effect of X on Y changes with Z.[18] We immediately see that so long as $\beta_7 \neq 0$, there is no single interaction effect between X and Z and that the modifying effect of Z on the marginal effect of X *depends on*, or is *conditional on*, the value of W. Importantly, we see that the coefficient on XZ, β_4, tells us the interaction effect between X and Z only when $W = 0$; it doesn't tell us the average or unconditional interaction effect between X and Z. Put differently, we see that the slope of the linear relationship between the marginal effect of X and Z varies with the value of W. Specifically, the slope of the relationship between the marginal effect of X and Z increases by β_7 for each unit increase in W. It follows from this that a simple t-test on β_7 can be used to determine whether the conditional relationship between X and Z depends on the value of W. We can see this more explicitly by taking the derivative of the modifying effect of Z with respect to W,

---

[18] Due to the symmetry of interactions, it also tells us how the marginal effect of Z changes with X.

$$\frac{\partial \left(\frac{\partial \left(\frac{\partial Y}{\partial X} \right)}{\partial Z} \right)}{\partial W} = \frac{\partial \left(\beta_4 + \beta_7 W \right)}{\partial W} = \beta_7. \tag{6.30}$$

This is a "second-order" interaction effect in that it describes the *modifying effect* of W on the *modifying effect* of Z on the marginal effect of X. It's this quantity that indicates whether our empirical results are consistent with a theoretical claim of "full interaction" between X, Z, and W.

⚠ **Important:** In an interaction model with variables X, Z, W, XZ, XW, ZW, and XZW, the coefficient on XZW represents a "second-order" interaction effect. It's this quantity that indicates whether the empirical results are consistent with a theoretical claim of "full interaction" between X, Z, and W.

The modifying effect of W on the marginal effect of X is

$$\frac{\partial \left(\frac{\partial Y}{\partial X} \right)}{\partial W} = \frac{\partial \left(\beta_1 + \beta_4 Z + \beta_5 W + \beta_7 ZW \right)}{\partial W} = \beta_5 + \beta_7 Z. \tag{6.31}$$

This is the "interaction effect" between X and W and tells us how the marginal effect of X on Y changes with W.[19] So long as $\beta_7 \neq 0$, we see that there's no single interaction effect between X and W and that the modifying effect of W on the marginal effect of X depends on the value of Z. Significantly, we see that the coefficient on XW, β_5, tells us the interaction effect between X and W only when $Z = 0$; it doesn't tell us the average or unconditional interaction effect between X and W. We also see that just as we can use a t-test on β_7 to determine whether the modifying effect of Z on the marginal effect of X depends on the value of W, we can use the same t-test to determine whether the modifying effect of W on the marginal effect of X depends on the value of Z. This follows from our earlier discussion of Figure 6.11 where we pointed out that the "second-order" interaction effects from a fully-interactive model are all identical. We can see this explicitly with respect to the modifying effect of Z on the modifying effect of W,

$$\frac{\partial \left(\frac{\partial \left(\frac{\partial Y}{\partial X} \right)}{\partial W} \right)}{\partial Z} = \frac{\partial \left(\beta_5 + \beta_7 Z \right)}{\partial Z} = \beta_7, \tag{6.32}$$

---

[19] Due to the symmetry of interactions, it also tells us how the marginal effect of W changes with X.

which is identical to the modifying effect of W on the modifying effect of Z on the marginal effect of X in Eq. 6.30.

⚠ **Important:** In an interaction model with variables X, Z, W, XZ, XW, ZW, and XZW, the coefficients on XZ, XW, and ZW do not represent unconditional or average interaction effects. Instead, these coefficients indicate the specified interaction effects only when the value of the third modifying variable is 0.

According to hypothesis $H_{X|Z,W}$, the marginal effect of X should always be positive. As a result, $\beta_1 + \beta_4 Z + \beta_5 W + \beta_7 ZW$ should be positive for all observed combinations of values for Z and W. It follows that β_1 should be positive. Hypothesis $H_{X|Z,W}$ also states that the magnitude of X's positive effect is expected to increase with Z but less so when W is high. Several predictions follow from this. The modifying effect of Z, $\beta_4 + \beta_7 W$, should be positive for all values of W. This, in turn, means that β_4 should be positive. Since the positive modifying effect of Z is expected to decline with higher values of W, β_7 should be negative. However, there is a constraint on how large the negative coefficient β_7 can be given that the modifying effect of Z should always be positive. Specifically, it must be the case that $\beta_4 + \beta_7 W_{max} > 0$. Solving for β_7 and combining with our prediction that β_4 is positive, we have $\frac{-\beta_4}{W_{max}} < \beta_7 < 0$. Finally, hypothesis $H_{X|Z,W}$ also states that the magnitude of X's positive effect is expected to increase with W when Z is low but decrease with W when Z is high. This part of the hypothesis speaks to the modifying effect of W, $\beta_5 + \beta_7 Z$. In what follows, we'll assume that the minimum value of Z is 0. Given that the modifying effect of W should be positive when Z is low, it follows that β_5 should be positive. Since we expect the modifying effect of W to be negative once Z is sufficiently large, β_7 and $\beta_5 + \beta_7 Z_{max}$ should both be negative. Given that β_5 is expected to be positive, $\beta_5 + \beta_7 Z_{max}$ will only be negative if the negative coefficient β_7 is sufficiently large, $\beta_7 < \frac{-\beta_5}{Z_{max}}$. This means that our prediction for the coefficient on β_7 is bounded such that $\frac{-\beta_4}{W_{max}} < \beta_7 < \frac{-\beta_5}{Z_{max}} < 0$.

As with the case where the modifying effects of Z and W are independent, it's not always possible to fully visualize the marginal effect of X with a 2-D marginal effect plot when the modifying effects of Z and W are dependent. Consider a 2-D marginal effect plot for X across the observed range of Z. As in the case where the modifying effects of Z and W are independent, Eq. 6.28 shows that the intercept of the marginal effect line $\beta_1 + \beta_5 W$ depends on the value of W. In other words, to draw a marginal effect line for X across the observed range of Z, we must choose a particular value of W. Note, though, that unlike the case where the

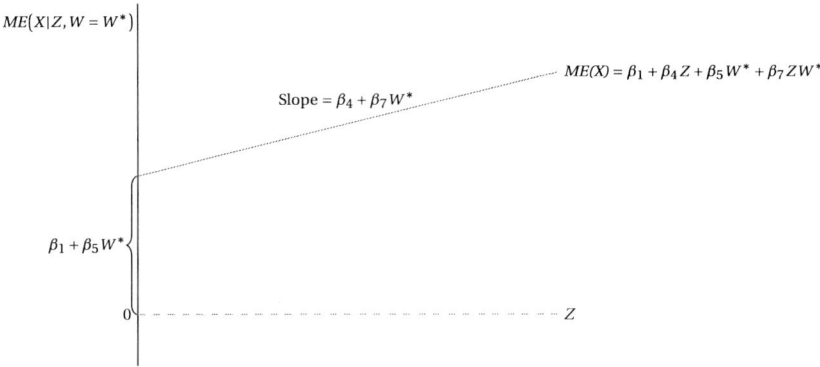

Figure 6.12 A different 2-D marginal effect plot for X when $W = W^*$

modifying effects of Z and W are independent, Eq. 6.29 shows that the slope of the marginal effect line with respect to Z, $\beta_4 + \beta_7 W$, also now depends on the chosen value of W. In Figure 6.12, we show what such a 2-D marginal effect plot might look like when $W = W^*$. This particular marginal effect plot is consistent with the predictions in hypothesis $H_{X|Z, W}$ in that the marginal effect of X is always positive and its magnitude increases with Z. A key point here is that the marginal effect line depicted here is just one of the infinitely many possible marginal effect lines for X across the observed range of Z. If we were to choose a different value of W, the intercept and the slope of the marginal effect line would both change. This is important because changes to the intercept and the slope can alter the sign of the marginal effect of X. In particular, we may find that that there are some values of W and Z at which the marginal effect of X isn't positive, thereby contradicting $H_{X|Z, W}$. Significantly, by including only one marginal effect line for X, the plot in Figure 6.12 provides no way to assess the prediction in hypothesis $H_{X|Z, W}$ that the positive modifying effect of Z decreases with W. Evaluating this particular prediction requires that we plot at least two marginal effect lines for X that correspond to two different values of W. A similar logic applies to constructing and evaluating a 2-D marginal effect plot for X across the observed range of W.

As in the case where the modifying effects of Z and W are independent, it can be helpful to construct a 2-D marginal effect plot showing the marginal effect lines for a few different substantively relevant values of W (or Z). This particular strategy is especially useful when one of the modifying variables is discrete. This is because the resulting 2-D marginal effect plot allows us to fully visualize the marginal effect of X on Y. In Figure 6.13, we show what a 2-D marginal effect plot for X that's consistent with hypothesis $H_{X|Z, W}$ might look like if Z is continuous and

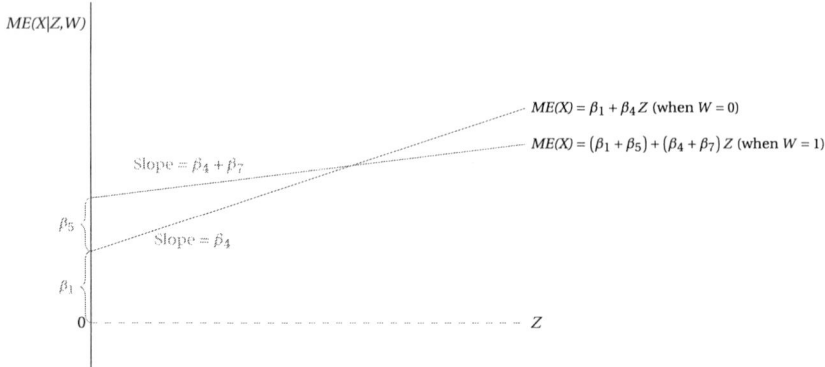

Figure 6.13 A different 2-D marginal effect plot for X consistent with hypothesis $H_{X|Z,\,W}$ when Z is continuous and W is dichotomous $(0/1)$

W is dichotomous $(0/1)$. We now have two marginal effect lines across the observed range of Z, one for when $W = 0$ and one for when $W = 1$. Both marginal effect lines are above the gray dashed zero line for all values of Z, indicating that the marginal effect of X is positive for all possible combined values of Z and W. Both marginal effect lines have a positive slope, indicating that the marginal effect of X always increases with Z. Although positive, the slope of the marginal effect line when $W = 1$ is less steep than the slope of the marginal effect line when $W = 0$. This captures the fact that W has a negative modifying effect on the positive modifying effect of Z. The marginal effect line when $W = 1$ is above the marginal effect line when $W = 0$ when Z is low but below the marginal effect line when $W = 0$ when Z is high. This indicates that the modifying effect of W is positive when Z is low but negative when Z is high.

When constructing an *estimated* marginal effect plot, scholars should include confidence intervals for each of the marginal effect lines. A potential problem arises in that the plot can become messy if the confidence intervals overlap. One solution is to "break up" the marginal effect plot for X into a *series* of marginal effect plots where each plot shows the marginal effect for X across the observed range of Z for a specific value of W. Scholars should also try to incorporate information about the distribution of the modifying variables. This may not be possible if both Z and W are continuous because there are likely to be few observations that have the specific *combinations* of values for Z and W depicted by the various points along each marginal effect line. Things are easier when one of the modifying variables, say W, is discrete. If W is dichotomous, for example, we can overlay a marginal effect plot like the one shown in Figure 6.13 with two histograms showing the percentage of observations at different values of Z when $W = 0$ and when

$W = 1$. If we were to present the two marginal effect lines for X shown in Figure 6.13 in separate plots, we could easily incorporate a histogram in each showing the frequency distribution for Z when $W = 0$ or $W = 1$.

It's important to note that while the marginal effect plot in Figure 6.13 allows us to fully visualize the marginal effect of X on Y, it doesn't provide us with enough information to evaluate all nine of the key predictions in hypothesis $H_{X|Z,W}$. As predicted, the slopes of the two marginal effect lines are positive, suggesting that there's a positive interaction effect between X and Z for both values of W. However, the information in the plot doesn't allow us to know whether these interaction effects are statistically significant. The confidence intervals that would be included in an estimated marginal effect plot can tell us whether the effect of X is significantly different from zero at different values of Z and W. However, they can't tell us whether the points along each marginal effect line are significantly different from each other. In other words, they can't tell us whether the slope of either marginal effect line is significantly different from zero. As a result, we recommend that scholars report the interaction effect between X and Z when $W = 0$ (β_4) and the interaction effect between X and Z when $W = 1$ ($\beta_4 + \beta_7$) somewhere in the plot along with their t-ratios or confidence intervals.

As predicted, the slope of the marginal effect line when $W = 1$ is less positive than the slope of the marginal effect line when $W = 0$. However, we can't actually tell from the plot whether the slopes are significantly different from each other. In other words, we can't see whether there's a second-order interaction effect where W significantly modifies the modifying effect of Z on the marginal effect of X. Again, the confidence intervals that would be included in an estimated marginal effect plot provide no insight into this issue. As a result, we recommend that scholars also report the second-order interaction effect (β_7) somewhere in the plot along with its t-ratio or confidence interval.

As expected, the "$W = 1$ marginal effect line" is above the "$W = 0$ marginal effect line" when Z is low but below the "$W = 0$ marginal effect line" when Z is high. This is consistent with our claim that the modifying effect of W is positive when Z is low but negative when Z is high. However, we can't actually tell from the plot if and when the modifying effect of W is ever statistically significant. Consider the earlier case depicted in Figure 6.3 where the modifying effects of Z and W are independent. In this case, the modifying effect of W on the marginal effect of X is unconditional and therefore constant across the observed range of values for Z. This is reflected in the fact that the marginal effect line for X when $W = 0$ is parallel to the marginal effect line for X when $W = 1$. The vertical distance between these two marginal effect lines captures the modifying effect of

W and is indicated by the coefficient β_5. In other words, we can see if the modifying effect of W is statistically significant simply by looking at coefficient β_5.

Things are more complicated in the scenario depicted in Figure 6.13 where the modifying effects of Z and W are dependent. Notably, the modifying effect of W on the marginal effect of X, which is visually represented by the vertical distance between the two marginal effect lines, now changes with the value of Z. Recall from Eq. 6.31 that the modifying effect of W is $\beta_5 + \beta_7 Z$. From this, and the intercept in Figure 6.13, we see that the modifying effect of W is equal to β_5 when $Z = 0$. This means that we can determine directly from the regression output whether the modifying effect of W is statistically significant when $Z = 0$. However, we have no way of knowing if and when the modifying effect of W is statistically significant for any other value of Z. In the particular case depicted in Figure 6.13, we know that the modifying effect of W starts off positive when Z is low, becomes zero as Z increases and the two marginal effect lines cross, and then becomes negative as Z increases further. This means that we know that there's going to be a range of values for Z where the modifying effect of W is statistically indistinguishable from zero. However, we don't know from Figure 6.13 what this range is or what percentage of the sample observations have values of Z where the modifying effect of W is statistically insignificant. To address this issue, we recommend combining the marginal effect plot for X shown in Figure 6.13 with an "interaction effect plot" for XW that graphs the modifying effect of W on the marginal effect of X along with its confidence interval across the observed range of Z. This interaction effect plot could also include a histogram showing the distribution of Z. The general point here is that we'll often need to combine a marginal effect plot for X with interaction effect plots for XZ and/or XW if we're to fully evaluate all nine of the predictions contained in a hypothesis like $H_{X|Z,W}$ that's derived from a theory where the modifying effects of Z and W are dependent.

> ⚠ **Important:** Suppose we have a fully interactive model with variables X, Z, W, XZ, XW, ZW, and XZW. When testing all nine of the key predictions contained in a typical hypothesis about the marginal effect of X, we will often need to combine a marginal effect plot for X with interaction effect plots for XZ and XW, as well as information about the second-order interaction effect between X, Z, an W.

Things are easier if both of the modifying variables are discrete. This is because it's possible to construct a single plot that conveys the necessary

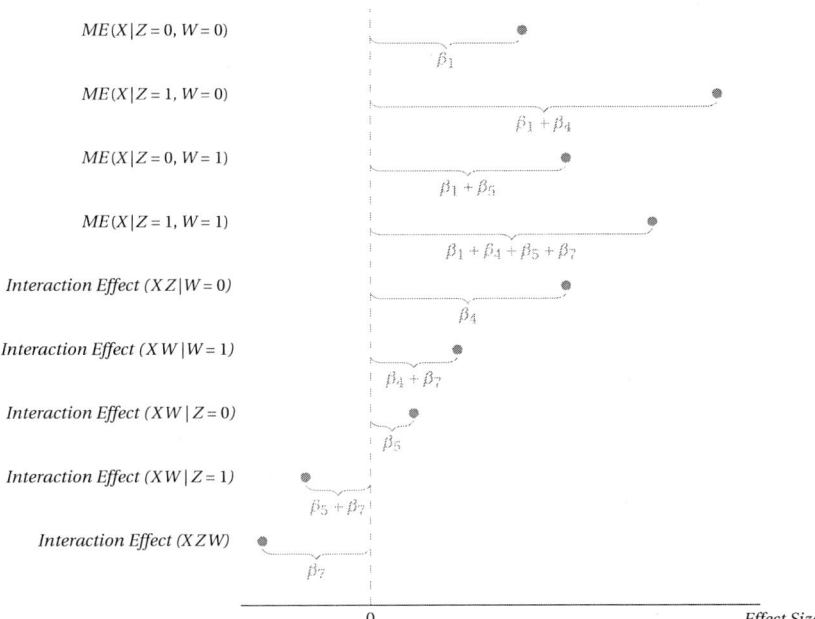

Figure 6.14 A different 2-D marginal effect plot for X consistent with hypothesis $H_{X|Z,W}$ when Z and W are both dichotomous

information to evaluate all of the key predictions contained in a hypothesis like $H_{X|Z,W}$. In Figure 6.14, we show what such a plot might look like if Z and W are both dichotomous (0/1). Again, we've drawn the plot to be consistent with hypothesis $H_{X|Z,W}$. As we noted earlier, hypothesis $H_{X|Z,W}$ contains nine key predictions that are all evaluated in Figure 6.14. Specifically, the plots shows the marginal effect of X for each of the four possible combinations of values for Z and W, the interaction effect between X and Z when $W = 0$ and when $W = 1$, the interaction effect between X and W when $Z = 0$ and when $Z = 1$, and the second-order interaction effect between X, Z, and W. The first four entries indicate that the marginal effect of X is positive for all four possible combinations of values for Z and W. The next two entries indicate that Z has a positive modifying effect on the marginal effect for X both when $W = 0$ and when $W = 1$, and that this positive modifying effect is smaller when $W = 1$. The next two entries indicate that W has a positive modifying effect on the marginal effect for X when $Z = 0$ but a negative modifying effect when $Z = 1$. The negative interaction effect for XZW at the very bottom speaks to whether there's a "fully interactive" relationship between X, Z, and W and indicates both that the positive modifying effect of Z on the marginal effect of X depends

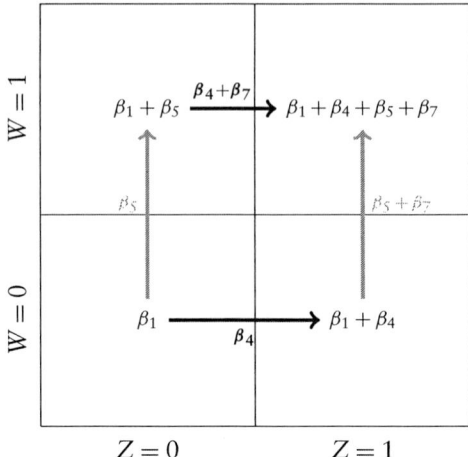

Figure 6.15 The marginal effect of X when Z and W are both dichotomous

Note: The modifying effect of Z is shown in bold and indicates how the marginal effect of X changes when we move from the left column to the right column. The modifying effect of W is shown in gray and indicates how the marginal effect of X changes when we move from the bottom row to the top row.

negatively on the value of W and that the modifying effect of W on the marginal effect of X depends negatively on the value of Z.

An alternative strategy when the modifying variables are both discrete is to present the information about the marginal effect of X in a tabular format rather than a marginal effect plot. In Figure 6.15, we show what this might look like if Z and W are both dichotomous (0/1). The marginal effect of X for each possible combination of values for Z and W is shown in the four cells. The modifying effects of Z are shown in bold and indicate how the marginal effect of X changes when we move from the left column ($Z = 0$) to the right column ($Z = 1$). How the modifying effect of Z changes with W can be evaluated by comparing the move from the left column to the right column in the bottom row (when $W = 0$) with the move from the left column to the right column in the top row (when $W = 1$). The modifying effects of W are shown in gray and indicate how the marginal effect of X changes when we move from the bottom row ($W = 0$) to the top row ($W = 1$). How the modifying effect of W changes with Z can be evaluated by comparing the move from the bottom row to the top row in the left column (when $Z = 0$) with the move from the bottom row to the top row in the right column (when $Z = 1$). According to hypothesis $H_{X|Z, W}$, the marginal effect of X reported in each cell should be positive. The magnitudes of these positive effects should grow as we increase the

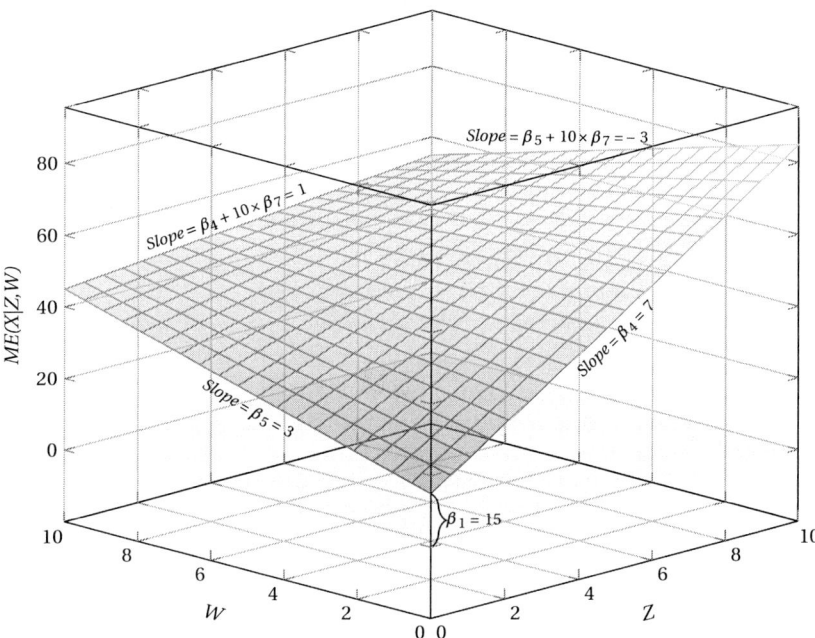

Figure 6.16 A 3-D plot of the marginal effect of X on Y consistent with hypothesis $H_{X|Z,W}$ (assuming $\beta_1 = 15$, $\beta_4 = 7$, $\beta_5 = 3$, and $\beta_7 = -0.6$)

Note: The shaded surface shows how the marginal effect of X on Y varies with Z and W. Darker colors indicate smaller values for the marginal effect of X. The solid colored "zero surface" is equivalent to the "zero line" we used in a 2-D marginal effect plot to indicate if and when the marginal effect of X is 0.

value of Z (move from the left column to the right column) but less so when we're in the bottom row. The magnitudes of these positive effects should increase when we increase the value of W (move from the bottom row to the top row) when $Z = 0$ (left column) but decrease when we increase the value of W (move from the bottom row to the top row) when $Z = 1$ (right column).

While it's not possible to fully visualize the marginal effect of X with a 2-D marginal effect plot or a table if both modifying variables are continuous, it is possible if we use a 3-D marginal effect plot. A 3-D marginal effect plot consistent with hypothesis $H_{X|Z,W}$ is shown in Figure 6.16. We've assumed arbitrarily that the values of Z and W both run between 0 and 10 and that $\beta_1 = 15$, $\beta_4 = 7$, $\beta_5 = 3$, and $\beta_7 = -0.6$. The shaded mesh surface indicates how the marginal effect of X varies with Z and W. Darker colors indicate smaller values for the marginal effect of X. The solid colored "zero surface" is equivalent to the "zero line" used in a 2-D marginal effect plot to help indicate if and when the marginal effect of X is 0. The shaded mesh surface in Figure 6.16 is always above the zero surface, indicating that the marginal effect of X is positive for all possible combinations of values for Z and W. The surface slopes up to the right, indicating that the marginal effect of X always increases with Z. However, the slope of the surface up to the right is steeper at the front where W is low than at the back where W is high. This indicates that the positive modifying effect of Z declines with higher values of W. The surface slopes up to the left at the front, indicating that the marginal effect of X increases with W when Z is low. However, the surface slopes down to the left at the back, indicating that the marginal effect of X declines with W when Z is high. Recall that the 3-D marginal effect surface for X in Figure 6.6, when we assumed that the modifying effects of Z and W were independent, was sloped but flat. In contrast, while the 3-D surface in Figure 6.16 remains smooth, it's no longer flat. The curved surface is indicative of the fact that the modifying effects of Z and W are now dependent. The darker colors in the bottom front corner of the surface show that the positive effect of X is smallest when Z and W are at their lowest values. The lighter colors in the top right corner show that the positive effect of X is largest when Z is at its highest value and W is at its lowest value.

When constructing an *estimated* 3-D marginal effect plot, we can color code or shade the surface to indicate the combinations of values for Z and W at which the marginal effect of X is statistically significant and use the axis label values to evaluate the magnitude of the marginal effect of X. While the 3-D marginal effect plot for X allows us to fully visualize the conditional effect of X on Y, it's important to remember again that it doesn't provide us with enough information to evaluate all of the key predictions in a typical hypothesis about the marginal effect of X from a theory positing that the modifying effects of Z and W are dependent. At most, the 3-D marginal effect plot can tell us the combinations of values for W and Z at which the marginal effect of X is positive, zero, or negative. In line with our earlier guidance, we recommend that if scholars employ a 3-D marginal effect plot for X, then they should also include two interaction effect plots showing how the modifying effect of Z on the marginal effect of X varies across the observed range of W and how the modifying effect of W on the marginal effect of X varies across the observed range of Z. They should also report the second-order interaction effect, β_7, along with its t-ratio or confidence interval.

In order to fully evaluate hypothesis $H_{X|Z, W}$ and its conditional claim about the marginal effect of X on Y,

$$\frac{\partial Y}{\partial X} = \beta_1 + \beta_4 Z + \beta_5 W + \beta_7 Z W, \tag{6.33}$$

we must calculate an appropriate measure of uncertainty for the marginal effect of X. Using the basic properties of variances shown in Appendix A, we calculate the variance of the marginal effect of X as

$$\text{var}\left(\frac{\partial Y}{\partial X}\right) = \text{var}(\beta_1) + Z^2 \text{var}(\beta_4) + W^2 \text{var}(\beta_5) + Z^2 W^2 \text{var}(\beta_7)$$
$$+ 2Z \text{cov}(\beta_1, \beta_4) + 2W \text{cov}(\beta_1, \beta_5)$$
$$+ 2ZW \text{cov}(\beta_1, \beta_7) + 2ZW \text{cov}(\beta_4, \beta_5)$$
$$+ 2Z^2 W \text{cov}(\beta_4, \beta_7) + 2ZW^2 \text{cov}(\beta_5, \beta_7). \quad (6.34)$$

Like the marginal effect of X on Y, the variance depends jointly on the values of Z and W. This means that each marginal effect of X and associated variance we calculate is for a specific *combination* of Z and W values. This is important to understand when making claims about when the marginal effect of X is or isn't statistically significant. As usual, we obtain the standard error for the marginal effect of X by taking the square root of the variance,

$$\text{se}\left(\frac{\partial Y}{\partial X}\right)$$
$$= \text{se}(\beta_1 + \beta_4 Z + \beta_5 W + \beta_7 ZW)$$
$$= \left\{ \text{var}(\beta_1) + Z^2 \text{var}(\beta_4) + W^2 \text{var}(\beta_5) + Z^2 W^2 \text{var}(\beta_7) \right.$$
$$+ 2Z \text{cov}(\beta_1, \beta_4) + 2W \text{cov}(\beta_1, \beta_5) + 2ZW \text{cov}(\beta_1, \beta_7) + 2ZW \text{cov}(\beta_4, \beta_5)$$
$$\left. + 2Z^2 W \text{cov}(\beta_4, \beta_7) + 2ZW^2 \text{cov}(\beta_5, \beta_7) \right\}^{\frac{1}{2}}. \quad (6.35)$$

We can use the standard error to conduct *t*-tests in the normal way to evaluate claims we might have about the marginal effect of X. We can also use it to construct confidence intervals for the marginal effect of X on Y at given combinations of values for Z and W. For example, a two-sided confidence interval for the effect of X on Y with confidence coefficient $1 - \alpha$ is

$$\beta_1 + \beta_4 Z + \beta_5 W + \beta_7 ZW \pm t_{n-k,\alpha/2} \times \text{se}\left(\frac{\partial Y}{\partial X}\right), \quad (6.36)$$

where $t_{n-k,\alpha/2}$ refers to the critical value from a *t*-distribution table for a particular α level of significance and $n - k$ degrees of freedom.

Having dealt with the marginal effect of X on Y, we now turn very briefly to the marginal effect of Z,

$$\frac{\partial Y}{\partial Z} = \beta_2 + \beta_4 X + \beta_6 W + \beta_7 XW. \quad (6.37)$$

From this, we see that the marginal effect of Z depends on the values of X and W. In particular, we see that the coefficient on the constitutive term Z in Eq. 6.27, β_2, tells us the effect of Z only when X and W are both 0. The modifying effect of X on the marginal effect of Z is

$$\frac{\partial \left(\frac{\partial Y}{\partial Z}\right)}{\partial X} = \frac{\partial \left(\beta_2 + \beta_4 X + \beta_6 W + \beta_7 X W\right)}{\partial X} = \beta_4 + \beta_7 W. \tag{6.38}$$

This is the interaction effect between Z and X. From this, we see that so long as $\beta_7 \neq 0$, the modifying effect of X on the marginal effect of Z depends on the value of W. As expected from the symmetry of interactions, how the modifying effect of X on the marginal effect of Z depends on the value of W is identical to how the modifying effect of Z on the marginal effect of X depends on the value of W shown earlier in Eq. 6.29. We also see that the coefficient on XZ, β_4, tells us the modifying effect of X on the marginal effect of Z only when $W = 0$. The modifying effect of W on the marginal effect of Z is

$$\frac{\partial \left(\frac{\partial Y}{\partial Z}\right)}{\partial W} = \frac{\partial \left(\beta_2 + \beta_4 X + \beta_6 W + \beta_7 X W\right)}{\partial W} = \beta_6 + \beta_7 X. \tag{6.39}$$

This is the interaction effect between Z and W. Again we see that, so long as $\beta_7 \neq 0$, the modifying effect of W on the marginal effect of Z depends on the value of X. We also see that the coefficient on ZW, β_6, tells us the modifying effect of W on the marginal effect of Z only when $X = 0$.

According to hypothesis $H_{Z|X, W}$, the marginal effect of Z should always be negative. As a result, $\beta_2 + \beta_4 X + \beta_6 W + \beta_7 X W$ should be negative for all combinations of observed values for X and W. It follows that β_2 should be negative. Hypothesis $H_{Z|X, W}$ also states that the magnitude of Z's negative effect is expected to decrease with X but less so when W is high. Several predictions follow from this. The modifying effect of X, $\beta_4 + \beta_7 W$, should be positive for all values of W. This, in turn, means that β_4 should be positive. Since the positive modifying effect of Z is expected to decline with higher values of W, β_7 should be negative. As we saw earlier, the negative value of β_7 is constrained such that $\frac{-\beta_4}{W_{max}} < \beta_7 < 0$. Finally, hypothesis $H_{Z|X, W}$ also states that the magnitude of Z's negative effect increases with W and increasingly so with higher values of X. This part of the hypothesis speaks to the modifying effect of W, $\beta_6 + \beta_7 X$. We see that this modifying effect should be negative for all values of X and hence that β_6 should be negative. Again, we see that β_7 should be negative.

The variance of the marginal effect of Z on Y is

$$\mathrm{var}\left(\frac{\partial Y}{\partial Z}\right) = \mathrm{var}\left(\beta_2\right) + X^2 \mathrm{var}\left(\beta_4\right) + W^2 \mathrm{var}\left(\beta_6\right) + X^2 W^2 \mathrm{var}\left(\beta_7\right)$$

$$+ 2X\mathrm{cov}\left(\beta_2, \beta_4\right) + 2W\mathrm{cov}\left(\beta_2, \beta_6\right) + 2XW\mathrm{cov}\left(\beta_2, \beta_7\right)$$

$$+ 2XW\mathrm{cov}\left(\beta_4, \beta_6\right) + 2X^2 W\mathrm{cov}\left(\beta_4, \beta_7\right) + 2XW^2 \mathrm{cov}\left(\beta_6, \beta_7\right). \tag{6.40}$$

We calculate the standard error and confidence intervals in the usual way.
Finally, we turn to the marginal effect of W on Y,

$$\frac{\partial Y}{\partial W} = \beta_3 + \beta_5 X + \beta_6 Z + \beta_7 XZ. \tag{6.41}$$

From this, we see that the marginal effect of W depends on the values of X and Z. In particular, we see that the coefficient on the constitutive term W in Eq. 6.27, β_3, tells us the effect of W only when X and Z are both 0. The modifying effect of X on the marginal effect of W is

$$\frac{\partial\left(\frac{\partial Y}{\partial W}\right)}{\partial X} = \frac{\partial\,(\beta_3 + \beta_5 X + \beta_6 Z + \beta_7 XZ)}{\partial X} = \beta_5 + \beta_7 Z. \tag{6.42}$$

This is the interaction effect between W and X. From this, we see that so long as $\beta_7 \neq 0$, the modifying effect of X on the marginal effect of W depends on the value of Z. As expected from the symmetry of interactions, how the modifying effect of X on the marginal effect of W depends on the value of Z is identical to how the modifying effect of W on the marginal effect of X depends on the value of Z shown earlier in Eq. 6.31. We also see that the coefficient on XW, β_5, tells us the modifying effect of X on the marginal effect of W only when $Z = 0$. The modifying effect of Z on the marginal effect of W is

$$\frac{\partial\left(\frac{\partial Y}{\partial W}\right)}{\partial Z} = \frac{\partial\,(\beta_3 + \beta_5 X + \beta_6 Z + \beta_7 XZ)}{\partial Z} = \beta_6 + \beta_7 X. \tag{6.43}$$

This is the interaction effect between W and Z. Again we see that, so long as $\beta_7 \neq 0$, the modifying effect of Z on the marginal effect of W depends on the value of X. We also see that the coefficient on ZW, β_6, tells us the modifying effect of Z on the marginal effect of W only when $X = 0$.

According to hypothesis $H_{W|X,Z}$, the marginal effect of W should be zero when X and Z are both at their minimum value, positive when X is at its maximum value and Z is at its minimum value, negative when X is at its minimum value and Z is at its maximum value, and negative when X and Z are both at their maximum values. For the purposes of our example, we'll assume as we did earlier that X, Z, and W are all continuous variables that run between 0 and 10. Given this, $H_{W|X,Z}$ predicts that $\beta_3 = 0$, $\beta_3 + 10\beta_5 > 0$, $\beta_3 + 10\beta_6 < 0$, and $\beta_3 + 10\beta_5 + 10\beta_6 + 100\beta_7 < 0$. Hypothesis $H_{W|X,Z}$ also states that the modifying effect of X is positive when Z is at its minimum value but negative when Z is at its maximum value. Several predictions follow from this. The modifying effect of X, $\beta_5 + \beta_7 Z$, should be positive when $Z = 0$ and negative when $Z = 10$. It follows that β_5 should be positive and that β_7 should be negative. As we saw earlier, the negative value of β_7 is constrained such that $\beta_7 < \frac{-\beta_5}{Z_{max}} < 0$. Finally, hypothesis $H_{W|X,Z}$ also states that the modifying effect of Z, $\beta_6 + \beta_7 X$, is always negative and increasingly so with higher values of X. It follows that β_6 and β_7 should both be negative.

The variance of the marginal effect of W on Y is

$$\text{var}\left(\frac{\partial Y}{\partial W}\right) = \text{var}(\beta_3) + X^2\text{var}(\beta_5) + Z^2\text{var}(\beta_6) + X^2 Z^2 \text{var}(\beta_7)$$
$$+ 2X\text{cov}(\beta_3, \beta_5) + 2Z\text{cov}(\beta_3, \beta_6) + 2XZ\text{cov}(\beta_3, \beta_7)$$
$$+ 2XZ\text{cov}(\beta_5, \beta_6) + 2X^2 Z\text{cov}(\beta_5, \beta_7) + 2XZ^2\text{cov}(\beta_6, \beta_7).$$
$$(6.44)$$

We calculate the standard error and confidence intervals in the usual way.

⚠ **Important:** Consider the following linear-interactive model:

$$Y = \beta_0 + \beta_1 X + \beta_2 Z + \beta_3 W + \beta_4 XZ + \beta_5 XW + \beta_6 ZW + \beta_7 XZW + \epsilon.$$

The marginal effect of X on Y is

$$\frac{\partial Y}{\partial X} = \beta_1 + \beta_4 Z + \beta_5 W + \beta_7 ZW,$$

and its estimated standard error is

$$\text{se}\left(\frac{\partial Y}{\partial X}\right) = \left\{\text{var}(\beta_1) + Z^2\text{var}(\beta_4) + W^2\text{var}(\beta_5) + Z^2 W^2\text{var}(\beta_7)\right.$$
$$+ 2Z\text{cov}(\beta_1, \beta_4) + 2W\text{cov}(\beta_1, \beta_5) + 2ZW\text{cov}(\beta_1, \beta_7)$$
$$\left. + 2ZW\text{cov}(\beta_4, \beta_5) + 2Z^2 W\text{cov}(\beta_4, \beta_7) + 2ZW^2\text{cov}(\beta_5, \beta_7)\right\}^{\frac{1}{2}}.$$

The marginal effect of Z on Y is

$$\frac{\partial Y}{\partial Z} = \beta_2 + \beta_4 X + \beta_6 W + \beta_7 XW,$$

and its estimated standard error is

$$\text{se}\left(\frac{\partial Y}{\partial Z}\right) = \left\{\text{var}(\beta_2) + X^2\text{var}(\beta_4) + W^2\text{var}(\beta_6) + X^2 W^2\text{var}(\beta_7)\right.$$
$$+ 2X\text{cov}(\beta_2, \beta_4) + 2W\text{cov}(\beta_2, \beta_6) + 2XW\text{cov}(\beta_2, \beta_7)$$
$$\left. + 2XW\text{cov}(\beta_4, \beta_6) + 2X^2 W\text{cov}(\beta_4, \beta_7) + 2XW^2\text{cov}(\beta_6, \beta_7)\right\}^{\frac{1}{2}}.$$

The marginal effect of W on Y is

$$\frac{\partial Y}{\partial W} = \beta_3 + \beta_5 X + \beta_6 Z + \beta_7 XZ,$$

and its estimated standard error is

$$\text{se}\left(\frac{\partial Y}{\partial W}\right) = \left\{ \text{var}(\beta_3) + X^2\text{var}(\beta_5) + Z^2\text{var}(\beta_6) + X^2 Z^2\text{var}(\beta_7) \right.$$

$$+ 2X\text{cov}(\beta_3,\beta_5) + 2Z\text{cov}(\beta_3,\beta_6) + 2XZ\text{cov}(\beta_3,\beta_7)$$

$$\left. + 2XZ\text{cov}(\beta_5,\beta_6) + 2X^2 Z\text{cov}(\beta_5,\beta_7) + 2XZ^2\text{cov}(\beta_6,\beta_7) \right\}^{\frac{1}{2}}.$$

6.2.1 An Alternative Interaction Model When the Modifying Variables Are All Discrete

As we've seen, we can evaluate the implications of a theory positing a fully interactive relationship between three independent variables X, Z, and W with the following "standard" interaction model:

$$Y = \beta_0 + \beta_1 X + \beta_2 Z + \beta_3 W + \beta_4 XZ + \beta_5 XW + \beta_6 ZW + \beta_7 XZW + \epsilon. \tag{6.45}$$

We can use this model specification irrespective of whether the independent variables are continuous or discrete. As was the case when we were examining theories positing an interactive relationship between variables X and Z, it's possible to use an alternative specification when the three interacting variables X, Z, and W are discrete. To estimate this alternative model, we must first break up X, Z, and W into three sets of logically exhaustive and mutually exclusive dichotomous variables that correspond to one of the interacting variables' distinct values. We then multiply each of the dichotomous variables obtained from X with each of the dichotomous variables from Z and W to create a series of interaction terms. If X has m distinct values, Z has n distinct values, and W has p distinct values, we'll end up with $K = m \times n \times p$ interaction terms that are themselves dichotomous variables. The new alternative interaction model will either include (i) all of the K dichotomous interaction terms and no constant or (ii) $K - 1$ of the dichotomous interaction terms and a constant.

To briefly see what's going on here, consider the case where X, Z, and W are all dichotomous. To provide some substantive context, we'll assume that X captures an individual's gender and indicates whether they're female (1) or male (0), Z captures an individual's race and indicates whether they're Black (1) or White (0), and W captures an individual's social class and indicates whether they're higher (1) or lower (0) class (Block, Golder and Golder, 2023). In this substantive context, the alternative model specification with a constant is

$$Y = \gamma_0 + \gamma_1 X_0 Z_0 W_1 + \gamma_2 X_0 Z_1 W_0 + \gamma_3 X_0 Z_1 W_1 + \gamma_4 X_1 Z_0 W_0$$
$$+ \gamma_5 X_1 Z_0 W_1 + \gamma_6 X_1 Z_1 W_0 + \gamma_7 X_1 Z_1 W_1 + \varepsilon, \qquad (6.46)$$

where X_0 is a dichotomous variable that equals 1 when $X=0$ and 0 otherwise, X_1 is a dichotomous variable that equals 1 when $X=1$ and 0 otherwise, Z_0 is a dichotomous variable that equals 1 when $Z = 0$ and 0 otherwise, Z_1 is a dichotomous variable that equals 1 when $Z = 1$ and 0 otherwise, W_0 is a dichotomous variable that equals 1 when $W = 0$ and 0 otherwise, W_1 is a dichotomous variable that equals 1 when $W = 1$ and 0 otherwise, and $X_0 Z_0 W_0$ is the omitted interaction term.

All of the interaction terms in the alternative model are, themselves, dichotomous variables. $X_0 Z_0 W_1$ is a dichotomous variable that equals 1 if an individual is a higher class White male, $X_0 Z_1 W_0$ is a dichotomous variable that equals 1 if an individual is a lower class Black male, $X_0 Z_1 W_1$ is a dichotomous variable that equals 1 if an individual is a higher class Black male, $X_1 Z_0 W_0$ is a dichotomous variable that equals 1 if an individual is a lower class White female, $X_1 Z_0 W_1$ is a dichotomous variable that equals 1 if an individual is a higher class White female, $X_1 Z_1 W_0$ is a dichotomous variable that equals 1 if an individual is a lower class Black female, $X_1 Z_1 W_1$ is a dichotomous variable that equals 1 if an individual is a higher class Black female, and the omitted interaction term $X_0 Z_0 W_0$ is a dichotomous variable that equals 1 if an individual is a lower class White male. Each of the dichotomous interaction terms captures a different category of observation, with the omitted interaction term determining the baseline or reference category against which the other categories are compared. In effect, each of the independent variables in the alternative model indicates membership in one of the eight possible identity groups that are defined by the possible combinations of dichotomous values for gender, race, and class. These identity groups correspond to the eight shaded cells we saw earlier in Figure 6.11. We can interpret the coefficients on the interaction terms as the difference in Y between the specified identity group associated with the interaction term and the baseline identity group. In other words, the coefficients capture differences in the mean value of Y across groups of observations.

The alternative model specification shown in Eq. 6.46 is often mistaken for an additive model because it comprises a series of dichotomous variables incorporated in an additive manner. This is especially the case when scholars use variable names that signal the different groups of observations captured by the interaction terms. In the context of our gender, race, and class example, the alternative model shown in Eq. 6.46 would likely appear to readers as

$$Y = \gamma_0 + \gamma_1 \text{Higher Class White Male} + \gamma_2 \text{Lower Class Black Male}$$
$$+ \gamma_3 \text{Higher Class Black Male} + \gamma_4 \text{Lower Class White Female}$$
$$+ \gamma_5 \text{Higher Class White Female} + \gamma_6 \text{Lower Class Black Female}$$
$$+ \gamma_7 \text{Higher Class Black Female.} + \varepsilon. \tag{6.47}$$

Although this might look like an additive model, it should be clear from our discussion that it is, in fact, an interaction model because each of the dichotomous variables indicating different identity groups are really "hidden" interaction terms.

⚠ **Important:** Scholars who posit a theory in which three interacting variables X, Z, and W are discrete may choose to adopt an "alternative" interactive specification. The alternative specification involves breaking X, Z, and W up into three sets of dichotomous variables that correspond to one of the interacting variables' distinct values and then multiplying each of the dichotomous variables from X with each of the dichotomous variables from Z and W to create a series of dichotomous interaction terms.

The alternative interaction model shown in Eq. 6.46 is identical to the standard interaction model shown in Eq. 6.45 in that it produces the exact same quantities of interest. To see this, note that we can simplify the alternative model specification by recognizing that $X_1 = X$, $Z_1 = Z$, and $W_1 = W$:

$$Y = \gamma_0 + \gamma_1 X_0 Z_0 W + \gamma_2 X_0 Z W_0 + \gamma_3 X_0 Z W + \gamma_4 X Z_0 W_0$$
$$+ \gamma_5 X Z_0 W + \gamma_6 X Z W_0 + \gamma_7 X Z W + \varepsilon. \tag{6.48}$$

Given that $X_0 = 1 - X$, $Z_0 = 1 - Z$, and $W_0 = 1 - W$, we can rewrite Eq. 6.48 as

$$Y = \gamma_0 + \gamma_1 (1-X)(1-Z)W + \gamma_2 (1-X)Z(1-W) + \gamma_3 (1-X)ZW$$
$$+ \gamma_4 X(1-Z)(1-W) + \gamma_5 X(1-Z)W + \gamma_6 XZ(1-W) + \gamma_7 XZW + \varepsilon. \tag{6.49}$$

Multiplying through, we have

$$Y = \gamma_0 + \gamma_1 W - \gamma_1 ZW - \gamma_1 XW + \gamma_1 XZW + \gamma_2 Z - \gamma_2 ZW - \gamma_2 XZ + \gamma_2 XZW$$
$$+ \gamma_3 ZW - \gamma_3 XZW + \gamma_4 X - \gamma_4 XW - \gamma_4 XZ + \gamma_4 XZW$$
$$+ \gamma_5 XW - \gamma_5 XZW + \gamma_6 XZ - \gamma_6 XZW + \gamma_7 XZW + \varepsilon. \tag{6.50}$$

And collecting terms, we have

$$Y = \gamma_0 + \gamma_4 X + \gamma_2 Z + \gamma_1 W + (\gamma_6 - \gamma_2 - \gamma_4) XZ$$
$$+ (\gamma_5 - \gamma_1 - \gamma_4) XW + (\gamma_3 - \gamma_1 - \gamma_2) ZW$$
$$+ (\gamma_1 + \gamma_2 - \gamma_3 + \gamma_4 - \gamma_5 - \gamma_6 + \gamma_7) XZW + \varepsilon. \tag{6.51}$$

We can now see that the alternative interaction model shown in Eq. 6.46 is just an algebraic transformation of the standard interaction model shown in Eq. 6.45, where $\beta_0 = \gamma_0$, $\beta_1 = \gamma_4$, $\beta_2 = \gamma_2$, $\beta_3 = \gamma_1$, $\beta_4 = \gamma_6 - \gamma_2 - \gamma_4$, $\beta_5 = \gamma_5 - \gamma_1 - \gamma_4$, $\beta_6 = \gamma_3 - \gamma_1 - \gamma_2$, and $\beta_7 = \gamma_1 + \gamma_2 - \gamma_3 + \gamma_4 - \gamma_5 - \gamma_6 + \gamma_7$. In effect, the two models are just different representations of the same interaction model. It follows that looking at how Y varies across different categories or groups of observations (alternative model) is exactly equivalent to looking at how the corresponding dimensions of difference interact to determine variation in Y (standard model).

⚠ **Important:** It is not possible to determine directly from the regression output if there is a second-order interaction effect between X, Z, and W when estimating the alternative interaction model,

$$Y = \gamma_0 + \gamma_1 X_0 Z_0 W_1 + \gamma_2 X_0 Z_1 W_0 + \gamma_3 X_0 Z_1 W_1 + \gamma_4 X_1 Z_0 W_0$$
$$+ \gamma_5 X_1 Z_0 W_1 + \gamma_6 X_1 Z_1 W_0 + \gamma_7 X_1 Z_1 W_1 + \varepsilon.$$

Instead, the researcher must formally test whether $\gamma_1 + \gamma_2 - \gamma_3 + \gamma_4 - \gamma_5 - \gamma_6 + \gamma_7 = 0$.

We summarize our comparison of the standard and alternative interaction models when all three interacting variables X, Z, and W are dichotomous in Table 6.2. The information in the table also links our discussion back to the nineteen key predictions that can typically be derived from a theory positing a fully interactive relationship between X, Z, and W by indicating the quantities of interest necessary for evaluating each prediction. The standard and alternative interaction models differ in how easy they make it to see particular quantities of interest. In this regard, the most notable difference has to do with how easy it is to evaluate support for a second-order interaction effect (P_{XZW}) and, hence, whether there's a fully interactive relationship between X, Z, and W. It's possible to directly identify whether there's a second-order interaction effect from the standard interaction model by looking at the coefficient on the interaction term XZW, β_7. In contrast, there's no way of knowing directly from the regression output with the alternative interaction model whether there's a fully interactive relationship. Instead, the researcher must formally test

Table 6.2 Comparing the standard and alternative interaction models when the modifying variables X, Z, and W are dichotomous: Nineteen key predictions

Key Prediction	Standard Interaction Model	Alternative Interaction Model
1. P_{XZW}	β_7	$\gamma_1 + \gamma_2 - \gamma_3 + \gamma_4 - \gamma_5 - \gamma_6 + \gamma_7$
2. $P_{XZ \mid W_{\min}}$	β_4	$\gamma_6 - \gamma_2 - \gamma_4$
3. $P_{XZ \mid W_{\max}}$	$\beta_4 + \beta_7$	$\gamma_1 - \gamma_3 - \gamma_5 + \gamma_7$
4. $P_{XW \mid Z_{\min}}$	β_5	$\gamma_5 - \gamma_1 - \gamma_4$
5. $P_{XW \mid Z_{\max}}$	$\beta_5 + \beta_7$	$\gamma_2 - \gamma_3 - \gamma_6 + \gamma_7$
6. $P_{ZW \mid X_{\min}}$	β_6	$\gamma_3 - \gamma_1 - \gamma_2$
7. $P_{ZW \mid X_{\max}}$	$\beta_6 + \beta_7$	$\gamma_4 - \gamma_5 - \gamma_6 + \gamma_7$
8. $P_{X \mid Z_{\min}, W_{\min}}$	β_1	γ_4
9. $P_{X \mid Z_{\min}, W_{\max}}$	$\beta_1 + \beta_5$	$\gamma_5 - \gamma_1$
10. $P_{X \mid Z_{\max}, W_{\min}}$	$\beta_1 + \beta_4$	$\gamma_6 - \gamma_2$
11. $P_{X \mid Z_{\max}, W_{\max}}$	$\beta_1 + \beta_4 + \beta_5 + \beta_7$	$-\gamma_3 + \gamma_7$
12. $P_{Z \mid X_{\min}, W_{\min}}$	β_2	γ_2
13. $P_{Z \mid X_{\min}, W_{\max}}$	$\beta_2 + \beta_6$	$\gamma_3 - \gamma_1$
14. $P_{Z \mid X_{\max}, W_{\min}}$	$\beta_2 + \beta_4$	$\gamma_6 - \gamma_2$
15. $P_{Z \mid X_{\max}, W_{\max}}$	$\beta_2 + \beta_4 + \beta_6 + \beta_7$	$\gamma_5 + \gamma_7$
16. $P_{W \mid X_{\min}, Z_{\min}}$	β_3	γ_1
17. $P_{W \mid X_{\min}, Z_{\max}}$	$\beta_3 + \beta_6$	$\gamma_3 - \gamma_2$
18. $P_{W \mid X_{\max}, Z_{\min}}$	$\beta_3 + \beta_5$	$\gamma_5 - \gamma_4$
19. $P_{W \mid X_{\max}, Z_{\max}}$	$\beta_3 + \beta_5 + \beta_6 + \beta_7$	$-\gamma_6 + \gamma_7$
Joint Effect of Z and W	$\beta_2 + \beta_3 + \beta_6$	γ_3
Joint Effect of X and W	$\beta_1 + \beta_3 + \beta_5$	γ_5
Joint Effect of X and Z	$\beta_1 + \beta_2 + \beta_4$	γ_6
Joint Effect of X, Z, and W	$\beta_1 + \beta_2 + \beta_3 + \beta_4 + \beta_5 + \beta_6 + \beta_7$	γ_7

whether $\gamma_1 + \gamma_2 - \gamma_3 + \gamma_4 - \gamma_5 - \gamma_6 + \gamma_7 = 0$ to determine if there's any evidence of a second-order interaction effect. One potential advantage of the alternative interaction model is that it allows us to see the effect of *jointly* changing some combination of X, Z, and W. In the context of our gender, race, and class example, for instance, γ_7 tells us the effect associated with changing someone's identity from lower class White male to higher class Black female. There's no equivalent coefficient for γ_7 in the standard interaction model. Instead, $\gamma_7 = \beta_1 + \beta_2 + \beta_3 + \beta_4 + \beta_5 + \beta_6 + \beta_7$. While it's certainly possible to calculate this quantity from the standard interaction model, it's not possible to read it directly from the regression output as it is with the alternative interaction model. No matter what model they employ, though, an analyst who wishes to fully evaluate a theory positing a fully interactive relationship between X, Z, and W and test all nineteen of the predictions we discussed earlier is going to have to make some post-estimation calculations. The regression output provided by both models isn't sufficient on its own to fully evaluate such a theory. Given this, the choice of model specification is largely a matter of taste.[20]

6.2.2 Substantive Application: The Impact of Demand, Supply, and Regime Type on Women's Legislative Representation

Why are there more women legislators in some countries than others? In the previous chapter, we replicated an analysis by Dhima (2022) looking at how demand-side and supply-side factors interact to influence the level of women's legislative representation around the world. Demand-side factors, as you'll recall, have to do with the preferences that people have for women legislators. Demand for women legislators is low when individuals hold traditional attitudes regarding gender roles. In contrast, supply-side factors determine the size of the pool of women with the experience and willingness to effectively compete for political office. When women have the resources and ambition to run for office, the "supply" for women's representation is high. As Dhima (2022) notes, there's an inherent conditionality built into the supply and demand theory of women's representation. An increase in the supply of qualified women candidates should have only a limited effect on women's representation when voters exhibit little desire for female

---

[20] That said, if we define "ease of interpretation" in terms of the number of coefficients that we need to examine in order to evaluate our nineteen key predictions, then the standard interaction model is clearly superior. There are seven key predictions where we need to evaluate fewer coefficients with the standard interaction model than the alternative interaction model, nine key predictions where we need to evaluate the same number of coefficients, and only three key predictions where we need to evaluate more coefficients with the standard interaction model.

legislators. Likewise, an increase in the demand for female legislators should have little effect on women's representation, at least in the short run, when the supply of qualified women candidates is low. In practice, we should expect high levels of women's representation only when supply and demand are both sufficiently high. This is precisely what we found in our earlier application.

In this new substantive application, we replicate part of a follow-up study by Dhima (2022) that suggests that the supply and demand framework that we've just outlined varies with regime type and is more applicable to democracies than dictatorships. The starting point here is that institutions like democracy and dictatorship define the *context* or *environment* in which voter preferences for female legislators (demand) and the size of the pool of qualified and willing female candidates (supply) interact to determine the observed level of women's representation. As such, it isn't appropriate to simply control for whether a country is a democracy or dictatorship when evaluating the supply and demand framework. Instead, we must recognize that the impact of supply-side and demand-side factors on women's legislative representation may depend on, or vary with, a country's regime type. This means incorporating a country's regime type into our model specification in an interactive, rather than an additive, manner.

Why might the supply and demand framework work differently in a democratic context than in an authoritarian one? While democratic institutions are designed, at least in principle, to be open and allow citizen preferences to be reflected in political outcomes, authoritarian institutions tend to be closed and are designed for cooptation and repression (Gandhi and Przeworski, 2006). Given this, we'd expect the level of women's representation in dictatorships to be much less responsive to supply-side and demand-side factors among the citizenry than in democracies. The level of women's legislative representation in dictatorships arguably has much more to do with whether political elites in these regimes want to have women in the legislature than with whether citizens demonstrate strong demand for female legislators. There are, after all, mechanisms by which authoritarian elites can, if they wish, increase the percentage of women in the legislature even when voter demand and/or the supply of qualified female candidates are low and, conversely, prevent the representation of women even if the supply of qualified candidates is high and citizens demand it. Several hypotheses can be derived from this reasoning.

Demand Hypothesis: An increase in mass demand for female representatives always has little effect on women's representation when the supply of qualified female candidates is low. However, it has an increasingly large positive effect as the supply of qualified female candidates grows so long as the regime is sufficiently democratic.

Supply Hypothesis: An increase in the supply of qualified female candidates always has little effect on women's representation when mass demand for female representatives is low. However, it has an increasingly large positive effect as the demand for female representatives grows so long as the regime is sufficiently democratic.

Democracy Hypothesis: Controlling for the level of elite demand in a country, an increase in the level of democracy has no effect on women's legislative representation when the supply of qualified female candidates and/or mass demand for women legislators is low. However, this effect becomes positive and increasingly large as the supply of qualified female candidates and mass demand for women legislators both increase.

These three hypotheses contain all nineteen of the predictions we recommend for a theory positing a fully interactive relationship like the one presented here.

As before, our dependent variable *Women's Representation* captures the percentage of female representatives in lower house legislatures around the world from 1990 through 2018 (Paxton, Kunovich and Hughes, 2007; Inter-Parliamentary Union, 2019). *Demand* captures the percentage of people in a country who disagree or strongly disagree with the claim that men make better political leaders than women (EVS, 2015; WVS, 2015). *Supply* captures the ratio of female to male labor force participation rates (EVS, 2015; WVS, 2015). *Democracy* runs from 0 to 20, with higher numbers indicating more democracy (Marshall, Gurr and Jaggers, 2019).[21] In an attempt to take account of the level of elite demand for women legislators in a country, we control for whether political actors have chosen to adopt an effective gender quota. *Effective Gender Quota* is a dichotomous variable that equals 1 if there's an effective gender quota and 0 otherwise (Hughes et al., 2017, 2019). Treating the dependent variable as continuous, we estimate an ordinary least squares regression with the following interactive specification:

$$
\begin{aligned}
\textit{Women's Representation} = \beta_0 &+ \beta_1 \textit{Demand} + \beta_2 \textit{Supply} + \beta_3 \textit{Democracy} \\
&+ \beta_4 \textit{Demand} \times \textit{Supply} \\
&+ \beta_5 \textit{Demand} \times \textit{Democracy} \\
&+ \beta_6 \textit{Supply} \times \textit{Democracy} \\
&+ \beta_7 \textit{Demand} \times \textit{Supply} \times \textit{Democracy} \\
&+ \beta_8 \textit{Effective Gender Quota} + \epsilon. \quad (6.52)
\end{aligned}
$$

---

[21] In our earlier application, *Democracy* ran from -10 to $+10$. We re-scaled the variable here so that a value of 0 would signify a "full dictatorship," making our substantive interpretation later a little easier.

While this model specification can be used to test all three of our hypotheses, we'll focus our upcoming discussion on how it can be used to evaluate the *Demand Hypothesis* and its predictions about the effect of mass demand for female representatives on women's representation. The marginal effect of *Demand* is

$$\frac{\partial\, Women's\ Representation}{\partial\, Demand} = \beta_1 + \beta_4 Supply + \beta_5 Democracy$$
$$+ \beta_7 Supply \times Democracy. \qquad (6.53)$$

According to the *Demand Hypothesis*, an increase in mass demand for female legislators should have little effect on women's representation when there's no supply of qualified women candidates and no democracy. As a result, β_1 should be close to 0. An increase in mass demand for female legislators should also have little effect when there's no supply of qualified women candidates. This means that $\beta_1 + \beta_5 Democracy$ should be close to 0 for all values of *Democracy*. Given that β_1 should be close to 0, this means that β_5 should also be close to 0. An increase in mass demand for female legislators should also have little effect in the absence of democracy. It follows that $\beta_1 + \beta_4 Supply$ should be close to 0 for all values of *Supply*. Given that β_1 should be close to 0, this means that β_4 should also be close to 0. An increase in mass demand for female legislators should have an increasingly large positive effect on women's representation when the supply of qualified female candidates and the level of democracy are both sufficiently high. This means that β_7 should be positive and that the marginal effect of *Demand* in Eq 6.53 should be positive when *Supply* and *Democracy* are both sufficiently high.

The results from the interaction model shown in Eq. 6.53 are presented in Table 6.3. We start by discussing how to interpret the coefficients for *Demand*, *Supply*, and *Democracy*. The coefficient on *Demand* is statistically insignificant and tells us that we can't reject the hypothesis that a one-unit increase in *Demand* has no effect on women's representation when *Supply* and *Democracy* are both 0. The coefficient on *Supply* is positive and statistically significant, indicating that a one-unit increase in *Supply* is associated with a 0.42 percentage point increase in women's representation when *Demand* and *Democracy* are both 0. Finally, the coefficient on *Democracy* is positive and statistically significant, telling us that a one-unit increase in *Democracy* is associated with a 3.08 percentage point increase in women's representation when *Supply* and *Democracy* are both 0. The important point here is that these coefficients don't indicate the unconditional, average, independent, or main effects of *Demand*, *Supply*, or *Democracy*. Moreover, not too much should be read into them as there are no real-world observations that satisfy the conditions associated with their interpretation.

Table 6.3 Demand, supply, regime type, and women's legislative representation I

Dependent Variable: *Women's Representation, 0–100*

	Model 1
Demand	0.26
	(0.43)
Supply	0.42**
	(0.20)
Democracy	3.08***
	(0.95)
Demand × Supply	−0.001
	(0.01)
Demand × Democracy	−0.05**
	(0.03)
Supply × Democracy	−0.05***
	(0.01)
Demand × Supply × Democracy	0.001**
	(0.0004)
Effective Gender Quota	9.04***
	(1.53)
Constant	−19.26
	(12.50)
Observations	186
R^2	0.56

Standard errors in parentheses. $^*p < 0.10$; $^{**}p < 0.05$; $^{***}p < 0.01$ (two-tailed)

We now move on to the coefficients on the first three interaction terms. The coefficient on *Demand × Supply* is statistically insignificant and indicates that there's no interaction effect between *Demand* and *Supply* when *Democracy* is 0. Consistent with our theory, this particular coefficient suggests that we can't reject the null hypothesis that there's no interaction between *Demand* and *Supply* in the absence of democracy. The coefficient on *Demand × Democracy* is negative and statistically significant, indicating that there's a negative and significant interaction effect between *Demand* and *Democracy* when *Supply* is 0. Put differently, it tells us that a one-unit increase in *Demand* (*Democracy*) reduces the marginal effect of *Democracy* (*Demand*) by −0.05 percentage points when *Supply* is 0. Not too much should be read into this particular interaction effect, though, because there are no real-world observations where *Supply* is 0. The coefficient on *Supply* × *Democracy* is negative and statistically significant, indicating that there's a negative and significant interaction effect between *Supply* and *Democracy* when *Demand* is 0. Put differently, it tells us that a one-unit increase in *Supply* (*Democracy*) reduces the marginal effect of *Democracy* (*Supply*) by −0.05 percentage points when *Demand* is 0. Again, we shouldn't read too much into this particular interaction effect because there are no real-world observations where *Demand* is 0. The important point again here is that the coefficients on these interaction terms don't capture unconditional, average, or independent interaction effects. So long as the coefficient on *Demand* × *Supply* × *Democracy* isn't equal to 0, there's no single interaction effect for *Demand* × *Supply*, *Demand* × *Democracy*, and *Supply* × *Democracy*. Whether these individual coefficients are substantively important will depend on whether it makes sense to interpret these interaction terms when the value of the third modifying variable is 0. In this particular application, this is true for only the coefficient on *Demand* × *Supply*.

We now turn to the coefficient on the last interaction term, *Demand* × *Supply* × *Democracy*. As predicted, this coefficient is positive and statistically significant. As such, it provides the important evidence of a fully interactive relationship between *Demand*, *Supply*, and *Democracy* and indicates, as Dhima (2022) argues, that the conditional nature of the supply and demand framework varies with regime type. To be more precise, the coefficient on this interaction term indicates that a one-unit increase in *Democracy* increases the interaction effect between *Demand* and *Supply* by 0.001 percentage points, that a one-unit increase in *Supply* increases the interaction effect between *Demand* and *Democracy* by 0.001 percentage points, and that a one-unit increase in *Demand* increases the interaction effect between *Supply* and *Democracy*. These second-order interaction effects are identical due to the symmetry of interactions.

Finally, in terms of the control variable, we see that countries with an effective gender quota have 9.04 percentage points more women legislators than those that don't have such a quota. This suggests that elite demand for female legislators plays an important role in determining a country's level of women's legislative representation even after taking into account the level of mass demand from the citizenry and the supply of qualified female candidates.

Having discussed how to interpret the individual coefficients from the regression output, we now turn our attention to evaluating the specific empirical support for the *Demand Hypothesis*. Given that our modifying variables are all continuous, we can fully visualize the conditional effect of *Demand* on women's legislative representation with a 3-D marginal effect plot like the one shown in panel (a) of Figure 6.17. The meshed surface shows how the effect of *Demand* varies across the different possible combinations of values for *Supply* and *Democracy*. The horizontal solid colored "zero surface" helps us to see when the marginal effect of *Demand* is positive, zero, and negative. The meshed surface is shaded darker when the marginal effect of *Demand* is statistically significant ($p < 0.05$, two-tailed). To help readers evaluate the substantive importance of the information contained in the marginal effect plot, we provide information about the distribution of the two modifying variables by placing circles on the base of the "cube" housing the plot that indicate the location of the sample observations in the two-dimensional space defined by *Supply* and *Democracy*. Although we've chosen to treat the level of democracy in a country as continuous, some scholars prefer to make a dichoto-mous distinction between democracies and dictatorships. As a result, we've made the circles representing "democratic" observations (*Democracy* > 15) darker than those representing "dictatorial" observations (*Democracy* < 16).

The 3-D plot in panel (a) allows us to evaluate the predictions about the sign of the marginal effect of *Demand* contained in the *Demand Hypothesis*. As predicted, we see that the marginal effect of *Demand* is positive and statistically significant only when the supply of qualified female candidates and the level of democracy in a country are both sufficiently high. This is indicated by the dark shaded region in the far corner of the plot. *Demand* never has a positive and statistically significant effect on women's representation when either *Supply* or *Democracy* is low. The small dark shaded region in the bottom left corner suggests that the marginal effect of *Demand* is associated with a statistically significant negative effect on women's representation when *Supply* is very low and the level of democracy in a country is relatively high. While the statistically significant negative effect of *Demand* in this region is inconsistent with the *Demand Hypothesis*, the distribution of the circles on the base indicates that very few real-world observations have the requisite combination of values for *Supply* and *Democracy* to be in the highlighted region. As a result, we're not overly concerned with the evidence of a significant negative effect for *Demand* in the small dark shaded region in the bottom left corner of the plot.

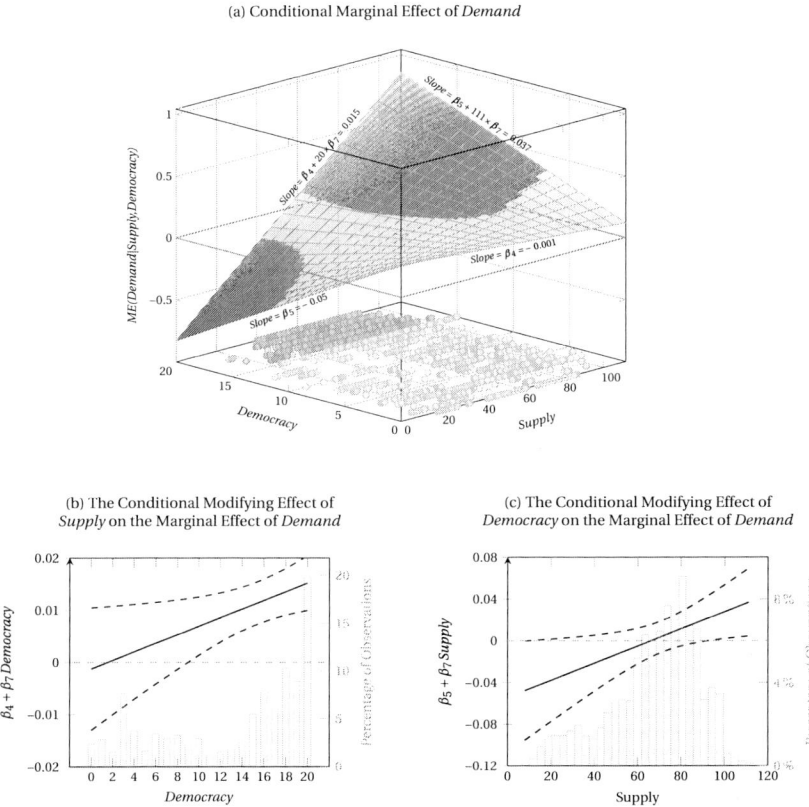

(a) Conditional Marginal Effect of *Demand*

(b) The Conditional Modifying Effect of *Supply* on the Marginal Effect of *Demand*

(c) The Conditional Modifying Effect of *Democracy* on the Marginal Effect of *Demand*

(d) Second-Order Interaction Effect: *Demand* × *Supply* × *Democracy*: 0.001 (0.0001, 0.002)

Figure 6.17 Evaluating *Demand Hypothesis* I

Note: In panel (a), the meshed surface shows how the conditional marginal effect of *Demand* varies with *Supply* and *Democracy*. The surface is shaded darker when the marginal effect is statistically significant ($p < 0.05$, two-tailed). The solid colored "zero surface" indicates when the marginal effect of *Demand* is positive and negative. The circles on the base of the "cube" indicate the location of the sample observations in the two-dimensional space defined by *Supply* and *Democracy*. The darker circles represent "democracies" (*Democracy* > 15), while the lighter circles represent "dictatorships" (*Democracy* < 16). The solid line in panel (b) shows how the conditional modifying effect of *Supply* on the marginal effect of *Demand* varies with the level of democracy in a country, while the solid line in panel (c) shows how the conditional modifying effect of *Democracy* on the marginal effect of *Demand* varies with *Supply*. The dashed lines represent two-tailed 95% confidence intervals. The histograms show the percentage of observations at different levels of *Democracy* (b) or different levels of *Supply* (c). The vertical axes for the histograms are shown on the right of each plot. The dependent variable *Women's Representation* is measured on a 0–100 scale. The plots in all three panels are based on the results shown in Table 6.3. The second-order interaction effect for *Demand* × *Supply* × *Democracy* and its confidence interval are shown at the bottom (d).

To some extent, the 3-D marginal effect plot also speaks to the predictions contained in the *Demand Hypothesis* regarding the interactive relationships between *Demand*, *Supply*, and *Democracy*. The fact that the meshed surface is not flat and perfectly parallel to the zero surface indicates that the effect of demand on women's legislative representation depends on, or varies with, the values of *Supply* and *Democracy*. The slope of the meshed surface to the right captures how the marginal effect of *Demand* varies with *Supply* (the interaction effect between *Demand* and *Supply*), while the slope of the meshed surface to the left captures how the marginal effect of *Demand* varies with *Democracy* (the interaction effect between *Demand* and *Democracy*). As we can see, the slope of the surface to the right changes with the value of *Democracy*, and the slope of the surface to the left changes with the value of *Supply*. The fact that the slope of the meshed surface isn't constant tells us that the interaction effects between *Demand* and *Supply* and between *Demand* and *Democracy* are dependent on the value of the other modifying variable and hence that we have a fully interactive relationship between *Demand*, *Supply*, and *Democracy*.

Although the 3-D plot in panel (a) fully visualizes the marginal effect of *Demand*, it's important to recognize that it can provide only *suggestive* evidence regarding the interactive relationships between *Demand*, *Supply*, and *Democracy*. While we can easily see that the slopes of the meshed surface to the right and left aren't always zero, it's much harder to see precisely when they're negative, zero, or positive. More importantly, we can't see if and when these slopes are statistically significant. Remember that the dark shaded regions on the meshed surface only indicate the points at which the marginal effect of *Demand* is statistically different from zero and not the points at which the *slope* of the surface is statistically different from zero.[22] In other words, the 3-D plot in panel (a) provides no evidence about if and when there's a statistically significant interaction effect between *Demand* and *Supply* or between *Demand* and *Democracy*. Nor does the 3-D plot allow us to determine whether these interaction effects are themselves conditional on the value of the other modifying variable. The fact that the slopes of the meshed surface to the left and to the right aren't constant is suggestive of a second-order interaction effect. However, we'd like to know if this second-order interaction effect is statistically significant. The main point here is that we need to move beyond the 3-D marginal effect plot shown in panel (a) if we are to evaluate any of the predictions regarding interactions in the *Demand Hypothesis*.

---

[22] In effect, the dark shaded regions indicate when the distance of a given point on the marginal effect surface to the zero surface is statistically significant. What we'd like to know here, though, is if and when the *slope* of the marginal effect surface is significantly different from zero.

The interaction effect plot in panel (b) shows the conditional interaction effect between *Demand* and *Supply* and thus how the modifying effect of *Supply* on the marginal effect of *Demand* varies with the level of democracy in a country, $\beta_4 + \beta_7 Democracy$. It's worth briefly seeing how the information contained in this plot relates to the information contained in the 3-D marginal effect plot shown in panel (a). The conditional modifying effect of *Supply* in panel (b) starts at -0.001 when *Democracy* is 0. This modifying effect corresponds to the almost flat slope to the right of the meshed surface on the front right wall of the 3-D plot in panel (a). As the plot in panel (b) indicates, the modifying effect of *Supply* becomes increasingly positive as the level of democracy in a country grows. When *Democracy* $= 20$, the modifying effect of *Supply* is 0.015. This modifying effect corresponds to the slope to the right of the meshed surface on the far left wall of the 3-D plot in panel (a). In effect, the modifying effect of *Supply* shown in panel (b) shows the slope to the right of the meshed surface in panel (a) across the different values of *Democracy*. The value of the interaction effect plot in panel (b) is that the dashed confidence intervals and horizontal zero line allow us to easily determine precisely if and when the modifying effect of *Supply* is positive and statistically significant. As we can see, the modifying effect of *Supply* is negative and statistically insignificant when *Democracy* < 1.6 (5.2%), positive and statistically insignificant when $1.6 < Democracy < 9$ (24.3%), and positive and statistically significant when *Democracy* > 9 (68.3%). As before figures in parentheses indicate the percentage of observations that fall within each of the specified ranges. The results here are consistent with the *Demand Hypothesis* and its prediction that the conditionality of the supply and demand framework is more applicable to democracies than dictatorships.

The interaction effect plot in panel (c) shows the conditional interaction effect between *Demand* and *Democracy* and thus how the modifying effect of *Democracy* on the marginal effect of *Demand* varies with the value of *Supply*, $\beta_5 + \beta_7 Supply$. The conditional modifying effect of *Democracy* is -0.05 when *Supply* is 0. This modifying effect corresponds to the slope of the meshed surface down to the left on the front left wall of the 3-D plot in panel (a). This particular modifying effect is not shown in panel (c) as the lowest observed value for *Supply* is 8.6. As the plot in panel (c) indicates, the negative modifying effect of *Democracy* shrinks in magnitude and eventually becomes positive with higher values of *Supply*. When *Supply* is at its maximum observed value of 111, the modifying effect of *Democracy* is 0.037. This corresponds to the slope of the meshed surface up to the left on the far right wall of the 3-D plot in panel (a). In effect, the modifying effect of *Democracy* shown in panel (c) corresponds to the slope to the left of the meshed surface in panel (a) across the different values of *Supply*. According to the interaction effect plot in panel (c), the modifying effect of *Democracy* is negative and statistically significant when *Supply* < 9.5 (0.08%), negative and statistically insignificant when 9.5 < *Democracy* < 66.1 (39.6%), positive and statistically insignificant when 66.1 < *Democracy* < 92.2 (51.4%), and positive and statistically significant when *Democracy* > 92.2 (9.1%). Figures in parentheses again indicate the percentage of observations that fall within each of the specified ranges. These results are consistent with the *Demand Hypothesis*, which predicts that the modifying effect of *Democracy* will be close to zero when *Supply* is low and positive when *Supply* is sufficiently high.

The symmetry of interactions means that the slopes of the two interaction effect lines in panels (b) and (c) are identical, $\beta_7 = 0.001$. Note that while we can determine if and when the interaction effects shown in panels (b) and (c) are statistically significant by looking at the dashed confidence intervals and the horizontal zero line, there's no way of actually determining directly from the plots whether the variable on the horizontal axis has a statistically significant modifying effect on the reported interaction effects. We can certainly see that the two interaction effect lines slope upwards, but there's no way of knowing directly from the plots whether these slopes are significantly different from zero. This is why we also report the coefficient on the interaction term *Demand* × *Supply* × *Democracy* along with its confidence interval in (d) at the bottom of Figure 6.17. It's this coefficient that defines the slope of the interaction effect lines in the two plots and shows whether there's any evidence of a second-order interaction effect or, equivalently, a fully interactive relationship between *Demand*, *Supply* and *Democracy*.

All four pieces of information reported in Figure 6.17 are necessary to fully evaluate the nine key predictions contained in the *Demand Hypothesis*. To be specific, the 3-D marginal effect plot in panel (a) allows us to evaluate the predictions about the sign of the marginal effect of *Demand* at different combinations of values for *Supply* and *Democracy*: $P_{Demand|Supply_{min}, Democracy_{min}}$, $P_{Demand|Supply_{min}, Democracy_{max}}$, $P_{Demand|Supply_{max}, Democracy_{min}}$, and $P_{Demand|Supply_{max}, Democracy_{max}}$. The interaction effect plot in panel (b) allows us to evaluate the predictions about the interaction effect between *Demand* and *Supply*: $P_{Demand×Supply|Democracy_{min}}$ and $P_{Demand×Supply|Democracy_{max}}$. The interaction effect plot in panel (c) provides the necessary information to evaluate the predictions about the interaction effect between *Demand* and *Democracy*: $P_{Demand×Democracy|Supply_{min}}$ and $P_{Demand×Democracy|Supply_{max}}$. Finally, the coefficient on the interaction term *Demand* × *Supply* × *Democracy* reported in (d) allows us to evaluate the prediction about a second order interaction effect: $P_{Demand×Supply×Democracy}$.

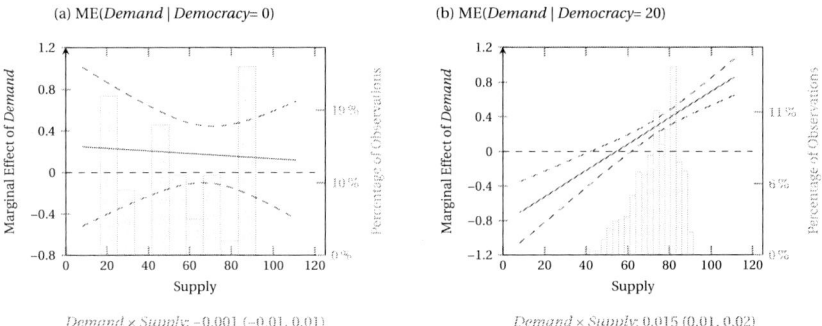

Figure 6.18 The marginal effect of *Demand*

Note: The solid lines in panels (a) and (b) show the marginal effect of *Demand* across the observed range of *Supply* when *Democracy* = 0 (a) and when *Democracy* = 20 (b). The interaction effect between *Demand* and *Supply* for the case where *Democracy* is 0 is shown below panel (a), while the interaction effect between *Demand* and *Supply* for the case where *Democracy* is 20 is shown below panel (b). The dashed lines in each panel represent two-tailed 95% confidence intervals. The histograms show the percentage of observations at different values of *Supply* at the specified level of *Democracy*. The vertical axes for the histograms are shown on the right of each plot. The dependent variable, *Women's Representation*, is measured on a 0–100 scale. The plots are based on the results shown in Table 6.3.

Some people find it difficult to interpret the information contained in a 3-D marginal effect plot like the one shown in panel (a) of Figure 6.17. An alternative strategy that serves a similar purpose is to present a series of 2-D marginal effect plots showing how the effect of *Demand* varies across the observed range of *Supply* for a few substantively relevant values of *Democracy*. In panels (a) and (b) in Figure 6.18, we provide such 2-D plots for the case where *Democracy* is 0 and the case where *Democracy* is 20. The marginal effect line in panel (a) is equivalent to the front right edge of the meshed surface shown in the 3-D plot in Figure 6.17, while the marginal effect line in panel (b) is equivalent to the far left edge. To help readers know if the slopes of the two marginal effect lines are statistically significant, we report the appropriate interaction effect between *Demand* and *Supply* and its confidence interval below each plot. These additional pieces of information are equivalent to the start and end points of the conditional interaction effect line for *Demand* × *Supply* reported in panel (b) of Figure 6.17.

As predicted, the plot in panel (a) of Figure 6.18 indicates that *Demand* never has a statistically significant effect on women's representation at any value of *Supply* when *Democracy* is at its lowest value of 0. Also as predicted, the plot in panel (b) indicates that *Demand* has a positive and statistically significant effect when *Supply* is sufficiently high and *Democracy* is at its maximum value of 20. In line with the dark shaded region in the bottom left corner of the 3-D plot in Figure 6.17, the 2-D plot in panel (b) indicates that *Demand* has a statistically significant negative effect when *Supply* is low and *Democracy* is high. As we noted previously, though, almost no real-world observations fall into this particular region of significance. Indeed, the histogram in panel (b) indicates that just 0.2% of the observations where *Democracy* = 20 have a *Supply* score low enough (<42.3) to see a statistically significant negative effect for *Demand*. Again as predicted, the slope of the marginal effect line in panel (b) is positive and statistically significant, whereas the slope of the marginal effect line in panel (a) is basically flat and statistically insignificant. This is consistent with the claim that the conditionality of the supply and demand framework is more applicable to democracies than dictatorships.

The disadvantage of using a series of 2-D marginal effect plots like this rather than the 3-D marginal effect plot in Figure 6.17 is that we can examine the marginal effect of *Demand* across the observed range of *Supply* for only a few (in this case, two) discrete values for *Democracy*. In contrast, the 3-D plot in Figure 6.17 allows us to see the marginal effect of *Demand* for all possible combinations of values for *Supply* and *Democracy*. We remind readers that the 2-D marginal effect plots in Figure 6.18 are a (partial) substitute for only the 3-D plot in panel (a) of Figure 6.17. Scholars who wish to fully evaluate all of the predictions contained in the *Demand Hypothesis* should still present the information contained in the other panels of Figure 6.17.

As always, we should consider the substantive significance of our results. In Figure 6.19, the shaded cells show how the effect of increasing *Demand* by 20 percentage points, which is slightly less than one standard deviation (20.71), varies across four different countries that differ in terms of their level of democracy and their supply of qualified female candidates. To be specific, our hypothetical countries differ in terms of whether their *Democracy* score is 0 (full dictatorship) or 20 (full democracy) and whether their *Supply* score is the observed mean of 67.88 or the observed maximum of 111.01. The modifying effects of changing the supply of qualified female candidates from its mean to its maximum observed value for each level of democracy are shown in the top two cells of the "Difference" column on the right. The modifying effects of changing the level of democracy from 0 to 20 for each value of *Supply* is shown in the first two cells of the "Difference" row at the bottom. The second-order interaction effect that captures how the specified change in *Democracy* modifies the modifying effect of *Supply*

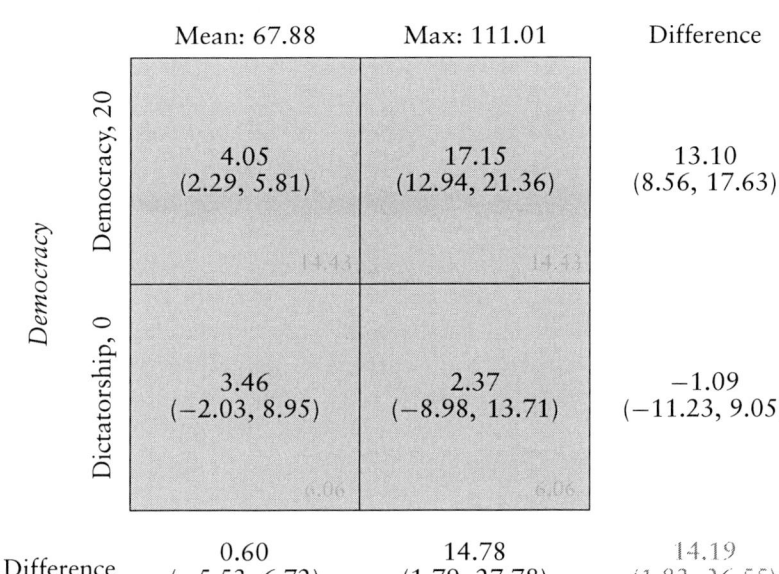

Figure 6.19 The substantive effect of *Demand* on women's legislative representation

Note: The shaded square shows the effect of a 20-unit increase in *Demand* along with 95% confidence intervals for four hypothetical countries that differ in terms of their *Democracy* score (0 or 20) and their *Supply* score (mean 67.88 or max 111.01). The modifying effect of changing the value of *Supply* from its mean to its maximum value for each level of *Democracy* is shown in the right "Difference" column. The modifying effect of changing their level of *Democracy* for each value of *Supply* is shown in the bottom "Difference" row. The small gray numbers in the bottom right corner of each cell indicate the mean level of women's legislative representation among countries with the specified level of *Democracy*. The dependent variable, *Women's Representation*, is measured on a 0–100 scale.

on the marginal effect of a 20-unit increase in *Demand* is shown in gray in the bottom right cell. Due to the symmetry of interactions, this second-order interaction effect also captures how the specified change in *Supply* modifies the modifying effect of *Democracy* on the marginal effect of a 20-unit increase in *Demand*.

To provide some kind of metric for evaluating whether the reported effects are substantively meaningful, we show the mean level of women's legislative representation at the specified level of democracy in small gray font in the bottom right corner of each cell. For example, the 6.06 written in gray in the bottom right corner of the bottom two cells indicates that the mean level of women's representation in countries whose *Democracy* score is 0 is 6.06%.[23]

As predicted, an increase in demand for female representatives is more proportionally translated into actual women's representation when the supply of qualified female candidates is high and political institutions are democratic. To be specific, a 20 percentage point increase in *Demand* is associated with a statistically significant 17.15 percentage point increase in women's legislative representation when *Supply* is at its maximum observed level and a country is fully democratic. This increase is equivalent to a 118.8% increase in the mean level of women's representation in a country that is fully democratic. In contrast, a 20 percentage point increase in *Demand* is associated with just a 3.46 percentage point increase in women's legislative representation when *Supply* is at its mean observed level and a country is fully dictatorial. While this increase is statistically insignificant, it does equate to a 57.1% increase in the mean level of women's representation in a country that is fully dictatorial. As predicted, an increase in *Supply* from its mean to its maximum observed value has little substantive impact on the effect of *Demand* in a fully dictatorial country. In contrast, this same increase in *Supply* has a substantively large impact on the effect of *Demand* in a fully democratic country. Indeed, the effect of *Demand* in a fully democratic country is 13.10 percentage points higher when *Supply* is at its maximum, as opposed to its mean, observed value. This "interaction effect" represents a 90.8% increase in the mean level of women's representation in a full democracy. Also as predicted, a change from a full dictatorship to a full democracy has little substantive impact on the effect of *Demand* when *Supply* is at its observed mean value. In contrast, this same increase in *Democracy* has a substantively large impact on the effect of *Demand* when *Supply* is at is maximum observed value. Indeed, the effect of *Demand* is 14.78 percentage points higher when *Supply* is at its maximum observed value in a full democracy than in a full dictatorship. This "interaction effect" represents a 102.4% increase in the mean global level of women's representation in the 1990–2018 time period.

---

[23] Ideally, we'd report the mean level of women's representation in a "baseline" scenario where countries have the specified levels of both *Democracy* and *Supply* corresponding to each of the four cells. Although there are multiple country observations that have the specified level of *Democracy* in each row, the more continuous nature of the *Supply* variable means that there are very few observations that are characterized by our two values of *Supply*. Rather than simply focus on those countries with the baseline level of *Democracy* as we do, an alternative strategy would be to indicate the mean level of women's representation for those countries that combine the specified level of democracy with a level of *Supply* that falls within a given range of the specified "baseline" value for *Supply*. A second alternative would be to report the *predicted* level of women's representation in the baseline scenarios.

We finish by noting that our application has focused on testing only the empirical support for the *Demand Hypothesis*. In practice, we would want to use similar techniques to those presented here to also test the empirical support for the *Supply Hypothesis* and the *Democracy Hypothesis*. Only by doing so can we be confident that the data support our particular theory of interaction between *Demand*, *Supply*, and *Democracy* rather than one of the other thousands of possible interactive relationships that could occur.

In this chapter, we began to look at some more theoretically complex forms of conditionality than we've seen previously. In particular, we examined theories that imply that the effect of an independent variable such as X on Y depends on the value of more than one other modifying variable. In the next chapter, we continue to look at more complex forms of conditionality by examining theories that imply that the effect of an independent variable such as X depends on its own value rather than the value of another variable.

6.3 EXERCISES

1. In the chapter, we examined how support for feminism was related to an individual's gender, education, and age. We estimated the following linear-interactive model specification:

$$Feminism = \beta_0 + \beta_1 Female + \beta_2 College\ Educated + \beta_3 Age$$
$$+ \beta_4 Female \times College\ Educated$$
$$+ \beta_5 Female \times Age$$
$$+ \beta_6 Controls + \epsilon. \tag{6.54}$$

The results from the analysis for our key variables are shown in Table 6.4. Part of the estimated variance-covariance matrix for the coefficients is shown below:

$$v(\beta) = \begin{pmatrix} 7.1072878 & & & & & \\ 0.53874128 & 1.4754821 & & & & \\ 0.05826377 & -0.00280156 & 0.00123011 & & & \\ -1.2959764 & -1.447081 & 0.00254588 & 2.8689833 & & \\ -0.11614581 & 0.00274064 & -0.00119335 & -0.0005735 & 0.0023413 & \\ \vdots & \vdots & \vdots & \vdots & \vdots & \vdots \end{pmatrix}. \tag{6.55}$$

a. You'll notice that we haven't reported the standard errors for the coefficients on our key independent variables in Table 6.4. Use the information in the variance-covariance matrix to identify the "missing" standard errors and put them in the appropriate empty parentheses in Table 6.4.

Table 6.4 Gender, educational attainment, age, and support for feminism II	
Dependent Variable: *Feminism*, 0–100	
	Model 1
Female	13.97506***
	( )
Education	2.411783**
	( )
Age	0.2223337***
	( )
Female × *Education*	3.460115**
	( )
Female × *Age*	−0.1782375***
	( )
Constant	74.79999
	(2.181709)
Controls	Yes
Observations	2,710
R^2	0.31

Standard errors in parentheses. $^*p < 0.10$; $^{**}p < 0.05$; $^{***}p < 0.01$ (two-tailed)

Note: *Controls* refers to the seven additional control variables that are shown in Table 6.1.

    b. Use the results in Table 6.4 and the variance-covariance matrix to calculate the effect of being female for a 30 year old college graduate. What are the variance, standard error, and *t*-statistic associated with this effect? What's the two-tailed 95% confidence interval for this effect?

2. In the chapter, we examined how women's legislative representation varied with the demand for female representatives (*Demand*), the supply of qualified female candidates (*Supply*), and regime type (*Democracy*). We estimated the following linear-interactive model specification:

Table 6.5 Demand, supply, regime type, and women's legislative representation II	
Dependent Variable: *Women's Representation, 0–100*	
	Model 1
Demand	0.2586052
	( )
Supply	0.4157726**
	( )
Democracy	3.08445***
	( )
Demand×Supply	−0.0012633
	( )
Demand×Democracy	−0.0543243**
	( )
Supply×Democracy	−0.0505661***
	( )
Demand×Supply×Democracy	0.0008223**
	( )
Observations	186

Standard errors in parentheses. $^*p < 0.10$; $^{**}p < 0.05$; $^{***}p < 0.01$ (two-tailed)

$$
\begin{aligned}
\textit{Women's Representation} = &\ \beta_0 + \beta_1 Demand + \beta_2 Supply + \beta_3 Democracy \\
&+ \beta_4 Demand \times Supply \\
&+ \beta_5 Demand \times Democracy \\
&+ \beta_6 Supply \times Democracy \\
&+ \beta_7 Demand \times Supply \times Democracy \\
&+ \beta_8 Controls + \epsilon.
\end{aligned} \tag{6.56}
$$

The results from the analysis for our key variables are shown in Table 6.5. Part of the estimated variance-covariance matrix for the coefficients is shown below:

$V(\beta)$

$$
=
\begin{pmatrix}
0.18704159 & & & & & & \\
0.06608508 & 0.0408126 & & & & & \\
0.28623153 & 0.15312129 & 0.90794061 & & & & \cdot \\
-0.00243877 & -0.00106831 & -0.00396534 & 0.00003546 & & & \\
-0.01087325 & -0.0040755 & -0.02192985 & 0.00014096 & 0.0007214 & & \\
-0.003947 & -0.0025801 & -0.01277234 & 0.00006256 & 0.00029574 & 0.00020294 & \\
-0.00014155 & 0.00006411 & 0.00029872 & -0.000002027 & 0.000009343 & 0.0000045 & 0.0000001318 \\
\vdots & \vdots & \vdots & \vdots & \vdots & \vdots & \vdots
\end{pmatrix}
$$

(6.57)

a. You'll notice that we haven't reported the standard errors for the coefficients on our key independent variables in Table 6.5. Use the information in the variance-covariance matrix to identify the "missing" standard errors and put them in the appropriate empty parentheses in Table 6.5.

b. Use the results in Table 6.5 and the variance-covariance matrix to calculate the effect of a 20 unit increase in *Demand* when *Supply* = 80 and *Democracy* = 15. What are the variance, standard error, and *t*-statistic associated with this effect? What's the two-tailed 95% confidence interval for this effect?

7 When an Independent Variable Interacts with Itself

In this chapter, we look at theories that imply that the effect of X depends in some way on its own value rather than the value of one or more other variables. We might refer to this type of interaction as a "self-interaction." In effect, we're interested in theories predicting a non-linear relationship between X and Y. All of the theories that we've looked at so far in this book assume a linear relationship between X and Y. A linear relationship implies that the marginal effect of X, holding the values of the other independent variables constant, is always the same irrespective of the level of X. In contrast, a non-linear relationship implies that the magnitude, and possibly the sign, of the marginal effect of X depends on, or varies with, the level of X. Non-linear relationships, such as between X and Y, can often be appropriately modeled using interaction models.

> ⚠ **Important:** A linear relationship between X and Y implies that the effect of X, holding the values of any other independent variables constant, is always the same irrespective of the level of X. In contrast, a non-linear relationship implies that the magnitude, and possibly the sign, of the effect of X on Y depends on the level of X.

Non-linear relationships can take many forms. Here, we focus on theories that predict one of two specific types of possible non-linear relationship between X and Y. First, we look at theories that predict some kind of smooth curvilinear relationship between X and Y. The conditional implications of these types of theories can often be evaluated using polynomial regression models. Although not always recognized as such, polynomial regression models are just a type of interaction model. As we demonstrate, polynomial regression models aren't just useful for modeling a non-linear relationship between X and Y. To this point in the book, we've focused on using interaction models to test the conditional implications of theories where the interaction effect between, say, X and

Z, is expected to be linear. It turns out that we can specify polynomial regression models that allow us to relax the linear interaction effect assumption. In other words, polynomial regression models allow us to test the conditional claims of theories predicting that, say, the marginal effect of X varies in a *non-linear* way with Z.

Second, we briefly look at theories that predict a non-linear relationship between X and Y that involves some kind of threshold effect. These theories might predict, for example, that X has one effect below some threshold $X = X^T$ and a different effect above this threshold. We focus our attention on piecewise linear and switching regression models, which allow us to capture situations where X^T represents a break in the linear relationship between X and Y. Although not always recognized as such, piecewise linear regression models and switching regression models are types of interaction model.

7.1 POLYNOMIAL REGRESSION MODELS

In Figure 7.1, we show three examples of possible non-linear relationships between X and Y that might be predicted by our theories. The predicted value plots on the left show how the value of Y changes across the observed range of values of X. To the right of each predicted value plot is a corresponding plot showing how the marginal effect of X on Y varies with the value of X. These marginal effect plots show how the slope of the line in the predicted value plot changes with X. The dashed gray zero line in these plots helps us to see when the marginal effect of X is positive, negative, and zero.

Each of the three predicted value plots indicates a different possible non-linear relationship between X and Y. In panel (a), the predicted value plot shows that Y increases with higher values of X but at a declining rate. The slope of the relationship between X and Y is positive and large when X is low. The magnitude of this positive slope declines steadily as X increases in value and eventually becomes zero when X is at its maximum value of 6. The changing nature of the slope relationship between X and Y is explicitly captured in the corresponding marginal effect plot on the right. From this, we see that the marginal effect of X is large and positive when X is low. We also see that the magnitude of this positive effect declines in a linear way with higher values of X until it becomes zero when X is 6. The fact that the marginal effect of X is a linear function of the modifying variable X tells us that, although the relationship between X and Y is non-linear, we continue to have a linear interaction effect; the modifying effect of X – the slope of the marginal effect line – is constant. Substantively, the plot shows that the marginal effect of X is associated with a 12 unit increase in Y when X is 0, a 6 unit increase in Y when X is 3, and no change in Y when X is 6.

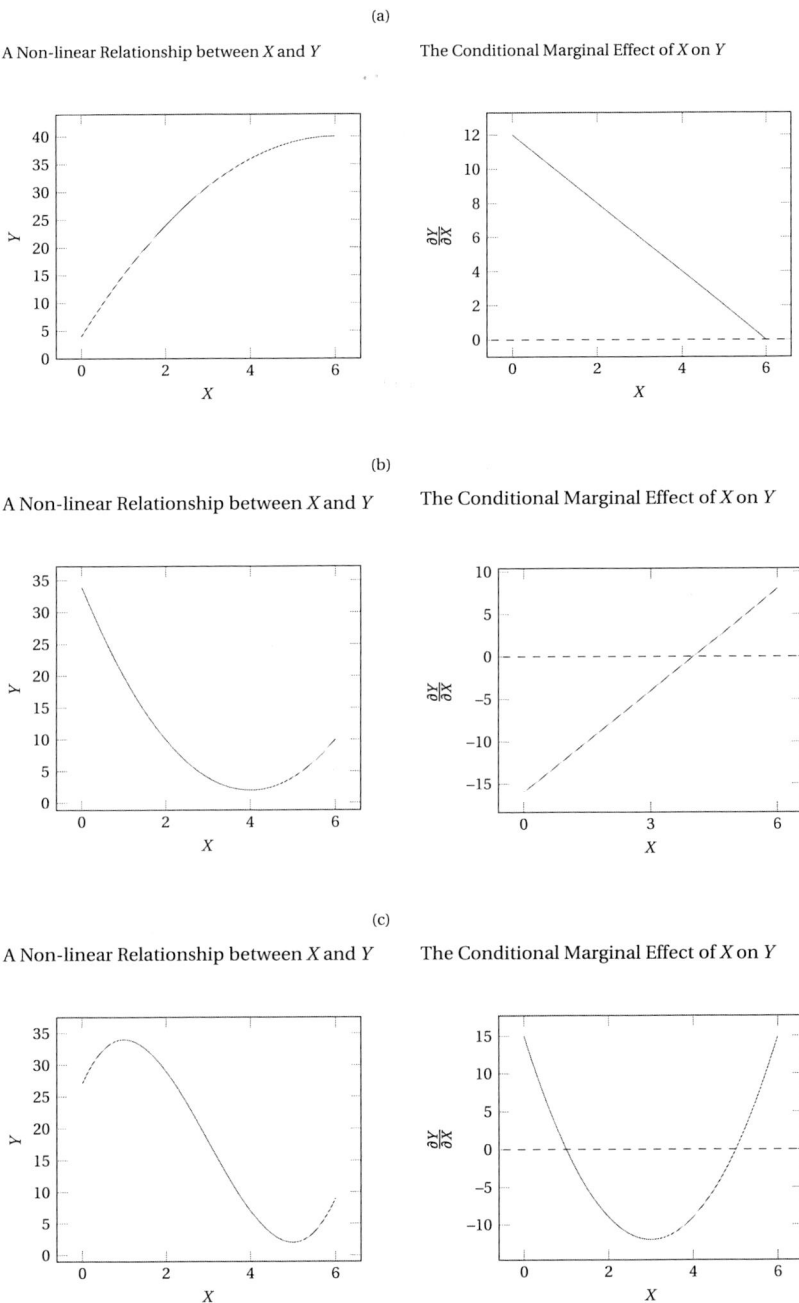

Figure 7.1 (a)–(c) Three possible non-linear relationships between X and Y

Note: The plots on the left show how the value of Y changes with X. To the right of these plots is a corresponding marginal effect plot showing how the marginal effect of X on Y changes with the value of X.

In panel (b), the predicted value plot shows that Y first decreases with X until X reaches a value of 4 and then increases with X. The slope of the relationship between X and Y is negative and large when X is low. The magnitude of this negative slope declines steadily as X increases in value and eventually becomes zero when X is 4. The slope of the relationship between X and Y then becomes positive and increasingly large as the value of X increases further. The changing nature of the slope relationship between X and Y is captured in the corresponding marginal effect plot on the right. The marginal effect of X is large and negative when X is low. The magnitude of this negative effect declines in a linear way with higher values of X and eventually becomes positive and increasingly large when X is larger than 4. The fact that the marginal effect of X is again a linear function of the modifying variable tells us that we continue to have a linear interaction effect where the modifying effect of X is constant. Substantively, the plot shows that the marginal effect of X is associated with a 16 unit reduction in Y when X is 0, no change in Y when X is 4, and an 8 unit increase in Y when X is 6.

In panel (c), we have a more complicated non-linear relationship in which Y first increases with X, then decreases with X, and then increases with X. The changing nature of the slope relationship between Y and X is again captured in the corresponding marginal effect plot on the right. The marginal effect of X is large and positive when X is low. The magnitude of this positive effect first declines at a decreasing rate with higher values of X and becomes negative when X is 1. The magnitude of this negative effect then increases with X at a decreasing rate until X is 3, after which the magnitude of the negative effect decreases with X at an increasing rate until X reaches 5. At this point, the marginal effect of X becomes positive and increasingly large with higher values of X. In this example, the marginal effect of X changes in a non-linear way with the modifying variable. In other words, we have a non-linear interaction effect where the modifying effect of X isn't constant.[1] Substantively, the plot shows that the marginal effect of X is associated with a 15 unit increase in Y when X is either 0 or 6, no change in Y when X is either 1 or 5, and a 12 unit reduction in Y when X is 3.

A common approach to modeling non-linear relationships between X and Y like those shown in Figure 7.1 involves using polynomial regression.[2]

---

[1] So far in the book, we've focused on theories that predict a linear interaction effect. This is our first example where the predicted interaction effect is non-linear.

[2] While we focus on polynomial regression models in what follows, we note that scholars sometimes use exponential growth or power models to capture particular types of non-linear relationships. We don't examine these models here as the non-linear relationships they capture don't arise from potentially interacting variables. In other words, they're not interaction models.

Polynomial regression allows for a smooth curvilinear relationship between X and Y and is especially appropriate when the slope of the relationship between X and Y changes sign. Polynomial regression involves modeling Y as a function of X and one or more higher powers of X. A generic polynomial regression model can be written as

$$Y = \beta_0 + \beta_1 X + \beta_2 X^2 + \beta_3 X^3 + \cdots + \beta_m X^m + \epsilon, \tag{7.1}$$

where m is the "degree" of the polynomial.[3] We sometimes give polynomial regression models specific names when the degree of the polynomial is small. For example, we often refer to a polynomial regression model of degree $m = 2$ as a *quadratic* regression model. And we typically refer to a polynomial regression model of degree $m = 3$ as a *cubic* regression model. It should be immediately obvious that polynomial regression models are just multiplicative interaction models that involve self-interaction. After all, X^2 is really just $X \times X$, X^3 is just $X \times X \times X$, and so on. This means that everything we've learned previously about interaction models transfers directly over to thinking about estimating and interpreting polynomial regression models.[4]

> ⚠ **Important:** Theories implying that the effect of X on Y depends in some way on the value of X predict a non-linear relationship between X and Y. A common way to model such a non-linear relationship involves using a polynomial regression model,
>
> $$Y = \beta_0 + \beta_1 X + \beta_2 X^2 + \beta_3 X^3 + \cdots + \beta_m X^m + \epsilon, \tag{7.2}$$
>
> where m is the chosen "degree" of the polynomial. Polynomial regression models allow for a smooth curvilinear relationship between X and Y and are multiplicative interaction models.

Polynomial regression models are able to capture many different types of non-linear relationship by allowing there to be "turning points" in the

---

[3] Although these models are used to capture possible non-linear relationships between X and Y, they remain *linear* regression models as they're *linear in the parameters*. Loosely speaking, a model that's linear in the parameters is one where (i) each term is either a constant or the product of a parameter and an independent variable and (ii) the terms are summed. Thus, while a polynomial regression model is non-linear in the independent variables due to the higher powers of X, it's linear in the parameters and therefore qualifies as a linear regression model.

[4] This isn't the only time that different terminology is used to describe methods that are fundamentally just multiplicative interaction models. Later in this chapter, for example, we'll look at complicated-sounding methods such as piecewise linear and switching regression models. All of these methods are, at their core, just interaction models.

line capturing the relationship between X and Y where the slope of the relationship, and hence the effect of X on Y, changes sign. In most cases, the number of turning points that occur with a polynomial regression model of degree m is $m-1$.[5] This means that we can use our theoretical predictions about the non-linear relationship between X and Y to determine the appropriate degree of our polynomial regression model. The predicted value plot in panel (b) of Figure 7.1 indicates that we're expecting a non-linear relationship between X and Y where there's one turning point. In this case, we'd want to use a polynomial regression model of degree $m=2$,

$$Y = \beta_0 + \beta_1 X + \beta_2 X^2 + \epsilon. \tag{7.3}$$

In contrast, the predicted value plot in panel (c) of Figure 7.1 indicates that we're expecting a non-linear relationship between X and Y where there are two turning points. In this case, we'd want to use a polynomial regression model of degree $m=3$,

$$Y = \beta_0 + \beta_1 X + \beta_2 X^2 + \beta_3 X^3 + \epsilon. \tag{7.4}$$

The values of X at which the turning points occur may or may not fall within the observed range of X. When they fall in the observed range, the slope relationship between X and Y is non-monotonic in that it changes sign. This is the case for the non-linear relationships depicted in panels (b) and (c) of Figure 7.1. In contrast, when they don't fall in the observed range, the slope relationship between X and Y is monotonic in that it doesn't change sign. This is similar to the case for the non-linear relationship shown in panel (a) of Figure 7.1 where the turning point occurs at the maximum observed value of $X=6$, and the slope relationship between X and Y is always non-negative.

As with all interaction models, we should include all of the constitutive elements of the interaction terms when specifying a polynomial regression model to test a hypothesis about the conditional effect of X on Y. This is because our model parameters are likely to be estimated with bias when one or more of the constitutive elements of an interaction term are omitted. This means, for example, that we should include X when we have a polynomial model of degree $m=2$ that includes X^2 and that we should include X and X^2 when we have a polynomial model of degree $m=3$ that includes X^3, and so on. The intuition here is the same as we saw earlier in Chapter 3. Consider the following quadratic regression model:

$$Y = \beta_0 + \beta_1 X + \beta_2 X^2 + \epsilon. \tag{7.5}$$

---

[5] Technically, a polynomial model of degree m has at most $m-1$ turning points where the slope relationship between X and Y changes sign.

The marginal effect of X is

$$\frac{\partial Y}{\partial X} = \beta_1 + 2\beta_2 X. \tag{7.6}$$

We see that the marginal effect of X is a linear function of X. Importantly, the coefficient on the constitutive term X, β_1, captures the intercept of the marginal effect line. If we were to omit this constitutive term from the quadratic model in Eq. 7.5, we'd be forcing the intercept of the marginal effect line for X to be 0, almost certainly introducing bias into our estimate of the coefficient on the interaction term X^2, β_2, and hence the estimated slope of the marginal effect line. We'd also be overestimating or underestimating the magnitude of the marginal effect of X if $\beta_1 \neq 0$. The same intuition applies to higher degree polynomial regression models. The bottom line is that scholars should include all of the constitutive elements of their interaction terms in a polynomial regression model.

In line with our previous recommendations, we encourage scholars who wish to evaluate claims from a theory predicting that the effect of X on Y varies with the value of X to look beyond any particular interaction effect. This is because any observed interaction effect is always consistent with a wide variety of ways in which X interacts with itself to determine Y, some of which may be inconsistent with their underlying theory. In particular, analysts should supplement a prediction about the way that the effect of X on Y varies with X with predictions about the sign of the effect of X at different values of X.[6] With a quadratic regression model, this means that scholars should supplement a prediction about the sign of the coefficient on the interaction term X^2 with predictions about the sign of the marginal effect of X when X is at its minimum and maximum observed values. In effect, scholars should make three key predictions when testing the implications from a theory positing a quadratic relationship between X and Y: P_{XX}, $P_{X|X_{min}}$, and $P_{X|X_{max}}$. The precise number of key predictions that a scholar needs to make to fully evaluate the conditional implications of their theory will vary with the nature of the predicted non-linear relationship between X and Y and hence the degree of the polynomial regression model.

As a hypothetical example, let's assume we have a theory predicting a quadratic relationship between X and Y such that an increase in X has a negative effect on Y when X is low but a positive effect on Y when X is high. This type of non-linear relationship is graphically presented in panel (b) of

---

[6] In what follows, we focus on the use of quadratic regression models. The exercise at the end of the chapter explores the specification and interpretation of cubic regression models.

Figure 7.1. We can easily capture our predictions in a single hypothesis about the marginal effect of X on Y and how this changes with X,

$H_{X|X}$: The effect of an increase in X on Y is negative when X is low but positive when X is high.

As recommended, hypothesis $H_{X|X}$ contains predictions about the sign of the marginal effect of X on Y when X is at its minimum and maximum observed values. Implied in hypothesis $H_{X|X}$ is also a prediction about the interaction effect; namely, that the negative effect of X declines in magnitude with higher values of X and eventually becomes positive when the value of X is sufficiently high.

We can test the implications contained in hypothesis $H_{X|X}$ using the quadratic regression model shown in Eq. 7.5. The marginal effect of X is shown in Eq. 7.6. We immediately see that the marginal effect of X depends on the value of X so long as $\beta_2 \neq 0$. This means that a simple t-test on the interaction term coefficient, β_2, can be used to evaluate the claim that the effect of X on Y depends on the value of X. We also see that in a quadratic regression model, the marginal effect of X is assumed to increase linearly with the value of X. As Eq. 7.6 indicates, the marginal effect of X is β_1 when X is 0. In other words, the coefficient on the constitutive term X in Eq. 7.5 doesn't tell us the average, unconditional, independent, or main effect of X; instead it tells us the marginal effect of X only when X is 0. Note that the marginal effect of X on Y increases by $2\beta_2$ for each unit increase in X. In other words, the marginal effect of X is β_1 when X is 0, $\beta_1 + 2\beta_2$ when X is 1, $\beta_1 + 4\beta_2$ when X is 2, $\beta_1 + 6\beta_2$ when X is 3, and so on. This means that the coefficient on the interaction term X^2 in Eq. 7.5 doesn't capture the "interaction effect"; it only captures half of the interaction effect. This can be confirmed by taking the derivative of the marginal effect of X with respect to X to identify the interaction effect and see how the value of X modifies the effect of X on Y:

$$\frac{\partial \left(\frac{\partial Y}{\partial X} \right)}{\partial X} = \frac{\partial \left(\beta_1 + 2\beta_2 X \right)}{\partial X} = 2\beta_2. \tag{7.7}$$

As we noted earlier, quadratic regression models allow us to capture non-linear relationships between X and Y that involve a single turning point. More specifically, they allow us to capture symmetric U-shaped or inverted U-shaped relationships between X and Y.[7] The sign of the coefficient on the interaction term X^2, β_2, indicates whether we have

---

[7] It's worth noting that the assumed symmetry of the non-linear relationship between X and Y about the turning point won't always be theoretically appropriate. As previously noted, it may also be the case that the assumed symmetry is not apparent in the range of the observed data as the turning points may fall outside of this range.

a U-shaped or inverted U-shaped relationship. We have a U-shaped or concave-up relationship when β_2 is positive and an inverted U-shaped or concave-down relationship when β_2 is negative. The magnitude of β_2 affects the curvature of the U-shaped or inverted U-shaped relationship. The larger the magnitude of β_2, the narrower the U-shaped or inverted U-shaped relationship. To find the value of X that minimizes Y when we have a U-shaped relationship or the value of X that maximizes Y in an inverted U-shaped relationship, we set the marginal effect of X in Eq. 7.6 – the slope of the relationship between X and Y – equal to 0 and solve for X,[8]

$$X^* = \frac{-\beta_1}{2\beta_2}. \tag{7.8}$$

In Figure 7.2, we graphically illustrate what the different possible combinations of signs for β_1 and β_2 imply about the non-linear relationship between X and Y and the associated marginal effect of X. In panel (a), we focus on those cases where β_2 is positive. This implies that the non-linear relationship between X and Y is U-shaped and that the marginal effect of X increases linearly with the value of X. The predicted value and marginal effect plots in the left column assume that $\beta_1 = 0$, the plots in the middle column assume that $\beta_1 < 0$, and those in the right column assume that $\beta_1 > 0$. Recall that β_1 indicates whether the slope of the U-shaped relationship is zero, positive, or negative when X is 0. This is equivalent to whether the marginal effect of X is zero, negative, or positive when X is 0. The equations used for Y and the associated marginal effect of X are shown above each plot. In panel (b), we focus on those cases where β_2 is negative. This implies that the non-linear relationship between X and Y has an inverted U-shape and that the marginal effect of X decreases linearly with the value of X. As before, the plots in the left column assume that $\beta_1 = 0$, those in the middle column assume that $\beta_1 < 0$, and those in the right column assume that $\beta_1 > 0$.

According to hypothesis $H_{X|X}$, the marginal effect of X should be negative when X is low but positive when X is high. This implies that $\beta_1 + 2\beta_2 X$ should be negative when X is sufficiently low but positive when X is sufficiently high. The predicted change in the sign of the marginal effect of X from negative to positive as the value of X increases implies a U-shaped relationship between X and Y and thus that β_2 should be positive. Whether we can make a precise prediction about the sign of β_1 depends on

---

[8] We can also derive X^* algebraically using Vieta's formulas. Vieta's formulas show that the sum of the roots (the values of X where $Y = 0$) of a generic quadratic polynomial $Y = aX^2 + bX + c$ is $-b/a$. Due to the symmetry of quadratic polynomials, the value of X that maximizes or minimizes Y is just the midpoint between, or the average of, the roots; that is $-b/2a$.

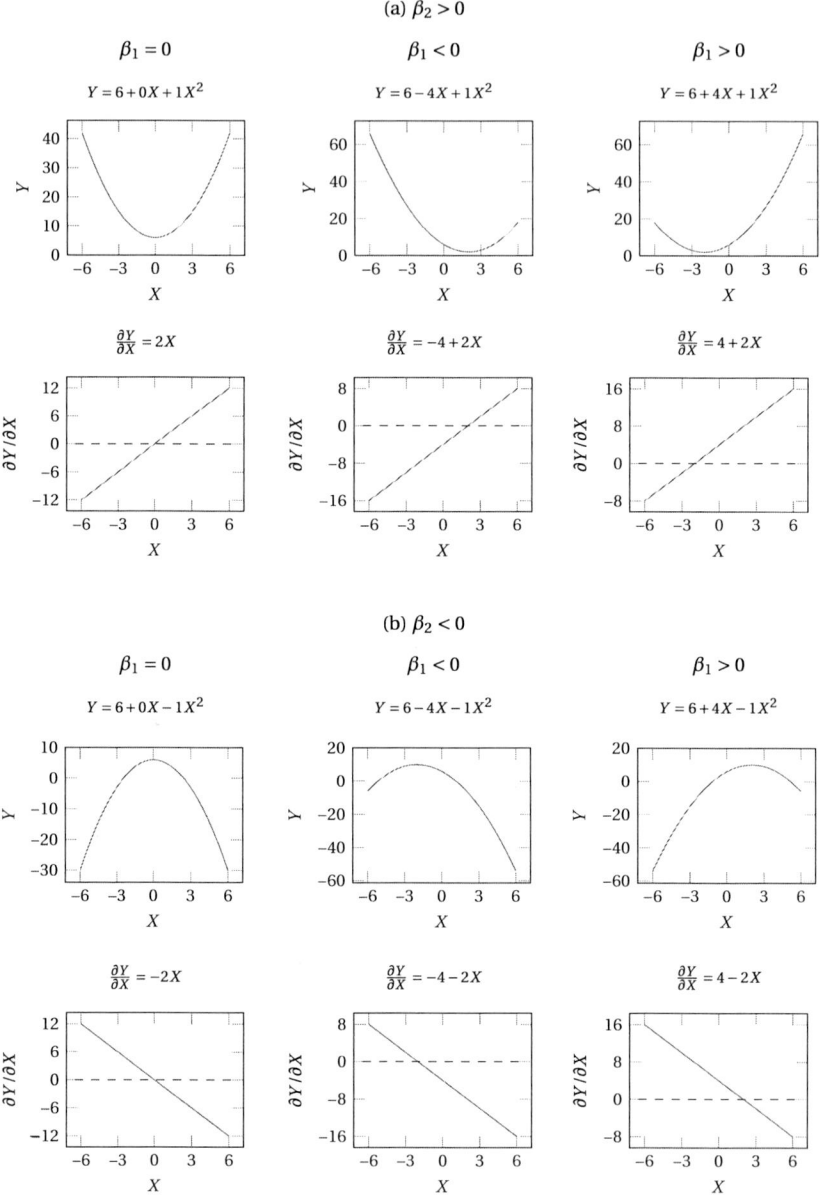

Figure 7.2 Different combinations of signs for β_1 and β_2 in a quadratic model

Note: Figure 7.2 shows predicted value plots along with their associated marginal effect plots for different combinations of signs for β_1 and β_2 from the quadratic regression model shown in Eq. 7.5. The plots in panel (a) all assume that β_2 is positive, while those in panel (b) all assume that β_2 is negative. Within each panel, the left column assumes that β_1 is 0, the middle column assumes that β_1 is negative, and the right column assumes that β_1 is positive.

the scale used to measure X. For example, if the minimum value of X is non-negative, then hypothesis $H_{X|X}$ predicts that β_1 should be negative. However, if the minimum value of X is negative, then it's unclear whether β_1 should be zero, negative, or positive. This is because hypothesis $H_{X|X}$, as currently stated, isn't sufficiently strong to predict whether the vertex of the U-shaped relationship – the minimum predicted value for Y – should be to the left of $X = 0$, to the right of $X = 0$, or at $X = 0$.

In order to fully evaluate hypothesis $H_{X|X}$ and its conditional claim about the effect of X on Y,

$$\frac{\partial Y}{\partial X} = \beta_1 + 2\beta_2 X, \tag{7.9}$$

we must calculate an appropriate measure of uncertainty for the marginal effect of X. Using the basic properties of variances shown in Appendix A, we calculate the variance of the marginal effect of X as

$$\text{var}\left(\frac{\partial Y}{\partial X}\right) = \text{var}(\beta_1) + 4X^2\text{var}(\beta_2) + 4X\text{cov}(\beta_1, \beta_2). \tag{7.10}$$

As usual, we obtain the standard error for the marginal effect of X by taking the square root of the variance,

$$\text{se}\left(\frac{\partial Y}{\partial X}\right) = \sqrt{\text{var}(\beta_1) + 4X^2\text{var}(\beta_2) + 4X\text{cov}(\beta_1, \beta_2)}. \tag{7.11}$$

We can use this standard error to conduct t-tests in the normal way to evaluate claims we might have about the marginal effect of X. We can also use it to construct confidence intervals for the marginal effect of X on Y at different values of X. For example, a two-sided confidence interval for the effect of X on Y with confidence coefficient $1 - \alpha$ is

$$\beta_1 + 2\beta_2 X \pm t_{n-k, \alpha/2} \times \text{se}\left(\frac{\partial Y}{\partial X}\right), \tag{7.12}$$

where $t_{n-k, \alpha/2}$ refers to the critical value from a t-distribution table for a particular α level of significance and $n - k$ degrees of freedom.

⚠ **Important:** Consider the following quadratic regression model:

$$Y = \beta_0 + \beta_1 X + \beta_2 X^2 + \epsilon.$$

The marginal effect of X on Y is

$$\frac{\partial Y}{\partial X} = \beta_1 + 2\beta_2 X,$$

and its estimated standard error is

$$\text{se}\left(\frac{\partial Y}{\partial X}\right) = \sqrt{\text{var}\left(\beta_1\right) + 4X^2\text{var}\left(\beta_2\right) + 4X\text{cov}\left(\beta_1, \beta_2\right)}.$$

When scholars present the results from a quadratic regression model, they often use a predicted value plot showing how the value of Y changes across the observed range of X. While these plots are often pretty and can help us to visualize the non-linear relationship between X and Y, they don't typically convey the necessary information to evaluate our hypotheses. In this sense, they're similar to the 3-D predicted value plots we've criticized previously in that they tend to place more emphasis on style rather than substance.

To examine this issue further, imagine that the results from a quadratic regression produce the predicted value plot shown in the middle column of panel (a) in Figure 7.2 where $\beta_0 = 6$, $\beta_1 = -4$, and $\beta_2 = 1$. Given that our plot is based on estimated results, it would make sense to add a confidence interval for the predicted value line and convey information about the location of the underlying data by incorporating a histogram showing the frequency distribution of X on the horizontal axis.[9] We can glean some useful information from a plot like this. For example, the inverted U-shaped relationship suggests that there's a non-linear relationship between X and Y and hence that the effect of X on Y depends on the value of X. We also see that the minimum predicted value of Y occurs when $X = 2$, suggesting that the marginal effect of X is negative when X is less than 2 and positive when X is greater than 2.

A significant problem with a predicted value plot like this, though, is that it doesn't convey all of the necessary quantities of interest and appropriate measures of uncertainty to evaluate the conditional claims from our theory. While the confidence interval that appears in a predicted value plot can tell us if and when the predicted value of Y is significantly different from 0, it can't tell us if and when the marginal effect of X on Y is

---

[9] Using the basic properties of variances shown in Appendix A, we see that the estimated variance of the predicted value of Y for the quadratic regression model in Eq. 7.5 is

$$\text{var}\left(\hat{Y}\right) = \text{var}\left(\beta_0 + \beta_1 X + \beta_2 X^2\right)$$
$$= \text{var}(\beta_0) + X^2\text{var}(\beta_1) + X^4\text{var}(\beta_2)$$
$$+ 2X\text{cov}(\beta_0, \beta_1) + 2X^2\text{cov}(\beta_0, \beta_2) + 2X^3\text{cov}(\beta_1, \beta_2). \quad (7.13)$$

We can use this to calculate the confidence interval for the predicted value line in the usual way.

statistically significant. In other words, it can't tell us if and when the slope relationship between X and Y is statistically significant. This is important because hypothesis $H_{X|X}$ speaks to the marginal effect of X and not the predicted value of Y. Significantly, we know that there must be a range of values of X at which the marginal effect of X isn't statistically significant because of the estimated U-shaped relationship. This follows from the fact that the slope relationship between X and Y is 0 at the vertex of the inverted U. There's simply no way of knowing from a predicted value plot how far the region of statistical insignificance for the marginal effect of X extends either side of the vertex of the inverted U. In addition, the predicted value plot can't tell us if the marginal effect of X depends on the value of X in a statistically significant way. We can certainly see that there appears to be a curvilinear relationship between X and Y. However, there's no way to discern whether this non-linear relationship is significantly different from a linear relationship. Whether the marginal effect of X depends on the value of X, and hence whether there's a non-linear relationship between X and Y, depends on whether β_2 is significantly different from 0. This information isn't contained in the typical predicted value plot.

In general, predicted value plots offer little leverage when it comes to testing the implications of a theory positing self-interaction. This is because they provide almost none of the information we need to evaluate the empirical support for our hypotheses. As a result, we strongly recommend that researchers use a marginal effect plot for X rather than a predicted value plot when testing conditional hypotheses with a polynomial regression model. Marginal effect plots, so long as they also indicate the sign and statistical significance of the interaction effect, provide all of the information necessary to evaluate the key predictions that can be derived from theories positing some form of interaction. One strategy that can sometimes be useful is to combine a marginal effect plot for X with a predicted value plot as we've done throughout this section.

⚠ **Important:** Predicted value plots are good for visualizing the non-linear relationship between X and Y when estimating a polynomial regression model. However, they offer little leverage when it comes to actually testing the implications of a theory positing self-interaction.

In Chapter 4, we noted that it's important to think about the effects of our variables in terms of derivatives and differences rather than coefficients. When there's a linear relationship between X and Y, we saw that the marginal effect of X, which can be calculated with a derivative, is identical to the effect of a one-unit increase in X, which is calculated with a difference. This equivalence is nice as it allows us to talk about the marginal effect of X and the effect of a one-unit increase in X interchangeably. This equivalence no longer holds, though, when we have a non-linear relationship between X and Y because the slope of the function that describes the relationship isn't constant. In the non-linear case, it will, therefore, almost always be the case that the marginal effect of X is different to the effect of a one-unit increase in X. This means that we need to be careful in how we describe our results when dealing with self-interaction and other forms of non-linear relationships.

To see what's going on here, consider Figure 7.3, which shows a quadratic relationship between X and Y where $Y = 15 + 0X - 5X^2$. The effect of X on Y changes depending on the value of X. Suppose we want to calculate the effect of X when $X = -1$. One approach would be to calculate the marginal effect of X on Y, $\frac{\partial Y}{\partial X} = -10X$. Recall that the marginal effect of X indicates how Y responds when we increase X by an infinitesimally small amount divided by the change in X. When $X = -1$, the marginal effect of X is $-10 \times -1 = 10$. This tells us that the slope of the relationship

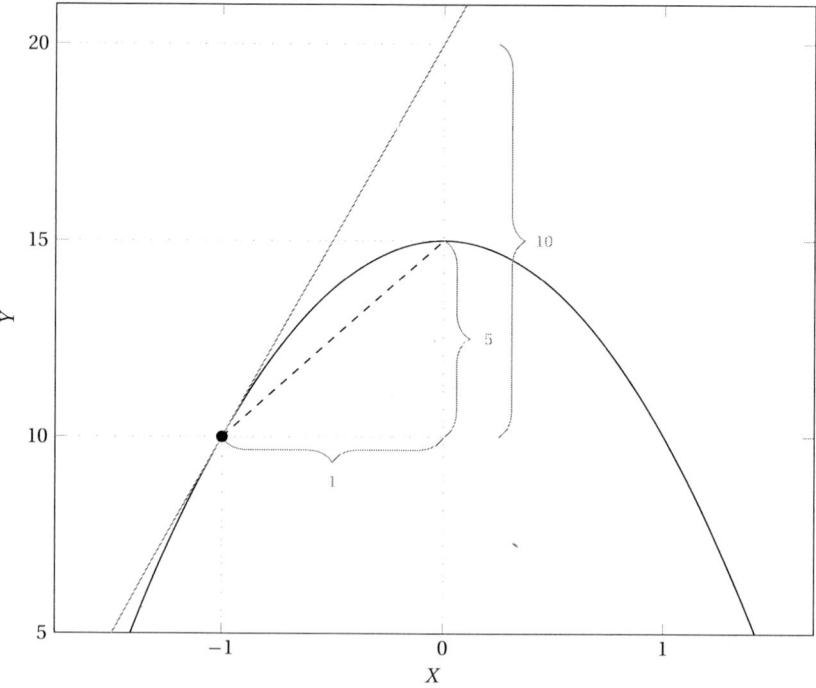

Figure 7.3 The difference between a derivative and a difference when the effect of X is non-linear

between X and Y when $X = -1$, which is indicated by the gray line in Figure 7.3, is 10. A different approach would be to calculate the effect of a one-unit increase in X on Y. Calculating this "difference" shows us that when we increase X by one-unit from -1 to 0, the value of Y increases by 5. This effect is captured by the slope of the dashed line in Figure 7.3. In this particular example, the effect of X when $X = -1$ when measured as a marginal effect is twice as large as the effect of X when measured as a one-unit difference. It should be clear that it's possible for the sign of the marginal effect of X to be different from the sign of the effect of a one-unit (or other discrete-unit) increase in X. For example, the marginal effect of X when $X = -0.5$ is 5, whereas the effect of increasing X by one unit from -0.5 to 0.5 is 0. And the marginal effect of X when $X = -0.25$ is 2.5, whereas the effect of increasing X by one unit from -0.25 to 0.75 is -2.5.

These differences are particularly important when it comes to discussing the substantive significance of the results from a polynomial regression or other form of non-linear model. In the real world, actors can typically increase X by some discrete amount but can't realistically increase X by an infinitesimally small amount. As a result, we encourage scholars to calculate differences rather than marginal effects when using counterfactuals to discuss the substantive significance of the results from a non-linear model. In line with these recommendations, scholars may also wish to produce "marginal effect" plots that show the effect of a small, but substantively meaningful, discrete-unit change in X across the observed range of X rather than the marginal effect of X.[10]

As always, scholars should provide an appropriate measure of uncertainty when reporting differences. Suppose we're interested in the effect of increasing X in the quadratic model shown in Eq 7.5 from some baseline value X_b to some counterfactual value X_c. This difference is calculated as

$$\hat{Y}_{X_c} - \hat{Y}_{X_b} = \left(\beta_0 + \beta_1 X_c + \beta_2 X_c^2\right) - \left(\beta_0 + \beta_1 X_b + \beta_2 X_b^2\right)$$
$$= \beta_1 (X_c - X_b) + \beta_2 \left(X_c^2 - X_b^2\right). \tag{7.14}$$

If we think in terms of a one-unit increase in X such that $X_c = X_b + 1$, then the difference shown in Eq. 7.14 simplifies to

---

[10] Scholars should focus on the effect of a *small*, but substantively meaningful, discrete change in X because we typically still want to mimic a marginal effect and identify what happens to Y when we start to increase X *from some specific value*. We can certainly calculate the effect of increasing X by a *large* discrete amount from some specific value, and this effect will be meaningful. However, doing so can give a potentially misleading impression as to exactly what happens to Y when we start to increase X from that value. This suggests one reason why we should be wary of calculating, say, the effect of changing X from its minimum value to its maximum value. This point applies to any situation in which we're interested in calculating the effect of a discrete change in an independent variable in a non-linear setting.

$$\hat{Y}_{X_b+1} - \hat{Y}_{X_b} = \beta_1 (X_b + 1 - X_b) + \beta_2 \left((X_b + 1)^2 - X_b^2 \right)$$
$$= \beta_1 + (1 + 2X_b)\beta_2. \tag{7.15}$$

Using the basic properties of variances shown in Appendix A, we see that the estimated variance of this difference is

$$\text{var}\left(\hat{Y}_{X_b+1} - \hat{Y}_{X_b} \right) = \text{var}(\beta_1 + (1 + 2X_b)\beta_2)$$
$$= \text{var}(\beta_1) + \left(1 + 4X_b + 4X_b^2 \right) \text{var}(\beta_2)$$
$$+ (2 + 4X_b)\text{cov}(\beta_1, \beta_2). \tag{7.16}$$

We calculate standard errors and confidence intervals in the usual way.[11]

> ⚠ **Important:** In polynomial regression and other non-linear models, the marginal effect of X on Y is not equivalent to the effect of a one-unit increase in X on Y. Scholars should calculate differences rather than marginal effects when using counterfactuals to discuss the substantive significance of the results from a non-linear model.

7.1.1   Substantive Application: The Impact of Party Ideology on Campaign Sentiment

What explains the type of election campaign run by political parties? In this application, we replicate an analysis by Crabtree et al. (2020) that examines the strategic use of campaign sentiment by political parties in Europe. The theory is situated in the retrospective voting literature, which assumes that individuals base their vote choice on how they perceive the state of the world at election time – something that's usually attributed to the performance of the incumbent government. The basic intuition is that people vote for the incumbent when the state of the world is above some threshold but switch to the opposition when this isn't the case. While there's obviously only one objective state of the world at any moment in time, political parties can influence how people perceive the state of the world through the emotive content of their campaign messages. Numerous studies have shown that campaign messages can trigger emotional responses that produce predictable changes in voter behavior. If this is the case, then political parties should be strategic about their use of campaign sentiment.

---

[11] When our non-linear model becomes more complicated, it can often be easier to calculate appropriate measures of uncertainty for quantities of interest like differences via simulation (King, Tomz and Wittenberg, 2000). We discuss how to do this in more detail in Chapter 8.

While campaign messages can engender different types of sentiment, such as fear, anxiety, sadness, optimism, and so on, Crabtree et al. (2020) focus on the use of positive and negative sentiment. Campaign messages that include positive sentiment are expected to stir optimism and encourage people to adopt a positive frame when evaluating the world around them. According to Crabtree et al. (2020), the level of positive sentiment exhibited by political parties in their campaign messages depends on their incumbency status, their policy position, and objective economic conditions. In what follows, we focus on how a party's position in a one-dimensional (left-right) policy space affects the use of positive sentiment in campaign messages. In general, we'd expect ideologically extreme parties to exhibit less positive sentiment than ideologically moderate parties. This is because voters are more likely to reject moderate parties and turn to more extreme parties when they perceive the state of the world to be particularly bad. Radical parties in Europe, for example, propose "root and branch" reform of the political and economic system, and many adopt populist rhetoric that holds all moderate parties responsible for society's ills (Mudde, 2007; Golder, 2016). These parties don't just want voters to punish the incumbent, they want voters to abandon the mainstream parties altogether. This is most likely to occur when the current state of affairs is considered especially bad. This argument suggests that there's a non-linear, specifically quadratic, relationship between the use of positive sentiment in campaign messages and the ideological positions of political parties.

Extreme Ideology Hypothesis: Ideologically extreme parties use lower levels of positive sentiment in their campaign messages than ideologically moderate parties.

As recommended, the *Extreme Ideology Hypothesis* contains all three of the key predictions that we recommend for a theory positing this form of interaction. If we think about the left-right ideological dimension, the hypothesis predicts that a shift to the right in a party's policy position increases the use of positive sentiment when the party is on the far left $(P_{X|X_{\min}})$ and reduces the use of positive sentiment when the party is on the far right $(P_{X|X_{\max}})$. Implied in the hypothesis is also a prediction about the interaction effect (P_{XX}); namely, that the positive effect of a rightward shift in a party's policy position declines in magnitude as a party's ideological position moves to the right and eventually becomes negative when its ideological position is sufficiently far to the right.

To examine positive sentiment in campaign messages, Crabtree et al. (2020) focus on the use of emotive language in over 400 party manifestos across eight European countries from 1980 to 2012. The dependent variable, *Positive Sentiment*, captures the percentage of positive emotional words in a manifesto minus the percentage of negative emotional words

(Pennebaker, Booth and Francis, 2007).[12] *Positive Sentiment* has a mean of 1.70 and standard deviation of 1.45. Our key independent variable *Left-Right* captures a party's left-right policy position on a 0–10 scale as identified by country experts (Döring and Manow, 2015). Given the nature of our dependent variable, we estimate the following interaction model – a quadratic regression model – using ordinary least squares:

$$Positive\ Sentiment = \beta_0 + \beta_1 Left\text{-}Right + \beta_2 Left\text{-}Right^2 + \epsilon. \qquad (7.17)$$

The marginal effect of increasing a party's left-right position on the level of positive sentiment in a party manifesto is

$$\frac{\partial Positive\ Sentiment}{\partial Left\text{-}Right} = \beta_1 + 2\beta_2 Left\text{-}Right. \qquad (7.18)$$

From this, we see that β_1 indicates the marginal effect of increasing a party's left-right position for those parties whose current left-right position is 0 (most left-wing). According to the *Extreme Ideology Hypothesis*, we expect an increase in a party's left-right position to be positive when a party is located on the extreme left, and so β_1 should be positive. The *Extreme Ideology Hypothesis* predicts that an increase in a party's left-right position should be negative when a party is located on the extreme right, and so $\beta_1 + 2\beta_2 Left\text{-}Right$ should be negative once *Left-Right* is sufficiently high. Given the predicted change in the sign of the marginal effect of *Left-Right* from positive to negative as we move across the left-right ideological dimension, β_2 should be negative.

The results from the interaction model shown in Eq. 7.18 are presented in Table 7.1. As predicted, the coefficient on *Left-Right* is positive and statistically significant, indicating that the marginal effect of a policy shift to the right is associated with an increase in the level of positive sentiment in campaign messages among far left parties where *Left-Right* is 0. Also as predicted, the coefficient on *Left-Right*^{2} is negative and statistically significant. The fact that this coefficient is statistically significant provides the important evidence of interaction and tells us that the effect of changing ideological position on the use of positive sentiment depends on a party's current policy position. The fact that it's negative indicates an inverted U-shaped relationship between *Positive Sentiment* and *Left-Right*. The policy position that maximizes the level of positive sentiment – the vertex of the inverted U-shaped relationship – is $\frac{-\beta_1}{2\beta_2} = \frac{-0.69}{2 \times -0.07} = 5.16$. This tells us that the marginal effect of a policy shift to the right is associated with an increase in the level of positive sentiment when parties are to the left

---

[12] The percentages of positive and negative words are calculated using the Linguistic Inquiry and Word Count (LIWC) program, which is a tool for conducting automatic sentiment analysis.

Table 7.1 Party left-right ideology and positive campaign sentiment

Dependent Variable: *Positive Sentiment*	
	Model 1
Left-Right	0.69***
	(0.15)
Left-Right^2	−0.07***
	(0.01)
Constant	0.35
	(0.33)
Observations	382
R^2	0.06

Standard errors in parentheses. $^*p < 0.10$; $^{**}p < 0.05$; $^{***}p < 0.01$ (two-tailed)

of 5.16 and a decrease in the level of positive sentiment when they're further to the right. The predicted non-linear relationship between *Left-Right* and *Positive Sentiment* is shown graphically in panel (a) of Figure 7.4.

It's not possible to know from the individual coefficients in Table 7.1 or the predicted value plot in panel (a) of Figure 7.4 the range of values for *Left-Right* at which the effect of a policy shift to the right is statistically significant. All we know from the coefficients is that the marginal effect of *Left-Right* is positive and statistically significant when *Left-Right* is 0 and that it's 0 and statistically insignificant when *Left-Right* is 5.16. The plot in panel (b) of Figure 7.4 shows how the marginal effect of *Left-Right*, $\beta_1 + 2\beta_2 Left\text{-}Right$, varies across the observed range of *Left-Right*. The confidence interval is calculated using the variance formula shown in Eq. 7.10. The plot indicates that the marginal effect of a shift in policy to the right is positive and statistically significant when *Left-Right* < 4.69 (50.3%) and is negative and statistically significant when *Left-Right* > 5.68 (47.4%). Figures in parentheses indicate the percentage of observations that fall within each of these ranges. The plot in panel (c) of Figure 7.4 shows how the effect of a one-unit increase in *Left-Right*, $\beta_1 + \beta_2 + 2\beta_2 Left\text{-}Right$, varies across the observed range of *Left-Right*. The confidence interval for this is calculated using the variance formula shown in Eq. 7.16. The plot indicates that a one-unit shift in policy to the right is

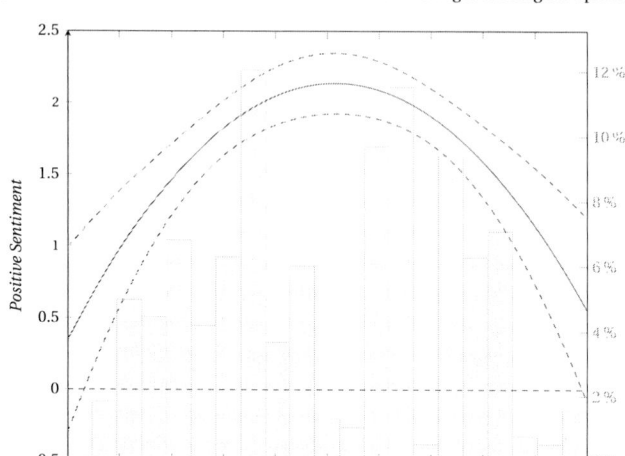

(a) Predicted Value of Positive Sentiment across the Left-Right Ideological Spectrum

(b) Marginal Effect of *Left-Right*

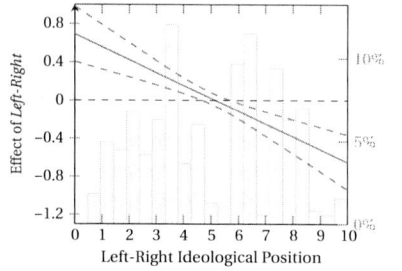

(c) Effect of a One-Unit Increase in *Left-Right*

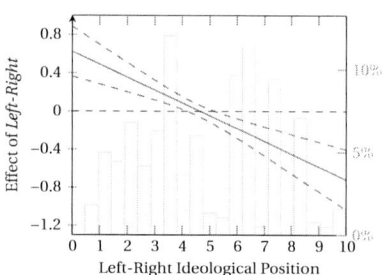

Figure 7.4 (a)–(c) Party left-right ideology and positive campaign sentiment

positive and statistically significant when *Left-Right* < 4.19 (44.2%) and is negative and statistically significant when *Left-Right* > 5.18 (48.4%).

As expected, the information from the plot in panel (b) showing the marginal effect of *Left-Right* is different, if only slightly, from the information contained in the plot in panel (c) showing the effect of a one-unit increase in *Left-Right*. In this application, we've shown both plots to highlight the conceptual and substantive difference between marginal effects and differences in the context of non-linear models such as those involving polynomial regression. In practice, we recommend that scholars show only one of these plots. Given that marginal effects don't necessarily capture realistic counterfactuals in a non-linear context, we encourage scholars to present plots that show the effect of a small, but substantively meaningful, discrete change in the variable of interest.

In terms of substantive significance, a one-unit shift in policy to the right, which equates to 42.3% of a standard deviation in party left-right policy positions, is associated with a 0.36 [0.20, 0.51] unit increase in *Positive Sentiment* among left-wing parties where *Left-Right* is 2. This is equivalent to a 33.5% increase in the mean level of positive sentiment (1.07) exhibited by left-wing parties at or to the left of *Left-Right* = 2. The same one-unit shift in policy position to the right has almost no effect on the level of positive sentiment, −0.04 [−0.11, 0.02], for parties located at the mean left-right policy position (4.98).

7.1.2 Using Polynomial Regression to Model Non-linear Interaction Effects

So far in this section, we've looked at how we can use polynomial regression (interaction) models to test the conditional implications of a theory positing that the effect of X depends on its own value rather than the value of one or more other variables. However, it should be clear that we can easily use polynomial regression models to test the conditional implications of theories positing that the effect of X depends on its own value *and* on the value of other variables. As we'll see, we can use these types of polynomial models to examine conditional claims where the interaction effect is expected to be non-linear.

Suppose we have a theory positing that the effect of X on Y depends on the value of X and Z.[13] This is essentially the same scenario as we examined in the previous chapter. We have a variable X whose effect on Y is predicted to depend on the value of *two* variables. The only difference now is that one of these two modifying variables happens to be X itself. As we saw in Chapter 6, we need to think carefully in these situations about whether our theory predicts that the interaction effects between X and each of its two modifying variables, in this case X and Z, are *independent* (unconditional) or *dependent* (conditional).

To examine this in more detail, let's assume that we're expecting a quadratic non-linear relationship between X and Y. In the case where the modifying effects of X and Z are expected to be independent, we'd estimate the following interaction model:

[13] We can, of course, imagine more complicated theories in which the posited effect of X on Y depends on the value of X and multiple other variables. All of the points we make in what follows generalize to these more complex scenarios.

$$Y = \beta_0 + \beta_1 X + \beta_2 X^2 + \beta_3 Z + \beta_4 XZ + \epsilon. \tag{7.19}$$

With this model, the marginal effect of X is

$$\frac{\partial Y}{\partial X} = \beta_1 + 2\beta_2 X + \beta_4 Z. \tag{7.20}$$

So long as $\beta_2 \neq 0$ and $\beta_4 \neq 0$, we see that the marginal effect of X on Y depends on the values of both X and Z. The modifying effect of X on the marginal effect of X is

$$\frac{\partial \left(\frac{\partial Y}{\partial X} \right)}{\partial X} = 2\beta_2, \tag{7.21}$$

and the modifying effect of Z on the marginal effect of X is

$$\frac{\partial \left(\frac{\partial Y}{\partial X} \right)}{\partial Z} = \beta_4. \tag{7.22}$$

As the marginal effect in Eq. 7.20 indicates, the modifying effects of X and Z are additive and therefore independent of each other. Put differently, and as Eq. 7.21 and Eq. 7.22 explicitly show, the modifying effect of X on the marginal effect of X doesn't depend on the value of Z, and the modifying effect of Z on the marginal effect of X doesn't depend on the value of X. It's in this sense that the modifying effects of X and Z are *unconditional*. This independence between the modifying effects is assumed in the interaction model shown in Eq. 7.19.

In the case where the modifying effects of X and Z are expected to be dependent, we'd estimate the following interaction model:

$$Y = \gamma_0 + \gamma_1 X + \gamma_2 X^2 + \gamma_3 Z + \gamma_4 XZ + \gamma_5 X^2 Z + \varepsilon. \tag{7.23}$$

With this model, the marginal effect of X is

$$\frac{\partial Y}{\partial X} = \gamma_1 + 2\gamma_2 X + \gamma_4 Z + 2\gamma_5 XZ. \tag{7.24}$$

From this, we again see that the the marginal effect of X is allowed to depend on the values of both X and Z. Unlike with the interaction model in Eq. 7.19, though, the marginal effect of X now also depends on the value of XZ. It's this that allows the modifying effects of X and Z to be *dependent* or *conditional*. The modifying effect of X on the marginal effect of X is

$$\frac{\partial \left(\frac{\partial Y}{\partial X} \right)}{\partial X} = 2\gamma_2 + 2\gamma_5 Z. \tag{7.25}$$

This is the "interaction effect" between X and X and tells us how the marginal effect of X changes with the value of X. So long as $\gamma_5 \neq 0$, there's no single interaction effect between X and X, and the modifying effect of X on the marginal effect of X depends on the value of Z. We can see exactly how Z modifies the interaction effect between X and X by taking the derivative of Eq. 7.25 with respect to Z,

$$\frac{\partial \left(\frac{\partial \left(\frac{\partial Y}{\partial X} \right)}{\partial X} \right)}{\partial Z} = 2\gamma_5. \tag{7.26}$$

The modifying effect of Z on the marginal effect of X is

$$\frac{\partial \left(\frac{\partial Y}{\partial X} \right)}{\partial Z} = \gamma_4 + 2\gamma_5 X. \tag{7.27}$$

This is the "interaction effect" between X and Z and tells us how the marginal effect of X changes with the value of Z. So long as $\gamma_5 \neq 0$, there's no single interaction effect between X and Z, and the modifying effect of Z on the marginal effect of X depends on the value of X. We can see exactly how X modifies the interaction effect between X and Z by taking the derivative of Eq. 7.27 with respect to X,

$$\frac{\partial \left(\frac{\partial \left(\frac{\partial Y}{\partial X} \right)}{\partial Z} \right)}{\partial X} = 2\gamma_5. \tag{7.28}$$

This is, of course, the same as Eq. 7.26 due to the symmetry of interactions. The various methods and strategies described in the previous chapter can be used to interpret and present the results from these models.

It turns out that the types of polynomial regression models we've just examined allow us to evaluate the conditional implications of theories that predict a non-linear interaction effect (Kam and Franzese, 2007; Berry, Golder and Milton, 2012). So far in this book, we've focused almost entirely on using interaction models to evaluate the implications of theories where the interaction effect is expected to be linear. The linearity of the interaction effect has been assumed in virtually all of the interaction models that we've examined.[14] As a reminder, a linear interaction effect means that the marginal effect of a variable, say X, on Y changes linearly, or at a constant rate, when we change the value of the modifying variable, say Z. Suppose we have the following interaction model:

$$Y = \beta_0 + \beta_1 X + \beta_2 Z + \beta_3 XZ + \epsilon. \tag{7.29}$$

---

[14] The one exception came earlier in this chapter when we very briefly discussed a situation where we had a non-linear relationship between X and Y that could be modeled with a cubic polynomial. As the plot on the right of panel (c) in Figure 7.1 indicates, the marginal effect of X in a cubic polynomial model isn't a linear function of X. We examine the non-linear interaction effect that arises with a cubic polynomial in more detail in Exercise 1 at the end of this chapter.

(a) The Conditional Marginal Effect of X on Y

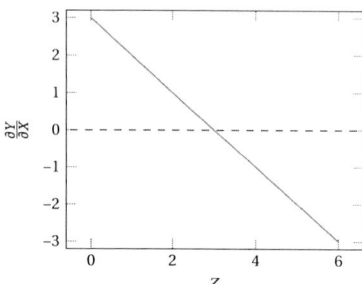

(b) The Modifying Effect of Z on the Marginal Effect of X on Y

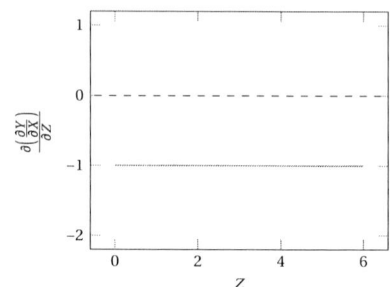

Figure 7.5 (a)–(b) An example of a linear interaction effect

This model assumes a linear interaction effect. To see this, note that the marginal effect of X,

$$\frac{\partial Y}{\partial X} = \beta_1 + \beta_3 Z, \tag{7.30}$$

is a *linear* function of Z, where each one unit increase in Z leads to a *constant* β_3 increase in the effect of X on Y. If we were to produce a marginal effect plot for X across the observed range of values for Z, we'd get a straight line with an intercept of β_1 and a slope of β_3. An example of such a marginal effect plot is shown in the left panel of Figure 7.5. In the right panel, we plot the corresponding interaction effect – the slope relationship between the marginal effect of X and Z – across the observed range of values for Z. As expected, this is just a flat line whose value is equal to β_3. In this particular example, each one unit increase in Z modifies the marginal effect of X on Y by $\beta_3 = -1$.

> ⚠ **Important:** A linear interaction effect means that the marginal effect of a variable, say X, changes linearly or at a constant rate with the value of its modifying variable, say Z.

But not all of our theories positing interaction predict that the interaction effect should be linear. Polynomial models often provide us with an appropriate and flexible way to evaluate the conditional implications of these types of theories while remaining in a traditional regression-based parametric framework.[15] This is because they allow for a smooth

[15] Hainmueller, Mummolo and Zu (2019) discuss other ways to relax the assumption of a linear interaction effect. In addition to various semi-parametric and

curvilinear relationship between the marginal effect of a variable such as X and its modifying variable.

Suppose we have a theory positing interaction between two variables X and Z. Our theory might predict that the marginal effect of X on Y varies in a smooth curvilinear way with Z. On the left-hand side of Figure 7.6, we show three possible ways in which our theory might predict that the marginal effect of X on Y varies across the observed range of values for Z. In panel (a), our theory predicts that the marginal effect of X is positive when Z is low or high but zero when Z is moderately high; it also predicts that the positive effect of X is greater when Z is high than when Z is low. We can easily see that we have a non-linear interaction effect because the marginal effect of X doesn't vary in a linear way with Z. To the right of the marginal effect plot is the corresponding "interaction effect" plot showing how the modifying effect of Z – the slope relationship between the marginal effect of X and Z – varies across the observed range of values for Z. The key thing to note is that the interaction effect line isn't flat, and so the interaction effect isn't constant. We see, for example, that the modifying effect of Z is negative when $Z < 2$ but positive when $Z > 2$. The fact that the modifying effect of Z isn't constant tells us that we're dealing with a non-linear interaction effect. In panel (b), our theory predicts that the marginal effect of X is always positive, but that the magnitude of this positive effect increases at an increasing rate with higher values of Z. We know that we have a non-linear interaction effect because the marginal effect of X doesn't vary in a linear way with Z, and the line in the corresponding "interaction effect" plot isn't flat. In panel (c), our theory predicts that the marginal effect of X is negative when Z is low but that it increases in value at a decreasing rate as Z gets larger; it also predicts that the marginal effect of X becomes positive once Z is sufficiently large. We again know we have a non-linear interaction effect because the marginal effect of X doesn't vary in a linear way with Z and the line in the corresponding "interaction effect" plot isn't flat.

Although they look quite different, note that all three of the cases shown in Figure 7.6 depict a smooth curvilinear relationship between the marginal effect of X and Z that involves a single "turning point". The "turning point" is entirely visible at $Z = 2$ in panel (a) but is partially obscured in panels (b) and (c) because it occurs at one end of the range of observed values for Z. Building on our discussion of polynomial regression models in this section, you might think that we could capture these

non-parametric methods, they also discuss a parametric binning estimator that's conceptually similar in some ways to the piece-wise linear regression models that we briefly examine in the next section.

(a)

The Conditional Marginal Effect of X on Y

The Modifying Effect of Z on the Marginal Effect of X on Y

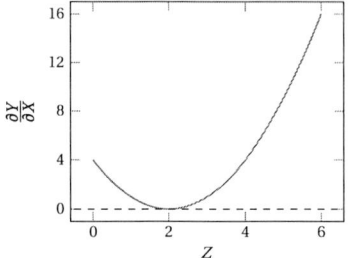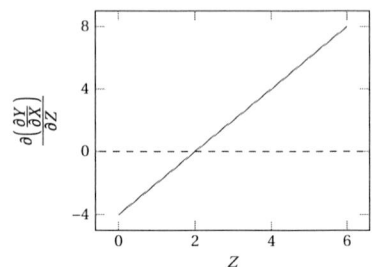

(b)

The Conditional Marginal Effect of X on Y

The Modifying Effect of Z on the Marginal Effect of X on Y

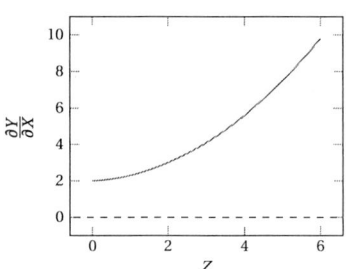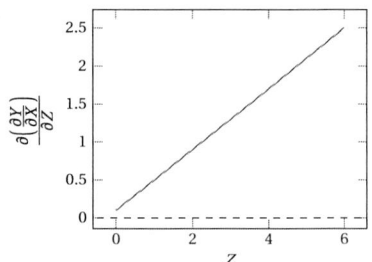

(c)

The Conditional Marginal Effect of X on Y

The Modifying Effect of Z on the Marginal Effect of X on Y

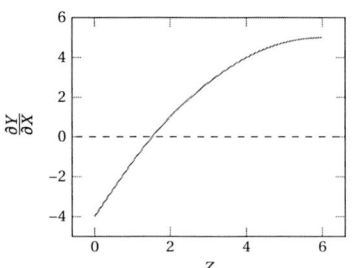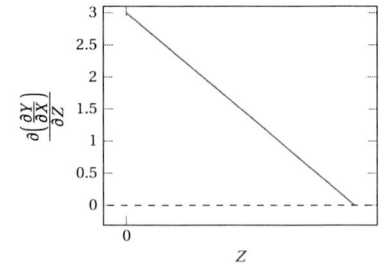

Figure 7.6 (a)–(c) Three possible non-linear relationships for how the marginal effect of X on Y varies with the value of Z

Note: The plots on the left show three possible ways in which the marginal effect of X could vary with the value of Z based on the interaction model shown in Eq. 7.31. The plots on the right show the corresponding interaction effect between X and Z and how this varies with Z.

types of non-linear interaction effects with some sort of quadratic model specification. And you'd be right.

It turns out that all three of the cases shown in Figure 7.6, as well as other cases where we expect some kind of smooth curvilinear relationship between the marginal effect of X and Z with only one turning point, can be modeled with the following polynomial (interaction) model:

$$Y = \beta_0 + \beta_1 X + \beta_2 Z + \beta_3 XZ + \beta_4 Z^2 + \beta_5 XZ^2 + \epsilon. \tag{7.31}$$

The marginal effect of X on Y is

$$\frac{\partial Y}{\partial X} = \beta_1 + \beta_3 Z + \beta_5 Z^2. \tag{7.32}$$

We see that the marginal effect of X is now a quadratic, rather than a linear, function of Z. As a quadratic function, we can interpret the coefficients in the marginal effect in the same way as we did earlier. For example, the sign of β_5 tells us whether the non-linear relationship between the marginal effect of X and Z has a U-shaped (positive) or inverted U-shaped (negative) relationship. The interaction effect between X and Z is

$$\frac{\partial\left(\frac{\partial Y}{\partial X}\right)}{\partial Z} = \beta_3 + 2\beta_5 Z. \tag{7.33}$$

From this we see that the modifying effect of Z on the marginal effect of X is no longer constant but is instead a linear function of Z. We also see, as expected, that β_3 only tells us the modifying effect of Z when $Z = 0$. Depending on the values for β_1, β_3, β_5, and the observed range of values for Z, the interaction model in Eq. 7.31 can capture many different ways in which the marginal effect of X on Y varies in a (one turning point) curvilinear way with Z. For example, we obtain the plots in panel (a) of Figure 7.6 if Z runs from 0 to 6 and $\beta_1 = 4$, $\beta_3 = -4$, and $\beta_5 = 1$. We obtain the plots in panel (b) if $\beta_1 = 2$, $\beta_3 = 0.1$, and $\beta_5 = 0.2$. And we obtain the plots in panel (c) if $\beta_1 = -4$, $\beta_3 = 3$, and $\beta_5 = -0.25$.

We can easily capture more complicated theoretical scenarios involving non-linear interaction effects where the predicted curvilinear relationship between the marginal effect of X and Z has more than one turning point by estimating a higher degree polynomial (interaction) model. For example, if we had a theory predicting a non-linear interaction effect where the curvilinear relationship between the marginal effect of X and Z has two turning points, we could estimate the following model:

$$Y = \beta_0 + \beta_1 X + \beta_2 Z + \beta_3 XZ + \beta_4 Z^2 + \beta_5 XZ^2 + \beta_6 Z^3 + \beta_7 XZ^3 + \epsilon. \tag{7.34}$$

Now the marginal effect of X on Y is

$$\frac{\partial Y}{\partial X} = \beta_1 + \beta_3 Z + \beta_5 Z^2 + \beta_7 Z^3. \tag{7.35}$$

This is a cubic function that, as we've seen, allows a curvilinear relationship, in this case between the marginal effect of X and Z, to have two turning points. The interaction effect between X and Z is now a quadratic function of Z,

$$\frac{\partial \left(\frac{\partial Y}{\partial X} \right)}{\partial Z} = \beta_3 + 2\beta_5 Z + 3\beta_7 Z^2. \tag{7.36}$$

In Figure 7.7, we show one of the many types of possible non-linear relationships for how the marginal effect of X on Y varies with Z that can be captured with the interaction model in Eq. 7.34. The specific example in Figure 7.7 is based on the case where $\beta_1 = 27$, $\beta_3 = 15$, $\beta_5 = -9$, and $\beta_7 = 1$. As you can see, we can specify polynomial (interaction) models that allow us to capture a wide range of non-linear interaction effects that might be predicted by our theories.[16]

(a) The Conditional Marginal Effect of X on Y

(b) The Modifying Effect of Z on the Marginal Effect of X on Y

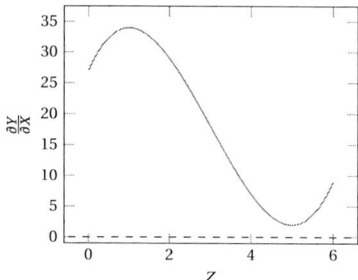
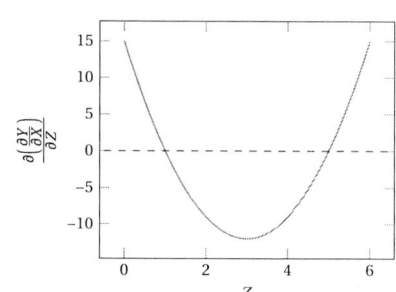

Figure 7.7 Another possible non-linear relationship for how the marginal effect of X on Y varies with the value of Z

Note: The plot on the left (a) shows a possible way in which the marginal effect of X on Y could vary with the value of Z based on the interaction model shown in Eq. 7.34. The plot on the right (b) shows the corresponding interaction effect between X and Z and how this varies with Z.

---

[16] As Berry, Golder and Milton (2012) show in their online appendix, we can also use these types of polynomial models to evaluate whether the inferences from an interaction model that assumes a linear interaction effect are robust to relaxing this assumption. For an alternative way to evaluate the robustness of the linear interaction effect assumption, see Hainmueller, Mummolo and Zu (2019).

⚠ **Important:** We can often specify polynomial regression models that allow us to test the conditional implications of theories that posit a non-linear interaction effect.

7.2 POSSIBLE THRESHOLD EFFECTS

In this chapter, we've been looking at theories predicting that the effect of X on Y depends on its own value. Such theories predict a non-linear relationship between X and Y. So far, we've examined how polynomial models can be used to capture many different types of non-linear relationships. However, polynomial models assume a smooth curvilinear relationship between X and Y and therefore aren't always appropriate. In what's left of this chapter, we'll very briefly highlight one situation where this is the case. It occurs when a theory predicts that X has one effect for some, say, low values of X but a different effect for other, say, high values of X. One way to think about these types of theory is that they predict a threshold relationship where X has one effect below some threshold $X = X^T$ and a different effect above this threshold. We might say that the threshold value X^T represents a break in the linear relationship between X and Y. The implications from these types of theories can often be evaluated with *piecewise linear regression models* or *switching regression models*. Although not always recognized as such, both of these types of models are interaction models. In what follows, we encourage scholars who are interested in testing these sorts of claims about how the effect of X on Y varies across different values of X to employ a switching regression model rather than a piecewise linear regression model. This is because piecewise linear regression models omit the constitutive elements of the interaction term, potentially introducing bias into the estimated results.

7.2.1 Piecewise Linear Regression Models

Piecewise linear regression models are often used to evaluate the implications from theories predicting non-linear relationships between X and Y like those seen in Figure 7.8. The predicted value plots on the left show the piecewise slope relationship between X and Y, while the corresponding marginal effect plots on the right show how the marginal effect of X on Y changes with the value of X. In panel (a), the slope of the relationship between X and Y is positive (3) in the interval where X is less than or equal to 3 and negative (-4) in the interval where X is greater than 3. In panel (b), the slope of the relationship between X and Y is small and positive (3) in the

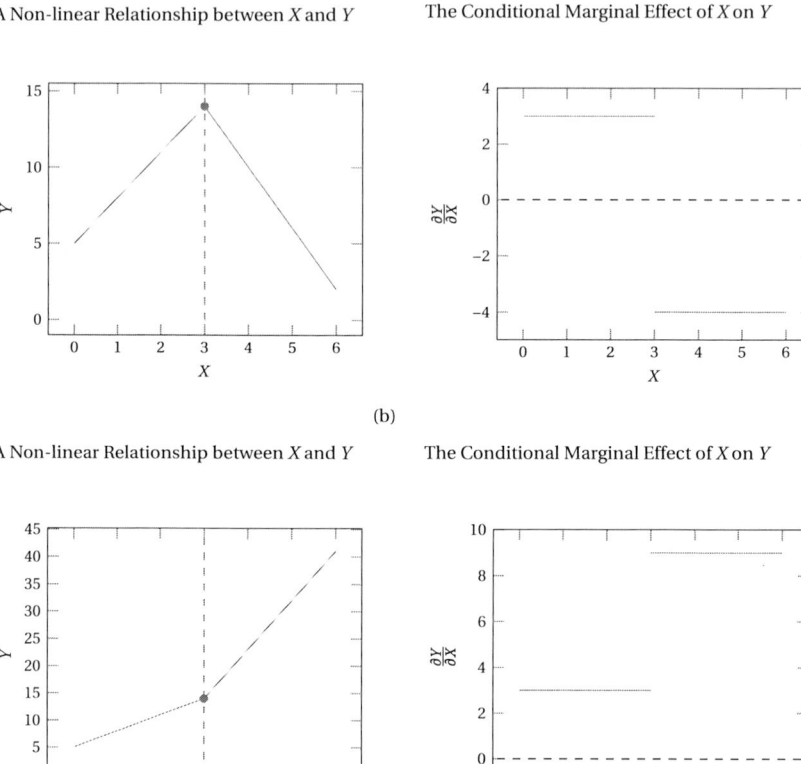

Figure 7.8 (a)–(b) Two possible continuous piecewise linear relationships between X and Y

interval where X is less than or equal to 3 and large and positive (9) in the interval where X is greater than 3. The break in the relationship between X and Y in both panels occurs when $X = 3$ and is indicated by the vertical dashed line in the two predicted value plots. The commonality in these types of scenarios is that the relationship between X and Y is continuous but comprises different linear "pieces" that are "joined" or "tied" together at some threshold value of X. The point at which the different linear pieces are tied together is sometimes referred to as a "knot" and is indicated by the circle at coordinates (3,14) in both panels. Both of the piecewise linear relationships between X and Y shown in Figure 7.8 change only once. However, we can obviously have more complicated theoretical stories where the piecewise linear relationship between X and Y changes multiple times. In these scenarios, we'd have multiple breaks and knots.

Piecewise linear relationships are special cases of a larger set of relationships called spline functions (Keele, 2008, 39–94). Spline functions capture relationships that are made up of distinct "pieces" or functions that are joined together in a continuous way. When the distinct pieces of the relationship are linear, as they are in Figure 7.8, we have a linear spline. Thus, a piecewise linear regression model is also a linear spline model.[17]

We can use the following piecewise linear regression model to evaluate the implications of a theory predicting a non-linear relationship between X and Y like those shown in Figure 7.8:

$$Y = \gamma_0 + \gamma_1 X + \gamma_2 \left(X - X^T\right) D + \epsilon, \tag{7.37}$$

where X^T is the threshold value of X at which the break occurs and D is a dichotomous variable that equals 1 if X is greater than X^T and 0 otherwise. As you can see, the piecewise linear regression model contains a multiplicative interaction term, $\left(X - X^T\right) D$, and is thus an interaction model.[18]

---

[17] The different pieces of the relationship between X and Y that are tied together at a knot don't have to be linear. In substantive applications, scholars often use spline models to allow for very complicated, but continuous, non-linear relationships between X and Y. In many of these applications, each distinct piece of the relationship between X and Y is modeled as a cubic polynomial. The cubic polynomials for each distinct piece of the relationship between X and Y are constrained to join up in a smooth way at the knots. These particular spline models are referred to as cubic splines. It should be clear how very complicated non-linear relationships between X and Y can be captured by having multiple cubic polynomials, each associated with a distinct interval of values for X, tied together with multiple knots.

[18] In many settings, the piecewise linear regression model is written as

$$Y = \gamma_0 + \gamma_1 X + \gamma_2 \left(X - X^T\right)_+ + \epsilon, \tag{7.38}$$

where $\left(X - X^T\right)_+$ is substituted in for the interaction term $\left(X - X^T\right) D$. The term $\left(X - X^T\right)_+$ is a *positive part function* where

$$\left(X - X^T\right)_+ = \begin{cases} X - X^T & \text{if } X - X^T > 0 \\ 0 & \text{if } X - X^T \leq 0, \end{cases} \tag{7.39}$$

or

$$\left(X - X^T\right)_+ = \max\{X - X^T, 0\}. \tag{7.40}$$

The positive part function $\left(X - X^T\right)_+$ is exactly equivalent to the interaction term $\left(X - X^T\right) D$. We prefer to use $\left(X - X^T\right) D$ here to highlight the fact that the piecewise linear regression model shown in Eq. 7.37 is an interactive, as opposed to an additive, model specification.

When X is less than or equal to X^T ($D = 0$), the model in Eq. 7.37 simplifies to

$$Y = \gamma_0 + \gamma_1 X + \epsilon. \tag{7.41}$$

From this, we see that γ_1 tells us the effect of a one-unit increase in X on Y when X is less than or equal to X^T. When X is greater than X^T ($D = 1$), the model in Eq. 7.37 simplifies to

$$\begin{aligned} Y &= \gamma_0 + \gamma_1 X + \gamma_2 \left(X - X^T\right) + \epsilon \\ &= \gamma_0 + \gamma_1 X + \gamma_2 X - \gamma_2 X^T + \epsilon \\ &= \left(\gamma_0 - \gamma_2 X^T\right) + (\gamma_1 + \gamma_2) X + \epsilon. \end{aligned} \tag{7.42}$$

From this, we see that $\gamma_1 + \gamma_2$ tells us the effect of a one-unit increase in X on Y when X is greater than X^T.[19] A simple t-test of $\gamma_2 = 0$ thus allows us to see whether there's a break in the linear relationship between X and Y – that is, whether the slope is different when $X \leq X^T$ than when $X > X^T$ – and hence whether the effect of X on Y varies with the value of X in a way that's consistent with the marginal effect plots shown in Figure 7.8.

To illustrate with specific examples, the predicted value plot in panel (a) of Figure 7.8 is based on the following piecewise linear regression results:

$$Y = 5 + 3X - 7(X - 3)D. \tag{7.43}$$

These results tell us that the effect of a one-unit increase in X is 3 when X is less than or equal to 3 and $3 - 7 = -4$ when X is greater than 3. These effects are shown in the corresponding marginal effect plot on the right. The predicted value plot in panel (b) is based on the following piecewise linear regression results:

$$Y = 5 + 3X + 6(X - 3)D. \tag{7.44}$$

These results tell us that the effect of a one-unit increase in X is 3 when X is less than or equal to 3 and $3 + 6 = 9$ when X is greater than 3. These effects are again shown in the corresponding marginal effect plot on the right.

The piecewise linear regression model shown in 7.37 can easily be generalized to deal with more than one break. For example, if we had a theory predicting that the effect of X on Y is piecewise linear and depends on whether the value of X is, say, low, medium, or high, we could estimate the following piecewise linear regression model,

$$Y = \gamma_0 + \gamma_1 X + \gamma_2 \left(X - X^{T_1}\right) D_1 + \gamma_3 \left(X - X^{T_2}\right) D_2 + \epsilon, \tag{7.45}$$

[19] The variance for the effect of a one-unit increase in X on Y when X is greater than X^T is $\text{var}(\gamma_1 + \gamma_2) = \text{var}(\gamma_1) + \text{var}(\gamma_2) + 2\text{cov}(\gamma_1, \gamma_2)$.

where X^{T_1} is the threshold value of X that determines the first break between low and medium values of X, X^{T_2} is the threshold value of X that determines the second break between medium and high values of X, D_1 is a dichotomous variable that equals 1 if X is greater than X^{T_1} and 0 otherwise, and D_2 is a dichotomous variable that equals 1 if X is greater than X^{T_2} and 0 otherwise. The predicted value of Y for each of the linear pieces is

$$Y = \begin{cases} \gamma_0 + \gamma_1 X \\ \left(\gamma_0 - \gamma_2 X^{T_1}\right) + (\gamma_1 + \gamma_2) X \\ \left(\gamma_0 - \gamma_2 X^{T_1} - \gamma_3 X^{T_2}\right) + (\gamma_1 + \gamma_2 + \gamma_3) X. \end{cases} \tag{7.46}$$

We see that the effect of a one-unit increase in X on Y is γ_1 when X is less than or equal to X^{T_1} (X is low), $\gamma_1 + \gamma_2$ when X is greater than X^{T_1} but less than or equal to X^{T_2} (X is medium), and $\gamma_1 + \gamma_2 + \gamma_3$ when X is greater than X^{T_2} (X is high).

One thing you'll have noticed is that the piecewise linear regression model shown in Eq. 7.37 doesn't include either of the constitutive components for the interaction term $(X - X^T) D$. Note that $X - X^T$ *can't* be included as there would be perfect multicollinearity with X and $(X - X^T) D$. But what about the constitutive component D? It turns out that D is deliberately excluded to make sure that the two linear regression pieces join together in a knot at the threshold value of $X = X^T$. We can see that the model in Eq. 7.37 constrains the two linear regression pieces to be tied together at $X = X^T$ by calculating the predicted value of Y when $X = X^T$. When we substitute $X = X^T$ into Eq. 7.41 for the first linear piece, we get

$$Y = \gamma_0 + \gamma_1 X^T. \tag{7.47}$$

We get the exact same thing when we substitute $X = X^T$ into Eq. 7.42 for the second linear piece,

$$\begin{aligned} Y &= \left(\gamma_0 - \gamma_2 X^T\right) + (\gamma_1 + \gamma_2) X \\ &= \gamma_0 - \gamma_2 X^T + \gamma_1 X^T + \gamma_2 X^T \\ &= \gamma_0 + \gamma_1 X^T. \end{aligned} \tag{7.48}$$

Ultimately, the omission of the constitutive component D forces the estimated piecewise linear relationship between X and Y to be continuous. In effect, the omission of D forces there to be no intercept shift in the relationship between X and Y when we transition from the first linear regression piece to the second one at the threshold $X = X^T$. The intuition here fits squarely with our earlier discussion of the consequences of omitting constitutive components in Chapter 3. Omitting Z is equivalent to assuming that the coefficient on D would be 0 if D were included in the model. To the extent that this assumption is incorrect, which would occur if there was any sort of discontinuity in the relationship between X and Y at the threshold $X = X^T$, the estimated results from the piecewise linear regression model will be biased. To avoid potential bias in the results, we encourage scholars to include D in their model specification. Doing so leads to a type of *switching regression model*.[20]

7.2.2 Switching Regression Models

A switching regression model can be used to evaluate the implications from theories predicting non-linear relationships between X and Y like those shown in Figure 7.9. As before, predicted value plots are shown on the left, while corresponding marginal effect plots are shown on the right. We're still talking about piecewise linear relationships between X and Y. The difference now is that these relationships aren't necessarily continuous because there's a potential discontinuity at the threshold point $X = X^T$. In other words, the linear pieces capturing the relationship between X and Y in a switching regression model are no longer constrained to join up at the threshold points. The types of non-linear relationship between X and Y shown in Figure 7.9 are often theorized in studies employing a regression discontinuity design.

We can use the following switching regression model to evaluate the implications of a theory predicting a non-linear relationship between X and Y like those shown in Figure 7.9:

$$Y = \beta_0 + \beta_1 X + \beta_2 D + \beta_3 \left(X - X^T \right) D + \epsilon, \tag{7.49}$$

where X^T is the threshold value of X at which the break occurs and D is a dichotomous variable that equals 1 if X is greater than X^T and 0 otherwise. This model is identical to the piecewise linear regresssion model shown in Eq. 7.37, except that it also includes the previously omitted constitutive term D.

[20] The general idea with switching regression models is that there are a finite (and usually small) number of possible "states" in which the value of the model parameters may change. In effect, switching models assume that the observations for a dependent variable Y are generated by two or more distinct regression equations that correspond to different "states." There are many different types of switching models (Goldfeld and Quandt, 1973). For example, regression discontinuity design models (Cattaneo, Idrobo and Titiunik, 2020) and Markov transitions models (Przeworski et al., 2000; Epstein et al., 2006) are types of switching model. We focus here on a switching model in which the transition between different states is deterministic and depends on comparing the value of some variable X to a conjectured threshold X^T.

(a)

A Non-linear Relationship between X and Y The Conditional Marginal Effect of X on Y

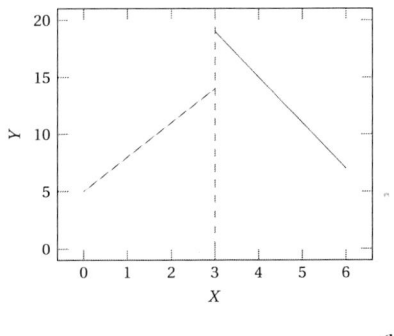 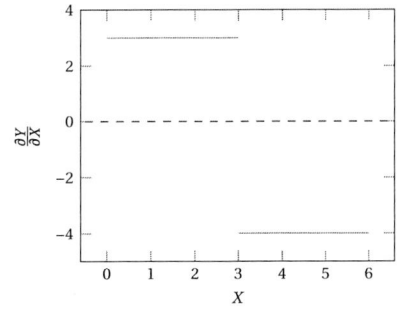

(b)

A Non-linear Relationship between X and Y The Conditional Marginal Effect of X on Y

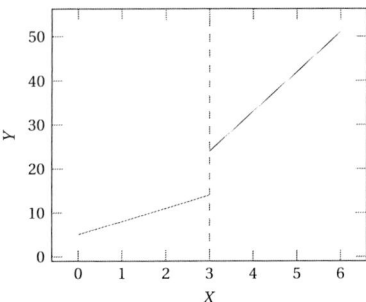 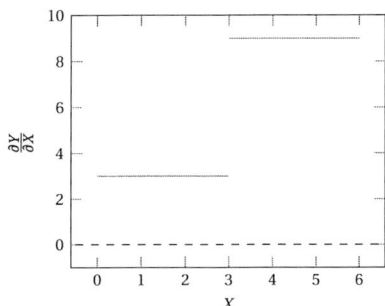

Figure 7.9 (a)–(b) Two non-continuous piecewise linear relationships between X and Y

When X is less than or equal to X^T ($D = 0$), the model in Eq. 7.49 simplifies to

$$Y = \beta_0 + \beta_1 X + \epsilon. \tag{7.50}$$

From this, we see that β_1 tells us the effect of a one-unit increase in X on Y when X is less than or equal to X^T. When X is greater than X^T ($D = 1$), the model in Eq. 7.49 simplifies to

$$
\begin{aligned}
Y &= \beta_0 + \beta_1 X + \beta_2 + \beta_3 \left(X - X^T\right) + \epsilon \\
&= \beta_0 + \beta_1 X + \beta_2 + \beta_3 X - \beta_3 X^T + \epsilon \\
&= \left(\beta_0 + \beta_2 - \beta_3 X^T\right) + (\beta_1 + \beta_3) X + \epsilon.
\end{aligned} \tag{7.51}
$$

From this, we see that $\beta_1 + \beta_3$ tells us the effect of a one-unit increase in X on Y when X is greater than X^T. A simple t-test of $\beta_3 = 0$ thus allows us to see whether there's a break in the linear relationship between X and Y and hence whether the effect of X on Y varies with the value of X in a way that's consistent with the marginal effect plots shown in Figure 7.9.

⚠ **Important:** Some theories predict that there is a piecewise linear relationship between X and Y, where X has, say, one effect when the value of X is low and a different effect when the value of X is high. One way to test the implications of such a theory is with a piecewise linear regression model,

$$Y = \gamma_0 + \gamma_1 X + \gamma_2 \left(X - X^T \right) D + \epsilon, \tag{7.52}$$

where X^T is the threshold value of X at which the relationship between X and Y changes and D is a dichotomous variable that equals 1 if X is greater than X^T and 0 otherwise. An alternative approach is to use a switching regression model,

$$Y = \beta_0 + \beta_1 X + \beta_2 D + \beta_3 \left(X - X^T \right) D + \epsilon. \tag{7.53}$$

Switching regression models have the advantage that they include the constitutive components of the interaction term and thus allow the piecewise linear relationship between X and Y to be non-continuous at the break points.

We can confirm that the switching regression model shown in Eq. 7.49 allows there to be an intercept shift when we transition from the first linear regression piece to the second one at $X = X^T$ by calculating the predicted value of Y when $X = X^T$. When we substitute $X = X^T$ into Eq. 7.50, we get

$$Y = \beta_0 + \beta_1 X^T. \tag{7.54}$$

However, when we substitute $X = X^T$ into Eq. 7.51, we get

$$\begin{aligned} Y &= \left(\beta_0 + \beta_2 - \beta_3 X^T \right) + (\beta_1 + \beta_3) X^T \\ &= \beta_0 + \beta_2 - \beta_3 X^T + \beta_1 X^T + \beta_3 X^T \\ &= \beta_0 + \beta_1 X^T + \beta_2. \end{aligned} \tag{7.55}$$

From this, we see that the predicted value of Y is β_2 higher at $X = X^T$ when $D = 1$ than when $D = 0$. In other words, the coefficient on D in Eq. 7.49 tells us the intercept shift when we transition from the first linear regression piece to the second one. By omitting the constitutive component

D, as happens in the piecewise linear regression model in Eq. 7.37, we're essentially assuming that β_2 in the switching regression model is 0 and hence that there's no intercept shift at the threshold value of X. It should now be clear why omitting D is potentially problematic. In terms of the panels in Figure 7.9, omitting D would force the two linear regression pieces in each of the predicted value plots to meet at $X = 3$. The only way for this to happen is for the slopes of the two linear regression pieces, and hence the marginal effects of X at different values of X, to change. Thus, omitting D introduces bias into the estimates of the conditional effect of X whenever β_2 isn't exactly 0.

One's theory may well predict that the piecewise linear relationship between X and Y is continuous and hence that a piecewise linear regression model is appropriate. However, this is probably something that we'd want to test rather than assume. We can do precisely this with the switching regression model, where we can look to see if the coefficient on D is, in fact, 0. This is why we recommend that scholars employ the (unconstrained) switching regression model in Eq. 7.49 rather than the (constrained) piecewise linear regression model in Eq. 7.37.

In the last two chapters, we've examined more theoretically complex forms of conditionality than we've seen previously. In the last chapter, we looked at theories positing that the effect of an independent variable X depends on the value of more than one other modifying variable. And in this chapter, we've looked at theories positing that the effect of an independent variable X depends not on the value of some other variable but on its own value.

So far in this book, we've focused on cases where our theory makes some kind of conditional claim in the context of a continuous dependent variable. In the next part of the book, we look at how to specify and interpret interaction models to test the conditional implications from theories where we have some form of limited dependent variable. The specific focus in the next chapter is on theories dealing with a dichotomous or binary dependent variable.

7.3 EXERCISES

1. In this chapter, we examined quadratic regression models in some detail. In this question, we're going to look more carefully at cubic regression models,

$$Y = \beta_0 + \beta_1 X + \beta_2 X^2 + \beta_3 X^3 + \epsilon. \tag{7.56}$$

a. Quadratic models allow there to be a single turning point in the non-linear relationship between X and Y where the effect of X on

Y changes sign. We saw earlier that the sign of the coefficient on the X^2 term indicated whether the estimated non-linear relationship between X and Y had a U-shaped or inverted U-shaped relationship. Cubic regression models allow there to be up to two turning points in the non-linear relationship between X and Y. An example of such a cubic relationship was shown earlier in the predicted value plot in panel (c) of Figure 7.1. Looking at the cubic regression model in Eq. 7.56, what does the sign of the coefficient on the X^3 term say about the shape of the non-linear relationship between X and Y? Draw a predicted value plot corresponding to a typical cubic regression model in which β_3 is negative and one in which β_3 is positive.

b. Calculate the marginal effect of X. How do we interpret the coefficient β_1?

c. Calculate the modifying effect of X on the marginal effect of X. How do we interpret the coefficient β_2?

d. In the main chapter, we found the value of X associated with the turning point in a quadratic model by setting the marginal effect of X equal to 0 and solving for X. Use the same method to identify the values of X at which the bends occur in the cubic regression model shown in Eq. 7.56.

e. It can be useful to think visually about how the cubic relationship between X and Y is associated with the marginal effect of X and the modifying effect of X on the marginal effect of X. Suppose we have the following results from a cubic regression model: $Y = 432 + 144X - 3X^2 - X^3$. Graph the predicted value of Y from $X = -14$ to $X = 14$. In the same plot, graph the marginal effect of X and the modifying effect of X on the marginal effect of X. Briefly discuss how these graphs are related.

f. Using the properties of variances in Appendix A, what is the variance for the marginal effect of X?

g. What is the variance for the modifying effect of X?

h. What is the effect of a one-unit increase in X? What is the variance for a one-unit increase in X?

i. How many *key predictions* are possible from a theory predicting that there's a cubic relationship between X and Y? What types of plots could be used to evaluate these key predictions?

2. In this chapter, we briefly looked at how we can specify polynomial regression models to evaluate the conditional implications from theories where the interaction effect is expected to be non-linear.

a. We noted that we could use the following polynomial regression model to capture the situation where we expect there to be a smooth curvilinear relationship between the marginal effect of X and Z with only one turning point:

$$Y = \beta_0 + \beta_1 X + \beta_2 Z + \beta_3 XZ + \beta_4 Z^2 + \beta_5 XZ^2 + \epsilon. \qquad (7.57)$$

Using the properties of variances in Appendix A, calculate the variances for the marginal effect of X and the interaction effect between X and Z.

b. We also noted that we could use the following polynomial regression model to capture the situation where we expect there to be a smooth curvilinear relationship between the marginal effect of X and Z with two turning points:

$$Y = \beta_0 + \beta_1 X + \beta_2 Z + \beta_3 XZ + \beta_4 Z^2 + \beta_5 XZ^2 + \beta_6 Z^3 + \beta_7 XZ^3 + \epsilon. \qquad (7.58)$$

Calculate the variances for the marginal effect of X and the interaction effect between X and Z.

PART III

Interactions and Limited Dependent Variables

8 Interactions and Dichotomous Dependent Variables

So far we've focused on theories that posit an interaction between two or more independent variables on a *continuous* dependent variable. However, not all of our theories with conditional implications deal with a continuous dependent variable. For example, we might have a theory that speaks to whether a country is at war or not, how individuals allocate their vote in a multiparty election, or someone's qualitative level of political interest. In each of these cases, we have some kind of limited dependent variable where the range of values is restricted in some important way and, as a result, ordinary least squares estimation is unlikely to be appropriate. Instead, we'll usually want to estimate some kind of maximum likelihood model that allows for limited dependent variables (King, 2001). In this chapter, we begin our transition to limited dependent variables by looking in detail at how to evaluate the conditional implications of our theories when we have a dichotomous or binary dependent variable. In the following chapters, we expand our discussion to look at interaction models in the context of other kinds of limited dependent variables.

8.1 THE LINEAR PROBABILITY MODEL

We often have theories with conditional implications that involve a dichotomous dependent variable Y that can take on the values 0 and 1. The dichotomous dependent variable might have to do with whether someone votes or not, whether a country is at war or not, whether someone is sentenced for a crime or not, and so on. In these circumstances, scholars typically employ a logit or probit model. Before we discuss these models in some detail, we briefly examine an alternative for dealing with a dichotomous dependent variable, the *linear probability model*.

The linear probability model (LPM) arises when we use ordinary least squares (OLS) estimation in the context of a dichotomous dependent variable. The expected value of Y is modeled as a linear function of some independent variables,

$$E[Y] = \beta_0 + \beta_1 X_1 + \beta_2 X_2 + \cdots + \beta_k X_k = X\beta, \tag{8.1}$$

where $X\beta$ is just the more succinct matrix notation for $\beta_0 + \beta_1 X_1 + \beta_2 X_2 + \cdots + \beta_k X_k$. With a dichotomous dependent variable, we have

$$\begin{aligned} E[Y] &= Pr(Y = 1) \times 1 + Pr(Y = 0) \times 0 \\ &= Pr(Y = 1), \end{aligned} \tag{8.2}$$

where $Pr(Y = 1)$ is the probability that $Y = 1$ and $Pr(Y = 0)$ is the probability that $Y = 0$. As a result, the linear probability model can be written as

$$Pr(Y = 1) = \beta_0 + \beta_1 X_1 + \beta_2 X_2 + \cdots + \beta_k X_k = X\beta. \tag{8.3}$$

The linear probability model is easy to estimate, and we can interpret the coefficients as we would from a normal ordinary least squares estimation. In other words, the coefficients in Eq. 8.3 indicate how a one-unit change in an independent variable affects the probability that $Y = 1$. We can easily test the conditional implications of a theory by incorporating interaction terms into our model. Everything that we've discussed previously about how to specify and interpret the results from an interactive linear regression model estimated by OLS applies to an interactive linear probability model. There's nothing new here if we adopt this approach to dealing with a dichotomous dependent variable.

However, there are reasons to think that the linear probability model may not be optimal for dealing with a dichotomous dependent variable. The first reason is that $X\beta$ is unbounded, meaning that we can obtain predicted probabilities that are greater than 1 and less than 0; this is obviously undesirable. The second reason is that the variance of Y depends on the values of the independent variables and the coefficients, meaning that the LPM is heteroskedastic by construction,

$$\begin{aligned} \text{var}(Y) &= E[Y](1 - E[Y]) \\ &= X\beta(1 - X\beta). \end{aligned} \tag{8.4}$$

The third reason is that the errors are not normally distributed due to the fact that they can take on only two values, $1 - X\beta$ or $-X\beta$, thereby creating issues for hypothesis testing. A fourth reason is that the linear functional form doesn't allow the marginal effect of an independent variable to exhibit diminishing returns, which is something that we'd expect theoretically as the probability that $Y = 1$ approaches 1 or 0. It should be noted that none of these issues create problems for our point estimates. In other words, the coefficients from the LPM remain unbiased. That said, arguably better and more theoretically appropriate methods for dealing with a dichotomous dependent variable are available. Many scholars, for example, prefer to estimate a logit or probit model.

> ⚠ **Important:** Potential reasons why a linear probability model (LPM) can be problematic when we have a dichotomous dependent variable:
>
> 1. Predicted probabilities greater than 1 and less than 0.
> 2. Heteroskedasticity by construction.
> 3. Non-normal errors.
> 4. Effects of the independent variables do not exhibit diminishing returns as $Pr(Y = 1)$ approaches 0 or 1.

8.2 THE BASIS FOR LOGIT AND PROBIT MODELS

Before looking at how we can use a logit or probit model to test the conditional implications of a theory, it's useful to first review where these models come from and how they work. We start by discussing the basis for logit and probit models.

Recognizing the issues with the linear probability model, scholars have adopted three broad approaches for dealing with a dichotomous dependent variable: (1) the *pure probability approach*, (2) the *latent variable approach*, and (3) the *random utility approach*. All three approaches can lead us to a logit or probit model. As we'll see, each approach gives us a slightly different way of thinking about how to interpret the results from a logit or probit model.

8.2.1 The Pure Probability Approach

With the pure probability approach, we start by looking for a probability distribution that can handle a dependent variable that takes on only two values, 1 and 0. The Bernoulli distribution can do this,

$$f(1) = \rho,$$
$$f(0) = 1 - \rho, \tag{8.5}$$

where the outcome $Y = 1$ or "success" occurs with probability ρ and the outcome $Y = 0$ or "failure" occurs with probability $1 - \rho$. The Bernoulli distribution would be an appropriate model for a dichotomous dependent variable if the outcome $Y = 1$ had the same chance of occurring for every observation in our sample. In practice, this is unlikely to be the case. For example, we wouldn't think that all individuals have the same probability of voting or that all countries are equally likely to go to war. Thus, the

Bernoulli distribution is too restrictive as it stands. Instead we'd like to let ρ vary across observations. However, we can't let each observation have its own ρ because our model wouldn't be identified. Instead, we assume that ρ is some function $g(\cdot)$ of a linear combination of independent variables; that is, $\rho = g(X\beta)$. This allows us to both reduce the number of parameters in our model and examine the effect of theoretically-relevant explanatory variables on the probability that $Y = 1$. We just need to find an appropriate function $g(\cdot)$ that can take the values of a linear combination of independent variables $X\beta$, which may be between $\pm\infty$, and convert them to the scale of a probability that runs between 0 and 1. We also want the function $g(\cdot)$ to return a 1 in the limit as the linear predictor $X\beta$ approaches $+\infty$ and a 0 in the limit when the linear predictor $X\beta$ approaches $-\infty$. While there are an infinite number of possible functions that satisfy these criteria, scholars have settled on a small number of "S-shaped" curves for $g(\cdot)$. The two most popular are the cumulative standard logistic distribution, $\Lambda(X\beta)$, which leads to the logit model, and the cumulative standard normal distribution, $\Phi(X\beta)$, which leads to the probit model.

The standard logistic cumulative distribution function (CDF) is

$$\Lambda(X\beta) = \frac{e^{X\beta}}{1 + e^{X\beta}} = \frac{1}{1 + e^{-X\beta}} = Pr(Y = 1) = \rho, \tag{8.6}$$

and the standard normal cumulative distribution function (CDF) is

$$\Phi(X\beta) = \int_{-\infty}^{X\beta} \frac{1}{\sqrt{2\pi}} e^{-\frac{(X\beta)^2}{2}} dX\beta = \Pr(Y = 1) = \rho. \tag{8.7}$$

In Figure 8.1, we visually show the cumulative standard logistic (smooth) and normal (dashed) distributions. We can now see how these specific S-shaped curves transfer unbounded values of $X\beta$ onto a probability scale that's bounded between 0 and 1. As an example, the standard logistic takes an $X\beta$ value of 1 and translates it into the probability $\Lambda(1) = 0.73$. The standard normal takes the same value for $X\beta$ and translates it into a probability of $\Phi(1) = 0.84$. The two S-shaped curves are quite similar and touch at $X\beta = 0$, with both curves mapping this particular value to a probability that $Y = 1$ of 0.5. The logit and probit curves share certain features. For example, their slopes are always steepest at $X\beta = 0$, meaning that the effect of any increase in an independent variable is always greatest among those cases where the probability that $Y = 1$ is 0.5.[1] The slopes of both curves decline as $X\beta$ departs from 0 or as $Pr(Y = 1)$ departs from 0.5 in either direction. Both slopes approach their minimum of 0 as $X\beta$ goes

---

[1] Those who wish to relax this particular assumption may be interested in using a scobit (skewed-logit) model (Nagler, 1994).

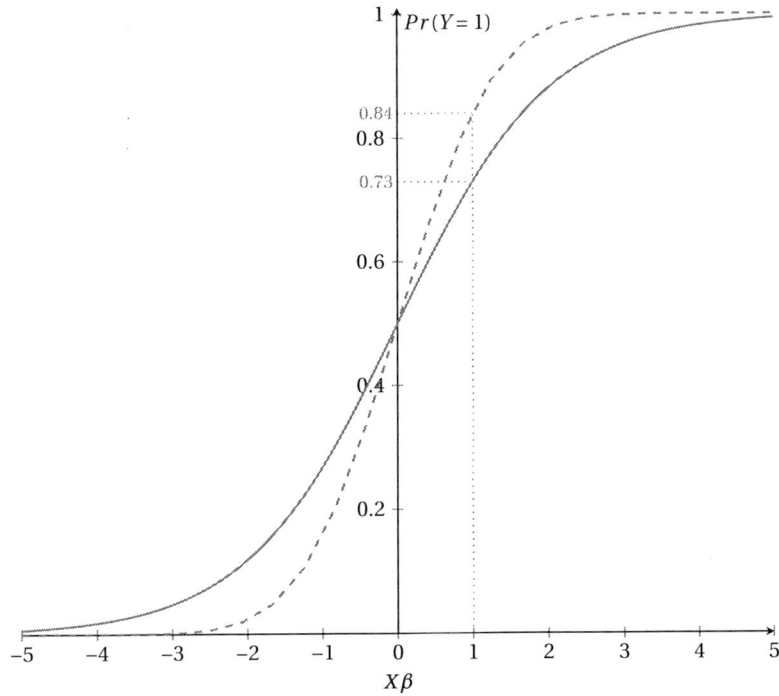

Figure 8.1 The cumulative standard logistic (smooth) and normal (dashed) distributions

to $\pm\infty$ or as $Pr(Y = 1)$ goes to either 1 or 0. It's worth noting that both curves are close to linear when $X\beta$ is 0 and when $X\beta$ is either very large or very small; however, they're both quite non-linear when $X\beta$ takes on other values. The coefficients from a logit model are larger than those from a probit model by a scale factor of about 1.6 to 1.8. Once the logit or probit coefficients are passed through the appropriate function $g(\cdot)$ to calculate substantive quantities of interest such as predicted probabilities and effect sizes, though, we typically obtain very similar results (Long, 1997, 83). It's because of this that the choice of whether to adopt a logit or probit model is usually treated as a "matter of taste or convenience" (Gelman and Hill, 2006, 199).

In our presentation of the pure probability approach, we looked for a function $g(\cdot)$ that takes the unbounded values of a linear combination of independent variables and translates them onto a probability scale that's bounded between 0 and 1, $\rho = g(X\beta)$. However, we could just as easily have done the opposite. In other words, we could have looked for a function $h(\cdot)$ that takes probabilities that are bounded between 0 and 1 and translates them onto an unbounded scale that runs from $-\infty$ to $+\infty$, $h(\rho) = X\beta$. Such a function, which is called the *link function*, is just the inverse of $g(\cdot)$.[2]

In the case of the logit model, the link function – the inverse of $\Lambda(X\beta)$ – is $ln\left(\frac{\rho}{1-\rho}\right)$, where ln refers to the natural log, and so we have

$$h(\rho) = ln(\Omega) = ln\left(\frac{\rho}{1-\rho}\right) = X\beta = \beta_0 + \beta_1 X_1 + \beta_2 X_2 + \cdots + \beta_k X_k.$$

$$(8.8)$$

The term $\Omega = \frac{\rho}{1-\rho}$ is known as the "odds" and indicates the ratio of the probability of a "success" ($Y = 1$) to the probability of a "failure" ($Y = 0$). As an example, if the probability of a success is 0.75, then the odds of a success is $\frac{0.75}{1-0.75} = \frac{0.75}{0.25} = 3$. We sometimes say that the odds of success are 3 to 1 or $3 : 1$. The term $ln(\Omega)$ is known as the "logit" of ρ or the "log-odds" of $Y = 1$. From this, we see that the log-odds is modeled as a linear combination of independent variables in a logit model. Thus, one interpretation of the coefficients in a logit model is that they tell us the effect of a one-unit increase in our independent variables on the log-odds of success.

8.2.2 The Latent Variable Approach

The latent variable approach is different in that it essentially treats dichotomous dependent variables as a measurement problem. The basic idea is that there exists some continuous underlying or "latent" variable Y^* that we haven't been able to fully measure. All we have is a dichotomous indicator Y of it. In this setup, the underlying model is

$$Y^* = X\beta + \epsilon,$$

$$(8.9)$$

where the error term ϵ has a mean of 0 and is distributed according to some symmetric distribution. Rather than observe the latent variable Y^*, all we get to observe is a dichotomous indicator of it,

$$Y = \begin{cases} 1 & \text{if } Y^* \geq 0 \\ 0 & \text{if } Y^* < 0. \end{cases}$$

---

[2] The pure probability approach essentially treats logit and probit models as a type of generalized linear model (Nelder and Wedderburn, 1972). A generalized linear model has three core components. The first is a *random component* μ specifying the conditional probability distribution for the dependent variable Y given the values of the independent variables in the model. The second is a *linear predictor* of the independent variables, $X\beta$. And the third is a smooth and invertible *link function* $h(\cdot)$ that "links" the random component to the linear predictor by transforming the expected value of the dependent variable to the linear predictor, $h(\mu) = X\beta$. Instead of specifying a generalized linear model in terms of the link function, though, we can, of course, use the inverse link function $g(\cdot)$ to transform the linear predictor to the expected value of the dependent variable, $\mu = g(X\beta)$.

One way to think about this setup is that Y^* indicates the *propensity* that some event such as casting a vote or starting a war happens. When the propensity for an event is sufficiently high (greater than 0), which depends on a linear combination of independent variables, the event happens.

From this setup, we see that

$$Pr(Y = 1) = Pr(Y^* \geq 0)$$
$$= Pr(X\beta + \epsilon \geq 0)$$
$$= Pr(\epsilon \geq -X\beta)$$
$$= Pr(\epsilon \leq X\beta), \tag{8.10}$$

where the last equality holds because we've assumed that the distribution for ϵ is symmetric. To calculate $Pr(\epsilon < X\beta)$, and hence $Pr(Y = 1)$, we just need to choose an appropriate distribution for ϵ. If we assume that ϵ has a standard logistic distribution with mean 0 and variance $\frac{\pi^2}{3}$, we get the logit model where

$$Pr(\epsilon < X\beta) = \Lambda(X\beta) = \frac{e^{X\beta}}{1 + e^{X\beta}} = \frac{1}{1 + e^{-X\beta}} = Pr(Y = 1). \tag{8.11}$$

And if we assume that ϵ has a standard normal distribution with mean 0 and variance 1, we get the probit model where

$$Pr(\epsilon < X\beta) = \Phi(X\beta) = \int_{-\infty}^{X\beta} \frac{1}{\sqrt{2\pi}} e^{-\frac{(X\beta)^2}{2}} dX\beta = Pr(Y = 1). \tag{8.12}$$

Based on the latent variable approach, we see that the coefficients from a logit or probit model can be interpreted as telling us the effect of a one-unit increase in our independent variables on the *propensity* of some event ($Y = 1$) occurring.

8.2.3 The Random Utility Approach

The random utility approach to modeling dichotomous dependent variables is again different and has its origins in microeconomic theories of choice and utility maximization. To get a sense of what's going on here, imagine a situation where someone is choosing among a set of alternatives or potential outcomes. For example, a country's leader may be deciding to go to war or not; an individual may be deciding whether to vote for Party A, Party B, or Party C; a jury may be deciding whether someone is guilty or not, and so on. The *random utility model* (RUM) assumes that the actor making the choice attaches a utility to each of the available alternatives and that these utilities are based on a systematic and stochastic component. It's assumed that the actor engages in utility maximizing behavior and chooses the alternative with the highest utility.

In this setup, the utility that actor i obtains from alternative m is

$$U_{im} = V_{im} + \epsilon_{im}, \tag{8.13}$$

where V_{im} indicates the systematic component of utility for actor i associated with alternative m, and ϵ_{im} indicates the stochastic component of utility for actor i associated with choice m. We can think that the stochastic component captures the idea that there may be some uncertainty about the precise utility of alternative m. You'll notice that e_{im} is subscripted by both i and m, meaning that there's a distinct disturbance for each actor for each alternative. In the random utility model, we assume that the systematic component of the actor's utility for alternative m is determined by a linear combination of independent variables,

$$V_{im} = X_{im}\beta_m, \tag{8.14}$$

and so the utility of actor i for alternative m can be written as

$$U_{im} = X_{im}\beta_m + \epsilon_{im}. \tag{8.15}$$

The fact that the independent variables are subscripted by both i and m means that they can include factors that are specific to the actor and/or the alternative. The coefficients indicate the relative importance of these factors. We assume that actor i chooses alternative m if and only if

$$U_{im} \geq U_{ij} \ \forall j \neq m. \tag{8.16}$$

In other words, we assume that actor i chooses alternative m if and only if the utility associated with alternative m is larger than or equal to the utility associated with all of the other available alternatives. What's nice about the random utility setup is that it's embedded in a specific micro-level theory of individual choice and allows us to look at situations where actors can choose between two or more alternatives.

We can easily apply the random utility approach to the situation where we have just two alternatives and we have a dichotomous dependent variable. In this case, the probability of choosing alternative 1 is just the probability that the utility from alternative 1 is larger than or equal to the utility from alternative 2,

$$P(Y_i = 1) = P(U_{i1} \geq U_{i2}). \tag{8.17}$$

If we now set actor i's utility from alternative 2 to some arbitrary value, such as 0, we get

$$\begin{aligned}
P(Y_i = 1) &= P(U_{i1} \geq 0) \\
&= P(V_{i1} + \epsilon_{i1} \geq 0) \\
&= P(X_{i1}\beta_1 + \epsilon_{i1} \geq 0) \\
&= P(\epsilon_{i1} \leq X_{i1}\beta_1). \tag{8.18}
\end{aligned}$$

This is essentially the same as what we had in the latent variable setup in Eq. 8.10. To calculate $P(\epsilon_{i1} \leq X_{i1}\beta_1)$, and hence $P(Y_i = 1)$, we need to choose an appropriate distribution for ϵ_{1i}. If we assume that ϵ_{1i} has a standard logistic distribution, we get the logit model, and if we assume that it has a standard normal distribution, we get the probit model. Based on the random utility approach, we see that the coefficients from a logit or probit model can be interpreted as telling us the effect of a one-unit increase in our independent variables on the utility of alternative 1 ($Y = 1$) relative to alternative 2 ($Y = 0$).

> ⚠ **Important:** There are three broad approaches to thinking about logit and probit models:
>
> 1. The pure probability approach.
> 2. The latent variable approach.
> 3. The random utility approach.

8.3 INTERPRETATION AND INTERACTION EFFECTS IN "ADDITIVE" LOGIT AND PROBIT MODELS

To start our discussion on interpretation, we first look at how to interpret the results from an "additive" logit or probit model that doesn't include any explicit interaction terms. This will provide a firm foundation and point of comparison for looking at the specific issues that arise when interpreting the results from an "interactive" logit or probit model that includes an explicit interaction term. As we'll see, it's possible to identify interactions even in additive logit or probit models. These interactions, which are referred to as "compression interactions" (Berry, DeMeritt and Esarey, 2010) or "structural interactions" (Bowen, 2012), can be substantively meaningful. As we'll see, scholars sometimes have theories that make predictions about these types of interactions.

In what follows, we adopt a latent variable setup for our logit and probit models. To provide some substantive context, we assume that the dichotomous dependent variable Y indicates whether someone votes (1) or not (0).

8.3.1 Coefficients

With the latent variable setup, we model the underlying variable Y^* as a linear combination of independent variables. If we have three independent variables X_1, X_2, and X_3, the underlying model is

$$Y^* = X\beta + \epsilon$$
$$= \beta_0 + \beta_1 X_1 + \beta_2 X_2 + \beta_3 X_3 + \epsilon. \tag{8.19}$$

The coefficients tell us the marginal effect of the independent variables on the *propensity* that someone votes. Given the linearity of the underlying model, they also tell us the effect of a one-unit increase in the independent variables on the propensity to vote. Thus, we see that a one-unit increase in X_1 increases the propensity of voting by β_1, a one-unit increase in X_2 increases it by β_2, and a one-unit increase in X_3 increases it by β_3. The coefficient β_0 indicates the propensity that someone votes when X_1, X_2, and X_3 are all 0. As we've seen, there are other ways of thinking about logit and probit models. In the random utility setup, for example, the coefficients indicate the effect of a one-unit increase in the independent variables on the relative *utility* of voting. In the pure probability setup, the coefficients from a logit model indicate the effect of a one-unit increase in the *log-odds* of voting. Providing a substantive interpretation of the coefficients from the *probit* model in a pure probability setup is a little harder as $X\beta$ has no obvious substantive meaning. The important thing to note is that we can interpret the coefficients from logit and probit models in any of these ways. In other words, we can interpret the magnitude, sign, and statistical significance of the logit and probit coefficients in terms of the *propensity*, *utility*, or *log-odds* (in the case of logit) of $Y = 1$.

Using the coefficients to talk about the effect of the independent variables on the latent dependent variable Y^* is meaningful, and doing so can throw light on the level of empirical support we have for the implications of our theories. In many cases, though, we'll also want to know the effect of our independent variables on the actual *probability* that $Y = 1$. In other words, we'll want to know how our variables affect the probability that, say, someone votes and not just the propensity, utility, or log-odds of voting. This is especially important when thinking about the substantive importance of our estimated effects as there's no meaningful scale for Y^*. So, what do the logit and probit coefficients tell us about the probability that $Y = 1$?

Recall that the probability that $Y = 1$ is $\Lambda(X\beta)$ in the logit model and $\Phi(X\beta)$ in the probit model. By visually showing how the value of $X\beta$ and hence Y^* is related to the probability that $Y = 1$ in logit and probit models, Figure 8.1 provides insight into how we can interpret the coefficients from these models. When a coefficient is positive, an increase in an independent variable X_k leads to an increase in $X\beta$. As Figure 8.1 indicates, an increase in $X\beta$ is associated with an increase in $Pr(Y = 1)$. When a coefficient is negative, an increase in an independent variable X_k leads to a decrease in $X\beta$, which, as Figure 8.1 indicates, leads to a decrease in $Pr(Y = 1)$. The non-linear S-shaped nature of the curves in Figure 8.1 means that the magnitude of the change in $Pr(Y = 1)$ associated with an increase in an independent variable depends on the level of $X\beta$ at which the increase occurs. For example, a given increase in X_k has a larger effect (indeed, its largest effect) on $Pr(Y = 1)$ when $X\beta$ is 0 than when, say, $X\beta$ is 3 or -3. What this means is that the sign and significance of a logit or probit coefficient indicates whether an increase in an independent variable is associated with a statistically significant increase (positive) or decrease (negative) in the probability that $Y = 1$ but not the substantive magnitude of the change in probability. This is one reason why we often choose to move beyond the coefficients when substantively interpreting the results from a logit or probit model.

8.3.2 Marginal Effects on Probabilities

One way to think about the effects of the independent variables on the probability that $Y = 1$ is in terms of derivatives and, hence, marginal effects. Recall that the probability that $Y = 1$, which we'll now refer to more simply as $Pr(Y)$, is $F(X\beta)$, where F could be the standard logistic cumulative distribution function Λ or the standard normal cumulative distribution function Φ. To find the marginal effect of some independent variable X_k on $Pr(Y)$ from an additive logit or probit model like the one shown in Eq. 8.19, we take the derivative of $Pr(Y)$ with respect to X_k,

$$\frac{\partial Pr(Y)}{\partial X_k} = \frac{\partial F(X\beta)}{\partial X_k} = \frac{\partial F(X\beta)}{\partial X\beta} \cdot \frac{\partial X\beta}{\partial X_k} = f(X\beta) \cdot \beta_k, \tag{8.20}$$

where $f(\cdot)$ is the standard logistic or standard normal *probability density function*.[3] So, in the case of the logit model, the marginal effect of X_k is

$$\frac{\partial Pr(Y)}{\partial X_k} = \frac{\partial \Lambda(X\beta)}{\partial X_k} = \lambda(X\beta) \cdot \beta_k, \tag{8.21}$$

where λ is the standard logistic probability density function,

$$\lambda(X\beta) = \Lambda(X\beta) \cdot [1 - \Lambda(X\beta)] = \frac{e^{X\beta}}{(1 + e^{X\beta})^2}. \tag{8.22}$$

And in the case of the probit model, the marginal effect of X_k is

$$\frac{\partial Pr(Y)}{\partial X_k} = \frac{\partial \Phi(X\beta)}{\partial x_k} = \phi(X\beta) \cdot \beta_k, \tag{8.23}$$

---

[3] Taking the derivative requires the use of the chain rule, which gives us the derivative of the outside function $\frac{\partial F(X\beta)}{\partial X\beta}$ multiplied by the derivative of the inside function $\frac{\partial X\beta}{\partial X_k}$. The derivative of the outside function is $f(X\beta)$ because the derivative of a cumulative density function F is just the associated probability density function f. The derivative of the additive inside function is the coefficient on the variable of interest, β_k.

where ϕ is the standard normal probability density function,

$$\phi(X\beta) = \frac{1}{\sqrt{2\pi}}e^{-\frac{(X\beta)^2}{2}}. \tag{8.24}$$

As an example, if $\beta_k = 2$ and $X\beta = 1$, then the marginal effect of X_k in the logit model is

$$\lambda(X\beta) \cdot \beta_k = \lambda(1) \cdot 2 = 0.197 \times 2 = 0.39, \tag{8.25}$$

and the marginal effect of X_k in the probit model is

$$\phi(X\beta) \cdot \beta_k = \phi(1) \cdot 2 = 0.242 \times 2 = 0.48. \tag{8.26}$$

When we discussed how to interpret logit and probit coefficients, we noted that the magnitude of any change in the probability that $Y = 1$ associated with an increase in an independent variable depends on the level of $X\beta$ at which the increase occurs. This is explicitly recognized in the equation for the marginal effect of X_k in Eq. 8.20, which shows that the coefficient on X_k is multiplied by the probability density function $f(X\beta)$. The non-linear effect of the independent variables on $Pr(Y)$ arises because of the S-shaped curve we need to use to take the unbounded values of $X\beta$ and map them appropriately onto a probability space that's bounded between 0 and 1. The key insight here is that the effect of any independent variable X_k on $Pr(Y)$ *always* depends on the values of *all* of the variables in the model including its own (so long as the coefficients aren't 0). In other words, the effects of the independent variables on $Pr(Y)$ are always conditional when we estimate a logit or probit model. This is true even when we estimate an underlying model for Y^* that's purely additive and doesn't include any interaction terms.[4]

We can obtain greater insight into the conditional nature of the marginal effect of X_k on $Pr(Y)$ by slightly rewriting the equation shown in Eq. 8.20:

$$\frac{\partial Pr(Y)}{\partial X_k} = \frac{\partial F(X\beta)}{\partial X\beta} \cdot \frac{\partial X\beta}{\partial X_k} = \frac{\partial Pr(Y)}{\partial Y^*} \cdot \frac{\partial Y^*}{\partial X_k} = f(X\beta) \cdot \beta_k. \tag{8.28}$$

---

[4] The marginal effect of X_k on $Pr(Y)$ in a logit model can be written as

$$\frac{\partial Pr(Y)}{\partial X_k} = \Lambda(X\beta) \cdot [1 - \Lambda(X\beta)] \cdot \beta_k. \tag{8.27}$$

As we noted earlier, the effect of an independent variable on $Pr(Y)$ is largest when $X\beta = 0$ or, equivalently, when $\Lambda(X\beta) = 0.5$. As a result, the largest that the marginal effect of X_k on $Pr(Y)$ will be is $0.5 \times 0.5 \times \beta_k = 0.25 \times \beta_k$. In other words, we can quickly identify what the *largest* marginal effect of a variable on $Pr(Y)$ in a logit model will be by dividing its coefficient by 4. The largest that the marginal effect of an independent variable will be on $Pr(Y)$ in a probit model is $\frac{1}{2\pi}\beta_k \approx 0.4\beta_k$ (Glasgow, 2022, 13).

From this, we see that there are two distinct components to the marginal effect of X_k on the probability that $Y = 1$. The first gray-colored term $\frac{\partial Pr(Y)}{\partial Y^*}$ is the marginal effect of the unbounded latent variable Y^* on the probability that $Y = 1$. This is equivalent to the slope of the logit or probit curves in Figure 8.1 evaluated at $X\beta$. The second gray-colored term $\frac{\partial Y^*}{\partial X_k}$ tells us the marginal effect of X_k on the unbounded latent variable Y^*. In effect, we see that an increase in X_k affects $Pr(Y)$ in a two-step process. The first step sees an increase in X_k change the value of the underlying latent variable Y^*. In the "additive" world where there's no interaction term, the marginal effect of X_k on Y^* is unconditional, β_k, and doesn't depend on the value of any other independent variable. In the second step, the change in Y^* produced in the first step is mapped via the S-shaped logit or probit curve into a change in the probability that $Y = 1$. It's in this second step that conditionality or interaction with all of the independent variables is introduced into the effect of X_k on $Pr(Y)$. Berry, DeMeritt and Esarey (2010) refer to the introduction of this conditionality as a "compression effect" because it results from "compressing" the unbounded values of Y^* into a bounded probability space. Bowen (2012) refers to it as a "structural effect" because it automatically follows from the S-shaped structure of the logit and probit curves. We'll discuss the substantive significance of these *compression* or *structural* interactions shortly.

The fact that the marginal effect of an independent variable on the probability that $Y = 1$ depends on the value of $X\beta$ raises a number of issues for presenting marginal effects and, as we'll see, other quantities of interest from logit and probit models. The most important to recognize is that any marginal effect we present is the marginal effect of X_k only for the specific scenario where all of the independent variables take on the values we choose for $X\beta = \beta_0 + \beta_1 X_1 + \beta_2 X_2 + \cdots + \beta_k X_k$. The values we choose for $X\beta$ essentially specify a particular baseline scenario for evaluating the marginal effect of X_k. We should think carefully about the specific baseline scenario we choose when calculating the marginal effect of X_k. What baseline scenario is most relevant for evaluating our theoretical hypotheses and demonstrating the substantive importance of our results?

⚠ **Important:** In a logit or probit model, the effect of any independent variable X_k on $Pr(Y)$ always depends on the values of all of the variables in the model, including its own. This means that an "additive" logit or probit model that does not include any explicit interaction terms is still an interactive model. The interaction that occurs in an "additive" model arises purely due to the compression effects that result from "compressing" the unbounded values of Y^* into a bounded probability space.

There are at least three different approaches we might adopt for specifying our baseline scenario. The first approach, which has historically been the most common, involves setting the values of the independent variables to their means if they're continuous and their modes if they're discrete. If the distribution of a continuous variable is heavily skewed, it may be preferable to use the median instead of the mean. The idea with this approach is that these measures of central tendency are one way of thinking about what it means for the values of the independent variables to be "typical." One potential problem with this approach, though, is that while the individual values for each independent variable may be "typical," the *combination* of these values may not correspond to a "typical" observation. In other words, there may be few actual real-world observations, if any, with this particular combination of values for the independent variables.

The second approach involves setting the values of the independent variables to particular values that are, for some reason, deemed "substantively interesting." These values may correspond, for example, to a specific empirical case in the sample, possibly one that was used to motivate the research in the first place. They may also correspond to a specific scenario that has particular theoretical importance. Depending on the purpose, one potential drawback of this second approach is that it may be unclear how "representative" the chosen baseline scenario is of real-world cases in general.

The third approach involves using the real-world combinations of values for the independent variables associated with each sample observation as different baseline scenarios (Hanmer and Kalkan, 2013). We then use each of these baseline scenarios to calculate the desired quantity of interest, such as the marginal effect of X_k, and we report the (weighted) average. Weighting may be necessary if we don't have a random sample and some sub-populations are under- or over-sampled. The idea with this approach is that the average represents, in some sense, the quantity of interest in the "typical" or "average" baseline scenario.

⚠ **Important:** Scholars should think carefully about how they select their baseline scenarios when evaluating the substantive significance of the effects of their variables on $Pr(Y)$. Common approaches involve setting the values of the independent variables (1) to some measure of central tendency, (2) to the values observed in some substantively interesting or representative case, or (3) to the values observed in each sample case and calculating the average effect across cases.

As we indicated in the previous chapter when discussing polynomial regression models, presenting the effects of the independent variables on $Pr(Y)$ in terms of derivatives and, hence, marginal effects may not be the most appropriate thing to do when we have a non-linear relationship between Y or $Pr(Y)$ and our independent variables. This is because the marginal effect of some variable X_k is no longer equivalent to the effect of a one-unit increase in X_k. Instead, it's the effect of an infinitesimally small increase in X_k divided by the change in X_k. It's hard to substantively interpret this kind of effect. In the real world, actors can typically increase X_k by some discrete amount but can't realistically increase X_k by an infinitesimally small amount. As a result, we reiterate our recommendation from earlier that scholars use differences rather than derivatives when presenting counterfactual effects to illustrate the substantive significance of the results from a non-linear model. In the context of a logit or probit model, this means looking at how the predicted probability that $Y = 1$ changes in response to some small, but substantively meaningful, discrete change in X_k. Before discussing differences in predicted probabilities, we first take a closer look at predicted probabilities.

8.3.3 Predicted Probabilities

Many scholars show predicted probabilities when presenting the results of a logit or probit model. To examine how this works, we return to our earlier underlying model,

$$Y^* = X\beta + \epsilon$$
$$= \beta_0 + \beta_1 X_1 + \beta_2 X_2 + \beta_3 X_3 + \epsilon, \tag{8.29}$$

where Y^* is, say, someone's propensity to vote. Suppose that X_1 is a continuous independent variable whose observed values run from 0 to 10, X_2 is a dichotomous variable whose mode is 0, and X_3 is some continuous variable whose mean is 0. Let's assume that when we estimate our model, we find that $\beta_0 = -0.4$, $\beta_1 = 0.2$, $\beta_2 = -0.6$, and $\beta_3 = -0.2$. To calculate the probability that someone votes, we need to pick a particular scenario by choosing specific values for X_1, X_2, and X_3. For some reason, say we're interested in the probability that someone votes when $X_1 = 4$, X_2 is at its mode of 0, and X_3 is at its mean of 0. In this scenario, $X\beta = -0.4 + 0.2 \times 4 - 0.6 \times 0 - 0.2 \times 0 = 0.4$ and so the probability that someone votes is $\Lambda(0.4) = 0.60$ if we estimated a logit model and $\Phi(0.4) = 0.66$ if we estimated a probit model.

Of course, this is just one possible scenario, and it may be more illustrative to calculate the probability that someone votes under different conditions. Scholars with hypotheses about the effect of an independent variable such as X_1 often choose to calculate $Pr(Y)$ across the observed range of values for X_1 while keeping the values of the other independent variables constant. As an example, we might calculate the probability of voting across the observed range of values for X_1 while holding the values of X_2 and X_3 constant at 0. In this case, where X_2 is dichotomous, we might actually think to calculate the probability of voting across the observed range of X_1 for each possible value of X_2 while holding the value of X_3 at 0. One way to present this sort of information is in the form of a predicted probability plot like the one shown in Figure 8.2. We've drawn this plot assuming that we estimated a logit model. It's important when presenting a plot like this that the reader knows the values that were used for the variables when calculating the reported predicted probabilities. This information, which is important for evaluating the substantive importance of the predicted probabilities, can often be presented most easily in a note beneath the plot. In a real application, the plot should also show confidence intervals for the predicted probability lines.

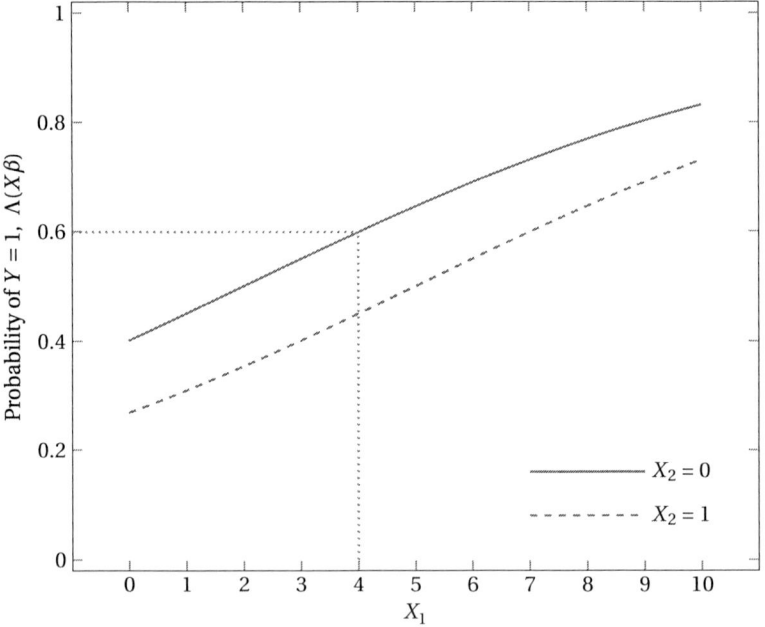

Figure 8.2 Predicted probability that $Y = 1$ from the logit model in Eq. 8.29 where $\beta_0 = -0.4$, $\beta_1 = 0.2$, $\beta_2 = -0.6$, and $\beta_3 = -0.2$.

Note: Figure 8.2 shows how the predicted probability that $Y = 1$ changes across the observed range of values for X_1 when $X_2 = 0$ (solid) and when $X_2 = 1$ (dashed), assuming that X_3 is held at its mean value of 0.

With respect to Figure 8.2, the dotted gray line confirms that the predicted probability that $Y = 1$ we calculated earlier for the particular scenario where $X_1 = 4$, $X_2 = 0$, and $X_3 = 0$ is 0.60. The two predicted probability lines are curved (if only slightly) as we'd expect from the non-linear logit model. The dashed predicted probability line for the case where $X_2 = 1$ is always below the solid line for the case where $X_2 = 0$, indicating that an increase in X_2 is associated with a lower probability of voting. The upward sloping nature of both lines indicates that the probability of voting increases with higher values of X_1.

As we've noted previously, predicted value plots such as the predicted probability plot in Figure 8.2 are good for visualizing the non-linear relationship between the dependent variable and the independent variables. However, they often fail to provide the necessary information to fully evaluate the implications of our theories. This is because our hypotheses are typically about the *effects* of the independent variables; that is, they're about how $Pr(Y)$ *changes* in response to an increase in our independent variables and not about the *level* of $Pr(Y)$. Note that we can't necessarily determine in Figure 8.2 whether the effects of X_1 and X_2 on the predicted probability of voting are statistically significant. Any confidence intervals that might be included with a plot like this would indicate only whether the predicted probabilities are significantly different from 0, which they will be by construction. This may not be too problematic in an "additive" logit or probit model like the one that we have here as we can determine the statistical significance of the effects of our independent variables directly from the statistical significance of the coefficients in our regression output. However, this won't be the case when we have an "interactive" logit or probit model. To capture whether an increase in, say, X_1 leads to a significant change in the probability of voting, we need to instead calculate a difference in predicted probability and examine the confidence interval around this quantity.

8.3.4 Differences in Predicted Probabilities

We recommend that scholars report differences in predicted probabilities when presenting the substantive effects of their independent variables. Suppose that we're using the same reported results from the underlying model shown in Eq. 8.29 and that we want to know the effect of a one-unit increase in X_1 on the probability of voting. It's important to remember that there's no single effect of a one-unit increase in X_1. Instead, it depends on the values of X_1, X_2, and X_3 at which the increase in X_1 occurs. So, suppose we want to know the effect of a one-unit increase in X_1 when $X_1 = 4$, and X_2 and X_3 are both 0. This effect is the change in the probability of voting between the counterfactual scenario where $X_1 = 5$ and $X_2 = X_3 = 0$ and the baseline scenario where $X_1 = 4$ and $X_2 = X_3 = 0$,

$$F(X_c\beta) - F(X_b\beta) = F(-0.4 + 0.2 \times 5 - 0.6 \times 0 - 0.2 \times 0) \qquad (8.30)$$
$$- F(-0.4 + 0.2 \times 4 - 0.6 \times 0 - 0.2 \times 0)$$
$$= F(0.6) - F(0.4), \qquad (8.31)$$

where $F(\cdot)$ is the relevant cumulative distribution function. With the logit model, we get

$$\Lambda(X_c\beta) - \Lambda(X_b\beta) = \Lambda(0.6) - \Lambda(0.4) = 0.65 - 0.60 = 0.05, \qquad (8.32)$$

and with the probit model, we get

$$\Phi(X_c\beta) - \Phi(X_b\beta) = \Phi(0.6) - \Phi(0.4) = 0.73 - 0.66 = 0.07. \qquad (8.33)$$

From this, we see that a one-unit increase in X_1 in the specified baseline scenario is associated with a 0.05 point increase in the probability of voting if we estimated a logit model and a 0.07 point increase if we estimated a probit model.

⚠ **Important:** We recommend that scholars report (percentage) changes in predicted probabilities rather than marginal effects when evaluating the substantive significance of their logit or probit model results.

In addition to reporting the absolute change in the probability that $Y = 1$ associated with a discrete change in some independent variable in some specified scenario, it can be helpful to also report the *percentage change in the probability* that $Y = 1$. Doing so can aid the reader in evaluating the substantive magnitude of any change in probability by making the change relative to the probability that $Y = 1$ in the baseline scenario. After all, even very small changes in probability can be substantively meaningful if the baseline probability that $Y = 1$ is low. And apparently large positive changes in probability may not be as substantively important as they first seem if the baseline probability that $Y = 1$ is already quite high. The percentage change in the probability of voting associated with a one-unit increase in X_1 when $X_1 = 4$ and $X_2 = X_3 = 0$ is

$$\frac{F(X_c\beta) - F(X_b\beta)}{F(X_b\beta)} \times 100 = \frac{F(0.6) - F(0.4)}{F(0.4)} \times 100. \qquad (8.34)$$

With the logit model we get

$$\frac{\Lambda(0.6) - \Lambda(0.4)}{\Lambda(0.4)} \times 100 = \frac{0.65 - 0.60}{0.60} \times 100 = 7.8. \qquad (8.35)$$

With the probit model we get

$$\frac{\Phi(0.6) - \Phi(0.4)}{\Phi(0.4)} \times 100 = \frac{0.73 - 0.66}{0.66} \times 100 = 10.7. \tag{8.36}$$

From this, we see that a one-unit increase in X_1 in the specified baseline scenario is associated with a 7.8% increase in the probability of voting if we estimated a logit model and a 10.7% increase if we estimated a probit model.

We've calculated the (percentage) change in the probability of voting associated with a one-unit increase in X_1 under just one possible scenario so far. We can use a plot like the one shown in Figure 8.3 to show how the (percentage) change in the probability of voting associated with a one-unit increase in X_1 varies across the range of values for X_1 and X_2 when the value of X_3 is held at its mean of 0. Panel (a) shows the *change* in the probability of voting, while panel (b) shows the *percentage change* in the probability of voting. In a substantive application, we would also show confidence intervals to allow us to see when the effect of a one-unit increase in X_1 is statistically significant. As expected from the positive coefficient on X_1, a one-unit increase in X_1 always leads to an increase in the probability of voting, with the magnitude varying with both the value of X_1 and the value of X_2. Panel (a) indicates that the magnitude of this increase in probability is sometimes larger when $X_2 = 0$ and sometimes larger when $X_2 = 1$, depending on the value of X_1 (whether X_1 is larger or smaller than 3) at which the increase happens. Despite this, panel (b) clearly indicates that the percentage change in the probability of voting associated with a one-unit increase in X_1 is always larger when $X_2 = 1$ than when $X_2 = 0$.

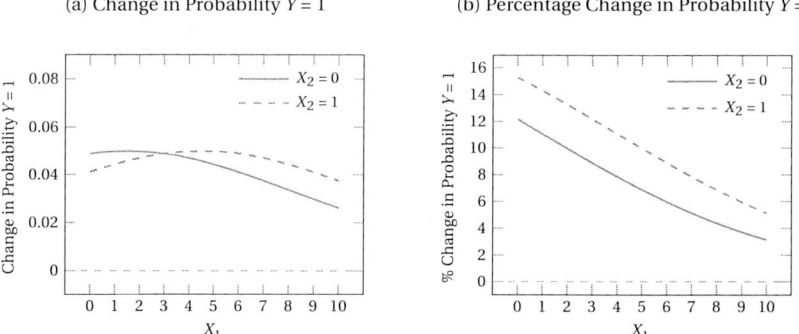

(a) Change in Probability $Y = 1$ (b) Percentage Change in Probability $Y = 1$

Figure 8.3 The effect of a one-unit increase in X_1 based on the logit model in Eq. 8.28 where $\beta_0 = -0.4$, $\beta_1 = 0.2$, $\beta_2 = -0.6$, and $\beta_3 = -0.2$

Note: Figure 8.3 shows how the change (a) or percentage change (b) in the probability that $Y = 1$ associated with a one-unit increase in X_1 varies across the observed range of values for X_1 when $X_2 = 0$ (solid) and when $X_2 = 1$ (dashed). In both panels, we've assumed that X_3 is held at its mean value of 0.

8.3.5 Odds Ratios

Scholars using a *logit* model can also examine the effects of their independent variables in terms of "odds ratios." Recall from the pure probability approach that the log-odds of $Y = 1$ is a linear combination of independent variables,

$$ln(\Omega) = ln\left(\frac{\rho}{1-\rho}\right) = X\beta = \beta_0 + \beta_1 X_1 + \beta_2 X_2 + \cdots + \beta_k X_k, \quad (8.37)$$

where $\Omega = \frac{\rho}{1-\rho}$ is known as the "odds" and indicates the ratio of the probability that $Y = 1$ to the probability that $Y = 0$. The effect of increasing an independent variable X_k by one unit on the log-odds of $Y = 1$, while holding the value of the other variables constant, is

$$ln(\Omega)\big| X_k + 1 - ln(\Omega)\big| X_k = \frac{\partial X\beta}{\partial X_k} = \beta_k. \quad (8.38)$$

This reminds us that the logit coefficients indicate the effect of a one-unit increase in the independent variables on the log-odds of $Y = 1$.

But most people don't think in terms of log-odds; instead, they think in terms of odds. It would be nice to be able to rewrite Eq. 8.38 in terms of odds instead of log-odds. Recall that a logarithm converts multiplication and division to addition and subtraction. It's inverse, exponentiation, converts addition and subtraction back to multiplication and division. We have a difference on the left-hand side of Eq. 8.38. Thus, if we exponentiate both sides of Eq. 8.38, we get

$$\frac{\Omega\big| X_k + 1}{\Omega\big| X_k} = e^{\beta_k}. \quad (8.39)$$

From this, we see that the exponentiated coefficients from a logit model tell us the effect of a one-unit increase in an independent variable on a ratio of odds that $Y = 1$. An odds ratio greater than 1 indicates that the one-unit increase in X_k increases the odds that $Y = 1$, while an odds ratio less than 1 indicates that it decreases the odds that $Y = 1$.[5]

To see how this works, let's return to our running example where the coefficient on X_1 is 0.2 and the coefficient on X_2 is -0.6. If we exponentiate these coefficients, we see that the odds that someone votes is $e^{0.2} = 1.22$ times larger when we increase X_1 by one unit and $e^{-0.6} = 0.55$ times as large when we increase X_2 by one unit. It's often preferable to report these changes in terms of percentage changes. Doing this, we see

---

[5] Odds ratios are sometimes confused with *risk ratios* (Glasgow, 2022). A risk ratio, which is also known as the *relative risk*, is the ratio of the probability that $Y = 1$ in the counterfactual scenario where $X_k = c$ and the probability that $Y = 1$ in the baseline scenario where $X_k = b$, $\frac{Pr(Y)|X_k = c}{Pr(Y)|X_k = b}$.

that a one-unit increase in X_1 leads to a $100 \times [e^{0.2} - 1] = 22\%$ increase in the odds that someone votes and that a one-unit increase in X_2 leads to a $100 \times [e^{-0.6} - 1] = 45\%$ reduction in the odds that someone votes. We obviously don't need to focus on the effect of a one-unit increase in an independent variable. The odds that $Y = 1$ is $e^{\beta_k \times \delta}$ times larger if we increase an independent variable X_k by some amount δ. Thus, a two-unit increase in X_1 is associated with a $100 \times [e^{0.2 \times 2} - 1] = 49\%$ increase in the odds that some votes.

Note that unlike the other substantive quantities of interest that we've examined, odds ratios don't depend on the value of the other independent variables or the value of X_k at which the increase occurs. A given increase in X_k always leads to the same factor or percentage change in the odds of $Y = 1$. Put differently, the effects of the independent variables in terms of odds ratios are unconditional in an additive logit model.

8.3.6 Interaction Effects

As we've noted, the effect of each independent variable on $Pr(Y)$ depends on the values of all of the variables in a logit or probit model. This is due to the compression effects that are introduced by the S-shaped curves in these models. This means that there are always going to be interaction effects even if we don't include any explicit interaction terms in our model specification. As a reminder, an interaction effect indicates how the effect of an independent variable on some outcome of interest such as the probability of voting varies with the value of another independent variable. We can also have a self-interaction where the effect of a variable depends on its own value. How do we identify the interaction effects in an additive logit or probit model? And are they substantively meaningful?

We can think of the compression-based interaction effects in an additive logit or probit model in terms of marginal effects or in terms of "differences in differences." For ease of exposition, we focus on identifying interaction effects in the context of the same additive logit model as before,

$$Y^* = X\beta + \epsilon$$
$$= \beta_0 + \beta_1 X_1 + \beta_2 X_2 + \beta_3 X_3 + \epsilon, \tag{8.40}$$

where Y^* captures someone's propensity to vote. Based on Eq. 8.21, the marginal effect of X_1 on the probability of voting is

$$\frac{\partial Pr(Y)}{\partial X_1} = \frac{\partial \Lambda(X\beta)}{\partial X_1} = \underbrace{\Lambda(X\beta) \cdot [1 - \Lambda(X\beta)]}_{\lambda(X\beta)} \cdot \beta_1$$

$$= \Lambda(X\beta) \cdot \beta_1 - [\Lambda(X\beta)]^2 \cdot \beta_1. \tag{8.41}$$

At this point, there are several interaction effects in which we may be interested. For example, we may be interested in how the marginal effect of X_1 varies with the value of X_1, X_2, or X_3. Suppose that we're interested in how X_2 modifies the marginal effect of X_1. To calculate this, we simply take the derivative of the marginal effect shown in Eq. 8.41 with respect to X_2,

$$
\frac{\partial \left(\frac{\partial Pr(Y)}{\partial X_1} \right)}{\partial X_2} = \frac{\partial \left(\Lambda(X\beta) \cdot \beta_1 - [\Lambda(X\beta)]^2 \cdot \beta_1 \right)}{\partial X_2}
$$

$$
= \lambda(X\beta) \cdot \beta_1 \cdot \beta_2 - 2 \cdot \Lambda(X\beta) \cdot \lambda(X\beta) \cdot \beta_1 \cdot \beta_2
$$

$$
= \lambda(X\beta) \cdot [1 - 2 \cdot \Lambda(X\beta)] \cdot \beta_1 \cdot \beta_2
$$

$$
= \Lambda(X\beta) \cdot [1 - \Lambda(X\beta)] \cdot [1 - 2 \cdot \Lambda(X\beta)] \cdot \beta_1 \cdot \beta_2. \tag{8.42}
$$

We can immediately see from this that there's no single interaction effect between X_1 and X_2. Just as any marginal effect on $Pr(Y)$ that we might calculate for an independent variable is specific to a particular scenario, so is any interaction effect we might calculate.

As an illustrative example, suppose we're interested in the marginal effect of X_1 on the probability of voting and the interaction effect between X_1 and X_2 for the baseline scenario where $X_1 = 1$, and X_2 and X_3 are both 0. As before, we'll assume that $\beta_0 = -0.4$, $\beta_1 = 0.2$, $\beta_2 = -0.6$, and $\beta_3 = -0.2$. In this scenario, $X\beta = -0.4 + 0.2 \times 1 - 0.6 \times 0 - 0.2 \times 0 = -0.2$, and $\Lambda(X\beta) = \Lambda(-0.2) = 0.45$. Based on Eq. 8.41, the marginal effect of X_1 is $0.45 \times 0.55 \times 0.2 = 0.05$. And based on Eq. 8.42, the interaction effect between X_1 and X_2 is $0.45 \times 0.55 \times 0.1 \times 0.2 \times -0.6 = -0.003$. Thus, we see that an increase in X_2 *reduces* the positive marginal effect of X_1 in this particular scenario. Suppose that we're now interested in the same scenario except that X_1 is now 4 instead of 1. In this scenario, $X\beta = 0.4$ and $\Lambda(0.4) = 0.6$. Based on Eq. 8.41, the marginal effect of X_1 is $0.6 \times 0.4 \times 0.2 = 0.048$. And based on Eq. 8.42, the interaction effect between X_1 and X_2 is $0.6 \times 0.4 \times -0.2 \times 0.2 \times -0.6 = 0.006$. In this new scenario, we see that an increase in X_2 *increases* the positive marginal effect of X_1.

As these two examples illustrate, the sign of the compression-based interaction effect between independent variables is not fixed and can change across different scenarios. We can be more specific about this by looking at the terms that comprise the interaction effect in Eq. 8.42. As probabilities, the terms $\Lambda(X\beta)$ and $1 - \Lambda(X\beta)$ are both non-negative. Whether the term $1 - 2 \cdot \Lambda(X\beta)$ is positive, negative, or zero depends on whether $Pr(Y) = \Lambda(X\beta)$ is larger than, smaller than, or equal to 0.5, respectively in the particular scenario under consideration. If $Pr(Y)$ is greater than 0.5, then this term is negative. If $Pr(Y)$ is less than 0.5, then this term is positive. And if $Pr(Y)$ is exactly 0.5, then this term is 0. One implication of this is that there'll be no interaction effect whenever $Pr(Y) = 0.5$. Whether the term $\beta_1 \cdot \beta_2$ is positive or negative depends on whether these coefficients share the same sign (positive) or have different signs (negative). If either of the two coefficients is 0, then $\beta_1 \cdot \beta_2$ will be 0, and there'll be no interaction effect. From all of this, we see that the sign of the compression-based interaction effect between X_1 and X_2 depends on (1) the probability that $Y = 1$ in the baseline scenario and (2) whether the coefficients on X_1 and X_2 are non-zero and share the same sign. The possible relationships are summarized in Figure 8.4.

The relationships shown in Figure 8.4 are fairly intuitive. Consider the case where the probability that $Y = 1$ is greater than 0.5. As Figure 8.1 indicates, the logit curve is rising here but at a decreasing rate with higher values of $X\beta$. Recall from Eq. 8.28 that the marginal effect of X_1 is the slope of the logit curve evaluated at $X\beta$ multiplied by the coefficient on X_1, β_1. We'll assume for illustrative purposes that β_1 is positive. We can usefully think of the interaction effect between X_1 and X_2 as the difference in the marginal effect of X_1 in a counterfactual scenario where X_2 is "high" compared to in a baseline scenario where X_2 is "low." If the coefficient on X_2, β_2, is positive, then the value of $X\beta$ in the counterfactual scenario is higher and the slope of the logit curve is less steep than in the baseline scenario. This means that the marginal effect of X_1 is smaller in the counterfactual scenario, indicating that there's a negative interaction

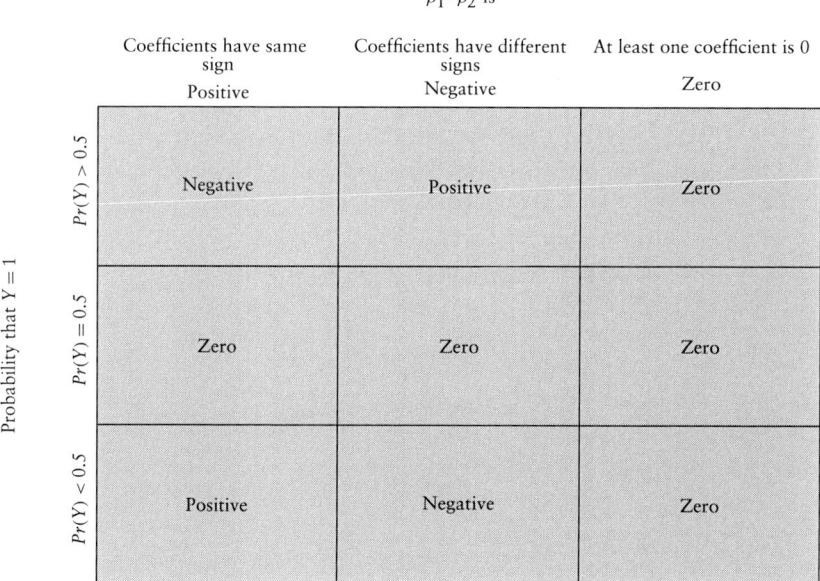

Figure 8.4 Different signs for the compression-based interaction effect between X_1 and X_2 on $Pr(Y)$ in the additive logit model shown in Eq. 8.40

effect between X_1 and X_2. If β_2 is negative, then the value of $X\beta$ in the counterfactual scenario is lower and the slope of the logit curve is steeper than in the baseline scenario. This means that the marginal effect of X_1 is larger in the counterfactual scenario, indicating that there's now a positive interaction effect between X_1 and X_2. And if β_2 is 0, then there's obviously no difference between the counterfactual and baseline scenarios, indicating that the interaction effect between X_1 and X_2 is zero.

The opposite relationships hold when the probability that $Y = 1$ is less than 0.5. In this scenario, the logit curve rises at an increasing rate with higher values of $X\beta$. Let's continue to assume that β_1 is positive. If the coefficient on X_2, β_2, is positive, then the value of $X\beta$ in the counterfactual scenario is higher and the slope of the logit curve is steeper than in the baseline scenario. This means that the marginal effect of X_1 is larger in the counterfactual scenario, indicating that there's a positive interaction effect between X_1 and X_2. If β_2 is negative, then the value of $X\beta$ in the counterfactual scenario is lower and the slope of the logit curve is less steep than in the baseline scenario. This means that the marginal effect of X_1 is smaller in the counterfactual scenario, indicating that there's now a negative interaction effect between X_1 and X_2. And if β_2 is 0, then there's no difference between the counterfactual and baseline scenarios, indicating that the interaction effect between X_1 and X_2 is zero.

The intuition as to why the interaction effect is always 0 when $Pr(Y) = 0.5$ or $X\beta = 0$ is a little trickier. The key thing to note here is that the logit (and probit) curve has an inflection point at $X\beta = 0$. This is where the curve switches from being concave up (U-shape) to being concave down (inverted U-shape). An inflection point occurs when the second derivative of a function, such as the logit or probit curve, changes signs; that is, when it's 0. Note that in order to calculate the interaction effect between X_1 and X_2, we have to take the derivative of the logit function twice, first with respect to X_1 and then with respect to X_2. Since the interaction effect, which is shown in Eq. 8.42, is a multiple of the second derivative, it follows that it will always be 0 at the inflection point where $Pr(Y) = 0.5$ no matter the signs of β_1 or β_2.

⚠ **Important:** There are compression-based interaction effects with respect to $Pr(Y)$ in an additive logit or probit model. The magnitudes, signs, and statistical significance of these interaction effects are not fixed and depend on the values of all of the variables in the model.

We can also think of compression-based interaction effects in an additive logit or probit model in terms of differences in differences. An interaction effect now captures how the effect of a discrete change in some independent variable on the probability of voting changes in response to a discrete change in another independent variable. For ease of exposition, we again focus on interaction effects in the context of the same logit model shown in Eq. 8.40. Suppose that we're interested in knowing how a discrete change in X_2 modifies the effect of a discrete change in X_1 on the probability of voting for a particular scenario. To identify this interaction effect, we must calculate three "differences." The first difference captures the *change* in the probability of voting associated with a discrete increase in X_1 from some baseline value X_{1b} to some counterfactual value X_{1c} when X_2 and all of the other independent variables are set at their own baseline values,

$$\Delta Pr(Y)_{X_{2b}} = \Lambda \left(\beta_0 + \beta_1 X_{1c} + \beta_2 X_{2b} + \beta_3 X_{3b} \right)$$
$$- \Lambda \left(\beta_0 + \beta_1 X_{1b} + \beta_2 X_{2b} + \beta_3 X_{3b} \right). \tag{8.43}$$

The second difference captures the *change* in the probability of voting associated with the same discrete increase in X_1 when X_2 is at its counterfactual value X_{2c} and all of the other independent variables remain at their own baseline values,

$$\Delta Pr(Y)_{X_{2c}} = \Lambda \left(\beta_0 + \beta_1 X_{1c} + \beta_2 X_{2c} + \beta_3 X_{3b} \right)$$
$$- \Lambda \left(\beta_0 + \beta_1 X_{1b} + \beta_2 X_{2c} + \beta_3 X_{3b} \right). \tag{8.44}$$

The third difference, which is the interaction effect between X_1 and X_2, is just the difference in these two differences,

$$Interaction\ Effect_{X_1, X_2} = \Delta Pr(Y)_{X_{2c}} - \Delta Pr(Y)_{X_{2b}}. \tag{8.45}$$

As an illustrative example, suppose that we're interested in a how a one-unit increase in X_2 modifies the effect of a one-unit increase in X_1 on the probability of voting in the scenario where $X_1 = 4$ and X_2 and X_3 are both 0. As before, we'll assume that $\beta_0 = -0.4$, $\beta_1 = 0.2$, $\beta_2 = -0.6$, and $\beta_3 = -0.2$. The change in $Pr(Y)$ associated with a one-unit increase in X_1 in the scenario where $X_2 = 0$ is

$$\Delta Pr(Y)_{X_{2b}} = \Lambda \left(-0.4 + 0.2 \times 5 - 0.6 \times 0 - 0.2 \times 0 \right)$$
$$- \Lambda \left(-0.4 + 0.2 \times 4 - 0.6 \times 0 - 0.2 \times 0 \right)$$
$$= \Lambda(0.6) - \Lambda(0.4)$$
$$= 0.646 - 0.599 = 0.047. \tag{8.46}$$

The change in $Pr(Y)$ associated with a one-unit increase in X_1 in the scenario where $X_2 = 1$ is

$$\Delta Pr(Y)_{X_{2c}} = \Lambda\left(-0.4 + 0.2 \times 5 - 0.6 \times 1 - 0.2 \times 0\right)$$
$$- \Lambda\left(-0.4 + 0.2 \times 4 - 0.6 \times 1 - 0.2 \times 0\right)$$
$$= \Lambda(0) - \Lambda(-0.2)$$
$$= 0.50 - 0.45 = 0.05. \tag{8.47}$$

The interaction effect between X_1 and X_2 in terms of a difference in differences is therefore

$$Interaction\ Effect_{X_1, X_2} = \Delta Pr(Y)_{X_{2c}} - \Delta Pr(Y)_{X_{2b}}$$
$$= 0.05 - 0.047 = 0.003. \tag{8.48}$$

Just like when we calculated the interaction effect for this particular scenario in terms of marginal effects, we see that the interaction effect on the probability of voting is positive when we think in terms of a difference in differences.[6] As expected, the magnitude of the interaction effect is different depending on whether we're thinking in terms of marginal effects for X_1 and X_2 (0.006) or a one-unit discrete change in these variables (0.003).

We now know how to calculate interaction effects in an additive logit or probit model. But are these compression-based interaction effects theoretically and substantively meaningful? Some scholars have argued that they're not, on the grounds that compression effects automatically arise in logit and probit models due to the nature of their S-shaped curves (Nagler, 1991, 1994). We believe, however, that it can sometimes make sense to interpret compression-based interaction effects as both theoretically and substantively meaningful.

It's important to recognize that there are theoretical reasons why we employ S-shaped curves to model the data-generating process behind dichotomous outcomes. As we noted earlier in the chapter, we need an appropriate function $g(\cdot)$ that can take the values of a linear combination of independent variables $X\beta$ that are unbounded $(-\infty, +\infty)$ and compress them onto a probability scale that's bounded $[0, 1]$. We want this function to return a 1 in the limit as $X\beta$ approaches $+\infty$ and a 0 when $X\beta$ approaches $-\infty$. Implied here is that the effect of an increase in $X\beta$ should be associated with a smaller and smaller increase in $Pr(Y)$ as $X\beta$ approaches $+\infty$ and that a reduction in $X\beta$ should be associated with a smaller and smaller decrease in $Pr(Y)$ as $X\beta$ approaches $-\infty$. The compression that results from using

---

[6] We could have predicted this based on the information in Figure 8.4. This is because the coefficients β_1 and β_2 have different signs and $Pr(Y) > 0.5$ in our baseline scenario.

an S-shaped curve to map the unbounded values of $X\beta$ into a bounded probability space is therefore desired by the analyst for theoretical reasons and isn't simply an artifact of the statistical model.

As Berry, DeMeritt and Esarey (2010, 255) note, it's possible to have a theoretical rationale for expecting a *pure* compression-based interaction effect on $Pr(Y)$ that's substantively meaningful. The example they give has to do with the factors that influence whether two countries go to war. Two factors that are thought to negatively affect the probability of conflict between a pair of countries or "dyad" are (1) the level of joint democracy in the dyad and (2) the geographic distance between the countries. According to "democratic peace" scholars, the probability that two countries go to war is close to zero when the level of joint democracy in a dyad is high (Oneal and Russett, 1997; Maoz, 1998). Suppose that we have a dyad involving two dictatorships and so the level of joint democracy is low. In this case, we'd expect the probability that war breaks out to be strongly and negatively influenced by how far the two countries are from each other. This is because geographic distance makes it less likely that grievances will arise and because it makes waging war more costly. Now suppose that we have a dyad involving two democracies, and so the level of joint democracy is high. According to the democratic peace literature, the probability of war will be close to 0. This is the case before we even think about any other characteristics of the dyad. It makes sense to think that geographic distance will still produce a large negative and unconditional effect on the unbounded *utility* or *propensity* of war, Y^*, in this dyad. However, the fact that probabilities are bounded below by 0 means that the magnitude of this negative effect on the *probability* of war, $Pr(Y)$, is constrained to be very small.[7] This compression, which means that the effect of geographic distance on the probability of war depends on a dyad's level of joint democracy, may be substantively large and is a direct implication of democratic peace theory.[8] There are many other cases like this where theory would lead us to expect that an extreme (high or low) value for one of our independent variables, such as joint democracy, would push $Pr(Y)$ close to one of its bounds and thereby restrict or "compress" the possible impact of the other independent variables.[9]

---

[7] In this theoretical framework, we might think to say that joint democracy is "nearly sufficient" for peace and that other factors, such as geographic distance, are "not necessary" for peace. For a discussion of how necessary and sufficient conditions are linked with interaction models, see Clark, Gilligan and Golder (2006).

[8] While we've referred to the effects discussed here as "compression effects," it should be clear that we could usefully think of them as "floor" or "ceiling" effects as well.

[9] It's worth noting that these types of theoretical implications about compression-based interaction effects, or floor and ceiling effects, can't be tested with a linear probability model.

> ⚠ **Important:** Compression-based interaction effects can be both theoretically and substantively meaningful.

We've seen that an additive logit or probit model that doesn't include an interaction term can capture a pure compression-based interaction effect on $Pr(Y)$. But does this mean that we can appropriately test a prediction about a substantively meaningful pure compression-based interaction effect with an additive specification? Testing such a prediction requires that we be able to potentially falsify this prediction. This is where things become a little tricky. Part of the issue with additive logit or probit models is that they don't allow us *not* to find evidence of non-zero compression-based interaction effects except under a very restrictive set of circumstances. In this sense, we might say that additive logit and probit models are biased toward finding non-zero compression-based interaction effects (Rainey, 2016). At the same time, though, additive logit or probit models don't guarantee that these interaction effects will always be *substantively large* or *statistically significant* (Berry, DeMeritt and Esarey, 2010). Thus, it *is* possible to falsify a claim of *non-trivial* compression-based interaction in a given application with an additive logit or probit model. What's not possible to test with an additive logit or probit model, though, is whether the interaction effect is a *pure* compression-based interaction effect where our variables have only an additive effect on Y^*. This is something that's *assumed* rather than *tested*. Only a logit or probit model that includes an explicit interaction term can test this particular prediction. It's largely because of this that we generally recommend that scholars estimate an interactive logit or probit model that includes an explicit interaction term even when they have a prediction about a pure compression-based interaction effect (Rainey, 2016).

There's a lot to unpack here. Consider again the compression-based interaction effect between X_1 and X_2 on the probability of voting that we calculated earlier in Eq. 8.40 based on an additive logit model,

$$\Lambda(X\beta) \cdot \left[1 - \Lambda(X\beta)\right] \cdot \left[1 - 2 \cdot \Lambda(X\beta)\right] \cdot \beta_1 \cdot \beta_2. \tag{8.49}$$

As confirmed by the information in Figure 8.4, it's impossible for this interaction effect to be 0 in a scenario where the baseline probability of voting is different from 0.5 and the effects of X_1 and X_2 are both non-zero. In effect, an additive model specification guarantees that we'll find a non-zero compression-based interaction effect, except under a very restrictive set of circumstances. Note, though, that while we're almost certainly guaranteed to find non-zero compression-based interaction effects when we estimate an additive model, this doesn't necessarily mean that these effects will be substantively large or statistically significant (Berry, DeMeritt and Esarey, 2010, 256–257). From a strict mathematical perspective, compression-based interaction effects of non-trivial magnitude are always possible in additive logit and probit models because these models are defined for all values of $X\beta$ from $-\infty$ to $+\infty$. In any given substantive application, though, the observed range of values for the independent variables will always be finite. This means that an additive model specification won't necessarily exhibit substantively large or statistically significant interaction within the range of values exhibited by real-world observations. This suggests that a theoretical claim of *non-trivial* compression-based interaction *can* be falsified with an additive logit or probit model.

Consider Figure 8.5, which shows the probability that $Y = 1$ for two different logit models with independent variables X_1 and X_2 whose real-world values are both constrained to the range $[0, 8]$. Panel (a) depicts the probability surface for a logit model where $Pr(Y) = \Lambda(-8 + X_1 + X_2)$, while panel (b) depicts the probability surface for a logit model where $Pr(Y) = \Lambda(-1 + 0.125X_1 + 0.125X_2)$. The plot in panel (a) indicates appreciable compression with the curved slope of the surface being steep when the values of X_1 and X_2 aren't at their extremes but quite shallow when they are. The left plot in panel (b) indicates that there's no appreciable compression when we restrict the probability surface to the observed $[0, 8]$ range of values for X_1 and X_2. The nearly flat probability surface sees the value of the marginal effect of X_1 on $Pr(Y)$ restricted to a very narrow range of 0.025 to 0.031 as we move across the real-world values for X_1 and X_2. In effect, there's no evidence of a non-trivial compression-based interaction effect between X_1 and X_2. Simulations with these models conducted by Berry, DeMeritt and Esarey (2010, 257) also find little evidence of a statistically significant interaction effect between X_1 and X_2 in the range of the observed data. In those cases where they do find a statistically significant compression-based interaction effect, the largest magnitude is a trivial 0.003.[10] Of course, we'll always find evidence of substantively significant compression if we extend the probability surface to a large enough range of values for X_1 and X_2. For example, the right plot in panel (b) indicates appreciable compression once we extend the range of values for X_1 and X_2 to $[-20, 20]$. The small solid gray region in the right plot corresponds to the probability surface depicted in the left plot, while the shaded meshed region indicates the probability surface for values of X_1 and X_2 that aren't observed in the real world. As we've noted previously,

---

[10] We'll discuss how to calculate measures of uncertainty for interaction effects and other quantities of interest from logit and probit models a little later in the chapter.

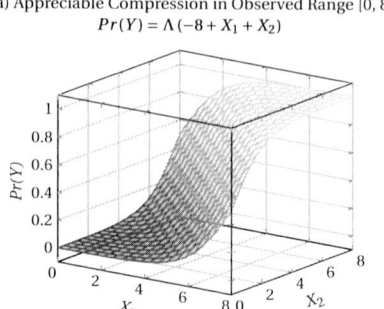

(a) Appreciable Compression in Observed Range [0, 8]
$$Pr(Y) = \Lambda(-8 + X_1 + X_2)$$

(b) No Appreciable Compression in Observed Range [0, 8]
$$Pr(Y) = \Lambda(-1 + 0.125X_1 + 0.125X_2)$$

Only Observed Range [0, 8]   Beyond the Observed Range [−20, 20]

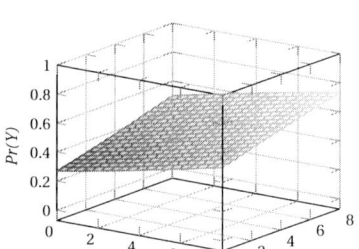
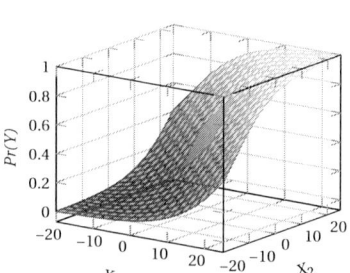

Figure 8.5 Logit models with and without appreciable compression in the observed range [0, 8] for X_1 and X_2

Note: Figure 8.5 shows the $Pr(Y)$ surface for two different additive logit models with independent variables X_1 and X_2 whose real-world values are both constrained to the range [0, 8]. The plot in panel (a) shows appreciable compression in the observed range of values for X_1 and X_2. The left plot in panel (b) is restricted to the observed range of values for X_1 and X_2 and shows no appreciable compression. The right plot in panel (b), which is based on the same logit model, shows appreciable compression but only when we look beyond the observed range of values for X_1 and X_2; the small solid gray region is identical to the surface shown in the left plot, while the shaded meshed region indicates $Pr(Y)$ based on values for X_1 and X_2 that are not observed in the real world.

though, we shouldn't draw inferences about counterfactuals that are far from the data (King and Zeng, 2006). The point here is that a theoretical prediction of a non-trivial compression-based interaction effect between independent variables on $Pr(Y)$ *can* be falsified and isn't guaranteed to receive empirical support with an additive logit or probit model.

While it's possible to falsify a claim of non-trivial interaction with an additive logit or probit model, it's not possible to falsify the claim that we only have *pure* compression-based interaction where our variables have only an additive effect on Y^*. This is something that's assumed rather than tested with an additive logit or probit model. If it turns out that our variables actually have an interactive, rather than an additive, effect on Y^*, then the estimate of the interaction effect from an additive logit or probit model will be biased and won't actually capture a pure compression-based interaction effect. We can only test a prediction of *pure* compression-based interaction by estimating an interactive logit or probit model that includes an explicit interaction term. For example, we can test a prediction of pure compression-based interaction between variables X_1 and X_2 with the following model that includes an explicit interaction term X_1X_2:

$$Y^* = X\beta + \epsilon$$
$$= \beta_0 + \beta_1 X_1 + \beta_2 X_2 + \beta_3 X_1 X_2 + \beta_4 X_3 + \epsilon. \qquad (8.50)$$

We can only have a *pure* compression-based interaction effect on $Pr(Y)$ if the coefficient on the interaction term, β_3, is 0. If this coefficient isn't 0, then our prediction of a pure compression-based interaction effect is falsified, and any interaction effect on $Pr(Y)$ that we find will come from two separate sources: (1) the interaction from the inclusion of the interaction term and (2) the interaction from the compression that occurs in the logit and probit models.[11]

So, what should we do? Should we estimate an additive or interactive model when we have a theory predicting a pure compression-based interaction effect? To some extent, we believe that scholars can reasonably disagree here. Note that in some ways there's nothing really unique here that's specific to dealing with a logit or probit functional form. Consider the case where we have a continuous dependent variable Y. We might have a theory that two variables X_1 and X_2 have independent and additive effects on Y. This would usually lead us to estimate an additive regression model. However, someone could criticize our choice to omit an interaction term between X_1 and X_2 based on the reasoning that despite our *a priori* theory, it would be better to include an interaction term so that we can test our at least implicit claim of no interaction rather than assume it. Of course, adding an additional interaction term that turns out to be unnecessary reduces the efficiency of our estimates of the effects of X_1 and X_2. Thus, there's a trade-off if we wish to test a prediction of no interaction. This same trade-off exists when we have a theory predicting a pure compression-based interaction and we must decide whether to

---

[11] We discuss these different sources of interaction in much more detail in the next section of this chapter when we look explicitly at interaction effects in interactive logit and probit models.

estimate an additive or interactive logit or probit model. Relevant here is the fact that including an unnecessary interaction term in a logit or probit model reduces our ability to detect compression-based interaction effects.[12] Note, though, that the comparison to the case where we have a continuous dependent variable isn't perfect. This is because, as we've seen, additive logit and probit models are "biased" in a sense toward finding non-zero compression-based interaction effects because of their functional form. While it's certainly true that their functional form doesn't guarantee that we'll find evidence of non-trivial interaction effects in the range of real-world data in a given application, this is always something to keep in mind. Simulations conducted by Rainey (2016), for example, suggest that additive logit or probit models often find evidence of compression-based interaction even when there's no such interaction in the data generating process. On the whole, we generally come down on the side of recommending that scholars use an interactive logit or probit model that includes an explicit interaction term to test the implications from a theory predicting a pure compression-based interaction effect.[13]

> ⚠ **Important:** Scholars sometimes have theories that predict the presence of a pure compression-based interaction effect on $Pr(Y)$. There are trade-offs to be considered when deciding whether to evaluate the implications of these theories with an additive as opposed to an interactive logit or probit model. We generally recommend using an interactive model.

To sum up our discussion here, scholars may well have theories predicting a pure compression-based interaction effect between independent variables such as X_1 and X_2 on the probability that $Y = 1$. In these theories, X_1 and X_2 are expected to each have a separate and additive effect on the *latent propensity* or *utility* of $Y = 1$ but an interactive effect on the *probability* that $Y = 1$ via compression. Scholars with theories like this are encouraged to be as explicit as possible about how and why variables like X_1 and X_2 have a different effect on $Pr(Y)$ as opposed to Y^*. They're also encouraged to use their predictions about the signs of the marginal

---

[12] In simulations, Rainey (2016, Appendix, p.1) finds that including an unnecessary interaction term in a logit model leads to a substantial drop in statistical power when the sample size is small (< 1000) *and* the amount of interaction is small. However, the drop in statistical power is minimal for larger sample sizes and large interaction effects.

[13] When efficiency and statistical power is of particular concern, it might be useful to present the results from both an additive and interactive model specification. That way, readers can decide for themselves how they wish to weigh the evidence.

effects of X_1 and X_2 to make a specific hypothesis about the sign of the expected compression-based interaction effects under different scenarios. There are trade-offs to be considered when deciding whether to evaluate the implications of theories predicting a pure compression-based interaction effect with an additive as opposed to an interactive logit or probit model. We generally recommend using an interactive logit or probit model. Given that any interaction effect on $Pr(Y)$ we might calculate necessarily varies across different scenarios, it's incumbent on us to indicate when (and how frequently) we can expect this effect to be statistically significant and substantively important.

While there can be situations in which scholars have theories predicting a pure compression-based interaction effect, we believe that it's much more common for them to have theories predicting *variable-specific* interaction (Nagler, 1994, 249). Compression-based interaction applies to *all* of the independent variables in a logit or probit model due to the functional form and S-shaped curves of these models. Our theoretical story about conflict and democratic peace, for example, predicts that there'll be compression-based interaction effects between joint democracy and all of the independent variables in the model and not just, say, geographic distance. In practice, most theories that make a conditional claim predict an interaction between specific independent variables rather than all of the independent variables. These theories predict variable-specific interaction. To test claims about variable-specific interaction, we *must* employ an interactive logit or probit model that includes an explicit interaction term. It's to these types of claims and models that we now turn.

⚠ **Important:** While *compression-based interaction* applies to all of the independent variables in a logit or probit model, *variable-specific* interaction applies only to specific independent variables. Variable-specific interaction can only be evaluated in an interactive logit or probit model.

8.4 INTERPRETATION AND INTERACTION EFFECTS IN "INTERACTIVE" LOGIT AND PROBIT MODELS

In this section, we look at how to interpret the results from an interactive logit or probit model that includes an explicit interaction term in order to evaluate the implications of a theory positing variable-specific interaction. To provide some substantive context, we'll continue to assume that our dichotomous dependent variable Y indicates whether someone votes (1) or not (0). As before, we'll adopt a latent variable setup. For illustrative purposes, we'll assume that the interaction predicted by our theory concerns the variables X_1 and X_2, and so the model for the underlying variable Y^* is

$$Y^* = X\beta + \epsilon$$
$$= \beta_0 + \beta_1 X_1 + \beta_2 X_2 + \beta_3 X_1 X_2 + \beta_4 X_3 + \epsilon, \tag{8.51}$$

where $X_1 X_2$ is an interaction term and X_3 is a control variable.

In situations like this, it's common for scholars to use their theory to derive a hypothesis, say about the effect of X_1 on $Pr(Y)$ and how this varies with X_2.[14] For example, the hypothesis might state that an increase in X_1 always increases the probability that someone votes but less so when the value of X_2 is high. We've seen hypotheses like this in previous chapters, and they haven't raised any particular problems. However, things are a bit more complicated when we have a dichotomous dependent variable. The main reason for this has to do with the implicit two-step process by which an increase in a variable like X_1 affects $Pr(Y)$.

In the first step, the increase in X_1 leads to a change in the unbounded latent variable Y^* that captures the propensity to vote. The effect of this increase in X_1 on Y^* is conditional on the value of X_2 due to the inclusion of the interaction term $X_1 X_2$. In the second step, the change in Y^* from the first step is mapped via the S-shaped logit or probit curve into a change in $Pr(Y)$. This means that there are two sources to any conditionality we find in the effect of X_1 on the probability of voting: (1) the conditionality from the interaction with X_2 on the propensity to vote and (2) the conditionality from the compression effect involving all of the independent variables that arises because of the S-shaped logit or probit curves. The introduction of compression effects in the second step means that although there's just one interaction effect between X_1 and X_2 on the *propensity* to vote, this isn't the case when it comes to the interaction between these variables on the *probability* of voting. We'll calculate a different interaction effect between X_1 and X_2 on the probability of voting for each baseline scenario in which we're interested. As we'll see, the two sources of conditionality that we've mentioned here can combine in complicated ways to determine the interaction effect on the probability of voting for a given scenario. Indeed, for reasons that we'll explain below, the sign and statistical significance of

---

[14] We note that the symmetry of interactions that we've discussed previously continues to hold when we have limited dependent variables. As a result, scholars often also specify a hypothesis about the effect of X_2 on $Pr(Y)$ and how this varies with the value of X_1.

the interaction effect between X_1 and X_2 on the probability of voting in a given scenario doesn't necessarily match the sign and statistical significance of the interaction effect between these variables on the propensity of voting.

> ⚠ **Important:** Any conditionality we find between X_1 and X_2 with respect to $Pr(Y)$ in an interactive logit or probit model that includes the interaction term $X_1 X_2$ has two non-separable sources: (1) the conditionality from the interaction between X_1 and X_2 on Y^* and (2) the conditionality from the compression effect involving all of the independent variables that arises due to the functional form of the logit and probit models.

All of this suggests that we should be wary of hypotheses like the one mentioned earlier stating that X_1 always has a positive effect on the probability of voting but less so when X_2 is high. Are our theories really sufficiently strong that we can make a prediction about how interaction between X_1 and X_2 on Y^* combines with compression effects to determine the sign of the interaction effect between these variables on $Pr(Y)$ for all possible scenarios in the real world? We suspect in most cases that our theories are actually only strong enough to predict how X_1 and X_2 interact to affect the propensity or utility of voting. If this is the case, then this should be explicitly recognized in our hypotheses, and we should evaluate the empirical support for a theoretical prediction of interaction by examining the sign and statistical significance of the effect of X_1 on Y^* and how this varies with the value of X_2. Of course, we'll eventually also want to know when and if the magnitude of the effects that we calculate are substantively meaningful. We can't do this by simply looking at the propensity to vote as there's no meaningful scale for Y^*. We can only evaluate the substantive significance of our results by looking at the effects of our variables on $Pr(Y)$. In this sense, we generally recommend a two-step process when it comes to evaluating the results from an interactive logit or probit model. In the first step, we should examine whether we observe the effects predicted by our theory with respect to Y^*. In the second step, we should examine whether these effects are substantively meaningful by looking at how they impact $Pr(Y)$. Each step provides slightly different information that can be used to reach an overall conclusion about the value of our theory.

In what follows, we look at different ways to interpret the results from an interactive logit or probit model. In doing so, we expand on some of the insights that we've just discussed.

8.4.1 Coefficients

We can interpret the coefficients from the interactive logit or probit model shown in Eq. 8.51 in exactly the same way as we've seen in previous interaction models with a continuous dependent variable if we focus on the propensity to vote, Y^*. For example, the marginal effect of X_1 on the propensity of voting is

$$\frac{\partial Y^*}{\partial X_1} = \beta_1 + \beta_3 X_2. \tag{8.52}$$

From this, we see that β_1 tells us the effect of a one-unit increase in X_1 on the propensity of voting when $X_2 = 0$. The marginal effect of X_1 on the propensity of voting increases by β_3 for each unit increase in X_2. In other words, the modifying effect of X_2 on the marginal effect of X_1 is β_3. We can confirm this by taking the derivative of the marginal effect of X_1 with respect to X_2,

$$\frac{\partial \left(\frac{\partial Y^*}{\partial X_1} \right)}{\partial X_2} = \frac{\partial \left(\beta_1 + \beta_3 X_2 \right)}{\partial X_2} = \beta_3. \tag{8.53}$$

This means that we can test for a statistically significant interaction between X_1 and X_2 on the propensity of voting by conducting a simple z-test that $\beta_3 = 0$.[15] The marginal effect of X_2 on the propensity of voting is

$$\frac{\partial Y^*}{\partial X_2} = \beta_2 + \beta_3 X_1. \tag{8.54}$$

From this, we see that β_2 tells us the effect of a one-unit increase in X_2 on the propensity of voting when $X_1 = 0$. As expected, due to the symmetry of interactions, the marginal effect of X_2 on the propensity of voting increases by β_3 for each unit increase in X_1. We can use the logit or probit coefficients to produce the same types of marginal effect plots or tables we've seen in previous chapters to show how the effect of X_1 (X_2) on the propensity of voting changes across the observed range of values for X_2 (X_1). There's absolutely nothing new here so long as we focus our interpretation on the propensity to vote, Y^*. The same is true if we adopt a different setup for our models and focus our interpretation on the utility of voting or, in the case of a logit model, on the log-odds of voting.

What do the coefficients tell us about the *probability* of voting? Recall that the probability of voting is $\Lambda(X\beta)$ in a logit model and $\Phi(X\beta)$ in a probit model. When the marginal effect of a variable like X_1 on the propensity of voting is positive, then we see an increase in $X\beta$ and, hence,

---

[15] We conduct a z-test rather than a t-test in this context because our coefficients are assumed to be distributed according to a normal distribution rather than a t-distribution when we estimate a logit or probit model by maximum likelihood.

an increase in the probability of voting. When the marginal effect of X_1 on the propensity of voting is negative, then we see a decrease in $X\beta$ and, thus, a decrease in the probability of voting. In effect, the sign and statistical significance of the marginal effect of a variable like X_1 on Y^* indicates whether an increase in X_1 is associated with a statistically significant increase (positive) or decrease (negative) in $Pr(Y)$. The S-shaped nature of the logit and probit curves means that the precise magnitude of the change in $Pr(Y)$ associated with an increase in X_1 depends on the level of $X\beta$ or Y^* at which the increase occurs. This is all very similar to how we interpret the effects of the variables in an additive logit or probit model.

Things are more complicated when it comes to the interaction effect between X_1 and X_2 on the probability of voting. This is because, as we'll see a little later, the sign and statistical significance of the interaction effect between these variables on the probability of voting can be different from the sign and statistical significance of the interaction effect between these variables on the propensity of voting. In other words, we can't infer anything from the sign and significance of the coefficient on the interaction term about the interaction effect between X_1 and X_2 on the probability of voting.

8.4.2 Marginal Effects on Probabilities

Let's look more explicitly at the marginal effects of the independent variables in our interactive logit or probit model on the probability of voting. Suppose that we're interested in finding the marginal effect of X_1 from an interactive logit or probit model like the one shown in Eq. 8.51. We do this by taking the derivative of $Pr(Y)$ with respect to X_1,

$$\frac{\partial Pr(Y)}{\partial X_1} = \frac{\partial F(X\beta)}{\partial X_1} = \frac{\partial F(X\beta)}{\partial X\beta} \cdot \frac{\partial X\beta}{\partial X_1} = \frac{\partial Pr(Y)}{\partial Y^*} \cdot \frac{\partial Y^*}{\partial X_1} = f(X\beta) \cdot [\beta_1 + \beta_3 X_2],$$
(8.55)

where $f(\cdot)$ is the standard logistic or standard normal probability density function. So, in the case of the logit model, the marginal effect of X_1 is

$$\frac{\partial Pr(Y)}{\partial X_1} = \frac{\partial \Lambda(X\beta)}{\partial X_1} = \lambda(X\beta) \cdot [\beta_1 + \beta_3 X_2]$$
$$= \Lambda(X\beta) \cdot [1 - \Lambda(X\beta)] \cdot [\beta_1 + \beta_3 X_2], \qquad (8.56)$$

and in the case of the probit model, the marginal effect of X_1 is

$$\frac{\partial Pr(Y)}{\partial X_1} = \frac{\partial \Phi(X\beta)}{\partial x_1} = \phi(X\beta) \cdot [\beta_1 + \beta_3 X_2]. \qquad (8.57)$$

The information in Eq. 8.55 indicates that an increase in X_1 affects the probability of voting in a two-step process. In the first step, an increase in

X_1 changes the value of the underlying latent variable Y^* by $\beta_1 + \beta_3 X_2$. We see that the magnitude of this effect depends on the value of X_2. In the second step, the change in Y^* from the first step is mapped via the S-shaped logit or probit curve into a change in the probability of voting. The precise mapping depends on the value of all of the independent variables specified in the baseline scenario of interest. The compression effects introduced in the second step mean that, just as with an additive logit or probit model, the effect of an independent variable like X_1 on $Pr(Y)$ always depends on the value of all of the independent variables in the model. The total conditionality between X_1 and X_2 on $Pr(Y)$, thus, results from a combination of the variable-specific interaction on Y^* in the first step *and* the compression-based interaction involving all of the independent variables in the second step. We'll discuss the nature of this conditionality in more detail shortly.

To illustrate how to calculate the marginal effect of X_1 on the probability of voting, suppose that we estimate a logit model using the specification shown in Eq. 8.51 and that we find that $\beta_0 = -0.4$, $\beta_1 = 0.4$, $\beta_2 = -0.6$, $\beta_3 = -0.2$, and $\beta_4 = 0.5$. The marginal effect of X_1 on the *propensity* of voting is $\beta_1 + \beta_3 X_2 = 0.4 - 0.2 X_2$ and so depends on the value of X_2. It's 0.4 when $X_2 = 0$, 0.2 when $X_2 = 1$, 0 when $X_2 = 2$, and so on. To calculate the marginal effect of X_1 on the *probability* of voting, we need to specify the baseline scenario in which the increase in X_1 occurs. Suppose that we're interested in the baseline scenario where $X_1 = 0$, $X_2 = 1$, and $X_3 = 0$. In this scenario, $X\beta = -0.4 + 0.4 \times 0 - 0.6 \times 1 - 0.2 \times 0 \times 1 + 0.5 \times 0 = -1$, $\lambda(X\beta) = \lambda(-1) = 0.197$, and $\beta_1 + \beta_3 X_2 = 0.4 - 0.2 \times 1 = 0.2$. Putting this information together, we see that the marginal effect of X_1 in our baseline scenario is $\lambda(X\beta) \cdot [\beta_1 + \beta_3 X_2] = 0.197 \times 0.2 = 0.039$. As expected, the sign of the marginal effect of X_1 on the propensity to vote (positive, 0.2) is the same as the sign of the marginal effect of X_1 on the probability of voting (positive, 0.039).

Of course, the marginal effect that we've calculated is for just one possible scenario, and it may be more illustrative to calculate it for a range of different scenarios. Scholars with hypotheses about how the effect of X_1 varies with the value of X_2 might choose, for example, to calculate the marginal effect of X_1 across the observed range of values for X_2 while keeping the other independent variables constant at their baseline values. Suppose that X_2 in our previous example is a continuous variable whose values run from 0 to 10. We might be interested in how the marginal effect of X_1 affects the probability of voting across the observed values for X_2 when X_1 and X_3 are both 0. One way to present this sort of information is in the form of a marginal effect plot like the one shown in panel (c) of Figure 8.6. As a point of comparison, we provide a plot showing the

(a) The Conditional Marginal Effect of X_1 on Y^*

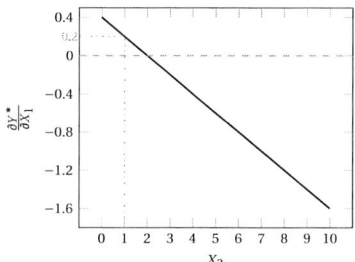

(b) The Unconditional Modifying Effect of X_2 on the Marginal Effect of X_1 on Y^*

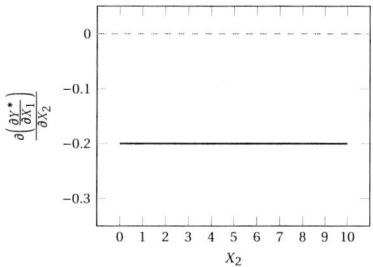

(c) The Conditional Marginal Effect of X_1 on $Pr(Y)$

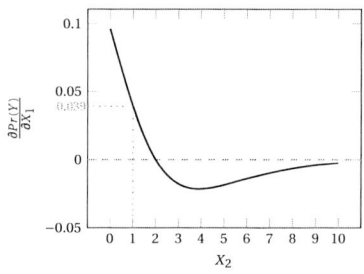

(d) The Conditional Modifying Effect of X_2 on the Marginal Effect of X_1 on $Pr(Y)$

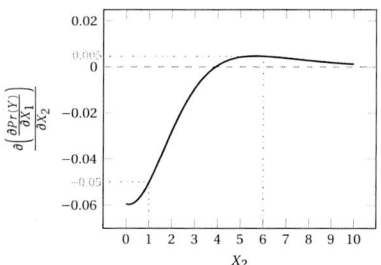

(e) The Conditional Probability that $Y = 1$

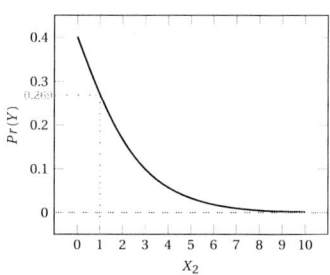

Figure 8.6 Different quantities from the logit model in Eq. 8.51 where $\beta_0 = -0.4$, $\beta_1 = 0.4$, $\beta_2 = -0.6$, $\beta_3 = -0.2$, and $\beta_4 = 0.5$

Note: Figure 8.6 shows different quantities of interest from the logit model in Eq. 8.51 for the baseline scenario where X_1 and X_3 are both 0. The plots in panels (a) and (b) show the marginal effect of X_1 (left) and the interaction effect between X_1 and X_2 (right) on the *propensity* to vote across the observed range of values for X_2. The plots in panels (c) and (d) show the marginal effect of X_1 (left) and the interaction effect between X_1 and X_2 (right) on the *probability* of voting across the observed range of values for X_2. The plot in panel (e) shows how the probability of voting changes across the observed range of values for X_2.

marginal effect of X_1 on the *propensity* of voting across the same range of values for X_2 in panel (a) directly above. In both plots, we've marked the marginal effects of X_1 we just calculated for the scenario where $X_2 = 1$. In a real application, the two plots should show confidence intervals for the marginal effect lines. They might also incorporate a histogram showing the frequency distribution for X_2.[16]

There are several things worth noting about these plots. First, the marginal effect of X_1 on the propensity of voting shown in panel (a) is only conditional on the value of X_2. As a result, the marginal effect plot shows the entire set of possible values for the marginal effect of X_1 on the propensity of voting. In contrast, the marginal effect of X_1 on the probability of voting shown in panel (c) is conditional on the value of all of the variables in the model. In other words, we see the marginal effect of X_1 on the probability of voting across the observed range of values for X_2 but only for the baseline scenario where $X_1 = 0$ and $X_3 = 0$. We'd obtain a different marginal effect plot if we chose a different baseline scenario. This reminds us that we need to be thoughtful in choosing the most appropriate baseline scenario(s) for evaluating the implications of our theory and demonstrating the substantive significance of our results.

Second, the sign of the marginal effect of X_1 on the propensity of voting shown in panel (a) is identical to the sign of the marginal effect of X_1 on the probability of voting shown in panel (c) for the same value of X_2. For example, we see that the marginal effect of X_1 on the propensity of voting in panel (a) is positive when X_2 is less than 2 and negative when X_2 is greater than 2. This is also true for the marginal effect of X_1 on the probability of voting in panel (c). This relationship always holds no matter what baseline scenario we're employing to calculate the marginal effect of X_1 on the probability of voting.

Third, the marginal effect of X_1 on the propensity of voting shown in panel (a) varies linearly with the value of X_2. Each one-unit change in the value of X_2 leads to a reduction of 0.2 in the marginal effect of X_1. This relationship results from our assumption of a linear interaction effect between X_1 and X_2 with respect to the underlying latent variable Y^* in our logit model. In this particular example, the marginal effect of X_1 ranges from a high of 0.4 when $X_2 = 0$ to a low of -1.6 when $X_2 = 10$. The magnitudes of these marginal effects are not substantively meaningful as there's no scale for Y^*. In contrast, the marginal effect of X_1 on the probability of voting varies in a non-linear way with the value of X_2.

---

[16] It's important to recognize, though, that any such frequency distribution would be unconditional and, therefore, not limited to the baseline scenario under consideration.

We first see the positive marginal effect of X_1 decline at a decreasing rate with higher values of X_2. When X_2 is greater than 2, the marginal effect becomes negative. The magnitude of this negative effect first increases at a decreasing rate with higher values of X_2 and then decreases at a decreasing rate. By the time X_2 is at its maximum value of 10, the marginal effect of X_1 on the probability of voting is almost 0. The non-linear relationship for the marginal effect of X_1 on $Pr(Y)$ across the observed values of X_2 results from compression effects that are introduced when mapping the unbounded $X\beta$ or Y^* into a bounded probability space.

As the plot in panel (c) indicates, the precise form of the non-linear relationship between the marginal effect of X_1 on $Pr(Y)$ across the observed range of values for X_2 can be rather complicated. We remind readers that this particular plot is for only one baseline scenario where $X_1 = X_3 = 0$. We suspect that most theories in the social sciences are only precise enough to make a prediction about the *sign* of the marginal effect of X_1 on $Pr(Y)$ for various values of X_2. Our theories aren't usually precise enough to tell us much about the exact non-linear way in which the magnitude of the marginal effect of X_1 on $Pr(Y)$ varies with X_2. This is important because the interaction effect between X_1 and X_2 on $Pr(Y)$ speaks precisely to the way in which the magnitude of the marginal effect of X_1 varies with the value of X_2 (and the values of the other independent variables in the model). We'll return to this point when we discuss interaction effects in more detail shortly.

As we've indicated previously, we recommend discussing the effects of the variables in an interactive logit or probit model in terms of differences rather than marginal effects. This means looking at how $Pr(Y)$ changes in response to some small, but substantively meaningful, discrete change in an independent variable. Before discussing differences in predicted probabilities, we first look at predicted probabilities.

8.4.3 Predicted Probabilities

Many scholars present predicted probabilities when showing the results of an interactive logit or probit model. To examine how this works, we return to our earlier underlying model,

$$Y^* = X\beta + \epsilon$$
$$= \beta_0 + \beta_1 X_1 + \beta_2 X_2 + \beta_3 X_1 X_2 + \beta_4 X_3 + \epsilon, \tag{8.58}$$

and continue to assume that $\beta_0 = -0.4$, $\beta_1 = 0.4$, $\beta_2 = -0.6$, $\beta_3 = -0.2$, and $\beta_4 = 0.5$. Suppose that we're interested in the probability of voting in the scenario where $X_1 = 0$, $X_2 = 1$, and $X_3 = 0$. In this scenario, $X\beta = -0.4 + 0.4 \times 0 - 0.6 \times 1 - 0.2 \times 0 \times 1 + 0.5 \times 0 = -1$, and so the probability that someone votes is $\Lambda(X\beta) = \Lambda(-1) = 0.269$ if we estimated a logit model and $\Phi(X\beta) = \Phi(-1) = 0.242$ if we estimated a probit model. Of course, this is just one possible scenario, and it may be more illustrative to calculate the probability that someone votes under different conditions. The plot in panel (e) at the bottom of Figure 8.6 shows how the logit-based probability of voting changes across the observed range of values for X_2 while keeping X_1 and X_3 at their baseline values of 0. Again, in a real application, we'd include a confidence interval. The dotted gray line confirms that the probability of voting is 0.269 when $X_2 = 1$.

The predicted probability plot in panel (e) of Figure 8.6 throws light on the particular nature of the non-linearity seen in the marginal effect plot for X_1 in panel (c) directly above. The plot in panel (a) indicates that the marginal effect of X_1 on the propensity of voting starts off positive when X_2 is less than 2 but becomes negative and increasingly large once X_2 is larger than 2. As we've noted, things are more complicated in panel (c) where we're focusing on the marginal effect of X_1 on the probability of voting. In line with the plot in panel (a), we see that the marginal effect of X_1 on the probability of voting switches from positive to negative when X_2 is larger than 2. However, the magnitude of this negative effect only grows for a short time as we raise the value of X_2 before it starts moving back toward 0. The reason for this difference between the two plots immediately becomes clear once we see the probability plot in panel (e). As the value of X_2 gets closer to its maximum value of 10, the probability of voting in the baseline scenario approaches 0. As a result, the magnitude of the negative marginal effect of X_1 on the probability of voting becomes increasingly constrained. Indeed, in the baseline scenario where the value of X_2 is 10, the probability of voting is almost 0, and so the negative marginal effect of X_1 is constrained to be almost 0 as well. This again highlights how compression effects are influencing our calculations of the effects of our variables on $Pr(Y)$.

8.4.4 Differences in Predicted Probabilities

We recommend that scholars report differences in predicted probabilities when presenting the effects of the independent variables in a logit or probit model. Suppose that we're interested in the effect of a one-unit increase in X_1 on the probability of voting when $X_1 = 0$, $X_2 = 1$, and $X_3 = 0$. This effect is the change in the probability of voting between the counterfactual scenario where $X_1 = 1$, $X_2 = 1$, and $X_3 = 0$ and the baseline scenario where $X_1 = 0$, $X_2 = 1$, and $X_3 = 0$,

$$
\begin{aligned}
F(X_c\beta) - F(X_b\beta) &= F(-0.4 + 0.4 \times 1 - 0.6 \times 1 - 0.2 \times 1 \times 1 + 0.5 \times 0) \\
&\quad - F(-0.4 + 0.4 \times 0 - 0.6 \times 1 - 0.2 \times 0 \times 1 + 0.5 \times 0) \\
&= F(-0.8) - F(-1),
\end{aligned}
\tag{8.59}
$$

where $F(\cdot)$ is the relevant cumulative distribution function. With the logit model, we get

$$\Lambda(X_c\beta) - \Lambda(X_b\beta) = \Lambda(-0.8) - \Lambda(-1) = 0.310 - 0.269 = 0.041, \tag{8.60}$$

and with the probit model, we get

$$\Phi(X_c\beta) - \Phi(X_b\beta) = \Phi(-0.8) - \Phi(-1) = 0.290 - 0.242 = 0.048. \tag{8.61}$$

From this, we see that a one-unit increase in X_1 in the specified baseline scenario is associated with a 0.041 point increase in the probability of voting if we estimated a logit model and a 0.048 point increase if we estimated a probit model.

As previously noted, it's often helpful to also report the *percentage change in the probability* that $Y = 1$ associated with a discrete change in an independent variable. The percentage change in the probability of voting associated with a one-unit increase in X_1 when $X_1 = 0$, $X_2 = 1$, and $X_3 = 0$ is

$$\frac{F(X_c\beta) - F(X_b\beta)}{F(X_b\beta)} \times 100 = \frac{F(-0.8) - F(-1)}{F(-1)} \times 100. \tag{8.62}$$

With the logit model we get

$$\frac{\Lambda(-0.8) - \Lambda(-1)}{\Lambda(-1)} \times 100 = \frac{0.310 - 0.269}{0.269} \times 100 = 15.3, \tag{8.63}$$

and with the probit model we get

$$\frac{\Phi(-0.8) - \Phi(-1)}{\Phi(-1)} \times 100 = \frac{0.290 - 0.242}{0.242} \times 100 = 19.7. \tag{8.64}$$

From this, we see that a one-unit increase in X_1 in the specified baseline scenario is associated with a 15.3% increase in the probability of voting if we estimated a logit model and a 19.7% increase if we estimated a probit model.

So far, we've calculated the (percentage) change in the probability of voting associated with a one-unit increase in X_1 under just one possible baseline scenario. Given the claim of conditionality between X_1 and X_2, it may be more illustrative to calculate it across the observed range of values for X_2. Figure 8.7 shows the change in the probability of voting (left panel) and the percentage change in the probability of voting (right panel) associated with a one-unit increase in X_1 from its baseline value of 0 across the observed range of values for X_2 when X_3 is held at its baseline value of 0. In a substantive application, these plots would also include confidence intervals and a histogram indicating the frequency distribution for X_2. The dotted gray lines in each plot confirm our earlier calculations that a one-unit change in X_1 from 0 to 1 is associated with a 0.041 unit or 15.3% increase in the probability of voting when $X_2 = 1$.

As expected, the plot in panel (a) is very similar, but not identical, to the marginal effect plot for X_1 with respect to the probability of voting shown earlier in panel (c) of Figure 8.6. When discussing the earlier marginal effect plot, we noted that the negative impact on the probability of voting associated with an increase in X_1 is very small in terms of raw magnitude when X_2 is high. We observe exactly the same thing here, with the black line being below, but almost touching, the zero-line when X_2 is high. As we mentioned previously, the reason for this has to do with compression effects whereby the magnitude of the negative impact of X_1 is severely constrained by the fact that the probability of voting is already almost 0 in the baseline scenario when X_2 is high. We might infer from this that X_1 has a substantively small impact on the probability of voting when X_2 is high. However, the plot in panel (b) showing the percentage change in the probability of voting suggests that this would likely be a mistake.

The plot in panel (b) of Figure 8.7 shows that the percentage change in the probability of voting associated with a one-unit increase in X_1 is positive when X_2 is less than 2 and negative when X_2 is greater than 2. This is consistent with our previous effect plots. Note, though, that the plot in panel (b) shows that the magnitude of the negative effect of X_1 when X_2 is greater than 2 continues to grow with higher values of X_2. Indeed, we see that a one-unit increase in X_1 is associated with a 31.7% reduction in the probability of voting when $X_2 = 4$ but a 79.8% reduction

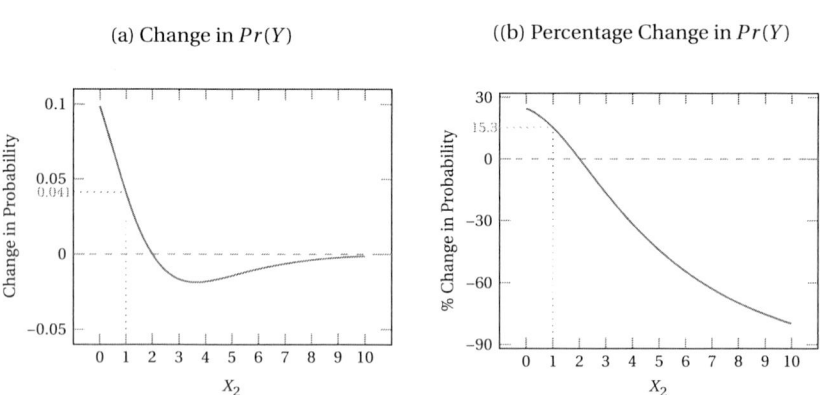

Figure 8.7 The effect of a one-unit increase in X_1 in terms of differences in probability

Note: Figure 8.7 shows how the change (a) or percentage change (b) in the probability that $Y = 1$ associated with a one-unit increase in X_1 varies across the observed range of values for X_2 in the baseline scenario when $X_1 = X_3 = 0$. The results are based on the logit model in Eq. 8.51 where $\beta_0 = -0.4$, $\beta_1 = 0.4$, $\beta_2 = -0.6$, $\beta_3 = -0.2$, and $\beta_4 = 0.5$.

when $X_2 = 10$. This is consistent with the pattern we saw in the marginal effect plot in panel (a) of Figure 8.6 that focused on the *propensity* to vote. What's going on here is that even though the raw magnitude of the effect of an increase in X_1 is small when X_2 is high, the baseline probability of voting is so low in these scenarios that even a small effect size translates into a very large percentage change in the probability of voting. This is substantively meaningful and reminds us that it's important when evaluating the substantive significance of our results to measure effect sizes relative to the value of some baseline scenario. It's for this reason that we strongly recommend that scholars report effects from logit and probit models in terms of the percentage change in $Pr(Y)$ in addition to, or instead of, in terms of the change in $Pr(Y)$.

> ⚠ **Important:** It is often helpful to combine information about the change in $Pr(Y)$ associated with a discrete change in some independent variable with information about the *percentage* change in $Pr(Y)$. Doing so can aid the reader in evaluating the substantive magnitude of any change in probability by making the change relative to $Pr(Y)$ in the baseline scenario.

We remind readers that although the plots shown in Figures 8.6 and 8.7 report the effects of X_1 on the probability of voting across a range of different baseline scenarios, they still only represent *some* of the possible effects of X_1 given the data. For example, the plot in panel (a) of Figure 8.7 tells us the effect of a one-unit increase in X_1 across the observed range of values for X_2 *only* for the case where $X_1 = X_3 = 0$. We'd get a different plot if we changed either of the baseline values for X_1 or X_3. We could, of course, create a 3-D plot where we plot the effect of X_1 across the observed range of values for, say, X_2 *and* X_1. However, this plot still only shows some of the possible effects for X_1. The issue obviously becomes harder to address the more independent variables there are in the model. Our point here is simply that it becomes incumbent on the analyst to explain exactly why and how the baseline scenarios captured in a plot are substantively important and how representative they are of possible baseline scenarios in general.

8.4.5 Odds Ratios

Scholars can also use odds ratios to examine the effects of the independent variables in an interactive logit model. The log-odds of voting in our logit model is

$$ln(\Omega) = ln\left(\frac{\rho}{1-\rho}\right) = X\beta = \beta_0 + \beta_1 X_1 + \beta_2 X_2 + \beta_3 X_1 X_2 + \beta_4 X_3.$$
$$(8.65)$$

The effect of increasing, say, X_1 by one unit on the log-odds of voting, while holding the value of the other variables constant, is

$$ln(\Omega)\big|\, X_k + 1 - ln(\Omega)\big|\, X_k = \frac{\partial X\beta}{\partial X_1} = \beta_1 + \beta_3 X_2. \qquad (8.66)$$

Thus, the effect of increasing X_1 by one unit on the ratio of the odds of voting in the counterfactual scenario rather than the baseline scenario is

$$\frac{\Omega\big|\, X_k + 1}{\Omega\big|\, X_k} = e^{\beta_1 + \beta_3 X_2}. \qquad (8.67)$$

In other words, we simply need to exponentiate the marginal effect of X_1 on the propensity to vote to find out how a one-unit increase in X_1 affects the odds ratio of voting. Continuing with our running example, we can say that a one-unit increase in X_1 increases the odds of voting by a factor of $e^{0.4-0.2\times1} = e^{0.2} = 1.22$ when $X_2 = 1$ and by a factor of $e^{0.4-0.2\times6} = e^{-0.8} = 0.45$ when $X_2 = 6$. Equivalently, we can say that a one-unit increase in X_1 increases the odds of voting by 22% when $X_2 = 1$ and decreases the odds of voting by 55% when $X_2 = 6$.

We can, of course, also calculate how a one-unit increase in X_1 affects the odds of voting across the observed range of values for X_2. Panel (a) in Figure 8.8 shows how a one-unit increase in X_1 affects the odds of voting. Values, or factor changes, greater than 1 indicate that the increase in X_1

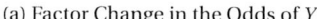

(a) Factor Change in the Odds of Y

(b) Percentage Change in the Odds of Y

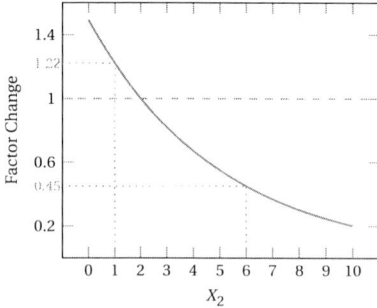

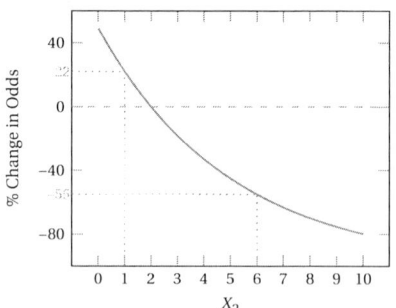

Figure 8.8 The effect of a one-unit increase in X_1 in terms of the odds of Y

Note: Figure 8.8 shows how the factor change in the odds of Y (a) or the percentage change in the odds of Y (b) associated with a one-unit increase in X_1 varies across the observed range of values for X_2. The results are based on the logit model in Eq. 8.51 where $\beta_0 = -0.4$, $\beta_1 = 0.4$, $\beta_2 = -0.6$, $\beta_3 = -0.2$, and $\beta_4 = 0.5$.

increases the odds of voting, while values, or factor changes, less than 1 indicate that it decreases the odds of voting. In line with our previous plots, we find that an increase in X_1 increases the odds of voting when X_2 is less than 2 but decreases the odds when X_2 is greater than 2. Panel (b) shows the percentage change in the odds of voting associated with a one-unit increase in X_1. Unlike other substantive quantities of interest such as predicted probabilities, which depend on the values of all of the variables in an interactive logit or probit model, odds ratios only depend on the value of the relevant modifying variable. This means that the two plots show all of the possible values for the effect of a one-unit increase in X_1 on the odds of voting.

We obviously don't need to focus on the effect of a one-unit increase in an independent variable. The odds of voting are $e^{(\beta_1 + \beta_3 X_2) \times \delta}$ larger if we increase X_1 by some amount δ. Thus, a two-unit increase in X_1 when $X_2 = 1$ is associated with a $e^{(0.4 - 0.2 \times 1) \times 2} = e^{0.4} = 1.492$ factor change in the odds of voting or, equivalently, a 49.2% increase in the odds of voting. The same two-unit increase in X_1 when $X_2 = 6$ is associated with a $e^{(0.4 - 0.2 \times 6) \times 2} = e^{-1.6} = 0.202$ factor change in the odds of voting or, equivalently, a 79.8% reduction in the odds of voting.

8.4.6 Interaction Effects

So far, we've focused on the effect of increasing, say, X_1 on the propensity, probability, or odds of voting. But what about how X_2 modifies the effect of X_1? In other words, what about interaction effects? There are typically two interaction effects in which we may be interested when we have an interactive logit or probit model. The first is the interaction effect between X_1 and X_2 with respect to the propensity of voting, Y^*. The second is the interaction effect between X_1 and X_2 with respect to the probability of voting, $Pr(Y)$. Scholars should think very carefully about which type of interaction effect is most relevant for testing their theoretical predictions. Significantly, the two interaction effects won't necessarily have the same magnitude, sign, or level of statistical significance.

In what follows, we examine interaction effects in the context of the same underlying model as we've examined previously,

$$Y^* = X\beta + \epsilon$$
$$= \beta_0 + \beta_1 X_1 + \beta_2 X_2 + \beta_3 X_1 X_2 + \beta_4 X_3 + \epsilon. \tag{8.68}$$

We'll continue to assume that we estimate a logit model, but the insights from our discussion apply equally well to a probit model. We'll also continue to assume that $\beta_0 = -0.4$, $\beta_1 = 0.4$, $\beta_2 = -0.6$, $\beta_3 = -0.2$, and $\beta_4 = 0.5$.

⚠ **Important:** There are two interaction effects in which we may be interested when we have an interactive logit or probit model. The first is the interaction effect between X_1 and X_2 on Y^*. The second is the interaction effect between X_1 and X_2 on $Pr(Y)$.

8.4.6.1 Interaction Effect on Y^*

Suppose that our theory makes a prediction about how X_2 modifies the effect of X_1 on the propensity or utility of voting. Such a prediction speaks to the interaction effect between X_1 and X_2 on Y^*. The marginal effect of X_1 on the propensity to vote is

$$\frac{\partial Y^*}{\partial X_1} = \beta_1 + \beta_3 X_2. \tag{8.69}$$

The interaction effect between X_1 and X_2 is just the slope relationship between this marginal effect and X_2,

$$\frac{\partial \left(\frac{\partial Y^*}{\partial X_1} \right)}{\partial X_2} = \frac{\partial (\beta_1 + \beta_3 X_2)}{\partial X_2} = \beta_3. \tag{8.70}$$

This interaction effect tells us how a one-unit increase in X_2 modifies the effect of a one-unit increase in X_1 on the propensity of voting. Due to the symmetry of interaction, it also tells us how a one-unit increase in X_1 modifies the effect of a one-unit increase in X_2 on the propensity of voting. As we can see, the sign, magnitude, and statistical significance of the interaction effect between X_1 and X_2 on the propensity to vote can be identified simply by looking at the coefficient on the interaction term β_3. The interaction effect with respect to Y^* doesn't depend on the value of any of the independent variables and is therefore unconditional. We show the interaction effect between X_1 and X_2 on the propensity of voting across the observed range of values for X_2 in the interaction effect plot in panel (b) of Figure 8.6. As expected, the interaction effect line is flat and has a constant value of -0.2, which is the slope of the marginal effect line shown to the left in panel (a).

8.4.6.2 Interaction Effect on $Pr(Y)$

Suppose that our theory makes a prediction about how X_2 modifies the effect of X_1 on the probability of voting. Such a prediction speaks to the interaction effect between X_1 and X_2 on $Pr(Y)$. We can think about this particular interaction effect in terms of marginal effects or in terms of "differences in differences." We'll start with marginal effects. Based on Eq. 8.56, the marginal effect of X_1 on the probability of voting is

$$\frac{\partial Pr(Y)}{\partial X_1} = \frac{\partial \Lambda(X\beta)}{\partial X_1} = \lambda(X\beta) \cdot [\beta_1 + \beta_3 X_2]$$

$$= \Lambda(X\beta) \cdot [1 - \Lambda(X\beta)] \cdot [\beta_1 + \beta_3 X_2]$$

$$= \Lambda(X\beta) \cdot (\beta_1 + \beta_3 X_2) - [\Lambda(X\beta)]^2 \cdot (\beta_1 + \beta_3 X_2). \tag{8.71}$$

As we can see, the marginal effect of X_1 on the probability of voting is conditional on the value of all of the independent variables in our logit model. As a result, we could calculate interaction effects between X_1 and any of our independent variables. However, we're interested here in the interaction effect between X_1 and X_2. This is just the slope relationship between the marginal effect of X_1 shown in Eq. 8.71 and X_2,

$$\frac{\partial \left(\frac{\partial Pr(Y)}{\partial X_1} \right)}{\partial X_2} = \frac{\partial \left[\Lambda(X\beta) \cdot (\beta_1 + \beta_3 X_2) - [\Lambda(X\beta)]^2 \cdot (\beta_1 + \beta_3 X_2) \right]}{\partial X_2}$$

$$= \lambda(X\beta) \cdot (\beta_1 + \beta_3 X_2) \cdot (\beta_2 + \beta_3 X_1) + \beta_3 \cdot \Lambda(X\beta)$$

$$- 2\Lambda(X\beta) \cdot \lambda(X\beta) \cdot (\beta_1 + \beta_3 X_2) \cdot (\beta_2 + \beta_3 X_1) - \beta_3 \cdot [\Lambda(X\beta)]^2$$

$$= \beta_3 \cdot \Lambda(X\beta) - \beta_3 \cdot [\Lambda(X\beta)]^2$$

$$+ \Lambda(X\beta) [1 - \Lambda(X\beta)] \cdot (\beta_1 + \beta_3 X_2) \cdot (\beta_2 + \beta_3 X_1)$$

$$- 2\Lambda(X\beta) \cdot [\Lambda(X\beta) [1 - \Lambda(X\beta)] \cdot (\beta_1 + \beta_3 X_2) \cdot (\beta_2 + \beta_3 X_1)]$$

$$= \Lambda(X\beta) [1 - \Lambda(X\beta)] \cdot \beta_3$$

$$+ \Lambda(X\beta) [1 - \Lambda(X\beta)] \cdot [1 - 2\Lambda(X\beta)] \cdot (\beta_1 + \beta_3 X_2) \cdot (\beta_2 + \beta_3 X_1). \tag{8.72}$$

It's sometimes useful to rearrange this interaction effect into two components, one that doesn't *directly* involve β_3 and one that's proportional to β_3:

$$\frac{\partial \left(\frac{\partial Pr(Y)}{\partial X_1} \right)}{\partial X_2} = \Lambda(X\beta) [1 - \Lambda(X\beta)] \cdot [1 - 2\Lambda(X\beta)] \cdot \beta_1 \cdot \beta_2$$

$$+ \beta_3 \cdot \Big[\Lambda(X\beta) [1 - \Lambda(X\beta)]$$

$$+ \Lambda(X\beta) [1 - \Lambda(X\beta)] [1 - 2\Lambda(X\beta)] (\beta_1 X_1 + \beta_2 X_2 + \beta_3 X_1 X_2) \Big]. \tag{8.73}$$

Due to the symmetry of interactions, this interaction effect tells us both how X_2 modifies the marginal effect of X_1 on the probability of voting and how X_1 modifies the marginal effect of X_2 on the probability of voting.

There are at least four important things to see here (Ai and Norton, 2003; Norton, Wang and Ai, 2004). First, there's no single interaction effect between X_1 and X_2 when it comes to the probability of voting. The interaction effect depends on the value of all of the variables in the model and is therefore conditional. As a result, we'll estimate a different interaction effect in each scenario we examine.

Second, the interaction effect between X_1 and X_2 on the probability of voting can be non-zero even if the coefficient on the interaction term β_3 is 0. The interaction effect on the probability of voting when $\beta_3 = 0$ is

$$\frac{\partial \left(\frac{\partial Pr(Y)}{\partial X_1} \right)}{\partial X_2} \bigg|_{\beta_3=0} = \Lambda(\overline{X\beta}) \left[1 - \Lambda(\overline{X\beta}) \right] \cdot \left[1 - 2\Lambda(\overline{X\beta}) \right] \cdot \beta_1 \cdot \beta_2, \quad (8.74)$$

where, in our case, $\overline{X\beta} = \gamma_0 + \gamma_1 X_1 + \gamma_2 X_2 + \gamma_3 X_3$. This is just the interaction effect between X_1 and X_2 on the probability of voting from an *additive* logit model that we saw earlier in Eq. 8.42. This interaction effect will be non-zero so long as $Pr(Y) = \Lambda(\overline{X\beta})$ isn't equal to 0.5 and neither β_1 nor β_2 is 0. This basically means that we can't infer that there's no interaction between X_1 and X_2 on the probability of voting when the coefficient on the interaction term is 0. We can, however, infer that any interaction that does exist in this case arises purely as a result of compression. Whether such interaction is substantively meaningful will, as we've discussed previously, depend on the application at hand and the nature of the underlying theory. It should also be clear that the interaction effect between X_1 and X_2 on the probability of voting can be zero even if the interaction term coefficient β_3 is non-zero. In other words, evidence of interaction with respect to the propensity to vote doesn't guarantee evidence of interaction with respect to the probability of voting. This situation arises when the interaction from compression exactly cancels out the effect of variable-specific interaction on Y^*.

Third, and following from the previous point, we can't infer whether there's a statistically significant interaction between X_1 and X_2 on the probability of voting by conducting a z-test on the interaction term coefficient. It's possible for the interaction effect between X_1 and X_2 on the probability of voting to not be statistically significant in a given scenario even if the interaction term coefficient is statistically significant. It's also possible for the interaction effect to be statistically significant in a given scenario even if the interaction term coefficient isn't statistically significant.

Fourth, the interaction effect between X_1 and X_2 on $Pr(Y)$ can have different signs depending on the values of the independent variables. This follows from the fact that the interaction effect in Eq. 8.72 comprises two additive terms that can each take different signs. The first term $\Lambda(X\beta) \left[1 - \Lambda(X\beta) \right] \cdot \beta_3$ will always be the same sign as β_3 because $\Lambda(X\beta)$ and $1 - \Lambda(X\beta)$ are both probabilities and hence non-negative. The sign for the second term $\Lambda(X\beta)\left[1-\Lambda(X\beta)\right]\cdot\left[1-2\Lambda(X\beta)\right]\cdot(\beta_1+\beta_3X_2)\cdot$ $(\beta_2+\beta_3X_1)$ depends on (1) the sign for the marginal effect of X_1 on the propensity to vote $\beta_1+\beta_3X_2$, (2) the sign for the marginal effect of X_2 on the propensity to vote $\beta_2+\beta_3X_1$, and (3) whether the probability of voting in the baseline scenario $\Lambda(X\beta)$ is greater or smaller than 0.5. It follows that we can't necessarily infer anything about the sign of the interaction effect between X_1 and X_2 on the probability of voting from the sign of the coefficient on the interaction term. The only times in which we can guarantee that the sign of the interaction effect will be the same as the sign of the coefficient on the interaction term are when (1) the marginal effect of X_1 on the propensity to vote is 0, (2) the marginal effect of X_2 on the propensity to vote is 0, or (3) the probability of voting is exactly 0.5.

The bottom line is that we can't typically draw valid inferences about the magnitude, sign, and statistical significance of the interaction effect between X_1 and X_2 on the probability of voting in a given scenario from the magnitude, sign, and statistical significance of the interaction term coefficient.

⚠ **Important:** It is not typically possible to draw valid inferences about the magnitude, sign, and statistical significance of the interaction effect between X_1 and X_2 on $Pr(Y)$ from the magnitude, sign, and statistical significance of the interaction term coefficient.

To see how things work, suppose that we're interested in the marginal effect of X_1 on the probability of voting and the interaction effect between X_1 and X_2 when $X_2 = 1$ and $X_1 = X_3 = 0$. Based on our running example, we have $X\beta = -0.4+0.4\times0-0.6\times1-0.2\times0\times1+0.5\times0 = -1$, $\Lambda(X\beta) =$ $\Lambda(-1) = 0.269$, $1-\Lambda(X\beta) = 1-0.269 = 0.731$, $1-2\Lambda(X\beta) = 1-0.269\times$ $2 = 0.462$, $\beta_3 = -0.2$, $\beta_1+\beta_3X_2 = 0.4-0.2\times1 = 0.2$, and $\beta_2+\beta_3X_1 =$ $-0.6-0.2\times0 = -0.6$. Drawing on Eq. 8.56, the marginal effect of X_1 on the probability of voting in our specified scenario is positive, $0.269\times$ $0.731\times0.2 = 0.039$. Drawing on Eq. 8.72, the interaction effect between X_1 and X_2 is *negative*, $-0.2\times0.269\times0.731+0.2\times-0.6\times0.269\times0.731\times$ $0.462 = -0.05$. This tells us that an increase in X_2 in our chosen scenario reduces the positive marginal effect of X_1 on the probability of voting. Suppose that we're now interested in the same scenario except that X_2 is 6 instead of 1. In this new scenario, $X\beta = -0.4+0.4\times0-0.6\times6-0.2\times0\times$ $6+0.5\times0 = -4$, $\Lambda(X\beta) = \Lambda(-4) = 0.018$, $1-\Lambda(X\beta) = 1-0.018 = 0.982$, $1-2\Lambda(X\beta) = 1-0.018\times2 = 0.964$, $\beta_3 = -0.2$, $\beta_1+\beta_3X_2 = 0.4-0.2\times6 =$ -0.8, and $\beta_2+\beta_3X_1 = -0.6-0.2\times0 = -0.6$. Drawing on Eq. 8.56, the marginal effect of X_1 is negative, $0.018\times0.982\times-0.8 = -0.014$. Drawing on Eq. 8.72, the interaction effect between X_1 and X_2 is *positive*, $-0.2 \times 0.018 \times 0.982 - 0.8 \times -0.6 \times 0.018 \times 0.982 \times 0.964 = 0.005$. This tells us that an increase in X_2 reduces the negative marginal effect of X_1. These particular examples illustrate how the sign of the interaction effect between X_1 and X_2 on the probability of voting can change across different scenarios.

We show how the interaction effect between X_1 and X_2 on the probability of voting varies across the observed range of values for X_2 when $X_1 = X_3 = 0$ in panel (d) of Figure 8.6. Unlike the interaction effect between X_1 and X_2 on the propensity to vote, which is shown in panel (b) directly above, the interaction effect on the probability of voting isn't constant, and it changes in a non-linear way with higher values of X_2. The plot indicates that the interaction effect on the probability of voting is negative when X_2 is a little smaller than 4 but positive when X_2 is larger than this. Since the interaction effect changes sign from negative to positive, we know that there'll be a range of values for X_2 at which the interaction effect is guaranteed to not be statistically significant. This will be the case even if the interaction term coefficient is statistically significant. The dotted gray lines confirm our calculations for the interaction effect when $X_2 = 1$ and when $X_2 = 6$.

The fact that the interaction effect plot in panel (d) of Figure 8.6 is placed to the right of the marginal effect plot for X_1 in panel (c) allows us to see exactly how these two plots are related. The interaction effect plot essentially indicates how the slope of the marginal effect line for X_1 in panel (c) changes with the value of X_2. This is why the interaction effect line is negative when the marginal effect line for X_1 is going down but positive when the marginal effect line is going up. The non-linearity of the interaction effect line follows directly from the non-linearity in the marginal effect line. As previously suggested, we suspect that few theories in the social sciences are sufficiently strong to make precise predictions about exactly how the marginal effect of X_1 on $Pr(Y)$ varies with X_2 beyond the sign of the marginal effect of X_1 at different values of X_2. This is important because predictions about the sign of the marginal effect of X_1 across the range of values for X_2 aren't enough on their own to make predictions about the sign of the interaction effect. For example, the interaction effect in panel (d) is sometimes negative and sometimes positive when the marginal effect of X_1 in panel (c) is negative. Much depends on the value of X_2 and the values of all of the other independent variables at which the interaction effect is examined. That the interaction effect can change sign depending on the values of the independent variables also suggests that it's inappropriate to present a hypothesis making an unconditional claim about the direction in which, say, X_2, modifies the marginal effect of X_1 on $Pr(Y)$. In effect, it requires a great deal of a theory to make clear predictions about the interaction effect between X_1 and X_2 on $Pr(Y)$.

The reason why it's so difficult to make a prediction about this particular interaction effect is because it depends on two distinct sources of interaction: (1) the variable-specific interaction that arises from the inclusion of the interaction term and (2) the compression-based interaction that arises from the S-shaped logit (or probit) curve. To make a prediction about the interaction effect between X_1 and X_2 on $Pr(Y)$, we'd need a theory strong enough to tell us exactly how these two sources of interaction combine under particular scenarios.

It would be nice if we could separate out the distinct contributions of these two sources of interaction. Unfortunately, it's not possible to fully do this. We might think to compare the interaction effect calculated from the interactive model in Eq. 8.68 that includes the interaction term X_1X_2 to the interaction effect calculated from the additive model in Eq. 8.40 that doesn't include this interaction term (Kaufman, 2019, 160). The intuition here is that the difference in these two interaction effects captures the incremental contribution of the interaction term to the pure compression-based interaction effect that exists in the "base" additive model. One problem with this approach, though, is that it involves comparing an interaction effect from a misspecified "additive" model to one from a correctly specified "interactive" model.

Avoiding this particular problem, Bowen (2012) proposes a strategy that involves comparing the interaction effect from an interactive logit model shown in Eq. 8.73 to the interaction effect shown in Eq. 8.74 from the same model where the coefficient on the interaction term is set to 0. Bowen (2012) refers to the first interaction effect as the "full" interaction effect of X_1 and X_2 on $Pr(Y)$ and to the second one as the "structural" interaction effect. He refers to the difference between these two interaction effects as the "secondary" interaction effect,

$$\begin{aligned}
\Big[&\Lambda(X\beta)\big[1-\Lambda(X\beta)\big]\cdot\big[1-2\Lambda(X\beta)\big] \\
&-\Big[\Lambda(\overline{X\beta})\big[1-\Lambda(\overline{X\beta})\big]\cdot\big[1-2\Lambda(\overline{X\beta})\big]\Big]\Big]\cdot\beta_1\cdot\beta_2 \\
&+\beta_3\cdot\Big[\Lambda(X\beta)\big[1-\Lambda(X\beta)\big] \\
&+\Lambda(X\beta)\big[1-\Lambda(X\beta)\big]\big[1-2\Lambda(X\beta)\big](\beta_1X_1+\beta_2X_2+\beta_3X_1X_2)\Big].
\end{aligned}$$
$$(8.75)$$

The idea again is that the secondary interaction effect captures the incremental contribution of the interaction variable to the "full" interaction effect beyond the contribution from the "structural" interaction effect "that exists in the 'base' model that excludes the interaction variable" (Bowen, 2012, 865). Unlike the full interaction effect shown in Eq. 8.73, the secondary interaction effect will be 0 if the coefficient on the interaction term β_3 is 0.[17] This means that we can conduct a simple z-test on the interaction term coefficient to see if there's any evidence of a statistically significant "secondary" interaction effect. Unfortunately, the secondary interaction effect is conditional on the values of the other independent variables, and we can't determine the sign of the secondary interaction effect simply by looking at the sign of β_3. To know this, we'd have to calculate the secondary interaction effect for each particular scenario in which we're interested.

One issue with the strategy proposed by Bowen (2012) has to do with variables that have no natural zero. As we noted in Chapter 3, it's possible to make the coefficients β_1 and β_2 on the constitutive terms take on any value we wish by arbitrarily re-scaling X_1 and X_2 when these variables have no natural zero. This means that the calculation of the structural interaction effect and, as a consequence, the secondary interaction effect will be dependent on the arbitrary decisions we make about how to scale our interacting variables. In particular, we can always re-scale X_1 and X_2 such that the structural interaction effect is 0. A second issue is that the structural interaction effect isn't equivalent to the interaction effect that we'd calculate in the "base" model that excludes the interaction term. This is because β_1 and β_2 capture the marginal effect of X_1 and X_2 on Y^* in the additive "base" model but they only capture the marginal effect of X_1 and X_2 on Y^* when the other modifying variable is 0 in the interactive model.

The bottom line is that the interaction effect between X_1 and X_2 on $Pr(Y)$ confounds two distinct sources of interaction that can't be fully separated. This has important implications when it comes to stating our theoretical hypotheses. In particular, we suspect that most theories in the social sciences implying conditionality are only precise enough to make predictions about the existence and sign of the interaction effect between X_1 and X_2 on Y^* and not about the existence and sign of the interaction effect between these variables on $Pr(Y)$. We encourage scholars to think carefully about whether their theoretical predictions regarding interaction apply to $Pr(Y)$, Y^*, or both and state their hypotheses accordingly. The same logic applies when it comes to testing our theoretical claims about interaction. In most cases, we'll want to evaluate whether there's empirical support for a theoretical claim about the existence and direction of interaction by focusing on Y^* and not $Pr(Y)$. This will mean

---

[17] This is easy to see once we recognize that $X\beta = \overline{X\beta}$ when $\beta_3 = 0$ and hence that
$$\Lambda(X\beta)\left[1 - \Lambda(X\beta)\right] \cdot \left[1 - 2\Lambda(X\beta)\right] - \left[\Lambda(\overline{X\beta})\left[1 - \Lambda(\overline{X\beta})\right] \cdot \left[1 - 2\Lambda(\overline{X\beta})\right]\right] = 0.$$

focusing on the sign and significance of the interaction term coefficient rather than the sign and significance of the interaction effect shown in 8.72. That said, we recommend that scholars also consider whether the magnitude of their estimated interaction effect is substantively meaningful. This requires looking at the interaction effect with respect to $Pr(Y)$ as there's no meaningful scale for the latent dependent variable Y^*. That the interaction effect on $Pr(Y)$ incorporates conditionality from compression effects isn't problematic when evaluating its substantive magnitude; compression effects are real and *should* be taken into account.

In this sense, we generally recommend a two-step process when it comes to evaluating interaction from an interactive logit or probit model. In the first step, we should examine whether we observe the effects predicted by our theory with respect to Y^*.[18] In the second step, we should examine whether the magnitude of these effects are substantively meaningful by looking at their impact on $Pr(Y)$. Each step provides slightly different information that can be used to reach an overall conclusion about the value of our theory.

> ⚠ **Important:** We generally recommend a two-step process when evaluating interaction in an interactive logit or probit model. First, examine whether we observe the effects predicted by our theory with respect to Y^*. Second, examine whether the effects are substantively meaningful by looking at how they affect $Pr(Y)$. Each step provides slightly different information that can be used to reach an overall conclusion about the value of our theory.

So far, we've looked at interaction effects with respect to $Pr(Y)$ in terms of marginal effects. However, the same basic insights apply if we think about them in terms of "differences-in-differences." An interaction effect now captures how the effect of a discrete change in some independent variable on the probability of voting changes in response to a discrete change in another independent variable. Suppose that we're interested in knowing how a discrete change in X_2 modifies the effect of a discrete change in X_1 on the probability of voting for a particular scenario. To identify this interaction effect, we must calculate three "differences." The first difference captures the *change* in the probability of voting associated with a discrete increase in X_1 from some baseline value X_{1b} to some counterfactual value X_{1c} when X_2 and all of the other independent variables are set at their own baseline values,

---

[18] Of course, if our theory is precise enough to (also) make predictions about the interaction effect on $Pr(Y)$, then we should evaluate these predictions (as well).

$$\Delta Pr(Y)_{X_{2b}} = \Lambda\,(\beta_0 + \beta_1 X_{1c} + \beta_2 X_{2b} + \beta_3 X_{1c} X_{2b} + \beta_4 X_{3b})$$
$$- \Lambda\,(\beta_0 + \beta_1 X_{1b} + \beta_2 X_{2b} + \beta_3 X_{1b} X_{2b} + \beta_4 X_{3b}). \quad (8.76)$$

The second difference captures the *change* in the probability of voting associated with the same discrete increase in X_1 when X_2 is at its counterfactual value X_{2c}, and all of the other independent variables remain at their own baseline values,

$$\Delta Pr(Y)_{X_{2c}} = \Lambda\,(\beta_0 + \beta_1 X_{1c} + \beta_2 X_{2c} + \beta_3 X_{1c} X_{2c} + \beta_4 X_{3b}) \quad (8.77)$$
$$- \Lambda\,(\beta_0 + \beta_1 X_{1b} + \beta_2 X_{2c} + \beta_3 X_{1b} X_{2c} + \beta_4 X_{3b}). \quad (8.78)$$

The third difference, which is the interaction effect between X_1 and X_2, is the difference in these two differences,

$$Interaction\ Effect_{X_1 X_2} = \Delta Pr(Y)_{X_{2c}} - \Delta Pr(Y)_{X_{2b}}. \quad (8.79)$$

To see how this works, suppose that we're interested in how a one-unit increase in X_2 modifies the effect of a one-unit increase in X_1 on the probability of voting in the scenario where $X_1 = 0$, $X_2 = 1$, and $X_3 = 0$. As before, we'll assume that $\beta_0 = -0.4$, $\beta_1 = 0.4$, $\beta_2 = -0.6$, $\beta_3 = -0.2$, and $\beta_4 = 0.5$. The change in $Pr(Y)$ associated with a one-unit increase in X_1 in the scenario where $X_2 = 1$ is

$$\Delta Pr(Y)_{X_{2b}} = \Lambda\,(-0.4 + 0.4 \times 1 - 0.6 \times 1 - 0.2 \times 1 \times 1 + 0.5 \times 0)$$
$$- \Lambda\,(-0.4 + 0.4 \times 0 - 0.6 \times 1 - 0.2 \times 0 \times 1 + 0.5 \times 0)$$
$$= \Lambda(-0.8) - \Lambda(-1)$$
$$= 0.310 - 0.269 = 0.041. \quad (8.80)$$

The change in $Pr(Y)$ associated with a one-unit increase in X_1 in the scenario where $X_2 = 2$ is

$$\Delta Pr(Y)_{X_{2c}} = \Lambda\,(-0.4 + 0.4 \times 1 - 0.6 \times 2 - 0.2 \times 1 \times 2 + 0.5 \times 0)$$
$$- \Lambda\,(-0.4 + 0.4 \times 0 - 0.6 \times 2 - 0.2 \times 0 \times 2 + 0.5 \times 0)$$
$$= \Lambda(-1.6) - \Lambda(-1.6)$$
$$= 0.168 - 0.168 = 0. \quad (8.81)$$

The interaction effect between X_1 and X_2 is therefore

$$Interaction\ Effect_{X_1 X_2} = \Delta Pr(Y)_{X_{2c}} - \Delta Pr(Y)_{X_{2b}}$$
$$= 0 - 0.041 = -0.041. \quad (8.82)$$

Just like when we calculated the interaction effect between X_1 and X_2 for this particular scenario in terms of marginal effects, we see that the interaction effect is also negative when we think in terms of a difference-in-differences with respect to probabilities. The magnitude of the interaction

(a) Marginal Effects   (b) Difference-in-Difference

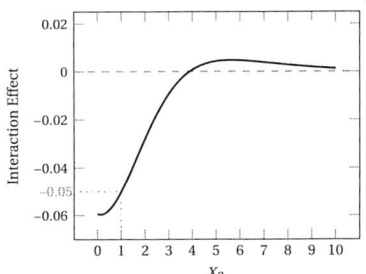
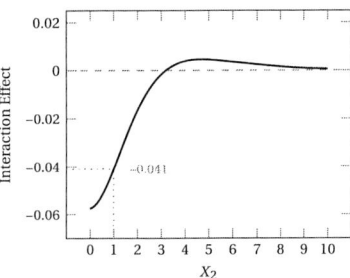

Figure 8.9 Interaction effects from the logit model in Eq. 8.51 where $\beta_0 = -0.4$, $\beta_1 = 0.4$, $\beta_2 = -0.6$, $\beta_3 = -0.2$, and $\beta_4 = 0.5$

Note: Figure 8.9 shows the interaction effect between X_1 and X_2 on $Pr(Y)$ across the observed range of values for X_2 in the baseline scenario where $X_1 = X_3 = 0$. Panel (a) calculates the interaction effect using marginal effects and Eq. 8.72, while panel (b) calculates the interaction effect as a difference-in-differences using Eq. 8.79.

effect is different, if only slightly, depending on whether we're thinking in terms of marginal effects (-0.05) or in terms of a difference-in-differences (-0.041). In Figure 8.9, we show how the interaction effect between X_1 and X_2 on $Pr(Y)$ varies across the range of values for X_2 in the baseline scenario where $X_1 = X_3 = 0$. The interaction effect in panel (a) is calculated using marginal effects and Eq. 8.72, while the one in panel (b) is calculated as a difference-in-differences using Eq. 8.79. While the two plots are very similar, they aren't identical. Among other things, we see that the interaction effect transitions from negative to positive at a slightly higher value of X_2 in panel (a) than in panel (b).

8.5 MEASURES OF UNCERTAINTY

As always, we should provide appropriate measures of uncertainty to go with the point estimates of our substantive quantities of interest. In this chapter, we've focused on logit and probit models, where the dependent variable is dichotomous. In the next two chapters, we'll turn to models designed to deal with ordered and unordered multichotomous dependent variables. There are, of course, many other types of limited dependent variables that we might also have to deal with such as counts or durations. The relevant substantive quantities of interest we'll want to calculate in these situations will often be different. With count models, for example, we're likely to want to calculate things like "incident rate ratios," expected counts, or differences in expected counts. With duration or survival models, we might want to present information about things like "hazard rates," expected durations, or differences in expected durations. No matter the setting, though, we should always provide appropriate measures of uncertainty to accompany the point estimates of our substantive quantities of interest. Analytically calculating measures of uncertainty in all of these cases can sometimes be quite challenging, if not impossible. In this section, we discuss three methods that can be used to calculate measures of uncertainty for the substantive quantities of interest that arise when dealing with limited dependent variable models: (1) the method of simulated coefficients, (2) the bootstrap method, and (3) the delta method (Glasgow, 2022). As we'll see, the method of simulated coefficients and the bootstrap method both leverage the power of computers to calculate measures of uncertainty via simulation. In contrast, the delta method is an analytic technique for calculating an approximate measure of uncertainty.

8.5.1 The Method of Simulated Coefficients

Most scholars estimate limited dependent variable models via maximum likelihood (King, 2001). While they're not typically unbiased, the estimated coefficients are consistent. Loosely-speaking, a consistent estimator is one that approaches the "correct" value as the sample size grows. This means that maximum likelihood estimators are large-sample estimators in that we need a "large enough" sample size to be confident that our estimates are good. The estimates we obtain for the coefficients in our model are uncertain. We obtain a single estimated coefficient for each of the independent variables in the model. However, these estimated coefficients are based on our particular sample of observations from the underlying population. If we'd drawn a different sample, we'd have obtained a slightly different set of coefficients for our independent variables. A nice property of maximum likelihood estimation is that we know the distribution of the estimated coefficients. Specifically, we know that the estimated coefficients will be distributed asymptotically multivariate normal, with the mean equal to the vector of population parameters

$$\beta = \begin{pmatrix} \beta_1 \\ \beta_2 \\ \vdots \\ \beta_k \\ \beta_0 \end{pmatrix} \tag{8.83}$$

and the variance equal to the variance-covariance matrix of the estimated coefficients (Cameron and Trivedi, 2005, 143)

$$V(\beta) = \begin{pmatrix} \text{var}(\beta_1) & \text{cov}(\beta_1, \beta_2) & \dots & \text{cov}(\beta_1, \beta_k) & \text{cov}(\beta_1, \beta_0) \\ \text{cov}(\beta_2, \beta_1) & \text{var}(\beta_2) & \dots & \text{cov}(\beta_2, \beta_k) & \text{cov}(\beta_2, \beta_0) \\ \vdots & \vdots & & \vdots & \vdots \\ \text{cov}(\beta_k, \beta_1) & \text{cov}(\beta_k, \beta_2) & \dots & \text{var}(\beta_k) & \text{cov}(\beta_k, \beta_0) \\ \text{cov}(\beta_0, \beta_1) & \text{cov}(\beta_0, \beta_2) & \dots & \text{cov}(\beta_0, \beta_k) & \text{var}(\beta_0) \end{pmatrix}.$$

$$(8.84)$$

This is true irrespective of the type of model we're estimating by maximum likelihood. This means that so long as our sample is large enough that we can rely upon the asymptotic properties of maximum likelihood, we know the distribution of the estimated coefficients. The method of simulated coefficients uses this knowledge to simulate measures of uncertainty for our substantive quantities of interest (King, Tomz and Wittenberg, 2000).[19]

The initial step in the method of simulated coefficients involves getting a computer to take M random draws from the asymptotic multivariate normal distribution of the estimated coefficients. Each draw consists of a "full set" of coefficients

$$\begin{pmatrix} \beta_{11} \\ \beta_{21} \\ \vdots \\ \beta_{k1} \\ \beta_{01} \end{pmatrix} \begin{pmatrix} \beta_{12} \\ \beta_{22} \\ \vdots \\ \beta_{k2} \\ \beta_{02} \end{pmatrix} \dots \begin{pmatrix} \beta_{1M} \\ \beta_{2M} \\ \vdots \\ \beta_{kM} \\ \beta_{0M} \end{pmatrix}, \qquad (8.85)$$

where β_{im} indicates the coefficient associated with the ith independent variable in the mth draw. The equivalent coefficients across the draws will all be slightly different from each other. In other words, the β_1s in each of the draws, $\beta_{11}, \beta_{12}, \dots, \beta_{1M}$, will all be slightly different to each other, with some larger and some smaller than the estimated β_1 in the original model. The same is true for the β_2s, and so on. As the number of draws increases, the means and standard deviations of the equivalent coefficients across the draws increasingly approximate the point estimates for the associated coefficients and standard errors in the original model. In other words, the mean and standard deviation of, say, the β_1s become closer and closer to the estimated β_1 and se(β_1). The goal in setting the number of draws, M, is to make sure that these means and standard deviations are close to the point estimates of the coefficients and the standard errors in the original model. In many cases, setting $M = 1,000$ will be sufficient. However, scholars should always check to see how close they are. If they're

[19] The method of simulated coefficients is also known as the Krinsky–Robb method (Krinsky and Robb, 1986) or the parametric bootstrap method (Efron and Tibshirani, 1994, 53–55).

not deemed to be sufficiently close, we can always increase the number of draws and check again.[20]

The next step is to use each of these M sets of coefficients to calculate M estimates of the desired quantity of interest. As an example, suppose we estimate the interactive logit model shown in Eq. 8.51 and that we're interested in calculating the predicted probability of voting in the (baseline) scenario where $X_2 = 1$ and $X_1 = X_3 = 0$,

$$\Lambda(X_b\beta) = \Lambda(\beta_0 + \beta_1 \times 0 + \beta_2 \times 1 + \beta_3 \times 0 \times 1 + \beta_4 \times 0). \qquad (8.86)$$

All we have to do is substitute each of our M sets of coefficients into this equation to calculate M lots of $\Lambda(X_b\beta)$. The process is visually shown in the first few columns of Table 8.1. For each draw of coefficients, we first calculate $X_b\beta$. We then run each of these M values of $X_b\beta$ through the standard logistic cumulative distribution function to get M values of $\Lambda(X_b\beta)$. These M values of $\Lambda(X_b\beta)$ provide information about the distribution for our desired quantity of interest. The standard deviation of these M values provides an estimate of the standard error associated with the probability of voting in our chosen scenario.

To construct a confidence interval for the predicted probability, we start by listing the M values of $\Lambda(X_b\beta)$ in order from smallest to largest. To obtain a two-tailed 95% confidence interval, we just need to identify the value of $\Lambda(X_b\beta)$ in the list that's associated with the 2.5 percentile (lower bound) and the 97.5 percentile (upper bound). If $M = 1,000$, this equates to the 25th and 975th largest values. To identify the bounds if we want a different level of confidence or if we want a one-sided, instead of two-sided, confidence interval, we just need to identify the values of $\Lambda(X_b\beta)$ in the list associated with the appropriate percentiles. Calculating confidence intervals in this way relaxes the assumption that they're symmetric. This

---

[20] One practical issue is that scholars should set a "seed" when conducting the random draws to aid with future replication. A seed specifies the starting point when a computer generates random draws. A seed is required because computers don't generate truly random draws; instead, they do so deterministically via some algorithm that needs a starting point. In this sense, computers only allow for "pseudo"-random draws. The computer will use a different seed each time we ask for M random draws. Since our quantities of interest are calculated based on the coefficients from these random draws, this means that we'll get a slightly different estimate of our quantity of interest each time we go through the process of making M random draws. We'll get the exact same estimate of the quantity of interest only if we use the same seed each time. By setting the seed ourselves, we guarantee that other scholars, as well as our future selves, will always be able to reproduce the exact same quantities of interest presented in our published research. We recommend that scholars always report the seed they use for their published quantities of interest in their replication files. Scholars should also report the version of the statistical package that was used for the published analysis as the seeds used in the process of making random draws may vary across versions.

is welcome as symmetric confidence intervals are likely to be problematic in some settings. As Glasgow (2022, 4) notes, "we would expect the confidence interval on a predicted probability to become asymmetric as the predicted probability approaches the boundary of 0 or 1, with the confidence interval becoming more narrow on the side of the predicted probability that is closer to the bound. Symmetric confidence intervals based on multiples of the standard error . . . will not capture this asymmetry, and in some cases can even produce confidence intervals that cross over the bounds (implying probabilities less than 0 or greater than 1)."

So far, we've discussed how we can use the method of simulated coefficients to obtain a measure of uncertainty around the predicted probability of voting when $X_2 = 1$ and $X_1 = X_3 = 0$. What about the point estimate for the predicted probability itself? We could use the mean of the M values of $\Lambda(X_b\beta)$ as our point estimate. However, we recommend that scholars just use the original estimated coefficients from the model to calculate $\Lambda(X_b\beta)$. Of course, the mean of the M values of $\Lambda(X_b\beta)$ should be very close to this predicted probability; if it isn't, we should start the simulation process again and increase the number of random draws M.

The method of simulated coefficients easily extends to different quantities of interest. As a brief example, suppose we want to calculate a measure of uncertainty to accompany an estimate that we've calculated for the change in the probability of voting associated with a one-unit increase in X_1 when $X_2 = 1$ and $X_1 = X_3 = 0$. First, we use the M draws of the set of coefficients to calculate M values of the probability of voting in the baseline scenario where $X_2 = 1$ and $X_2 = X_3 = 0$, i.e., $\Lambda(X_b\beta)$. This is what we just did. Next, we use the same process and the same M draws of the set of coefficients to calculate M values of the probability of voting in the counterfactual scenario where $X_1 = 1$, $X_2 = 1$, and $X_3 = 0$, i.e., $\Lambda(X_c\beta)$. We then subtract each value of $\Lambda(X_b\beta)$ from the value of $\Lambda(X_c\beta)$ that was calculated using the same draw of coefficients. This gives us M values for the change in the probability of voting associated with a one-unit increase in X_1 when $X_2 = 1$ and $X_1 = X_3 = 0$, i.e., $\Lambda(X_c\beta) - \Lambda(X_b\beta)$. As before, we can then use the distribution of the M values of $\Lambda(X_c\beta) - \Lambda(X_b\beta)$ to calculate a standard error or confidence interval for our change in the probability of voting. This whole process is visually shown across the columns of Table 8.1.

The method of simulated coefficients can be used to calculate appropriate measures of uncertainty for the relevant quantities of interest from any model estimated by maximum likelihood. The method is fairly fast, and we can easily increase the accuracy of any measure of uncertainty that we wish to calculate by increasing the number of random draws M that we use in the simulation process.

Table 8.1 Calculating quantities of interest via simulation

Draw	Coefficients					Baseline Scenario	Baseline Probability	Counterfactual Scenario	Counterfactual Probability	Change in Probability
1	β_{01}	β_{11}	β_{21}	β_{31}	β_{41}	$\beta_{01} + \beta_{11} \times 0 + \beta_{21} \times 1$ $+\beta_{31} \times 0 \times 1 + \beta_{41} \times 0 = (X_b\beta)_1$	$\Lambda(X_b\beta)_1$	$\beta_{01} + \beta_{11} \times 1 + \beta_{21} \times 1$ $+\beta_{31} \times 1 \times 1 + \beta_{41} \times 0 = (X_c\beta)_1$	$\Lambda(X_c\beta)_1$	$\Lambda(X_c\beta)_1 - \Lambda(X_b\beta)_1$
2	β_{02}	β_{12}	β_{22}	β_{32}	β_{42}	$\beta_{02} + \beta_{12} \times 0 + \beta_{22} \times 1$ $+\beta_{32} \times 0 \times 1 + \beta_{42} \times 0 = (X_b\beta)_2$	$\Lambda(X_b\beta)_2$	$\beta_{02} + \beta_{12} \times 1 + \beta_{22} \times 1$ $+\beta_{32} \times 1 \times 1 + \beta_{42} \times 0 = (X_c\beta)_2$	$\Lambda(X_c\beta)_2$	$\Lambda(X_c\beta)_2 - \Lambda(X_b\beta)_2$
3	β_{03}	β_{13}	β_{23}	β_{33}	β_{43}	$\beta_{03} + \beta_{13} \times 0 + \beta_{23} \times 1$ $+\beta_{33} \times 0 \times 1 + \beta_{43} \times 0 = (X_b\beta)_3$	$\Lambda(X_b\beta)_3$	$\beta_{03} + \beta_{13} \times 1 + \beta_{23} \times 1$ $+\beta_{33} \times 1 \times 1 + \beta_{43} \times 0 = (X_c\beta)_3$	$\Lambda(X_c\beta)_3$	$\Lambda(X_c\beta)_3 - \Lambda(X_b\beta)_3$
$\cdots$	$\cdots$					$\cdots$	$\cdots$	$\cdots$	$\cdots$	$\cdots$
M	β_{0M}	β_{1M}	β_{2M}	β_{3M}	β_{4M}	$\beta_{0M} + \beta_{1M} \times 0 + \beta_{2M} \times 0 \times 1$ $+\beta_{3M} \times 0 \times 1 + \beta_{4M} \times 0 = (X_b\beta)_M$	$\Lambda(X_b\beta)_M$	$\beta_{0M} + \beta_{1M} \times 1 + \beta_{2M} \times 1$ $+\beta_{3M} \times 1 \times 1 + \beta_{4M} \times 0 = (X_c\beta)_M$	$\Lambda(X_c\beta)_M$	$\Lambda(X_c\beta)_M - \Lambda(X_b\beta)_M$

8.5.2 The Bootstrap Method

Scholars may also use the bootstrap method to simulate measures of uncertainty (Efron and Tibshirani, 1994). The basic intuition behind bootstrapping is fairly straightforward. Suppose we want to calculate a measure of uncertainty for some quantity of interest, such as a predicted probability or a change in predicted probability. If we had, say, $1,000$ random samples from the population, we could estimate our model on each sample and calculate our quantity of interest $1,000$ times. Then, just as we did before, we could use the distribution of these $1,000$ values of our quantity of interest to calculate a measure of uncertainty, such as a standard error or a confidence interval. The problem is that we normally have only one sample from the population available to us.

The bootstrap method involves generating multiple samples by re-sampling from the current sample with replacement. For example, our sample of data might have, say, 800 observations. We could re-sample from this dataset with replacement until we had a second sample of 800 observations. We could repeat this, say, M times, until we had M different samples that each have 800 observations. We can then estimate our model on each of these "simulated" samples and use the M sets of coefficients to calculate M quantities of interest. At this point, we're in the same situation we were in with the method of simulated coefficients. We simply use the distribution of these M quantities of interest to identify an appropriate measure of uncertainty for our quantity of interest. Whereas the method of simulated coefficients involves estimating our maximum likelihood model once and then re-sampling on the estimates, the bootstrap method described here involves re-sampling on the data and estimating our model multiple times.

The bootstrap method is less restrictive than the method of simulated coefficients because it doesn't rely on the assumption of asymptotic normality for the estimated coefficients. This difference helps to explain why the method of simulated coefficients is sometimes called *parametric bootstrapping*, whereas the bootstrap method is sometimes called *non-parametric bootstrapping*. The downside of the bootstrap method is that it can be quite time consuming because it requires us to estimate our model multiple times.

8.5.3 The Delta Method

The delta method is an analytic technique that can be used to calculate an approximate variance for nonlinear functions of random variables that follow some known distribution (Oehlert, 1992; Ver Hoef, 2012). This method is useful for us because the substantive quantities in which we're interested, such as predicted probabilities and odds ratios, are all nonlinear functions of our asymptotically normally distributed model coefficients $S(\beta)$. The delta method works by using a first-order Taylor series expansion to linearly approximate $S(\beta)$. According to the delta method, the analytic variance of the linear approximation of $S(\beta)$ is

$$V(S(\beta)) \approx \left[\frac{\partial S(\beta)}{\partial \beta}\right]' V(\beta) \left[\frac{\partial S(\beta)}{\partial \beta}\right], \tag{8.87}$$

where $\left[\frac{\partial S(\beta)}{\partial \beta}\right]$ is a $k \times 1$ column vector containing the derivatives of our substantive quantity of interest $S(\beta)$ with respect to each of the k model coefficients β_k, $\left[\frac{\partial S(\beta)}{\partial \beta}\right]'$ is a $1 \times k$ row vector that's equal to the transpose of $\left[\frac{\partial S(\beta)}{\partial \beta}\right]$, and $V(\beta)$ is the estimated $k \times k$ variance-covariance matrix of the model coefficients. In effect, it's the square of the gradient or slope of $S(\beta)$ with respect to β multiplied by the variance of β. Taking the square root of this variance gives us the standard error. We can use the standard error to calculate a (two-tailed) confidence interval,

$$S(\beta) \pm z_{n-k,\alpha/2} \times se(S(\beta)). \tag{8.88}$$

To see how this works, imagine that we've estimated a logit model in which the underlying latent variable Y^* is modeled in the following way:

$$\begin{aligned} Y^* &= X\beta + \epsilon \\ &= \beta_0 + \beta_1 X_1 + \beta_2 X_2 + \beta_3 X_1 X_2 + \epsilon. \end{aligned} \tag{8.89}$$

Suppose that we're interested in calculating the predicted probability of voting in some scenario. As we've seen, this predicted probability, which is our substantive quantity of interest, is calculated as

$$S(\beta) = Pr(Y) = \frac{e^{X\beta}}{1 + e^{X\beta}}. \tag{8.90}$$

Using the delta method, the variance of $Pr(Y)$ is

$$
\begin{aligned}
V(S(\beta)) &= V(Pr(Y)) \\[6pt]
&\approx \left[\frac{\partial Pr(Y)}{\partial \beta}\right]' V(\beta) \left[\frac{\partial Pr(Y)}{\partial \beta}\right] \\[6pt]
&\approx \left[\frac{\partial Pr(Y)}{\partial \beta_1} \; \frac{\partial Pr(Y)}{\partial \beta_2} \; \frac{\partial Pr(Y)}{\partial \beta_3} \; \frac{\partial Pr(Y)}{\partial \beta_0}\right]
\begin{bmatrix}
\text{var}(\beta_1) & \text{cov}(\beta_1,\beta_2) & \text{cov}(\beta_1,\beta_3) & \text{cov}(\beta_1,\beta_0) \\
\text{cov}(\beta_2,\beta_1) & \text{var}(\beta_2) & \text{cov}(\beta_2,\beta_3) & \text{cov}(\beta_2,\beta_0) \\
\text{cov}(\beta_3,\beta_1) & \text{cov}(\beta_3,\beta_2) & \text{var}(\beta_3) & \text{cov}(\beta_3,\beta_0) \\
\text{cov}(\beta_0,\beta_1) & \text{cov}(\beta_0,\beta_2) & \text{cov}(\beta_0,\beta_3) & \text{var}(\beta_0)
\end{bmatrix}
\begin{bmatrix}
\frac{\partial Pr(Y)}{\partial \beta_1} \\
\frac{\partial Pr(Y)}{\partial \beta_2} \\
\frac{\partial Pr(Y)}{\partial \beta_3} \\
\frac{\partial Pr(Y)}{\partial \beta_0}
\end{bmatrix},
\end{aligned}
\tag{8.91}
$$

where

$$\frac{\partial Pr(Y)}{\partial \beta_1} = \frac{\partial \left(\frac{e^{X\beta}}{1+e^{X\beta}} \right)}{\partial \beta_1} = \frac{e^{X\beta}}{\left(1+e^{X\beta} \right)^2} \cdot X_1, \tag{8.92}$$

$$\frac{\partial Pr(Y)}{\partial \beta_2} = \frac{\partial \left(\frac{e^{X\beta}}{1+e^{X\beta}} \right)}{\partial \beta_2} = \frac{e^{X\beta}}{\left(1+e^{X\beta} \right)^2} \cdot X_2, \tag{8.93}$$

$$\frac{\partial Pr(Y)}{\partial \beta_3} = \frac{\partial \left(\frac{e^{X\beta}}{1+e^{X\beta}} \right)}{\partial \beta_3} = \frac{e^{X\beta}}{\left(1+e^{X\beta} \right)^2} \cdot X_1 X_2, \tag{8.94}$$

and

$$\frac{\partial Pr(Y)}{\partial \beta_0} = \frac{\partial \left(\frac{e^{X\beta}}{1+e^{X\beta}} \right)}{\partial \beta_0} = \frac{e^{X\beta}}{\left(1+e^{X\beta} \right)^2}. \tag{8.95}$$

To calculate a specific predicted probability and associated variance, we need only set the independent variables X_1 and X_2 to the values desired in our chosen scenario. The same basic steps can be applied to calculate measures of uncertainty for other substantive quantities that we might be interested in, such as odds ratios or differences in predicted probabilities.

The analytic basis of the delta method means that it's quicker for a computer to implement than the method of simulated coefficients or the bootstrap method. The delta method only provides an approximation of the variance for a substantive quantity of interest. While it's possible to increase the accuracy of this approximation by using a higher-order Taylor series expansion, this can be analytically challenging. In comparison, it's much easier to increase the accuracy of the method of simulated coefficients or the bootstrap method as doing so simply involves increasing the number of random draws in the simulation process. A potential concern with the delta method in some settings is that it produces symmetric confidence intervals (Glasgow, 2022, 6). As Eq. 8.88 indicates, this is because of the way that the estimated standard error is used in their construction.

8.6 SUBSTANTIVE APPLICATION: DETERMINANTS OF PRE-ELECTORAL COALITION FORMATION

What factors influence the formation of pre-electoral coalitions? In this substantive application, which is based on a partial replication of Golder (2006a,b), we examine how electoral system disproportionality, party system polarization, and legislative size affect the likelihood of pre-electoral coalition formation.

In most parliamentary democracies, parties that wish to exercise executive power are forced to enter some kind of coalition. This is because parties rarely control a legislative majority on their own. Parties have two basic options when thinking about forming a coalition. The first is to compete independently at election time and hope to be part of any government coalition that forms after the election. The second is to form a pre-electoral coalition (PEC) with one or more other parties prior to the election in the hopes of winning enough legislative seats that they can govern together afterward. In this framework, "a pre-electoral coalition exists when multiple parties choose to co-ordinate their electoral strategies rather than run for office alone" (Golder, 2006*a*, 195).

The formation of a pre-electoral coalition is the result of a bargaining process between party leaders who care about policy, office, and votes. Party leaders will form a pre-electoral coalition if the expected utility from forming a coalition is larger than the expected utility from running for office alone. A key feature of pre-electoral coalitions is that they can influence the probability that a party wins legislative seats and gets into government. This is especially the case in disproportional electoral systems where being a large party is particularly advantageous for winning legislative seats. Small parties that are disadvantaged in their ability to win legislative seats by a disproportional electoral system have an incentive to form a coalition with another party to compete together at election time. This incentive to form a pre-electoral coalition is much weaker in more proportional electoral systems where even small parties have a realistic hope of winning legislative seats and getting into government if they compete on their own.

The desire to "win" the election and get into government is especially strong when the party system is polarized. This is because electoral defeat risks having policy set by an "extreme" government from the other side of the ideological spectrum. This suggests two conditional claims. The first is that the positive incentive to form a pre-electoral coalition created by a disproportional electoral system will be stronger when the party system is polarized. The second is that we can expect party system polarization to be positively associated with pre-electoral coalition formation once the electoral system is sufficiently disproportional that these types of coalitions are electorally beneficial.

Disproportionality Hypothesis: Electoral system disproportionality always increases the utility or propensity of pre-electoral coalition formation. The magnitude of this positive effect grows with greater party system polarization.

Polarization Hypothesis: Party system polarization increases the utility or propensity of pre-electoral coalition formation when the electoral system is sufficiently disproportional.

Following our recommendations in this chapter, our hypotheses are specified in terms of the *utility* or *propensity* of pre-electoral coalition formation rather than the *probability* of pre-electoral coalition formation. This corresponds to the underlying bargaining model between party leaders presented by Golder (2006b, 42–54), which predicts variable-specific interaction between electoral system disproportionality and party system polarization on the *utility* of pre-electoral coalition formation. Following our recommendations in previous chapters, our hypotheses contain all five of the key predictions we recommend for a theory positing an interactive relationship like the one presented here. For example, the *Disproportionality Hypothesis* predicts that an increase in electoral system disproportionality will have a positive effect on the propensity of pre-electoral coalition formation at both the minimum and maximum values of party system polarization. The *Polarization Hypothesis* predicts that party system polarization will have little effect on the propensity of pre-electoral coalition formation when electoral system disproportionality is at its minimum value but a positive effect when electoral system disproportionality is at its maximum value. Implicit in both hypotheses is the prediction that there'll be a positive interaction between electoral system disproportionality and party system polarization.

Continuing with the theory, we note that the likelihood of a party entering government tends to increase with the size of its legislative representation (Glasgow, Golder and Golder, 2011, 2012). This would seem to suggest that the utility of pre-electoral coalition formation should grow with the expected size of the coalition. As Golder (2006a, 198) notes, though, if the coalition becomes sufficiently large, then at least one of the coalition members will start to think that it has a realistic chance of winning and entering government on its own. This reasoning suggests another conditional claim in which an increase in expected coalition size increases the utility of pre-electoral coalition formation when the expected coalition size is small but decreases it when the expected coalition size is large. This conditional claim predicts a variable-specific "self-interaction" with respect to expected coalition size.

Coalition Size Hypothesis: The utility or propensity of pre-electoral coalition formation first increases and then decreases with the expected size of the coalition.

Finally, the utility of pre-electoral coalition formation should be higher when the members of the proposed coalition share similar ideological positions. One reason for this is that party leaders have to make fewer concessions when agreeing to a coalition policy in these circumstances than if the coalition members have stronger policy differences.

Ideological Incompatibility Hypothesis: The utility or propensity of pre-electoral coalition formation declines as the ideological distance between potential coalition members grows.

We test these hypotheses using data on pre-electoral coalitions in 292 legislative elections in 20 advanced industrialized parliamentary democracies between 1946 and 1998. The data are organized in dyadic format, meaning that each observation represents a potential two-party coalition. The dataset contains 4,460 potential two-party pre-electoral coalitions, of which 237 actually formed. The dependent variable, *PEC*, is coded 1 if a pre-electoral coalition forms and 0 otherwise. Given the dichotomous nature of our dependent variable, we choose to employ a logit model with a latent variable setup to test our hypotheses. The latent dependent variable PEC^* measures the underlying propensity of party leaders in a dyad to form a pre-electoral coalition. The propensity to form a pre-electoral coalition is modeled with the following linear-interactive specification:

$$
\begin{aligned}
PEC^* = \beta_0 &+ \beta_1 Disproportionality + \beta_2 Polarization \\
&+ \beta_3 Disproportionality \times Polarization \\
&+ \beta_4 Coalition\ Size + \beta_5 Coalition\ Size^2 \\
&+ \beta_6 Ideological\ Incompatibility + \epsilon,
\end{aligned}
\tag{8.96}
$$

where PEC^* is assumed to be less than 0 when we don't observe a pre-electoral coalition ($PEC = 0$) and greater than or equal to 0 when we do ($PEC = 1$).

In terms of our independent variables, *Disproportionality* measures electoral system disproportionality in terms of something called the effective electoral threshold (Lijphart, 1994; Taagepera, 1998a,b). Higher effective thresholds equate to greater disproportionality. *Polarization* is calculated as the absolute ideological distance between the largest left-wing and right-wing parties in a country. Data on the policy positions of parties comes from the Manifesto Research Group (Budge et al., 2001), which evaluates parties on a left-right policy dimension that runs from -100 (extreme left) to $+100$ (extreme right). We focus on these particular parties because governments almost always contain either the main party on the left or the main party on the right. This means that when party leaders are worried about an "extreme" government coming to power, they're primarily concerned with the ideological positions of these parties. *Disproportionality* × *Polarization* is an interaction term that's included to test the conditionality of the *Disproportionality Hypothesis* and the *Polarization Hypothesis*. *Coalition Size* captures the expected size of the pre-electoral coalition in the current election. It's calculated as the percentage of legislative seats won by the two parties in the dyad in the previous election. *Coalition Size*^{2} is an interaction term that's included to test the conditionality of the *Coalition Size Hypothesis*. Finally, *Ideological Incompatibility* measures the absolute ideological distance between the parties in the dyad. Data on the ideological positions of the parties again comes from the Manifesto Research Group (Budge et al., 2001).

The marginal effect of *Disproportionality* on the propensity of pre-electoral coalition formation PEC^* is

$$\frac{\partial PEC^*}{\partial Disproportionality} = \beta_1 + \beta_3 Polarization. \tag{8.97}$$

According to the *Disproportionality Hypothesis*, an increase in electoral system disproportionality should always increase the propensity of pre-electoral coalition formation. Thus, $\beta_1 + \beta_3 Polarization$ should be positive for all values of *Polarization*. It follows that β_1 should also be positive. The positive effect of *Disproportionality* is expected to grow with *Polarization*, and so β_3 should be positive.

The marginal effect of *Polarization* on the propensity of pre-electoral coalition formation is

$$\frac{\partial PEC^*}{\partial Polarization} = \beta_2 + \beta_3 Disproportionality. \tag{8.98}$$

According to the *Polarization Hypothesis*, an increase in party system polarization should have little effect on the propensity of pre-electoral coalition formation when *Disproportionality* is low. Thus, β_2 should be close to 0. However, the effect of party system polarization should become positive and increasingly large as *Disproportionality* increases. It follows that $\beta_2 + \beta_3 Disproportionality$ should become positive once *Disproportionality* is sufficiently high and β_3 should be positive.

The marginal effect of *Coalition Size* on the propensity of pre-electoral coalition formation is

$$\frac{\partial PEC^*}{\partial Coalition\ Size} = \beta_4 + 2\beta_5 Coalition\ Size. \tag{8.99}$$

According to the *Coalition Size Hypothesis*, the propensity of pre-electoral coalition should first increase with the expected size of the potential pre-electoral coalition and then decrease. It follows that β_4 should be positive, β_5 should be negative, and $\beta_4 + 2\beta_5 Coalition\ Size$ should be negative once *Coalition Size* is sufficiently large.

Finally, the *Ideological Incompatibility Hypothesis* predicts that the marginal effect of *Ideological Incompatibility* on the propensity of pre-electoral coalition formation, β_6, should be negative.

The results from the interaction model shown in Eq. 8.96 are shown in Table 8.2. The first column presents the estimated coefficients, while the second column presents the exponentiated coefficients or odds ratios.

Table 8.2 Determinants of pre-electoral coalition formation I

Dependent Variable: *Pre-Electoral Coalition (PEC)*

	Coefficients	Odds Ratios
Disproportionality	0.03**	1.03**
	(0.01)	(0.01)
Polarization	−0.002	0.998
	(0.01)	(0.01)
Disproportionality×Polarization	0.001**	1.001**
	(0.0003)	(0.0003)
Coalition Size	0.04***	1.04***
	(0.01)	(0.01)
Coalition Size^2	−0.001***	0.999***
	(0.0002)	(0.0002)
Ideological Incompatibility	−0.01**	0.99**
	(0.004)	(0.004)
Constant	−3.60***	0.03***
	(0.29)	(0.01)
Observations	3,523	3,523
Log Likelihood	−694.73	−694.73

Standard errors in parentheses. $^*p < 0.10$; $^{**}p < 0.05$; $^{***}p < 0.01$ (two-tailed)

We'll focus first on the column of estimated coefficients. As predicted, the coefficient on *Disproportionality* is positive and statistically significant, indicating that electoral system disproportionality increases the propensity of pre-electoral coalition formation when there's no party system polarization. Also as predicted, the coefficient on the interaction term *Disproportionality×Polarization* is positive and statistically significant, indicating that the positive effect of electoral system disproportionality grows with increased party system polarization. In panel (a) of Figure 8.10, we show how a one-unit increase in *Disproportionality* affects the propensity of pre-electoral coalition formation across the observed range of values for *Polarization*. The histogram indicates the percentage of elections at different values of *Polarization*. The marginal effect plot provides strong support for the *Disproportionality Hypothesis*.

(a) The Conditional Effect of *Disproportionality*

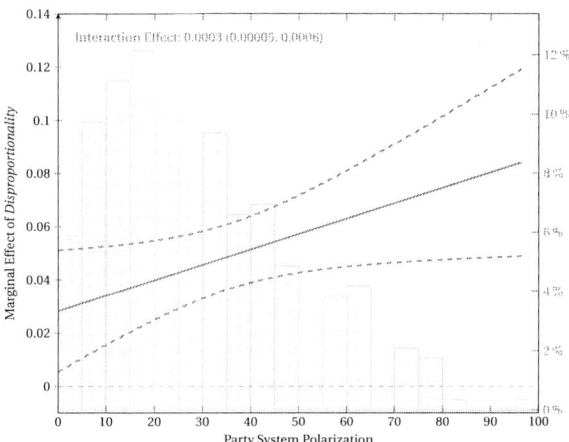

(b) The Conditional Effect of *Polarization*

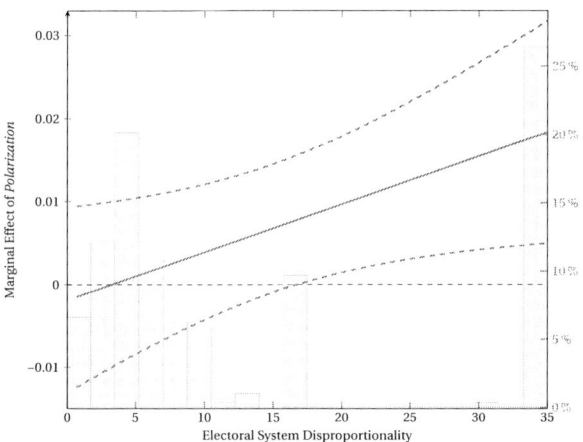

Figure 8.10 The conditional effects of *Disproportionality* and *Polarization* on the propensity of pre-electoral coalition formation

Note: The plot in panel (a) shows the effect of a one-unit increase in *Disproportionality* on the propensity of pre-electoral coalition formation across the range of observed values for *Polarization*. It also shows the interaction effect between *Disproportionality* and *Polarization*. The plot in panel (b) shows the effect of a one-unit increase in *Polarization* on the propensity of pre-electoral coalition formation across the range of observed values for *Disproportionality*. The curved dashed lines indicate two-tailed 95% confidence intervals. The histograms show the percentage of elections at different values of either *Polarization* or *Disproportionality*. The two plots are based on the results from the interaction model shown in the first column of Table 8.2.

As predicted, the coefficient on *Polarization* is close to 0 and not statistically significant, indicating that party system polarization has little effect on the propensity of pre-electoral coalition formation when there's no electoral system disproportionality. The positive and statistically significant coefficient on the interaction term *Disproportionality×Polarization* means that the effect of party system polarization grows with increased electoral system disproportionality. In panel (b) of Figure 8.10, we show how a one-unit increase in *Polarization* affects the propensity of pre-electoral coalition formation across the observed range of values for *Disproportionality*. The histogram this time indicates the percentage of elections at different levels of *Disproportionality*. In line with the *Polarization Hypothesis*, we see that *Polarization* only has a positive and statistically significant effect once *Disproportionality* is sufficiently high. To be specific, party system polarization increases the propensity of pre-electoral coalition formation in a statistically significant way but only when *Disproportionality* is greater than 16.74, which is the case in 32.2% of the elections in our sample.

As predicted, the coefficient on *Coalition Size* is positive and statistically significant, and the coefficient on *Coalition Size²* is negative and statistically significant. This pattern of signs for these coefficients is consistent with the expected inverted U-shaped relationship between the propensity of pre-electoral coalition formation and the expected size of the coalition. This inverted U-shaped relationship is graphically depicted in panel (b) of Figure 8.11 for the baseline scenario where *Disproportionality*, *Polarization*, and *Ideological Incompatibility* are all set to their mean observed values. The histogram indicates the percentage of potential coalition dyads at different values of *Coalition Size*. Following the insights from our discussion of quadratic relationships in Chapter 6, the expected coalition size that maximizes the propensity of pre-electoral coalition formation is $\frac{-\beta_4}{2\beta_5} = 37.4\%$. This tells us that an increase in expected coalition size increases the propensity of pre-electoral coalition formation when *Coalition Size* is less than 37.4% and decreases it when *Coalition Size* is greater than this. While somewhat informative, the predicted values plot in panel (b) of Figure 8.11 can't tell us if and when the effect of expected coalition size is statistically significant. This is why we plot the effect of a one-unit increase in *Coalition Size* across the observed range of values for *Coalition Size* in panel (a) of Figure 8.11.[21] This plot indicates that a one-unit increase in *Coalition Size* has a positive and statistically significant effect on the propensity of pre-electoral coalition formation when *Coalition*

---

[21] We remind the reader that the effect of a one-unit increase in *Coalition Size* is not identical to the marginal effect of *Coalition Size* given the non-linear relationship between the propensity of pre-electoral coalition formation and expected coalition size.

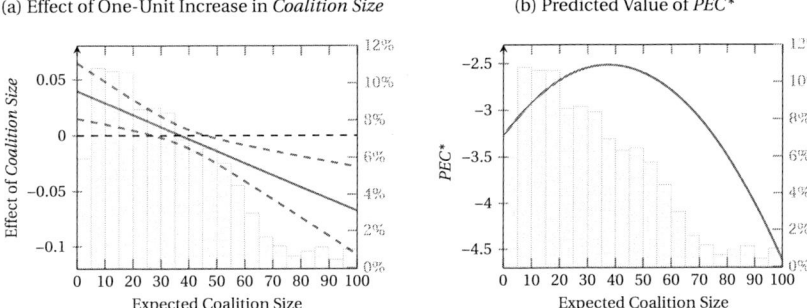

Figure 8.11 Expected coalition size and the propensity of pre-electoral coalition formation

Note: The plot in panel (a) shows the effect of a one-unit increase in *Coalition Size* on the propensity of pre-electoral coalition formation across the range of observed values for *Coalition Size*. The curved dashed lines indicate two-tailed 95% confidence intervals. The histogram shows the percentage of dyads at different values of *Coalition Size*. The self-interaction effect for *Coalition Size* is −0.0002 (0.0001). The plot in panel (b) shows the predicted propensity of pre-electoral coalition formation *PEC** across the observed range of values for *Coalition Size*. The two plots are based on the results from the interaction model shown in Table 8.2.

Size is less than 27.3% (49.3%) and that it has a negative and statistically significant effect when *Coalition Size* is greater than 45.9% (24.7%). The numbers in parentheses indicate the percentage of potential coalition dyads that satisfy each of these conditions.

Finally, the coefficient on *Ideological Compatibility* is negative and statistically significant as predicted. This tells us that an increase in the ideological distance between potential coalition partners reduces the propensity of pre-electoral coalition formation.

The information that we've provided speaks directly to the directional nature of the theoretical claims made by Golder (2006*a*,*b*) about pre-electoral coalition formation. As we've seen, the empirical evidence is entirely consistent with these claims. What we haven't determined so far is whether the magnitude of the conditional effects of electoral system disproportionality, party system polarization, and expected coalition size are substantively significant. This is because there's no meaningful scale for the latent dependent variable *PEC** that captures the propensity of pre-electoral coalition formation. To examine whether our results are substantively significant, we need to calculate effects in terms of quantities of interest that have a meaningful scale.

Suppose that we're interested in the substantive effect of electoral system disproportionality. As we've seen, there are different ways of evaluating this. One way is to look at how a discrete increase in *Disproportionality* changes the probability of pre-electoral coalition formation. To do this, though, we need to specify a particular baseline scenario in which the increase occurs. One common approach is to select a "representative" baseline scenario by setting the independent variables to their mean, median, or modal values. We also need to select a substantively realistic size for our proposed increase in *Disproportionality*.

In what follows, we examine the effect of a one standard deviation (13.3) increase in *Disproportionality* across the observed range of values for *Polarization* in the baseline scenario where *Disproportionality* (14.2), *Coalition Size* (31.4), and *Ideological Incompatibility* (24.4) are all set at their mean values. The plot in panel (a) of Figure 8.12 shows the change in the probability of pre-electoral coalition formation, while the plot in panel (b) shows the percentage change in the probability of pre-electoral coalition formation.[22] The histograms indicate the percentage of elections at different values of *Polarization*. One advantage of the plot in panel (b) is that it allows us to evaluate the magnitude of our effect relative to the probability of pre-electoral coalition formation in the baseline scenario. In line with the plot in panel (a) of Figure 8.10 that focused on the propensity of pre-electoral coalition formation, we see that our increase in electoral system disproportionality always has a positive and statistically significant effect on the probability of pre-electoral coalition formation and especially so when party system polarization is high. We can now be more specific about the magnitude of this effect, though. For example, we see that a one standard deviation increase in *Disproportionality* is associated with a 52.2% [20.8, 87.2] increase in the probability of pre-electoral coalition formation when *Polarization* is one standard deviation below its mean (10), a 72.4% [48.1, 98.7] increase (small circle) when *Polarization* is at its mean (29.5), and a 94.6% [64.8, 128.1] increase when *Polarization* is one standard deviation above its mean (49).

What about the interaction effect between *Disproportionality* and *Polarization*? In other words, what about how *Polarization* modifies the effect of a one standard deviation increase in *Polarization*? This has to do with whether the slopes of the solid lines in panels (a) and (b) are zero or not. That these lines always slope upwards means that the modifying effect of *Polarization* is always positive in the scenarios depicted in our plots. As expected, the lines aren't straight, indicating that the modifying effect of *Polarization* varies with the value of *Polarization* due to compression effects. The magnitude of the interaction effect between *Disproportionality* and *Polarization* is substantively meaningful. To illustrate this, note that the

---

[22] The confidence intervals in the two plots were constructed using the method of simulated coefficients described previously.

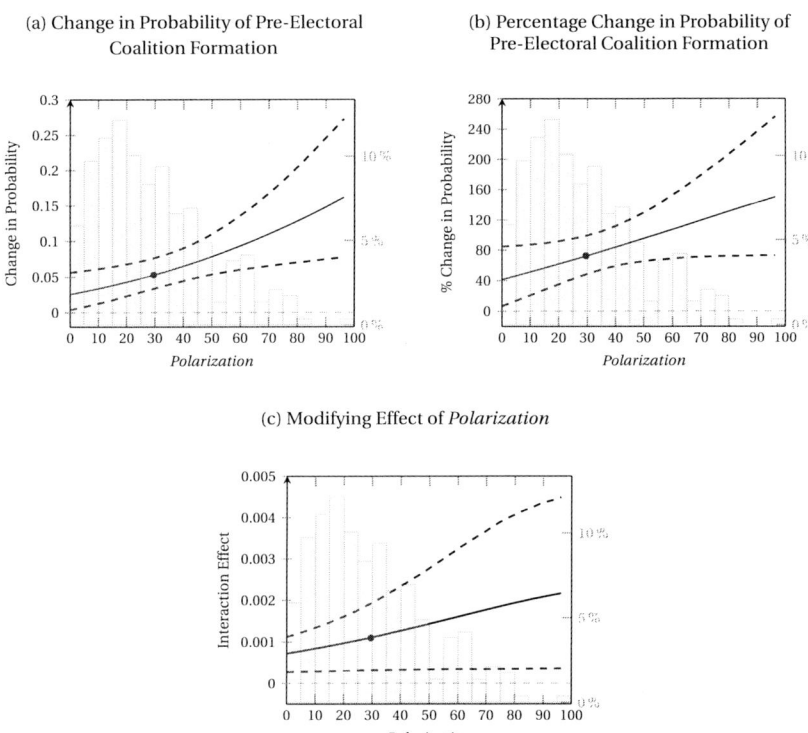

Figure 8.12 (a)–(c) The effect of *Disproportionality* on the probability of pre-electoral coalition formation

Note: The top two plots in Figure 8.12 indicate how the change (left) and the percentage change (right) in the probability of pre-electoral coalition formation associated with a one standard deviation (13.3) increase in *Disproportionality* varies across the observed range of values for *Polarization*. Values are calculated for the scenario where *Disproportionality* (14.2), *Coalition Size* (31.4), and *Ideological Incompatibility* (24.4) are set to their mean values. The bottom plot indicates how a one unit increase in *Polarization* modifies the change in the probability of pre-electoral coalition formation associated with a one standard deviation increase in *Disproportionality*. The small circles in each plot correspond to values when *Polarization* is at its mean (29.5). The three plots are based on the results shown in the first column of Table 8.2.

percentage change in the probability of pre-electoral coalition formation associated with a one standard deviation increase in *Disproportionality* is 30.6% larger when *Polarization* is one standard deviation above its mean (94.6%) than when it's at its mean (72.4%).

We can infer all of this from the top two plots in Figure 8.12. What we can't infer from these plots, though, is if and when the the modifying effect of *Polarization* is statistically significant. This is why we accompany our two "effect plots" with an "interaction effect plot" in panel (c) showing how a one-unit increase in *Polarization* modifies the change in the probability of pre-electoral coalition formation associated with a one standard deviation increase in *Disproportionality*. We see that the interaction effects shown in panel (c) are always positive and statistically significant, and that their magnitudes increase slightly with the value of *Polarization* due to compression effects. Together, the plots in Figure 8.12 provide compelling evidence that the empirical support for the *Disproportionality Hypothesis* is both statistically significant *and* substantively meaningful.

The strategy that we just employed to examine the substantive impact of electoral system disproportionality on the probability of pre-electoral coalition formation involved looking at how the effect of a one standard deviation increase in *Disproportionality* (13.3) varied across the observed range of values for *Polarization* in a "representative" baseline scenario. The representative baseline scenario in our particular example was one where *Disproportionality*, *Coalition Size*, and *Ideological Incompatibility* were all set to their mean values. Rather than employ a "representative" baseline scenario, an alternative strategy is to examine the effect of a one standard deviation increase in *Disproportionality* in *each* of the baseline scenarios defined by the values of the independent variables in our sample observations (potential coalition dyads) (Hanmer and Kalkan, 2013). This allows us to calculate the effect of the increase in *Disproportionality* on the probability of pre-electoral coalition formation for each "real-world" potential coalition dyad in the sample. On average across these dyads, a one standard deviation increase in *Disproportionality* is associated with a 77.05% (25.82) increase in the probability of pre-electoral coalition formation. The standard deviation is shown in parentheses.[23] The percentage increase in the probability of pre-electoral coalition formation ranges from a low of 38.04% to a high of 191.09% and is always statistically significant. The average percentage change in the probability of pre-electoral coalition formation across the sample observations is similar, but not identical, to

---

[23] To calculate a confidence interval around the average change in predicted probability, we could use the method of simulated coefficients. For example, we could take, say, 1,000 draws from the multivariate normal distribution defined by the model coefficients and the variance-covariance matrix. We could then use these draws to calculate 1,000 different changes in probability for each of our sample observations. From here, we could calculate 1,000 average changes in probability across our sample observations, one for each draw of coefficients. We could then use the standard deviation of these 1,000 average changes in probability to construct a confidence interval for our average change in probability. Note that the standard deviation of the 1,000 average changes in probability that we calculate using the method of simulated coefficients will be different from the standard deviation we obtain when we calculate the changes in probability for each of our sample observations using just the original set of model coefficients.

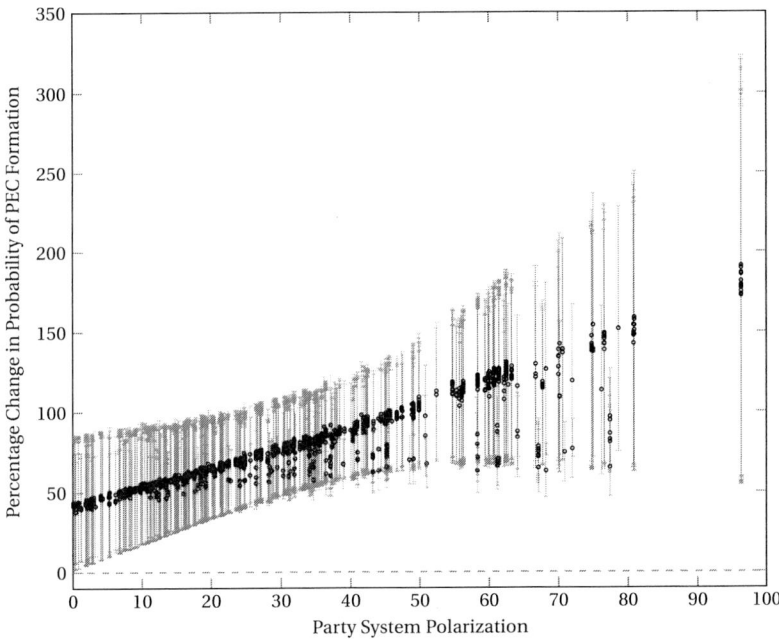

Figure 8.13 The conditional effect of *Disproportionality* on the probability of pre-electoral coalition formation

Note: The black circles show the percentage change in the probability of pre-electoral coalition formation associated with a one standard deviation increase in *Disproportionality* for each potential coalition dyad in our sample. The vertical gray bars indicate the associated two-tailed 95% confidence intervals.

the change in the probability of pre-electoral coalition formation that we calculated earlier when the independent variables were set to their mean values (72.4%); these quantities will generally be different.

In Figure 8.13, we show how the effect of a one standard deviation increase in *Disproportionality* on the probability of pre-electoral coalition formation (black circles) varies with the value of *Polarization* across the sample observations. The vertical gray lines indicate two-tailed 95% confidence intervals that were calculated using the delta method. The fact that the confidence intervals are all above the zero line confirms that increasing electoral system disproportionality is expected to have a positive and statistically significant effect on the probability of pre-electoral coalition formation regardless of the level of party system polarization. The upward trend in the black circles indicates that the positive effect of electoral system disproportionality tends to increase with higher levels of party system polarization even after taking account of the observed variation in the values of the other independent variables.[24] The modifying effect of party system polarization is quite substantial. If we restrict ourselves to elections in the lowest quartile of party system polarization (< 14.66), the one standard deviation increase in *Disproportionality* leads, on average, to a 50.1% increase in the probability of pre-electoral coalition formation. The average percentage increase is 2.26 times larger (113.3%) if we focus on those elections in the highest quartile of party system polarization (> 41.55).

We can examine the modifying effect of party system polarization more closely by calculating the interaction effect between *Disproportionality* and *Polarization* for each of our sample observations. On average, a one-unit increase in *Polarization* increases the effect of a one-standard deviation increase in *Disproportionality* on the probability of pre-electoral coalition formation by 0.001 (0.001). The standard deviation is shown in parentheses. The magnitude of the interaction effect ranges from a low of 0.0001 to a high of 0.003. The average interaction effect across the sample observations is again very similar, but not identical, to the interaction effect that we calculated earlier in panel (c) of Figure 8.12, where the independent variables were set to their mean values (0.001).

One helpful way to present the information about the interaction effect between *Disproportionality* and *Polarization* is graphically. In Figure 8.14, we plot how a one-unit increase in *Polarization* modifies the effect of a one-standard deviation increase in *Disproportionality* on the probability of pre-electoral coalition formation for each of our sample observations (black circles) along with two-tailed 95% confidence intervals (vertical gray bars). We have plotted these "interaction effects" against the probability of pre-electoral coalition formation in the baseline scenario so that they're ranked in a convenient manner. In comparison to the very large number of potential pre-election coalition dyads that could form in our sample, very few actually do form. This helps to explain why the probability of pre-electoral coalition formation shown on the horizontal axis ranges from a low of 0.006 to a high of 0.376. We immediately see from the fact that the black circles are all above the zero line in Figure 8.14 that the interaction effect between *Disproportionality* and *Polarization* is always positive, providing strong evidence that greater party system polarization increases the positive effect of electoral system disproportionality on the probability of pre-electoral coalition formation. The fact that very few of the confidence intervals cross the horizontal zero line indicates that the

---

[24] The black circles are aligned in vertical columns at different values of party system polarization. This is because party system polarization is a characteristic of elections, and multiple potential coalition dyads exist in each election.

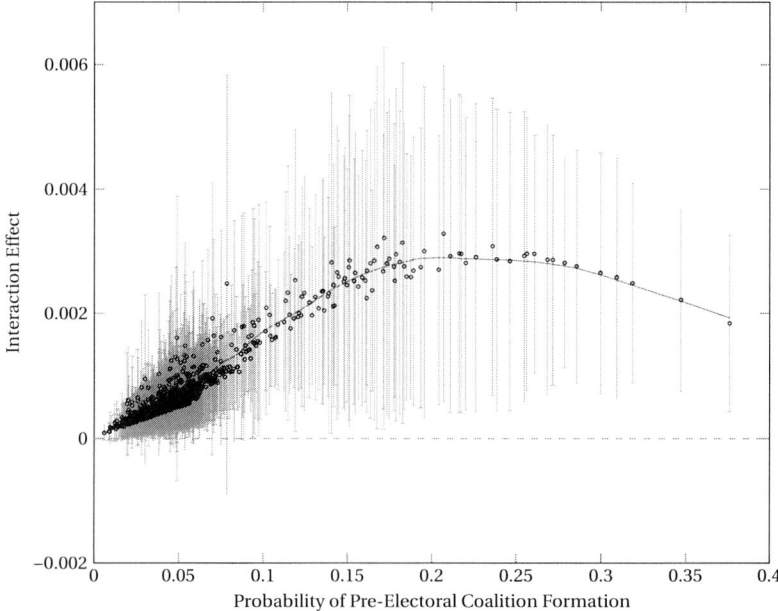

Figure 8.14 The modifying effect of party system polarization on the effect of electoral system disproportionality on the probability of pre-electoral coalition formation

Note: The black circles show the "interaction effect" calculated for each potential coalition dyad. These interaction effects indicate how a one unit increase in *Polarization* modifies the effect of a one standard deviation increase in *Disproportionality* on the probability of pre-electoral coalition formation. The vertical gray bars indicate the 95% confidence intervals associated with each of the interaction effects. The solid line indicates a lowess curve (bandwidth 0.03) summarizing the interaction effects.

positive interaction effect is almost always statistically significant. Indeed, the interaction effect is statistically significant in 95.4% of our sample observations.

Whether we look at a representative baseline scenario or across each of the baseline scenarios defined by our sample observations, we find that the empirical support for the *Disproportionality Hypothesis* is both statistically significant *and* substantively meaningful.

We could adopt one of these same approaches to examine the substantive effect of party system polarization on pre-electoral coalition formation. For illustrative purposes, though, we'll examine it in terms of odds ratios. Recall that we presented the results from our logit model in terms of odds ratios (exponentiated coefficients) in the second column of Table 8.2. The odds ratio associated with *Polarization* indicates that a one-unit increase in *Polarization* increases the odds of pre-electoral coalition formation by

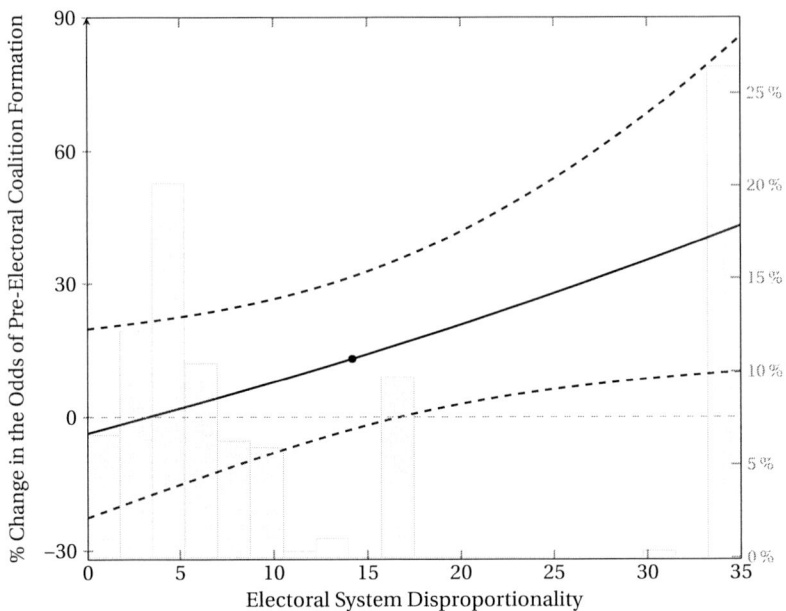

Figure 8.15 The effect of a one-standard deviation increase in *Polarization* on the odds of pre-electoral coalition formation

Note: Figure 8.15 shows how the percentage change in the odds of pre-electoral coalition formation associated with a one-standard deviation (19.5) increase in party system polarization varies across the observed range of electoral system disproportionality. The small circle corresponds to the values when *Disproportionality* is at its mean (14.2). The figure is based on the results shown in Table 8.2.

a factor of 0.998 when *Disproportionality* is 0. Put differently, a one-unit increase in *Polarization* reduces the odds of pre-electoral coalition formation by a substantively small 0.002% in the absence of electoral system disproportionality.

Suppose that we're interested in a more substantial change in party system polarization across different values of electoral system disproportionality. In Figure 8.15, we plot how the percentage change in the odds of pre-electoral coalition formation associated with a one-standard deviation increase in *Polarization* (19.5) varies across the observed range of values for *Disproportionality*. The histogram indicates the percentage of elections at different values of *Disproportionality*. In line with the plot in panel (b) of Figure 8.10, we see that our increase in party system polarization has almost no impact on the odds of pre-electoral coalition formation when *Disproportionality* is low but a statistically significant positive and growing impact when *Disproportionality* is greater than 16.7. From Figure 8.15, we see that a one-standard deviation increase in party system polarization increases the odds of pre-electoral coalition formation by 13.1% [−2.7, 31.6] when *Disproportionality* is at its mean value (14.2) and by 31.5% [7.5, 60.8] when *Disproportionality* is one standard deviation higher (27.5). This means that the percentage change in the odds of pre-electoral coalition formation associated with a one standard deviation increase in *Polarization* is 139.8% larger when *Disproportionality* is one standard deviation above its mean than when it's at its mean. Overall, Figure 8.15 provides strong evidence that the empirical support for the *Polarization Hypothesis* is both statistically *and* substantively meaningful.

What about the substantive impact of expected coalition size on pre-electoral coalition formation? In what follows, we examine the effect of a 10 unit (about half a standard deviation) increase in the expected size of a potential coalition on the probability of pre-electoral coalition formation across the observed range of values for *Coalition Size* in the baseline scenario where *Disproportionality* (14.2), *Polarization* (29.5), and *Ideological Incompatibility* (24.4) are all set at their mean values. The plot in panel (a) of Figure 8.16 shows the change in the probability of pre-electoral coalition formation, while the plot in panel (b) shows the percentage change in the probability of pre-electoral coalition formation. The histograms indicate the percentage of potential coalition dyads at different values of *Coalition Size*. We see that our 10 unit increase in expected coalition size has a positive and statistically significant effect on the probability of pre-electoral coalition formation in our baseline scenario when *Coalition Size* is less than 22.7 (41.2%) and a negative and statistically significant effect when *Coalition Size* is greater than 41.5 (30.2%). The numbers in parentheses again indicate the percentage of potential coalition dyads that satisfy the stated conditions. These results are consistent with our claim that an increase in expected coalition size has a positive effect on the probability of pre-electoral coalition formation when the potential coalition dyad is small but a negative effect when the potential coalition dyad is large. We can be more specific about the magnitude of this effect. As the plot in panel (b) indicates, a 10 unit increase in expected coalition size is associated with a statistically significant 25.8% [7.4, 46.9] increase in the probability of pre-electoral coalition formation when *Coalition Size* is one standard deviation below its mean (10.2), a statistically insignificant 1.08% [−6.4, 9.3] increase when *Coalition Size* is at its mean (31.4), and a statistically significant −18.3% [−29.0, −6.5] reduction in the probability of pre-electoral coalition formation when *Coalition Size* is one standard deviation above its mean (52.5).

What about the substantive magnitude of the interaction effect between *Coalition Size* and *Coalition Size*? In other words, how does

(a) Change in Probability of Pre-Electoral
Coalition Formation

(b) Percentage Change in Probability of
Pre-Electoral Coalition Formation

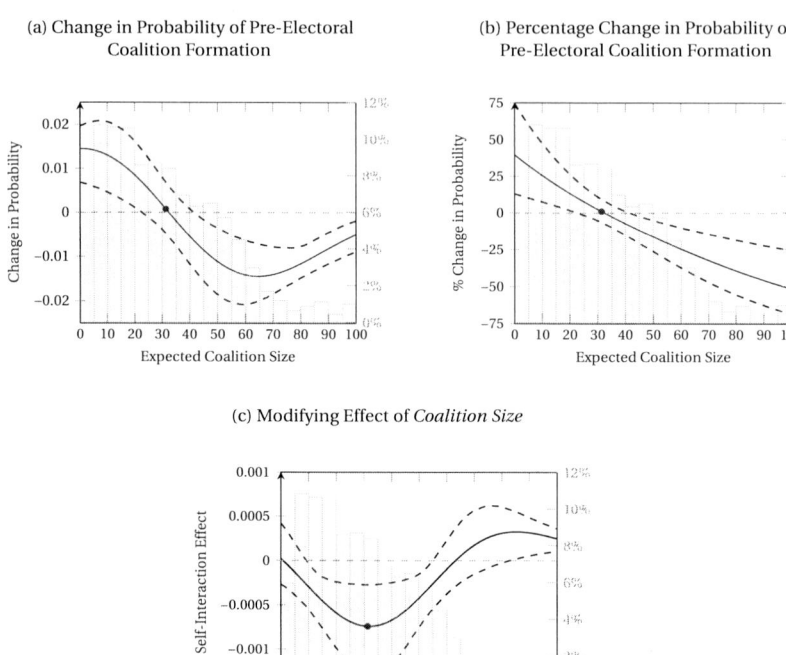

(c) Modifying Effect of *Coalition Size*

Figure 8.16   (a)–(c) The effect of *Coalition Size* on the probability of pre-electoral coalition formation

Note: The top two plots in Figure 8.16 indicate how the change (left) and the percentage change (right) in the probability of pre-electoral coalition formation associated with a ten unit increase in *Coalition Size* varies across the observed range of values for *Coalition Size*. Values are calculated for the baseline scenario where *Disproportionality* (14.2), *Polarization* (29.5), and *Ideological Incompatibility* (24.4) are set to their means. The bottom plot indicates how a one unit increase in *Coalition Size* modifies the change in the probability of pre-electoral coalition formation associated with a ten unit increase in *Coalition Size*. The small circles in each plot correspond to values when *Coalition Size* is at its mean (31.4). The three plots are based on the results shown in the first column of Table 8.2.

the effect of an increase in *Coalition Size* on the probability of pre-electoral coalition formation vary with the starting size of the potential coalition. This has to do with whether the slopes of the solid lines in panels (a) and (b) are zero or not. Note that the slopes of the lines in these two panels are quite different. The slope of the effects line in panel (b) is always negative, indicating that an increase in *Coalition Size* always has a negative modifying effect on how our 10 unit increase in expected coalition size affects the percentage change in the probability of pre-electoral coalition formation. The substantive magnitude of this negative modifying effect is large. As an example, the percentage change in the probability of pre-electoral coalition formation associated with a 10 unit increase in *Coalition Size* switches from positive 25.8% when *Coalition Size* is one standard deviation below its mean to negative 18.3% when it's one standard deviation above.

Things are more complicated in panel (a), where the slope of the effects line switches from negative to positive as *Coalition Size* increases. From this, we know that the modifying effect of *Coalition Size* on how our 10 unit increase in expected coalition size affects the change in the probability of pre-electoral coalition formation also switches from negative to positive. To illustrate this, panel (c) shows how a 1 unit increase in *Coalition Size* modifies how our 10 unit increase in expected coalition size changes the probability of pre-electoral coalition formation. We see that the self-interaction effect is negative and statistically significant when *Coalition Size* is greater than 9.2 and less than 55 (70.8%) and positive and statistically significant when *Coalition Size* is greater than 82.8 (2.9%). In other words, while the modifying effect of *Coalition Size* is always negative when we're thinking of effects in terms of the percentage change in the probability of pre-electoral coalition formation, this is not the case when we're thinking of them in terms of the change in the probability of pre-electoral coalition formation.

Overall, we see that the empirical results provide strong support for our hypotheses about the determinants of pre-electoral coalition formation.

In previous chapters, we've focused on theories that posit an interaction between two or more independent variables on a *continuous* dependent variable. We've now begun to transition to look at these types of theories in the context of limited dependent variables. In this chapter, we focused in great detail on how to evaluate conditional implications in the context of a dichotomous dependent variable. In the next two chapters, we transfer this new knowledge to look at interactions in two other types of limited dependent variable models, starting with those designed for ordered dependent variables.

8.7 EXERCISES

1. Our substantive application in the chapter looked at the determinants of pre-electoral coalition formation. Here, we revisit this application in order to practice how to calculate various quantities of interest. We modify our analysis somewhat by ignoring the role of pre-electoral coalition size. Thus, the propensity to form a pre-electoral coalition PEC^* is now modeled as

Table 8.3 Determinants of pre-electoral coalition formation II

Dependent Variable: *Pre-Electoral Coalition (PEC)*

	Coefficients	Odds Ratios
Disproportionality	0.0280***	1.028***
	(0.0103)	(0.0106)
Polarization	0.0037	1.0037
	(0.0050)	(0.0050)
Disproportionality × Polarization	0.0005**	1.0005**
	(0.0002)	(0.0002)
Ideological Incompatibility	−0.0105***	0.9896**
	(0.0038)	(0.0037)
Constant	−3.2367***	0.0393***
	(0.1868)	(0.0073)
Observations	4,395	4,395
Log Likelihood	−869.48	−869.48

Standard errors in parentheses. $^{*}p < 0.10$; $^{**}p < 0.05$; $^{***}p < 0.01$ (two-tailed)

$$PEC^{*} = \beta_0 + \beta_1 Disproportionality + \beta_2 Polarization$$
$$+ \beta_3 Disproportionality \times Polarization$$
$$+ \beta_4 Ideological\ Incompatibility + \epsilon, \qquad (8.100)$$

where PEC^{*} is assumed to be less than 0 when we don't observe a pre-electoral coalition ($PEC = 0$) and greater than or equal to 0 when we do ($PEC = 1$). The results from our slightly modified analysis are shown in Table 8.3.

a. Based on the results in Table 8.3, what are the marginal effects of *Disproportionality* and *Polarization* on the propensity of pre-electoral coalition formation?

b. Based on the coefficients for the first three variables listed in Table 8.3, what can we say about how *Disproportionality* and *Polarization* affect the *probability* of pre-electoral coalition formation?

c. Calculate the marginal effect of *Ideological Incompatibility* on the probability of pre-electoral coalition formation in the scenario where *Disproportionality* (5), *Polarization* (24.68), and

Ideological Incompatibility (19.51) are all set at their median values. What is the largest magnitude that the marginal effect of *Ideological Incompatibility* can be on the probability of pre-electoral coalition formation?

d. Calculate the marginal effects of *Disproportionality* and *Polarization* on the probability of pre-electoral coalition formation in the same scenario as before.

e. Calculate the interaction effect of *Disproportionality* and *Polarization* on the probability of pre-electoral coalition formation in the same scenario. In other words, calculate how *Polarization* (*Disproportionality*) modifies the marginal effect of *Disproportionality* (*Polarization*) on the probability of pre-electoral coalition formation in the same scenario.

f. Calculate the change and percentage change in the probability of pre-electoral coalition formation when we increase *Polarization* by 10 units in the scenario where *Polarization* = 15, *Disproportionality* = 2, and *Ideological Incompatibility* = 24. Now calculate the change and percentage change in the probability of pre-electoral coalition formation when we increase *Polarization* by 10 units for the same scenario except that *Disproportionality* is now 10. What is the modifying effect of increasing *Disproportionality* from 2 to 10 on the change in the probability of pre-electoral coalition formation associated with increasing *Polarization* by 10 units in our chosen scenario?

g. How do we interpret the odds ratios reported in Table 8.3 for *Ideological Incompatibility*, *Disproportionality*, and *Polarization*?

h. Calculate the change and percentage change in the odds of pre-electoral coalition formation associated with increasing *Disproportionality* by 5 units when *Polarization* is 10 and when *Polarization* is 30. What is the modifying effect of increasing *Polarization* from 10 to 30 on the change in the odds of pre-electoral coalition formation when we increase *Disproportionality* by 5 units?

i. Suppose that we re-estimated our logit model but omitted the interaction term *Disproportionality* × *Polarization*. The results from this purely additive model are shown in Table 8.4. As discussed in the chapter, additive models allow for pure compression-based interaction effects between the various independent variables. Given that the probability of pre-electoral coalition formation is very low in most scenarios, what can we infer about the compression-based interaction effect between *Disproportionality* and *Polarization* and the compression-based

Table 8.4 Determinants of pre-electoral coalition formation III

Dependent Variable: *Pre-Electoral Coalition (PEC)*

	Coefficients	Odds Ratios
Disproportionality	0.05***	1.05***
	(0.01)	(0.01)
Polarization	0.01***	1.01***
	(0.003)	(0.003)
Ideological Incompatibility	−0.01***	0.99***
	(0.004)	(0.004)
Constant	−3.497***	0.03***
	(0.15)	(0.005)
Observations	4,395	4,395
Log Likelihood	−871.84	−871.84

Standard errors in parentheses. $^*p < 0.10$; $^{**}p < 0.05$; $^{***}p < 0.01$ (two-tailed)

interaction effect between *Disproportionality* and *Ideological Incompatibility*? Under what conditions won't we observe a compression-based interaction effect between pairs of our independent variables?

9 Interactions and Ordered Dependent Variables

In the previous chapter, we began to look at interaction models in the context of limited dependent variables. So far, we've focused on theories positing interaction when our limited dependent variable is dichotomous and can take on only two values. Our discussion involved a detailed look at how to evaluate the conditional implications of a theory when estimating a logit or probit model. In the final two chapters, we take a look at theories positing interaction in the context of two types of discrete choice models (Train, 2009; Glasgow, 2022).[1]

Discrete choice models are those in which a choice is made from a finite set of mutually exclusive and collectively exhaustive choice alternatives. All of the discrete choice models that we'll examine have the same general form (Glasgow, 2022, 2). There's a probability that individual i chooses alternative j from a choice set consisting of J alternatives, p_{ij};[2] there's a linear function of independent variables and coefficients that describe the individual and/or the choice alternative, $X_{ij}\beta$; and there's a probability function that translates $X_{ij}\beta$ into choice probabilities. As you'll realize, the binary logit and probit models we examined in Chapter 8 are examples of discrete choice models where there are just two choice alternatives. In this chapter, we extend our discussion of discrete choice models to include situations in which the set of choice alternatives includes more than two *ordered* choices. And in the next chapter, we'll look at discrete choice models designed to handle situations where the set of choice

---

[1] Unlike in the previous chapter where we spent time discussing the motivation behind, and different ways of thinking about, logit and probit models, we'll assume that readers have a certain degree of familiarity with how to estimate and interpret the different types of models discussed in the next two chapters. Readers who wish to refresh their knowledge of these types of models might wish to consult Long (1997) and Cameron and Trivedi (2005). Our focus will be on how to substantively interpret the results from these models to evaluate theories positing interaction.

[2] Although discrete choice models are often discussed in terms of an identifiable actor making a choice, this isn't necessary. We can also discuss these models when there's no identifiable actor making a choice by simply assuming that there's some probability that each of the alternatives in the choice set might happen.

alternatives includes more than two *unordered* choices. As with logit and probit models, we'll see that we can substantively interpret the results of these other discrete choice models in terms of marginal effects, differences in predicted probabilities, and odds ratios.

> ⚠ **Important:** Discrete choice models are those in which a choice is made from a finite set of mutually exclusive and collectively exhaustive choice alternatives. An ordered choice model is appropriate when the available choice alternatives are ordered on some scale.

9.1 A LATENT VARIABLE APPROACH

Ordered choice models are appropriate when our dependent variable indicates the alternative chosen from a set of discrete choice alternatives that are ordered on some scale. Examples of ordered choices often arise in survey research where a respondent is asked to rate an item such as presidential approval or environmental concern on some ordinal scale. In a generic ordered choice model, we assume that the J values of the dependent variable Y are ordered such that 1 represents the lowest ranked alternative in the available choice set and J represents the highest ranked alternative. As with the binary choice models we examined in the last chapter, the probability that a particular alternative is chosen is a function of our independent variables and estimated coefficients $X\beta$ such that higher values of $X\beta$ increase the probability of choosing a higher ranked alternative.

One way to think about ordered choice models is in terms of an unobserved continuous latent dependent variable Y^* that's divided into J observed choice alternatives by $J-1$ ordered "thresholds" or "cutpoints." Each region of Y^* delineated by the thresholds corresponds to one of the values for our observed dependent variable Y. This approach to ordered choice models is graphically illustrated in Figure 9.1 for the case where we have $J=4$ choice alternatives and $J-1=3$ thresholds, $\tau_1 - \tau_3$ (Long, 1997, 117). For this particular case, we model the unobserved latent dependent variable Y^* in the usual way,

$$Y^* = X\beta + \epsilon, \tag{9.1}$$

and what we actually observe is

$$Y = \begin{cases} 1 & \text{if } Y^* \leq \tau_1 \\ 2 & \text{if } \tau_1 < Y^* \leq \tau_2 \\ 3 & \text{if } \tau_2 < Y^* \leq \tau_3 \\ 4 & \text{if } \tau_3 \leq Y^*. \end{cases}$$

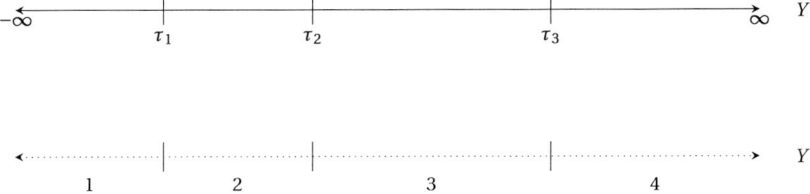

Figure 9.1 Mapping the unobserved latent dependent variable Y^* onto the observed ordered choice dependent variable Y

When estimating an ordered choice model, we estimate not only the coefficients on our independent variables but also the values of the various thresholds.

We can get insight into the latent variable approach by thinking about a survey question where respondents are asked their opinion on some issue, such as their concern for the environment (Greene, 2003, 736). The intensity of feelings that respondents have on the issue depends on several observable factors measured by the independent variables X and certain unobservable factors captured in ϵ. In principle, respondents could give a precise indicator of the intensity of their feelings Y^* on the issue. However, the survey question perhaps only provides, say, four possible answers, and so respondents are forced to choose the answer or category (choice alternative) that most closely captures the intensity of their feelings.

The probability of choosing a particular choice alternative or category is just the probability that Y^* falls between particular thresholds. For example, the probability of choosing alternative 1 is just the probability that Y^* is less than or equal to τ_1,

$$Pr(Y = 1) = Pr(Y^* \leq \tau_1) = Pr(X\beta + \epsilon \leq \tau_1) = Pr(\epsilon \leq \tau_1 - X\beta). \quad (9.2)$$

The probability of choosing alternative 2 is just the probability that Y^* is less than or equal to τ_2 but greater than τ_1. This is the same as the probability that Y^* is less than or equal to τ_2 minus the probability that Y^* is less than or equal to τ_1,

$$Pr(Y = 2) = Pr(\tau_1 < Y^* \leq \tau_2) = Pr(\epsilon \leq \tau_2 - X\beta) - Pr(\epsilon \leq \tau_1 - X\beta). \quad (9.3)$$

We can make similar calculations to find the probability of choosing the other alternatives. In order to calculate each of these probabilities, we need to choose an appropriate distribution for ϵ. If we assume that ϵ has a standard logistic distribution with mean 0 and variance $\frac{\pi^2}{3}$, we get the *ordered logit model*; and if we assume that ϵ has a standard normal distribution with mean 0 and variance 1, we get the *ordered probit model*.

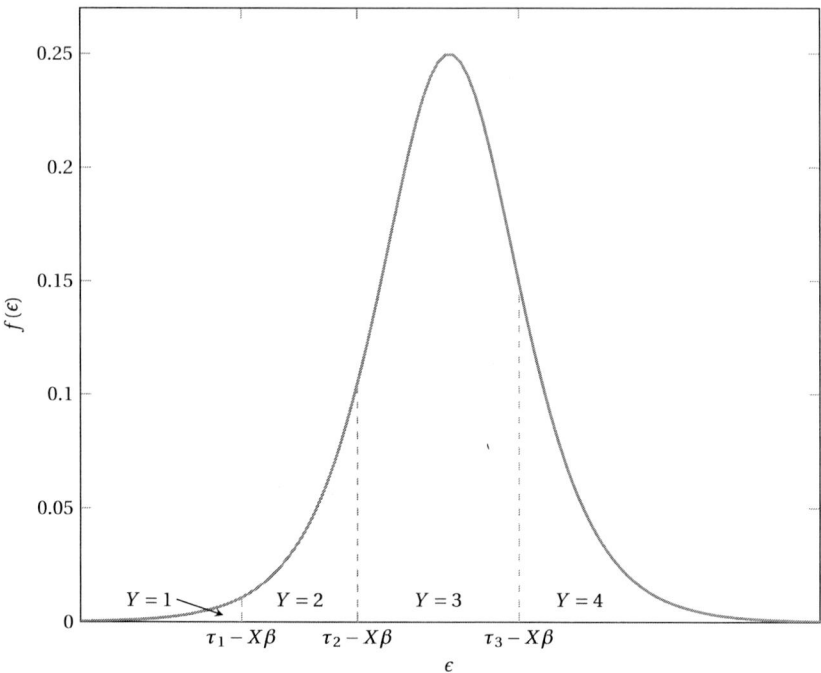

Figure 9.2 Probabilities in the ordered logit model

The implications of this setup are graphically illustrated in Figure 9.2 for the case where we have an ordered logit model.

More generally, the probability that individual i chooses alternative j in an ordered choice model is

$$P_{ij} = Pr(Y = j) = F(\tau_j - X\beta) - F(\tau_{j-1} - X\beta), \tag{9.4}$$

where $F(\cdot)$ represents either the cumulative standard logistic distribution $\Lambda(\cdot)$ or the cumulative standard normal distribution $\Phi(\cdot)$, τ_j refers to the threshold that delineates between alternative j and higher ranked alternatives, $\tau_0 = -\infty$, and $\tau_J = \infty$. The fact that $\tau_0 = -\infty$ means that the probability of choosing the lowest ranked alternative is just $p_{i1} = Pr(Y = 1) = F(\tau_1 - X\beta)$ because $F(\tau_0 - X\beta) = F(-\infty - X\beta) = 0$. And the fact that $\tau_J = \infty$ means that the probability of choosing the highest ranked alternative is $1 - F(\tau_{J-1} - X\beta)$ because $F(\tau_J - X\beta) = F(\infty - X\beta) = 1$. So in our example where there are four choice alternatives, we have

$$P_{i1} = Pr(Y = 1) = F(\tau_1 - X\beta)$$
$$P_{i2} = Pr(Y = 2) = F(\tau_2 - X\beta) - F(\tau_1 - X\beta)$$

$$P_{i3} = Pr(Y = 3) = F(\tau_3 - X\beta) - F(\tau_2 - X\beta)$$
$$P_{i4} = Pr(Y = 4) = 1 - F(\tau_3 - X\beta), \tag{9.5}$$

where the βs amd τs are to be estimated.[3]

9.2 INTERPRETATION

We can examine conditional claims about ordered choices using the same techniques related to marginal effects, differences in predicted probabilities, and odds ratios that we saw when discussing binary choice models in Chapter 8. The only real difference is that we now have to take account of the fact that we have more than two choice alternatives. To provide some substantive context, we'll assume in what follows that the ordered dependent variable Y captures how much a survey respondent approves of the performance of their country's political leader, a president. In particular, we'll assume that respondents can indicate that they disapprove strongly (1), disapprove not strongly (2), approve not strongly (3), and approve strongly (4). For illustrative purposes, we'll assume that the variable-specific interaction predicted by our theory concerns X_1 and X_2, and that our model for the underlying latent dependent variable Y^* is

$$Y^* = \beta_0 + \beta_1 X_1 + \beta_2 X_2 + \beta_3 X_1 X_2 + \beta_4 X_3 + \epsilon, \tag{9.6}$$

where $X_1 X_2$ is an interaction term and X_3 is a control variable.

9.2.1 Coefficients

We can interpret the coefficients from our interactive ordered choice model in a largely similar manner to how we interpreted the coefficients from an interactive binary choice model. If we focus on the latent dependent variable Y^*, we can again interpret the coefficients in the same way as we've seen in previous interaction models with a continuous dependent variable. The marginal effect of X_1 on the "propensity" to approve of the president is

$$\frac{\partial Y^*}{\partial X_1} = \beta_1 + \beta_3 X_2. \tag{9.7}$$

We see that β_1 tells us the effect of a one-unit increase in X_1 on the propensity to approve of the president when $X_2 = 0$. The marginal effect

---

[3] One identification issue that arises with ordered choice models has to do with how we pin down the location of the thresholds τ if the underlying scale for Y^* is unknown. To identify the thresholds, we constrain β_0 to be 0. This pins the distribution down at a reference point, and we estimate the model relative to that point. Note that this means that we won't see an estimate for the constant term in our regression output. An alternative, but equivalent, approach is to set $\tau_1 = 0$ and estimate a constant term.

of X_1 increases by β_3 for each unit increase in X_2. In other words, the modifying effect of X_2 on the marginal effect of X_1 is β_3. This means that we can test for a statistically significant interaction between X_1 and X_2 on the propensity to approve of the president by conducting a z-test that $\beta_3 = 0$. The marginal effect of X_2 on the propensity to approve of the president is

$$\frac{\partial Y^*}{\partial X_2} = \beta_2 + \beta_3 X_1. \tag{9.8}$$

We see that β_2 tells us the effect of a one-unit increase in X_2 on the propensity to approve of the president when $X_1 = 0$. Due to the symmetry of interactions, the marginal effect of X_2 on the propensity to approve of the president increases by β_3 for each unit increase in X_1. We can use the model coefficients to produce the same types of marginal effect plots or tables we've seen in previous chapters to show how the effect of X_1 (X_2) on the propensity to approve of the president changes across the observed range of values for X_2 (X_1). There's nothing new here if we focus our interpretation on the propensity to approve of the president, Y^*.

What do the coefficients tell us about the *probability* of approving of the president? As in binary choice models, when the marginal effect of a variable like X_1 on the propensity to approve of the president is positive, we see an increase in Y^* and, hence, an increase in the probability of observing a higher ranked choice alternative and a decrease in the probability of observing a lower ranked choice alternative. When the marginal effect on the propensity to approve of the president is negative, we see the opposite effects.

What about the probabilities of specific choice categories? The sign of the marginal effect of a variable like X_1 on the propensity to approve of the president tells us *only* about how the probabilities of observing the lowest and highest ranked presidential approval categories change. When the marginal effect on Y^* is positive, we know that the probability of observing the lowest approval category goes down and that the probability of observing the highest approval category goes up. And when the marginal effect on Y^* is negative, we know that the probability of observing the lowest approval categories goes up and that the probability of observing the highest approval category goes down. Without additional calculations, though, we can't know whether the probabilities of observing any of the intermediate approval categories increase or decrease. The intuition behind this is shown in Figure 9.3. An increase in an independent variable like X_1, while holding the relevant cutpoints constant, is equivalent to shifting our chosen probability distribution to the left or right (Greene, 2003, 738–739). In Figure 9.3, we show the case where the marginal effect of X_1 on

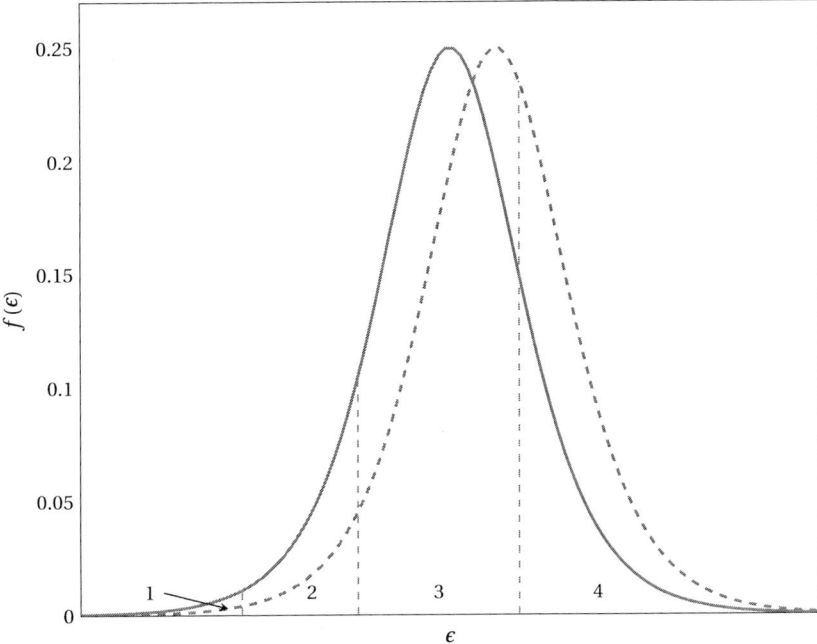

Figure 9.3 Effect of increasing X_1 on the probabilities from an ordered logit model when the marginal effect of X_1 on Y^* is positive

Note: Figure 9.3 shows how the probabilities from an ordered logit model change when we increase X_1 from some baseline value (solid curve) to some counterfactual value (dashed curve), assuming that the marginal effect of X_1 on Y^* is positive. The dashed vertical gray lines indicate the relevant cutpoints separating the choice alternatives or categories.

the propensity to approve of the president is positive, and so an increase in X_1 shifts the probability distribution rightward from the solid curve to the dashed curve. The effect of this shift is to unambiguously move some probability mass out of the lowest choice category (the area to the left of the first dashed vertical gray line shrinks) and some probability mass into the highest choice category (the area to the right of the last dashed vertical gray line increases). What happens with the intermediate choice categories is ambiguous. Some probability mass is moving in from the left, but some is also moving out to the right. Whether more probability mass is moving into an intermediate choice category than is moving out depends on the baseline scenario, the magnitude of the shift, and the location of the thresholds.

As before, we can't make any inference about the precise magnitude of any change in the probability of observing particular presidential approval categories from the magnitude of the marginal effect of an independent variable like X_1 on the propensity to approve of the president. This is one reason why we often choose to move beyond the coefficients and marginal effects with respect to Y^* when substantively interpreting the results from an ordered logit or probit model.

> ⚠ **Important:** In an ordered choice model, the sign of the marginal effect of an independent variable on the latent dependent variable Y^* tells us *only* about how the probabilities of observing the lowest and highest ranked choice categories change. Without additional calculations, we cannot know whether the probabilities of observing any of the intermediate choice categories increase or decrease.

The same issues arise when it comes to interpreting the interaction effect of X_1 and X_2 on the probability of the different presidential approval categories as we saw when discussing binary choice models in the previous chapter. In other words, we can't infer anything from the sign and significance of the coefficient on the interaction term about the sign and significance of the interaction effect of X_1 and X_2 on the probabilities of the different choice alternatives.

In addition to the coefficients, we'll also obtain estimates of the various thresholds separating our choice categories when we estimate an ordered choice model. In our particular example, where there are four presidential approval categories, this means that we'll estimate three thresholds. When we estimate an ordered choice model, we're assuming that the ordered choice alternatives are truly different. We should check this by testing whether the threshold parameters are significantly different from each other. If they're not, this suggests that the observed choice categories aren't truly different, and we might want to collapse the indistinguishable categories into a single category. On the whole, the thresholds are generally treated as nuisance parameters and aren't interpreted in a substantive way. Depending on the application, though, we might infer that substantially unequal widths between choice categories signal that the meanings of some categories are more expansive than others or that there are differences in the semantic distinctiveness of the adjectives used to label the various choice categories.

9.2.2 Marginal Effects on Probabilities

In an ordered choice model, we can calculate the marginal effect of an independent variable on the probability of choosing each of the choice alternatives or categories. In other words, there'll be a separate marginal effect when it comes to $Pr(Y)$ for each of the J choice alternatives.

Suppose that we're interested in the marginal effects of a variable like X_3 in Eq. 9.6 that's included *additively* in an ordered choice model. We can find the marginal effect of X_3 on the probability of each choice alternative by taking the derivative of P_{ij} in Eq. 9.4 with respect to X_3,

$$\frac{\partial P_{ij}}{\partial X_3} = \frac{\partial Pr(Y=j)}{\partial X_3} = \frac{\partial \left(F(\tau_j - X\beta) - F(\tau_{j-1} - X\beta)\right)}{\partial X_3}$$

$$= f(\tau_j - X\beta)(-\beta_4) - f(\tau_{j-1} - X\beta)(-\beta_4)$$

$$= \left[f(\tau_j - X\beta) - f(\tau_{j-1} - X\beta)\right] \cdot -\beta_4$$

$$= \left[f(\tau_{j-1} - X\beta) - f(\tau_j - X\beta)\right] \cdot \beta_4, \qquad (9.9)$$

where, recall, β_4 is the coefficient on X_3, and $f(\cdot)$ is either the standard logistic probability density function $\lambda(\cdot)$ in an ordered logit model or the standard normal probability density function $\phi(\cdot)$ in an ordered probit model. In our presidential approval example, the marginal effect of X_3 on each of our four choice probabilities is

$$\frac{\partial P_{i1}}{\partial X_3} = \frac{\partial Pr(Y=1)}{\partial X_3} = \left[f(\tau_1 - X\beta)\right] \cdot -\beta_4$$

$$\frac{\partial P_{i2}}{\partial X_3} = \frac{\partial Pr(Y=2)}{\partial X_3} = \left[f(\tau_1 - X\beta) - f(\tau_2 - X\beta)\right] \cdot \beta_4$$

$$\frac{\partial P_{i3}}{\partial X_3} = \frac{\partial Pr(Y=3)}{\partial X_3} = \left[f(\tau_2 - X\beta) - f(\tau_3 - X\beta)\right] \cdot \beta_4$$

$$\frac{\partial P_{i4}}{\partial X_3} = \frac{\partial Pr(Y=4)}{\partial X_3} = \left[f(\tau_3 - X\beta)\right] \cdot \beta_4. \qquad (9.10)$$

There are a couple of things to note here. The first is that because $f(\cdot)$ is necessarily positive, the marginal effect of X_3 on the probability of choosing the lowest ranked alternative will always have the opposite sign to that of the variable's coefficient. In other words, a positive (negative) coefficient for β_4 indicates that an increase in X_3 always reduces (increases) the probability of observing the lowest choice category. Similarly, the marginal effect on the probability of choosing the highest ranked alternative will always have the same sign as the variable's coefficient. In other words, a positive (negative) coefficient for β_4 indicates that an increase in X_3 always increases (reduces) the probability of observing the highest choice category. The sign of the marginal effect of X_3 on the probability of observing the intermediate choice categories always depends on the relative sizes of $f(\tau_{j-1} - X\beta)$ and $f(\tau_j - X\beta)$. These insights just confirm our earlier claim that we can only draw inferences about the sign of the change in the probability of observing the lowest and highest ranked choice alternatives from the sign of an independent variable's marginal effect on Y^*.

The second thing to note is that, as in the binary choice models from the last chapter, the marginal effect of an independent variable on the various choice probabilities depends on the value of $X\beta$ at which the increase takes place. In other words, the effect of an independent variable on the choice probabilities always depends on the values of *all* of the variables in the model, including its own. This tells us that the effects of the independent variables on the choice probabilities are always conditional when we estimate an ordered choice model, even when the variable of interest is included additively. This conditionality arises from the compression that occurs when we take the unbounded values of Y^* and map them onto a probability space that's bounded between 0 and 1.

Suppose now that we're interested in the marginal effect of a variable like X_1 in Eq. 9.6 that's included *interactively* in an ordered choice model. We again find the marginal effect on the probability of each choice alternative by taking the derivative of P_{ij} in Eq. 9.4 with respect to X_1,

$$\frac{\partial P_{ij}}{\partial X_1} = \frac{\partial Pr(Y=j)}{\partial X_1} = \frac{\partial\left(F(\tau_j - X\beta) - F(\tau_{j-1} - X\beta)\right)}{\partial X_1}$$
$$= f(\tau_j - X\beta)(-\beta_1 - \beta_3 X_2) - f(\tau_{j-1} - X\beta)(-\beta_1 - \beta_3 X_2)$$
$$= \left[f(\tau_j - X\beta) - f(\tau_{j-1} - X\beta)\right]\cdot - (\beta_1 + \beta_3 X_2)$$
$$= \left[f(\tau_{j-1} - X\beta) - f(\tau_j - X\beta)\right]\cdot(\beta_1 + \beta_3 X_2). \tag{9.11}$$

In our presidential approval example, the marginal effect of X_1 on each of our four choice probabilities is

$$\frac{\partial P_{i1}}{\partial X_1} = \frac{\partial Pr(Y=1)}{\partial X_1} = \left[f(\tau_1 - X\beta)\right]\cdot - (\beta_1 + \beta_3 X_2)$$
$$\frac{\partial P_{i2}}{\partial X_1} = \frac{\partial Pr(Y=2)}{\partial X_1} = \left[f(\tau_1 - X\beta) - f(\tau_2 - X\beta)\right]\cdot(\beta_1 + \beta_3 X_2)$$
$$\frac{\partial P_{i3}}{\partial X_1} = \frac{\partial Pr(Y=3)}{\partial X_1} = \left[f(\tau_2 - X\beta) - f(\tau_3 - X\beta)\right]\cdot(\beta_1 + \beta_3 X_2)$$
$$\frac{\partial P_{i4}}{\partial X_1} = \frac{\partial Pr(Y=4)}{\partial X_1} = \left[f(\tau_3 - X\beta)\right]\cdot(\beta_1 + \beta_3 X_2). \tag{9.12}$$

As in the binary choice models from the last chapter, an increase in X_1 affects our choice probabilities in a two-step process. In the first step, an increase in X_1 changes the value of the underlying latent variable Y^* by $\beta_1 + \beta_3 X_2$. The magnitude of this effect depends on just the value of X_2. In the second step, this change in Y^* is mapped into a change in each of our choice probabilities. The precise mapping depends on the value of all of the independent variables specified in the baseline scenario of interest. The compression effects introduced in the second step mean that, just as with an independent variable that's included additively, the effect of X_1 on the choice probabilities always depends on the values of all of the independent variables in the model. Thus, the eventual conditionality we see between X_1 and X_2 on the choice probabilities results from a combination of the variable-specific interaction on Y^* in the first step and the compression-based interaction involving all of the independent variables in the second step.

Scholars with hypotheses about how the effect of X_1 varies with the value of X_2 might wish to calculate the marginal effect of X_1 on each of their choice probabilities across the observed range of values for X_2 while keeping the other independent variables constant at their chosen baseline values. One way to present this type of information is in the form of a marginal effect plot. In our particular example, where we have four presidential approval categories, this would mean producing four separate marginal effect plots.

9.2.3 Differences in Predicted Probabilities

Scholars can also report differences in each of the predicted choice probabilities when presenting the effects of the independent variables in an ordered logit or probit model. The effect of increasing X_1 from some baseline value X_{1b} to some counterfactual value X_{1c} on each of the choice probabilities when the other independent variables are each set at their baseline values is

$$\Delta P_{ij} = \Delta Pr(Y = j) = \left[F(\tau_j - X_c\beta) - F(\tau_{j-1} - X_c\beta) \right]$$
$$- \left[F(\tau_j - X_b\beta) - F(\tau_{j-1} - X_b\beta) \right], \qquad (9.13)$$

where

$$X_c\beta = \beta_0 + \beta_1 X_{1c} + \beta_2 X_{2b} + \beta_3 X_{1c} X_{2b} + \beta_4 X_{3b} \qquad (9.14)$$

and

$$X_b\beta = \beta_0 + \beta_1 X_{1b} + \beta_2 X_{2b} + \beta_3 X_{1b} X_{2b} + \beta_4 X_{3b}. \qquad (9.15)$$

In our presidential approval example, we have

$$\Delta P_{i1} = \Delta Pr(Y = 1) = F(\tau_1 - X_c\beta) - F(\tau_1 - X_b\beta)$$
$$\Delta P_{i2} = \Delta Pr(Y = 2) = \left[F(\tau_2 - X_c\beta) - F(\tau_1 - X_c\beta) \right] - \left[F(\tau_2 - X_b\beta) - F(\tau_1 - X_b\beta) \right]$$
$$\Delta P_{i3} = \Delta Pr(Y = 3) = \left[F(\tau_3 - X_c\beta) - F(\tau_2 - X_c\beta) \right] - \left[F(\tau_3 - X_b\beta) - F(\tau_2 - X_b\beta) \right]$$
$$\Delta P_{i4} = \Delta Pr(Y = 4) = \left[1 - F(\tau_3 - X_c\beta) \right] - \left[1 - F(\tau_3 - X_b\beta) \right]. \qquad (9.16)$$

Rather than calculate the absolute change in each of the choice probabilities, it's often preferable, for the reasons that we've discussed previously, to calculate the percentage change.

Scholars with hypotheses about how the effect of X_1 varies with the value of X_2 might wish to calculate the (percentage) change in each of the choice probabilities associated with the specified increase in X_1 across the observed range of values for X_2 while keeping the other independent variables constant at their baseline values. One way to present this type of information is in the form of a "difference" plot. In our particular example, where we have four presidential approval categories, this would mean producing four separate "difference" plots.

9.2.4 Odds Ratios

If we use an ordered logit model, we can also present the effects of the independent variables in terms of odds ratios (Long, 1997, 138–140). Note that the *cumulative* probability of observing a choice alternative less than or equal to k is

$$Pr(Y \leq k) = \sum_{j=1}^{k} Pr(Y = j).$$
(9.17)

It follows that the odds of observing a choice alternative less than or equal to k versus a choice alternative greater than k is

$$\Omega_{\leq k} = \frac{Pr(Y \leq k)}{1 - Pr(Y \leq k)} = \frac{Pr(Y \leq k)}{Pr(Y > k)}.$$
(9.18)

In our presidential approval example, for instance, we could use Eq. 9.18 to calculate, say, the odds of observing choice categories 1, 2, or 3 ($k \leq 3$) versus choice category 4. In an ordered logit model, the odds of observing a choice alternative less than or equal to k versus a choice alternative greater than k is

$$\Omega_{\leq k} = \frac{Pr(Y \leq k)}{Pr(Y > k)} = e^{\tau_k - X\beta}.$$
(9.19)

Suppose that we're interested in evaluating the effect of a one-unit increase in an independent variable like X_3 in Eq. 9.6 that's included *additively* in an ordered logit model. One way to evaluate this effect is to calculate an odds ratio where we compare the odds of observing a choice alternative less than or equal to k when the value of X_3 is one unit higher than some baseline value ($X_3 + 1$) with the odds when the value of X_3 is at its baseline value (X_3), while holding the values of the other independent variables constant:

$$\frac{\Omega_{\leq k} \mid X_3 + 1}{\Omega_{\leq k} \mid X_3} = e^{-\beta_4}.$$
(9.20)

Recall that β_4 is the coefficient on X_3 and captures the marginal effect of X_3 on Y^*. From this, we see that a one-unit increase in X_3 leads to a multiplicative increase in the odds of $Y \leq k$ of $e^{-\beta_4}$. As an example, if, say, $\beta_4 = 0.2$, then the odds of observing a choice category less than or equal to k versus a choice category greater than k changes by a factor of $e^{-0.2} = 0.82$. Since the odds ratio is less than 1, we see that a one-unit increase in X_3 is associated with a reduction in the odds of $Y \leq k$. We obviously don't need to focus on the effect of a one-unit increase in X_3. If we increase X_3 by some amount δ, then the odds of $Y \leq k$ changes by a factor of $e^{-\beta_4 \times \delta}$. Thus, if we increase X_3 by two units, then the odds of observing a choice category less than or equal to k changes by a factor of $e^{-0.2 \times 2} = e^{-0.4} = 0.67$. We can also calculate the percentage change in the odds of $Y \leq k$ as $\left[e^{-\beta_4 \times \delta} - 1 \right] \times 100$. Thus, a two-unit increase in X_3 is associated with a $\left[e^{-0.4} - 1 \right] \times 100 = 33\%$ reduction in the odds of observing $Y \leq k$.

Note that the odds of observing a choice category greater than k is just the inverse of the odds of observing a choice category less than or equal to k. As a result, the factor change in the odds of $Y > k$ associated with a δ increase in X_3 is just the inverse of the factor change in the odds of $Y \leq k$ associated with the same δ increase; that is, $\frac{1}{e^{-\beta_4 \times \delta}} = e^{\beta_4 \times \delta}$. Thus, if we increase X_3 by two units, then the odds of observing a choice category greater than k changes by a factor of $e^{0.2 \times 2} = e^{0.4} = 1.49$. Put differently, a two-unit increase in X_3 is associated with a 49% increase in the odds of observing $Y > k$.

As with the binary choice models we examined in Chapter 8, Eq. 9.20 indicates that an increase in a variable like X_3 leads to a constant factor change in the odds of $Y \leq k$ that doesn't depend on the values of any of the independent variables, including its own, or the value of the threshold τ_k. The fact that it doesn't depend on the value of the threshold τ_k is important because this implies that the multiplicative change in the odds of $Y \leq k$ (or the odds of $Y > k$) will be the same for any choice category k. In the context of our presidential approval example, this means that a two-unit increase in X_3 reduces the odds of an approval score of 1 versus a combined approval score of 2, 3, or 4 by 33%, holding the values of all of the other variables constant. The same increase in X_3 reduces the odds of an approval score of 1 or 2 versus an approval score of 3 or 4 by 33% as well. And it also reduces the odds of an approval score of 1, 2, or 3 versus an approval score of 4 by 33%. This property is known as the *proportional odds* or *parallel regression* assumption.

Suppose now that we're interested in evaluating the effect of an increase in an independent variable like X_1 in Eq. 9.6 that's included *interactively* in an ordered logit model. Everything remains the same except that the odds ratio where we compare the odds of observing a choice alternative less than or equal to k when the value of X_1 is one unit higher than some baseline value $(X_1 + 1)$ with the odds when the value of X_1 is at its baseline value (X_1) is now

$$\frac{\Omega_{\leq k} | X_1 + 1}{\Omega_{\leq k} | X_1} = e^{-(\beta_1 + \beta_3 X_2)}, \tag{9.21}$$

where $\beta_1 + \beta_3 X_2$ is the marginal effect of X_1 on Y^*. From this, we see that the effect of a one-unit increase in X_1 on the odds of $Y \leq k$ depends on the value of X_2. As an example, if, say, $\beta_1 = 0.4$ and $\beta_3 = -0.2$, then the odds of observing a choice category less than or equal to k versus a choice category greater than k changes by a factor of $e^{-(0.4 - 0.2 \times 0)} = e^{-0.4} = 0.67$ when $X_2 = 0$; it changes by a factor of $e^{-(0.4 - 0.2 \times 1)} = e^{-0.2} = 0.82$ when $X_2 = 1$; it changes by a factor of $e^{-(0.4 - 0.2 \times 2)} = e^0 = 1$ when $X_2 = 2$; it changes by a factor of $e^{-(0.4 - 0.2 \times 3)} = e^{0.2} = 1.22$ when $X_2 = 3$, and so on. If we increase X_1 by some amount δ, then the odds of $Y \leq k$ changes by a factor of $e^{-(\beta_1 + \beta_3 X_2) \times \delta}$. And the percentage change in the odds of $Y \leq k$ is calculated as $\left[e^{-(\beta_1 + \beta_3 X_2) \times \delta} - 1 \right] \times 100$. If we're interested in the odds of observing a choice category greater than k, then the factor change in the odds of $Y > k$ associated with an increase of δ in X_1 is $\frac{1}{e^{-(\beta_1 + \beta_3 X_2) \times \delta}} = e^{(\beta_1 + \beta_3 X_2) \times \delta}$. Scholars with hypotheses about how the effect of X_1 varies with the value of X_2 might wish to plot the factor or percentage change in the odds of $Y \leq k$ (or the odds of $Y > k$) across the observed range of values for X_2.

9.2.5 Interaction Effects

We're generally interested in two interaction effects when we have an interactive ordered logit or probit model. The first is the interaction effect between X_1 and X_2 with respect to the propensity to approve of the president Y^*. The second is the interaction effect between X_1 and X_2 with respect to the probability of each of the choice alternatives $Pr(Y = j)$. Scholars should think very carefully about which type of interaction effect is most relevant for testing their theoretical predictions. In what follows, we look at interaction effects in the context of the same underlying model as we've examined previously,

$$Y^* = X\beta + \epsilon$$
$$= \beta_0 + \beta_1 X_1 + \beta_2 X_2 + \beta_3 X_1 X_2 + \beta_4 X_3 + \epsilon. \tag{9.22}$$

We'll continue to assume that we estimate an ordered logit model, but the insights from our discussion apply equally well to the ordered probit model.

9.2.5.1 Interaction Effect on Y^*

Suppose that our theory makes a prediction about how X_2 modifies the effect of X_1 on the propensity to approve of the president. Such a prediction speaks to the interaction effect between X_1 and X_2 on Y^*. The marginal effect of X_1 on the propensity to approve of the president is

$$\frac{\partial Y^*}{\partial X_1} = \beta_1 + \beta_3 X_2. \tag{9.23}$$

The interaction effect between X_1 and X_2 is just the slope of the line that describes the relationship between this marginal effect and X_2,

$$\frac{\partial \left(\frac{\partial Y^*}{\partial X_1}\right)}{\partial X_2} = \frac{\partial \left(\beta_1 + \beta_3 X_2\right)}{\partial X_2} = \beta_3. \tag{9.24}$$

This interaction effect tells us how a one-unit increase in X_2 modifies the effect of a one-unit increase in X_1 on the propensity to approve of the president. Due to the symmetry of interaction, it also tells us how a one-unit increase in X_1 modifies the effect of a one-unit increase in X_2 on the propensity to approve of the president. As we can see, the sign, magnitude, and statistical significance of the interaction effect between X_1 and X_2 on the propensity to approve of the president can be identified simply by looking at the coefficient on the interaction term β_3. The interaction effect with respect to Y^* doesn't depend on the value of any of the independent variables and is therefore unconditional.

9.2.5.2 Interaction Effect on $Pr(Y = j)$

Suppose that our theory makes a prediction about how X_2 modifies the effect of X_1 on the probability of selecting one or more of the choice categories. Such a prediction speaks to the interaction effect between X_1 and X_2 on $Pr(Y = j)$. We can think about this type of interaction effect in terms of marginal effects or in terms of "differences in differences."

We'll start with marginal effects. Based on Eq. 9.11, the marginal effect of X_1 on the probability of each choice alternative is

$$\frac{\partial Pr(Y = j)}{\partial X_1} = \frac{\partial \left(\Lambda(\tau_j - X\beta) - \Lambda(\tau_{j-1} - X\beta)\right)}{\partial X_1}$$

$$= \left[\lambda(\tau_{j-1} - X\beta) - \lambda(\tau_j - X\beta)\right] \cdot (\beta_1 + \beta_3 X_2). \tag{9.25}$$

If we let $\Lambda_j = \Lambda(\tau_j - X\beta)$, $\Lambda_{j-1} = \Lambda(\tau_{j-1} - X\beta)$, $\lambda_j = \lambda(\tau_j - X\beta)$ and $\lambda_{j-1} = \lambda(\tau_{j-1} - X\beta)$, then we can rewrite Eq. 9.25 as

$$\frac{\partial Pr(Y = j)}{\partial X_1} = \frac{\partial \left(\Lambda_j - \Lambda_{j-1}\right)}{\partial X_1}$$

$$= \left[\lambda_{j-1} - \lambda_j\right] \cdot (\beta_1 + \beta_3 X_2). \tag{9.26}$$

This, in turn, can be rewritten as

$$\frac{\partial Pr(Y=j)}{\partial X_1} = \left[\lambda_{j-1} - \lambda_j\right] \cdot (\beta_1 + \beta_3 X_2)$$

$$= \lambda_{j-1} \cdot (\beta_1 + \beta_3 X_2) - \lambda_j \cdot (\beta_1 + \beta_3 X_2)$$

$$= \Lambda_{j-1}\left[1 - \Lambda_{j-1}\right] \cdot (\beta_1 + \beta_3 X_2) - \Lambda_j\left[1 - \Lambda_j\right] \cdot (\beta_1 + \beta_3 X_2)$$

$$= \Lambda_{j-1} \cdot (\beta_1 + \beta_3 X_2) - \Lambda_{j-1}^2 \cdot (\beta_1 + \beta_3 X_2) - \Lambda_j \cdot (\beta_1 + \beta_3 X_2)$$

$$+ \Lambda_j^2 \cdot (\beta_1 + \beta_3 X_2). \tag{9.27}$$

To calculate the interaction effect between X_1 and X_2 on $Pr(Y=j)$, we just take the derivative of Eq. 9.27 with respect to X_2,

$$\frac{\partial\left(\frac{\partial Pr(Y=j)}{\partial X_1}\right)}{\partial X_2} = \frac{\partial\left[\Lambda_{j-1} \cdot (\beta_1 + \beta_3 X_2) - \Lambda_{j-1}^2 \cdot (\beta_1 + \beta_3 X_2) - \Lambda_j \cdot (\beta_1 + \beta_3 X_2) + \Lambda_j^2 \cdot (\beta_1 + \beta_3 X_2)\right]}{\partial X_2}$$

$$= \left[\Lambda_{j-1}\left(1 - \Lambda_{j-1}\right) - \Lambda_j\left(1 - \Lambda_j\right)\right] \cdot \beta_3$$

$$+ \left[\Lambda_{j-1}\left(1 - \Lambda_{j-1}\right) \cdot \left(1 - 2\Lambda_{j-1}\right) - \Lambda_j\left(1 - \Lambda_j\right) \cdot \left(1 - 2\Lambda_j\right)\right] \cdot$$

$$(\beta_1 + \beta_3 X_2) \cdot (-\beta_2 - \beta_3 X_1). \tag{9.28}$$

Due to the symmetry of interactions, this interaction effect tells us both how X_2 modifies the marginal effect of X_1 on $Pr(Y=j)$ and how X_1 modifies the marginal effect of X_2 on $Pr(Y=j)$.

The interaction effect in Eq. 9.28 has the same basic form as the interaction effect we saw earlier in Eq. 8.72 when we were dealing with a binary logit model. It also has the same basic characteristics. We immediately see that there's no single interaction effect between X_1 and X_2 when it comes to $Pr(Y=j)$. The interaction effect is conditional and depends not only on the choice category but also on the values of all of the independent variables in the model. The interaction effect can be non-zero even if the coefficient on the interaction term β_3 is 0. It's not possible to infer whether there's a statistically significant interaction effect between X_1 and X_2 on $Pr(Y=j)$ by conducting a z-test on the interaction term coefficient. The sign of the interaction effect varies depending on the choice category and the values of the independent variables. In sum, we can't typically draw valid inferences about the magnitude, sign, and statistical significance of the interaction effect between X_1 and X_2 on $Pr(Y=j)$ from the magnitude, sign, and statistical significance of the interaction term coefficient.

⚠ **Important:** In an ordered choice model, we cannot typically draw valid inferences about the magnitude, sign, and statistical significance of the interaction effect between X_1 and X_2 on $Pr(Y=j)$ from the magnitude, sign, and statistical significance of the interaction term coefficient.

We offer the same advice when it comes to interaction effects in ordered choice models as we did when we discussed interaction effects in binary choice models. We suspect that most theories in the social sciences are only strong enough to make predictions about the existence and the sign of the interaction effect between X_1 and X_2 on Y^* and not about the existence and sign of the interaction effects between X_1 and X_2 on $Pr(Y = j)$. To the extent that this is the case, we recommend a two-step process when it comes to evaluating interaction in an interactive ordered choice model. In the first step, we should examine whether we observe the effects predicted by our theory with respect to Y^*. In the second step, we should examine whether the magnitudes of these effects are substantively meaningful by looking at their impact on $Pr(Y = j)$.

So far, we've looked at interaction effects with respect to $Pr(Y = j)$ in terms of marginal effects. However, the same basic insights apply if we think about them in terms of "differences-in-differences." Suppose that we're interested in knowing how a discrete change in X_2 modifies the effect of a discrete change in X_1 on $Pr(Y = j)$ for a particular scenario. To identify this interaction effect, we must calculate three "differences." The first difference captures the *change* in $Pr(Y = j)$ associated with a discrete increase in X_1 from some baseline value X_{1b} to some counterfactual value X_{1c} when X_2 and all of the other independent variables are set at their own baseline values,

$$\Delta Pr(Y=j)_{X_{2b}} = \left[\Lambda\left(\tau_j - \beta_0 - \beta_1 X_{1c} - \beta_2 X_{2b} - \beta_3 X_{1c} X_{2b} - \beta_4 X_{3b}\right) \right.$$
$$- \Lambda\left(\tau_{j-1} - \beta_0 - \beta_1 X_{1c} - \beta_2 X_{2b} - \beta_3 X_{1c} X_{2b} - \beta_4 X_{3b}\right) \bigg]$$
$$- \left[\Lambda\left(\tau_j - \beta_0 - \beta_1 X_{1b} - \beta_2 X_{2b} - \beta_3 X_{1b} X_{2b} - \beta_4 X_{3b}\right) \right.$$
$$- \Lambda\left(\tau_{j-1} - \beta_0 - \beta_1 X_{1b} - \beta_2 X_{2b} - \beta_3 X_{1b} X_{2b} - \beta_4 X_{3b}\right) \bigg].$$
(9.29)

The second difference captures the *change* in $Pr(Y = j)$ associated with the same discrete increase in X_1 when X_2 is at its counterfactual value X_{2c} and all of the other independent variables remain at their own baseline values,

$$\Delta Pr(Y=j)_{X_{2c}} = \left[\Lambda\left(\tau_j - \beta_0 - \beta_1 X_{1c} - \beta_2 X_{2c} - \beta_3 X_{1c} X_{2c} - \beta_4 X_{3b}\right) \right.$$
$$- \Lambda\left(\tau_{j-1} - \beta_0 - \beta_1 X_{1c} - \beta_2 X_{2c} - \beta_3 X_{1c} X_{2c} - \beta_4 X_{3b}\right) \bigg]$$
$$- \left[\Lambda\left(\tau_j - \beta_0 - \beta_1 X_{1b} - \beta_2 X_{2c} - \beta_3 X_{1b} X_{2c} - \beta_4 X_{3b}\right) \right.$$
$$- \Lambda\left(\tau_{j-1} - \beta_0 - \beta_1 X_{1b} - \beta_2 X_{2c} - \beta_3 X_{1b} X_{2c} - \beta_4 X_{3b}\right) \bigg].$$
(9.30)

The third difference, which is the interaction effect between X_1 and X_2, is the difference in these two differences,

$$Interaction\ Effect_{X_1 X_2} = \Delta Pr(Y = j)_{X_{2c}} - \Delta Pr(Y = j)_{X_{2b}}. \qquad (9.31)$$

9.3 SUBSTANTIVE APPLICATION: IDEOLOGY, RACE, AND PRESIDENTIAL APPROVAL OF BARACK OBAMA

How do ideology and race affect presidential approval in the United States? Here, we modify a substantive application that we first saw in Chapter 5 in which we examined how an individual's ideology and race interacted to influence their support for Barack Obama at the time of the 2012 presidential election in the United States. In the original application, we operationalized support for Obama in terms of a continuous feeling thermometer survey question in which respondents were asked how they feel about Obama on a 0–100 scale. In our modified application, we operationalize support for Obama, *Obama Approval*, in terms of a survey question about presidential approval that's measured on an ordered and discrete 1–4 scale, where 1 indicates strong disapproval (SD), 2 indicates disapproval (D), 3 indicates approval (A), and 4 indicates strong approval (SA).

Our theoretical story remains largely the same. Spatial theories of politics predict that individuals will exhibit more support for candidates who share their ideology than those who don't (Downs, 1957; Black, 1958). Given that Obama was a liberal, we'd expect conservatives to approve of Obama less than liberals. In terms of race, theories of descriptive representation predict that Blacks will approve of Obama more than Whites because of their shared racial background (Pitkin, 1967; Phillips, 2014; Parker, 2016; Tillery, 2019). There are reasons to think that ideology and race will interact to influence how much someone approves of Obama. Given the low level of descriptive representation for Blacks in the United States, Black voters are expected to trade off ideological considerations for higher levels of descriptive representation. This implies that ideology is likely to weigh less for Blacks than Whites when it comes to evaluating their approval of Obama. Put differently, we expect the magnitude of the negative effect of increased conservatism on the part of the respondent to be smaller among Blacks than Whites.

Conservative Ideology Hypothesis: An increase in conservative ideology always reduces the level of approval for Obama. The magnitude of this negative impact will be smaller among Blacks than Whites.

Our theory also implies that the positive gap in the approval of Obama between Blacks and Whites is expected to grow with increased conservative ideology. This is because conservative Blacks will "punish" Obama less for his liberal ideology than conservative Whites.

Black Hypothesis: Blacks always exhibit greater approval for Obama than Whites. The magnitude of this positive effect will grow with the extent to which someone holds a conservative ideology.

Note that our theory and hypotheses refer to the level of approval given to Obama. Theoretically, respondents could give a precise indicator of their level of approval. In our upcoming empirical analysis, respondents are only given the opportunity to indicate their level of presidential approval on an ordered and discrete 1–4 scale. The important point to recognize here is that this is an operationalization issue and not a theoretical one. In other words, it would be wrong to discuss our theory and state our hypotheses in terms of how ideology and race affect the probability of choosing some ordered category of presidential approval.[4] This speaks directly to the quantities of interest that we should examine in our empirical analyses. When evaluating our hypotheses, we should examine how ideology and race interact to affect the underlying propensity to approve of Obama, *Obama Approval**.[5] Of course, there's no meaningful scale for the propensity to approve of Obama. Thus, to evaluate whether and when the results with respect to ideology and race are substantively meaningful, we'll also want to examine their effects on the probability of different approval categories.

In passing, we'll note that our hypotheses contain all five of the key predictions we recommend for a theory positing an interactive relationship like the one presented here. For example, the *Conservative Ideology Hypothesis* predicts that an increase in conservative ideology will have a negative effect on the propensity to approve of Obama when race is at both its minimum (White) and maximum (Black) values. The *Black Hypothesis* predicts that Blacks will exhibit higher approval of Obama than Whites when conservative ideology is at both its minimum and maximum values. Implicit in both hypotheses is the prediction that there'll be a positive interaction between conservative ideology and race.

We test our hypotheses using data from the 2016 release of the American National Election Studies 2012 Time Series Study (American

---

[4] Obviously, we want the operationalization process to produce empirical measures that match our theoretical concepts as closely as possible. Our argument here is simply that our theoretical concepts should drive our choice of empirical measures. Our choice of empirical measure shouldn't drive how we talk about our theory.

[5] Note that our theory isn't sufficiently precise given the complicated ways that variable-specific and compression-based interaction combine to make clear predictions about how ideology and race interact to affect the probability of each of the approval categories under all scenarios anyway.

National Election Studies, 2014). We've already described our dependent variable *Obama Approval*. In terms of our key independent variables, *Liberal-Conservative* is measured on a 1–7 scale, where 1 is extremely liberal, 4 is moderate or middle of the road, and 7 is extremely conservative. *Black* is a dichotomous variable that equals 1 if an individual self-identifies as Black and 0 if they self-identify as White. In terms of our control variables, *Worse Off* measures whether an individual self-identifies as worse off than a year ago on a 1–5 scale, where 1 indicates that they're much better off and 5 indicates that they're much worse off. *Female* is a dichotomous independent variable that equals 1 if the respondent self-identifies as female and 0 if they self-identify as male. *Age* captures the respondent's age in years. To test our hypotheses, we estimate an ordered logit model where the propensity to approve of President Obama, *Obama Approval**, is modeled as

$$Obama\ Approval^* = \beta_0 + \beta_1 Liberal\text{-}Conservative + \beta_2 Black$$
$$+ \beta_3 Liberal\text{-}Conservative \times Black$$
$$+ \beta_4 Worse\ Off + \beta_5 Female + \beta_6 Age + \epsilon.$$
$$(9.32)$$

The marginal effect of conservative ideology on the propensity to approve of Obama is

$$\frac{\partial Obama\ Approval^*}{\partial Liberal\text{-}Conservative} = \beta_1 + \beta_3 Black. \qquad (9.33)$$

According to our *Conservative Ideology Hypothesis*, an increase in conservative ideology on the part of the respondent should always reduce the propensity to approve of Obama. As a result, β_1 (for Whites) and $\beta_1 + \beta_3$ (for Blacks) should both be negative. We expect the negative effect of conservative ideology to be smaller for Blacks, and so β_3 should be positive.

The effect of being Black as opposed to White on the propensity to approve of Obama can be calculated as

$$\frac{\partial Obama\ Approval^*}{\partial Black} = \beta_2 + \beta_3 Liberal\text{-}Conservative. \qquad (9.34)$$

We see that β_2 indicates the effect of being Black when *Liberal-Conservative* is 0. Note that *Liberal-Conservative* is never 0 because it's measured on a 1–7 scale. According to our *Black Hypothesis*, Blacks should always exhibit greater approval of Obama than Whites. As a result, $\beta_2 + \beta_3 Liberal\text{-}Conservative$ should be positive for all observed values of *Liberal-Conservative*. Given this, we have no prediction about the sign of β_2; it could be positive, zero, or negative.[6] We expect the positive effect of being Black to grow with the conservative ideology of the respondent, and so β_3 should be positive.

The results from our ordered logit model are shown in Table 9.1. The first column shows the estimated coefficients, while the second column shows the exponentiated coefficients or odds ratios. We'll focus first on the column of coefficients. As predicted, the coefficient on *Liberal-Conservative* is negative and statistically significant, indicating that an increase in conservative ideology on the part of the respondent reduces the propensity to approve of Obama among Whites. To be specific, a one-unit increase in *Liberal-Conservative* is associated with a -1.01 $[-1.07, -0.95]$ unit reduction in the propensity to approve of Obama among Whites. Confidence intervals of 95% are shown in parentheses. From this negative effect, we can infer that an increase in *Liberal-Conservative* increases the probability that a White respondent chooses the lowest category of approval for Obama and reduces the probability that they choose the highest category. We can't infer how it will affect the probability that they choose one of the intermediate categories of approval. Nor can we infer the magnitude of these effects. Also as predicted, the coefficient on the interaction term *Liberal-Conservative×Black* is positive and statistically significant, indicating that this negative effect is smaller among Blacks. To be specific, a one-unit increase in *Liberal-Conservative* is associated with a $-1.01 + 0.78 = -0.22$ $[-0.35, -0.09]$ unit reduction in the propensity to approve of Obama among Blacks. From this, we can infer that an increase in *Liberal-Conservative* among Blacks is associated with a smaller increase in the probability that they choose the lowest category of approval for Obama than Whites and a smaller decrease in the probability that they choose the highest category. These results with respect to the propensity to approve of Obama, which are shown graphically in panel (a) of Figure 9.4, are consistent with our previous analysis in Chapter 5 and suggest that Blacks place significantly less weight on ideological concerns than Whites when evaluating Obama.

The coefficient on *Black* tells us the effect of being Black instead of White on the propensity to approve of Obama when *Liberal-Conservative* is 0. Not too much should be read into this particular coefficient because *Liberal-Conservative* is never 0; recall that the scale for *Liberal-Conservative* runs from 1 to 7. In panel (b) of Figure 9.4, we show the effect of being Black instead of White on the propensity to approve of Obama across the *observed* range of values for *Liberal-Conservative*.

---

[6] We can predict, though, from the fact that $\beta_2 + \beta_3 Liberal\text{-}Conservative$ should be positive once *Liberal-Conservative* is 1 or higher, that $\beta_3 > -\beta_2$.

Table 9.1 Ideology, race, and presidential approval for Barack Obama in 2012

	Dependent Variable: *Obama Approval, 1−4*	
	Coefficients	**Odds Ratios**
Liberal-Conservative	−1.01***	0.37***
	(0.03)	(0.01)
Black	−0.44	0.64
	(0.31)	(0.20)
Liberal-Conservative×Black	0.78***	2.19***
	(0.07)	(0.16)
Controls		
Worse Off	−0.53***	0.59***
	(0.03)	(0.02)
Female	0.24***	1.27***
	(0.07)	(0.09)
Age	0.01***	1.01***
	(0.002)	(0.002)
Thresholds		
Disapprove Strongly/Disapprove	-5.78	
	(0.19)	
Disapprove/Approve	-4.93	
	(0.18)	
Approve/Strongly Approve	-3.44	
	(0.17)	
Observations	4,000	4,000
Log Likelihood	−3,669.03	−3,669.03

Standard errors in parentheses. $^*p < 0.10$; $^{**}p < 0.05$; $^{***}p < 0.01$ (two-tailed)

The plot includes two histograms showing the percentage of Black (dark gray) and White (light gray) respondents at the different possible values of *Liberal-Conservative*. The effect of being Black is always positive and is statistically significant when *Liberal-Conservative* is greater than 1.1; 96.6% of respondents fall in the region of statistical significance. From this positive effect, we can infer that Black respondents almost always

(a) The Conditional Effect of *Liberal-Conservative*

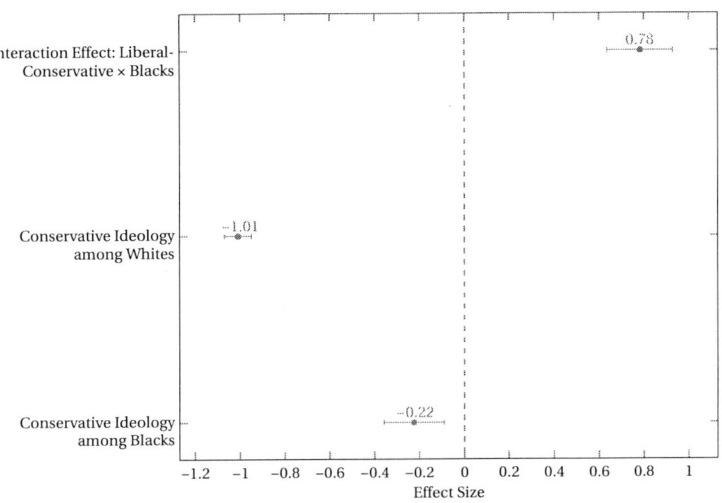

(b) The Conditional Effect of *Black*

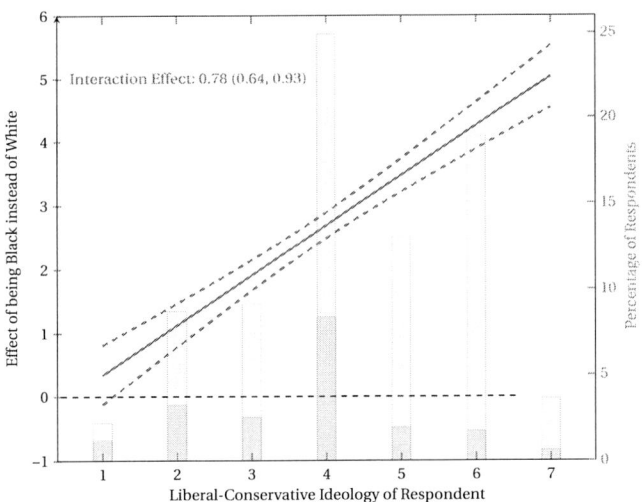

Figure 9.4 The conditional effects of *Liberal-Conservative* and *Black* on *Obama Approval**

Note: The plot in panel (a) shows the effects of a one-unit increase in the conservative ideology of the respondent among Whites and Blacks along with 95% confidence intervals. It also shows the interaction effect between conservative ideology and race. The plot in panel (b) shows the effect of being Black for individuals who vary in their level of conservative ideology along with 95% confidence intervals. The histograms show the percentage of Whites (light gray) and Blacks (dark gray) at different levels of conservative ideology. The two plots are based on the results shown in the first column of Table 9.1.

have a significantly lower probability of choosing the lowest category of approval for Obama than Whites and a higher probability of choosing the highest category, no matter what their level of conservative ideology. As expected, the positive gap between Blacks and Whites grows with the level of conservative ideology. This is indicated by the positive and statistically significant coefficient on the interaction term *Liberal-Conservative* × *Black*. Overall, these results, which are consistent with our previous analysis in Chapter 5, provide strong support for our *Black Hypothesis*.

In terms of the control variables, we see that respondents who felt that they were worse off today compared to a year ago were significantly less likely to approve of Obama. Female respondents and older respondents were significantly more likely to approve of Obama. Statistical tests indicate that all of the pairs of neighboring threshold parameters are significantly different and, thus, that it is reasonable to treat our four approval categories as different.

The information we've provided speaks to the directional nature of our theoretical claims. What we haven't examined so far is whether the magnitude of the conditional effects of conservative ideology and race are substantively meaningful. This is because there's no meaningful scale for *Obama Approval**. To examine whether our results are substantively significant, we need to calculate effects in terms of quantities of interest that have a meaningful scale.

Suppose that we're interested in the substantive effect of conservative ideology. As we've seen, there are different ways of evaluating this. One way is to look at how an increase in *Liberal-Conservative* changes the probability that White and Black respondents choose each of our categories of presidential approval. In Figure 9.5, we show how the marginal effect of *Liberal-Conservative* changes the probability of each of the approval categories for Obama among Whites and Blacks in the scenario where the respondent is a slightly liberal 45 year old woman who is doing about the same financially this year as last year. As expected, increasing the respondent's level of conservatism reduces the probability that our hypothetical respondent "strongly approves" of Obama by 0.24 [−0.25, −0.22] if they're White and by 0.03 [−0.05, −0.02] if they're Black. The modifying effect of race with respect to this particular approval category is substantively large. To be specific, the reduction in the probability of "strongly approving" of Obama is 7.56 [4.31, 15.17] times larger for our hypothetical respondent if they're White than if they're Black. Also as expected, increasing the respondent's level of conservatism increases the probability of "strongly disapproving" of Obama by 0.12 [0.11, 0.13] among Whites and by 0.005 [0.002, 0.01] among Blacks. We again see that ideology matters more for Whites than Blacks. The increase in the

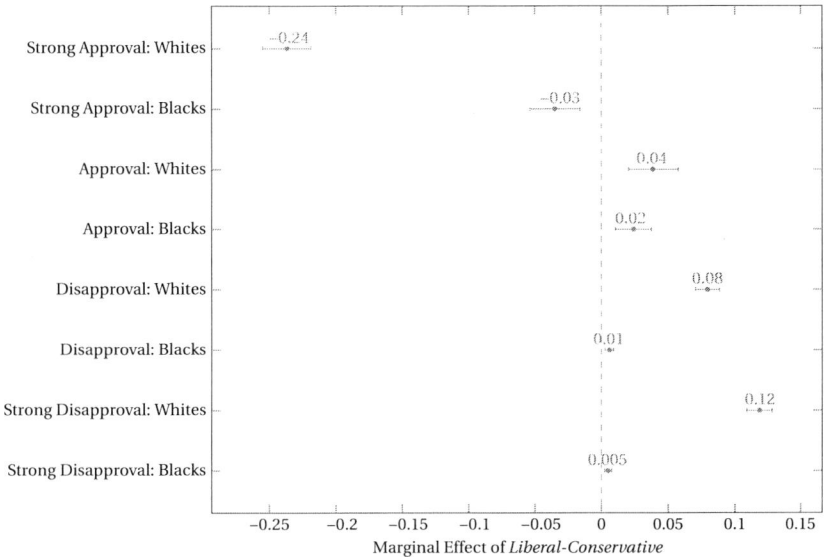

Figure 9.5 The marginal effect of *Liberal-Conservative* on the probability of different approval categories for Obama among Whites and Blacks

Note: The marginal effects and confidence intervals are calculated for a slightly liberal (*Liberal-Conservative* = 3) 45 year old woman who's doing about the same financially as in the previous year (*Worse Off* = 3).

probability of "strongly disapproving" of Obama is 26.40 [15.81, 50.37] times larger for our hypothetical respondent if they're White than if they're Black. What we can now see from Figure 9.5 that we couldn't see before is the effect of increasing the conservatism of someone's ideology on the probability of choosing the intermediate approval categories. In particular, we now see that an increase in the respondent's level of conservatism increases the probabilities that they'll choose to "approve" and "disapprove" of Obama. This is true irrespective of whether they're White or Black, although the effects are larger if they're White. The increase in the probability of "approving" of Obama is 1.81 [0.72, 4.17] times larger if they're White rather than Black, and the increase in the probability of "disapproving" of Obama is 14.81 [8.71, 28.34] times larger.

Rather than focus on marginal effects, we recommend that scholars calculate differences in predicted probabilities when substantively interpreting the results from a non-linear model like an ordered choice model. In Table 9.2, we present information about the effect of a one-unit increase in *Liberal-Conservative* on the probability of each of the approval categories for Obama among Whites and Blacks in the same scenario as before where the respondent is a slightly liberal 45 year old woman who's doing about the same financially this year as last year. We believe that a tabular format allows us to provide more substantive information in this context than a graphical format. For each approval category, the first two rows report the predicted probabilities in the baseline and counterfactual scenarios if our respondent is White and if they're Black. The "Difference" column for these two rows reports the effect of being Black instead of White in our two scenarios. The next two rows in each approval category report "effects." To be specific, the third and fourth rows report the change and percentage change in the predicted probability of observing the specified approval category associated with the one-unit increase in *Liberal-Conservative*. The "Difference" column associated with these last two rows reports "interaction effects." Specifically, it tells us how the change and percentage change in the predicted probability of observing the specified approval category change if the respondent is Black instead of White. Probabilities and "effects" are shown in gray, "interaction effects" are shown in bold, and two-tailed 95% confidence intervals calculated using the method of simulated coefficients are shown in brackets.

As expected, increasing our hypothetical respondent's level of conservatism reduces the probability that they strongly approve of Obama by a substantively large 51.9% [−54.3, −49.3] if they're White but by a substantively small 4.7% [−7.3, −2.1] if they're Black. In terms of the interaction effect, we see that the effect of ideology on this particular approval category is 91.0% [−85.9, −95.9] smaller for Blacks than Whites. Also as expected, increasing our hypothetical respondent's level of conservatism increases the probability that they strongly disapprove of Obama by a huge 121.4% [108.7, 134.2] if they're White but by a much smaller 24.6% [9.7, 41.9] if they're Black. This time the effect of ideology is 79.7% [−64.8, −92.2] smaller for Blacks than Whites. The results in Table 9.2, where we're dealing with a one-unit increase in *Liberal-Conservative*, are slightly different from those in Figure 9.5, where we're looking at the marginal effect of *Liberal-Conservative*. This is most noticeable when we examine how an increase in a White respondent's level of conservatism affects the probability that they "approve" of Obama. When we examined this effect in terms of marginal effects, we saw that our hypothetical White respondent increased their probability (0.04) of approving of Obama. In contrast, we see that a one-unit increase in *Liberal-Conservative* lowers their probability (−0.04) of approving of Obama. As we can see from Table 9.2, the effects of the independent variables and the interaction effects vary across the different approval categories. For example, the interaction effects range from −54.3% to 282.3%. All of these quantities would obviously also change if we altered the baseline scenario describing our hypothetical respondent.

Table 9.2 The effect of a one-unit increase in *Liberal-Conservative* on the probability of different approval categories for Obama among Whites and Blacks

	White	Black	Difference
Strongly Approve			
Baseline: *Liberal-Conservative* = 3	0.38 [0.35, 0.41]	0.80 [0.76, 0.84]	0.42 [0.38, 0.47]
Counterfactual: *Liberal-Conservative* = 4	0.18 [0.16, 0.20]	0.77 [0.73, 0.80]	0.58 [0.55, 0.62]
Change 3 → 4	−0.20 [−0.21, −0.18]	−0.04 [−0.06, −0.02]	**0.16** [0.13, 0.19]
Percentage Change 3 → 4	−51.9% [−54.3, −49.3]	−4.7% [−7.3, −2.1]	**−91.0%** [−85.9, −95.9]
Approve			
Baseline: *Liberal-Conservative* = 3	0.35 [0.33, 0.37]	0.14 [0.12, 0.17]	−0.21 [−0.24, −0.18]
Counterfactual: *Liberal-Conservative* = 4	0.31 [0.30, 0.34]	0.17 [0.15, 0.19]	−0.14 [−0.18, −0.12]
Change 3 → 4	−0.04 [−0.05, −0.02]	0.03 [0.01, 0.04]	**0.06** [0.04, 0.08]
Percentage Change 3 → 4	−10.3% [−14.3, −5.9]	17.8% [7.0, 30.6]	**282.3%** [167.4, 479.6]
Disapprove			
Baseline: *Liberal-Conservative* = 3	0.13 [0.12, 0.15]	0.03 [0.02, 0.04]	−0.10 [−0.12, −0.09]
Counterfactual: *Liberal-Conservative* = 4	0.20 [0.18, 0.22]	0.04 [0.03, 0.04]	−0.16 [−0.18, −0.15]
Change 3 → 4	0.07 [0.06, 0.08]	0.01 [0.003, 0.01]	**−0.06** [−0.07, −0.05]
Percentage Change 3 → 4	50.7% [42.6, 60.3]	23.0% [9.1, 39.2]	**−54.3%** [−20.2, −83.0]
Strongly Disapprove			
Baseline: *Liberal-Conservative* = 3	0.14 [0.12, 0.15]	0.02 [0.02, 0.03]	−0.11 [−0.13, −0.10]
Counterfactual: *Liberal-Conservative* = 4	0.30 [0.28, 0.33]	0.03 [0.02, 0.04]	−0.27 [−0.29, −0.25]
Change 3 → 4	0.17 [0.15, 0.18]]	0.01 [0.002, 0.01]	**−0.16** [−0.17, −0.15]
Percentage Change 3 → 4	121.4% [108.7, 134.2]	24.6% [9.7, 41.9]	**−79.7%** [−64.8, −92.2]

Note: The quantities reported in Table 9.2 are calculated for a slightly liberal (*Liberal-Conservative* = 3) 45 year old woman who's doing about the same financially as in the previous year (*Worse Off* = 3). Probabilities and "effects" are shown in gray. "Interaction effects" are shown in bold, and two-tailed 95% confidence intervals are shown in brackets.

What about the substantive effect of race? For illustrative purposes, we'll look at the conditional effect of race in terms of odds ratios rather than differences in predicted probabilities. Recall that we presented the results from our ordered logit model in terms of odds ratios (exponentiated coefficients) in the second column of Table 9.1. In this example, the odds ratios are calculated to give the factor change in the odds of being over a choice threshold (more approval) versus below that threshold (more disapproval). In other words, the odds ratio associated with *Black* indicates that being Black as opposed to White changes the odds of an approval category *greater* than some category k versus lower approval categories by a factor of 0.64 when *Liberal-Conservative* is 0. Put differently, being Black reduces the odds of "strong approval" versus "approval," "disapproval," or "strong disapproval" by a factor of 0.64 when *Liberal-Conservative* is 0; it reduces the odds of "strong approval" or "approval" versus "disapproval" or "strong disapproval" by a factor or 0.64 when *Liberal-Conservative* is 0; and it reduces the odds of "strong approval," "approval," or "disapproval" versus "strong disapproval" by a factor of 0.64 when *Liberal-Conservative* is 0. Again, not too much should be read into this particular odds ratio because the scale for *Liberal-Conservative* runs from 1 to 7, and so a value of 0 is never observed.

In Figure 9.6, we plot the factor change associated with being Black instead of White in the odds of an approval category greater than some category k versus a lower approval category across the observed range of values for *Liberal-Conservative*. The histograms show the percentage of Whites (light gray) and Blacks (dark gray) at different levels of conservative ideology. The reported factor change is always positive, indicating that the odds of choosing a higher approval category for Obama versus a lower one is greater for Blacks than Whites. This factor change is statistically significant when the value of *Liberal-Conservative* is greater than 1.1, a result that mirrors what we saw earlier in panel (b) of Figure 9.4. As predicted, the difference between Blacks and Whites grows with the respondent's level of conservatism. When the respondent is extremely liberal (*Liberal-Conservative* = 1), the factor change in the odds is 1.41 [0.88, 2.25] times larger for Blacks than Whites. In contrast, when the respondent is extremely conservative (*Liberal-Conservative* = 7), the factor change in the odds is 154.09 [93.42, 254.15] times larger for Blacks than Whites. The modifying effect of the respondent's level of conservatism is thus quite substantial. To be specific, the factor change in the odds is 109.5 times larger among the extremely conservative as opposed to the extremely liberal.

Overall, we see that the empirical results provide strong support for our hypotheses about how race and ideology interact to determine presidential approval for Obama.

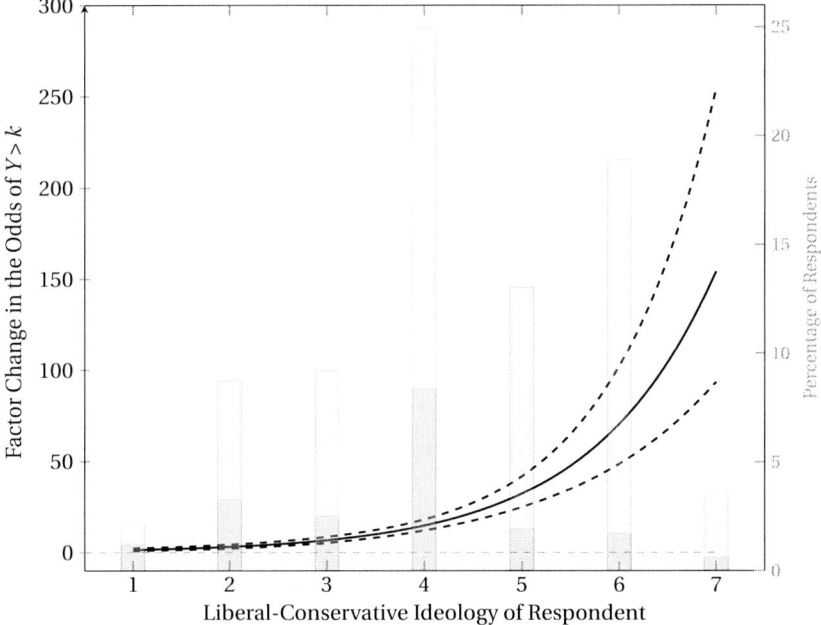

Figure 9.6 The effect of being Black on the odds of an approval category greater than some category k versus lower approval categories

Note: The plot shows the factor change associated with being Black instead of White in the odds of an approval category greater than some category k versus a lower approval category across the observed range of values for *Liberal-Conservative*. The histograms show the percentage of Whites (light gray) and Blacks (dark gray) at different levels of conservative ideology.

In this chapter, we've extended our discussion of theories positing interaction in the context of discrete choice models. Specifically, we've looked at how to evaluate conditional implications in discrete choice models where the choice alternatives are ordered. In the next chapter, we discuss how to evaluate conditional implications in discrete choice models where the choice alternatives are unordered.

9.4 EXERCISES

1. In a 2021 article in the *British Journal of Political Science*, Newman et al. (2021) examine whether explicitly racial and inflammatory speech by political elites emboldens prejudiced individuals to openly express their prejudices. Society can be divided up into citizens who

are prejudiced and those who aren't. Prejudiced citizens don't always express their prejudices openly. Newman et al. (2021) claim that citizens who aren't prejudiced won't express prejudice even in the presence of prejudiced elite speech. In contrast, they claim that prejudiced citizens will constrain the expression of their prejudice in the absence of prejudiced elite speech but feel emboldened to express it in its presence, especially if the speech is tacitly condoned by other elites. In this sense, prejudiced elite speech is expected to break "norms of equality" and allow prejudice to be openly expressed. Racial and inflammatory speech by political elites is expected to increase expressed prejudice but only among those individuals who already harbor prejudiced attitudes and beliefs.

To evaluate their theoretical claims, Newman et al. (2021) conduct a panel survey experiment. In the first wave, they use survey questions to measure, among other things, an individual's level of prejudicial attitudes and beliefs (*Prejudice*). *Prejudice* is a continuous measure that runs from 0 to 1, where higher values indicate greater levels of prejudice. The experimental component of the research design occurs in the second wave where individuals are exposed to some kind of speech by political elites. This speech is either not prejudiced (control) or is prejudiced (treatment). Individuals are then asked to rate the extent to which the discriminatory behavior that's described in a vignette is acceptable or not (*Expressed Prejudice*). *Expressed Prejudice* is, thus, the dependent variable and is measured on an ordinal 1 to 5 scale, where 1 indicates that the discriminatory behavior is completely unacceptable, 3 indicates that it's neither good nor bad, and 5 indicates that it's completely acceptable. The authors actually employ five treatment conditions, but for the sake of this exercise we're only interested in a treatment where the elite speech involves prejudiced speech by Trump on the issue of immigration reform that receives bipartisan support. We'll refer to this treatment as *Prejudiced Speech* and to the other treatments as *Treatments 2–5*.

a. Write down hypotheses that speak to the predicted effect of *Prejudice* and the predicted effect of *Prejudiced Speech*. Based on these hypotheses, what predictions can we make about the signs of some of our model coefficients? Explain.

Given the ordinal nature of their dependent variable, Newman et al. (2021) employ an ordered logit model where the propensity to express prejudice is specified in the following way:

Expressed Prejudice[*]

$$= \beta_0 + \beta_1 Prejudice$$
$$+ \beta_2 Prejudiced\ Speech + \beta_3 Prejudice \times Prejudiced\ Speech$$
$$+ \beta_4 Treatment2 + \beta_5 Prejudice \times Treatment2$$
$$+ \beta_6 Treatment3 + \beta_7 Prejudice \times Treatment3$$
$$+ \beta_8 Treatment4 + \beta_9 Prejudice \times Treatment4$$
$$+ \beta_{10} Treatment5 + \beta_{11} Prejudice \times Treatment5 + \epsilon. \tag{9.35}$$

The results from their analysis are shown in Table 9.3.

a. Based on the results in Table 9.3, what are the marginal effects of *Prejudice* and *Prejudiced Speech* on the propensity to express prejudice? To the extent possible, indicate whether the results are consistent with the theoretical predictions of Newman et al. (2021).

b. Based on the coefficients for the first three variables listed in Table 9.3, what can we say about how *Prejudice* and *Prejudiced Speech* affect the *probability* of openly expressing prejudice?

c. Calculate the marginal effects of *Prejudice* for those in the control group and those who received the *Prejudiced Speech* treatment on the probability of observing each of the five possible values of *Expressed Prejudice* in the scenario where *Prejudice* $= 0.3$. How does prejudiced elite speech modify the effect of *Prejudice* on the probability of openly expressing prejudice? Is the modifying effect of *Prejudiced Speech* always positive?

d. Calculate the change in the probability of choosing each of the five possible values of *Expressed Prejudice* associated with receiving the *Prejudiced Speech* treatment rather than the control for someone whose *Prejudice* score is 0.25 and for someone whose *Prejudice* score is 0.75. How does increasing someone's *Prejudice* score from 0.25 to 0.75 modify the effect of the *Prejudiced Speech* treatment on the probability of openly expressing prejudice?

e. Calculate the effect of a 0.5-unit increase in *Prejudice* on the odds of observing a value of 2 or less versus any higher value for *Expressed Prejudice* when *Prejudiced Speech* $= 0$ and when *Prejudiced Speech* $= 1$. Calculate the percentage change in the odds of observing a value of 3 or more versus any lower value for *Expressed Prejudice* associated with receiving the *Prejudiced Speech* treatment rather than the control when someone's *Prejudice* score is 0.25 and when it's 0.75.

Table 9.3 Prejudicial attitudes, elite speech, and expressed prejudice

Dependent Variable: *Expressed Prejudice*	
	Coefficients
Prejudice	−0.078
	(0.920)
Treatment 1: Prejudiced Speech	−0.597
	(0.412)
Prejudice × Prejudiced Speech	3.04**
	(1.21)
Treatment 2	−0.262
	(0.422)
Prejudice × Treatment 2	1.04
	(1.25)
Treatment 3	−0.609
	(0.438)
Prejudice × Treatment 3	1.64
	(1.27)
Treatment 4	−0.180
	(0.409)
Prejudice × Treatment 4	1.78
	(1.21)
Treatment 5	−0.289
	(0.415)
Prejudice × Treatment 5	1.76
	(1.20)
Thresholds	
Completely Unacceptable/Unacceptable	0.094
	(0.300)
Unacceptable/Neither Good Nor Bad	2.58
	(0.317)
Neither Good Nor Bad/Acceptable	3.79
	(0.356)
Acceptable/Completely Acceptable	4.46**
	(0.401)
Observations	997
Log Likelihood	−971.93

Standard errors in parentheses. $^*p < 0.10$; $^{**}p < 0.05$; $^{***}p < 0.01$ (two-tailed)

10 Interactions and Unordered Dependent Variables

In this chapter, we look at how to evaluate the conditional implications of a theory in the context of discrete choice models where the alternatives in the choice set aren't intrinsically ordered on some scale. Examples of unordered choices include an individual's vote choice in multi-party elections (Alvarez and Nagler, 1998) or the choice of a government out of the set of potential governments in the parliamentary government formation process (Glasgow, Golder and Golder, 2012). As with the other discrete choice models we've examined, we'll see that we can substantively interpret the results of an unordered choice model in terms of marginal effects, differences in predicted probabilities, and odds ratios.

> ⚠ **Important:** Discrete choice models are those in which a choice is made from a finite set of mutually exclusive and collectively exhaustive choice alternatives. An unordered choice model is appropriate when the available choice alternatives are not intrinsically ordered on some scale.

In an unordered choice model, the probability that a particular alternative is chosen by some actor is a function of our independent variables and estimated coefficients. Our independent variables can come in two types. The first type are referred to as *individual-specific* or *alternative-invariant* variables. These are variables that vary across our individual actors or cases but not across the choice alternatives. They typically capture characteristics of the actor making the choice, such as their age, gender, and race. The second type are commonly referred to as *alternative-specific* or *choice-specific* variables. These are variables that vary across the individual actors *and* choice alternatives; they provide information about the individual making the choice *relative* to each alternative.[1] A common example

---

[1] We recognize that the standard terminology of "alternative-specific" or "choice-specific" variables can be confusing as these variables vary across both choice

is a variable that captures the ideological distance between a voter and each of the political parties (choice alternatives) competing in an election. Unordered choice models can be estimated with only individual-specific variables, only alternative-specific variables, or both individual-specific and alternative-specific variables. When it comes to testing conditional claims, we can include interaction terms between alternative-specific variables, between individual-specific variables, or between alternative-specific and individual-specific variables.

In what follows, we look at unordered choice models in the context of multinomial logit (MNL) models. Unfortunately, the nomenclature surrounding unordered choice models can be confusing. In political science, for example, scholars sometimes distinguish between multinomial logit models, which include only individual-specific variables, conditional logit models, which include only alternative-specific variables, and "mixed logits," which include both individual-specific and alternative-specific variables. The problem is that there are other statistical models with these same names. To remove any confusion, we follow Glasgow (2022) and refer to the unordered choice model discussed in this chapter as a multinomial logit irrespective of the types of independent variables that it includes.

> ⚠ **Important:** In unordered choice models, we can have two types of independent variables. *Individual-specific* or *alternative-invariant* variables vary across our individual actors or cases but not across the choice alternatives. *Alternative-specific* or *choice-specific* variables vary across our cases and across the alternatives.

10.1   A RANDOM UTILITY APPROACH

A common way to think about unordered choice models is in a random utility framework (Train, 2009). The idea here is that there's some actor i who needs to make a choice from among J available alternatives. The actor derives a certain utility from each alternative, which depends on both a systematic and a stochastic component. We assume that the actor chooses the alternative with the highest utility. So, in this setup, the utility that actor i obtains from alternative k is

$$U_{ik} = V_{ik} + \epsilon_{ik}, \tag{10.1}$$

alternatives *and* individual decision-makers. In this sense, they're actually specific to the interaction of a choice alternative and an individual actor.

where V_{ik} indicates the systematic component of utility for actor i associated with choice alternative k, and ϵ_{ik} indicates the stochastic component of utility for actor i associated with choice k. We assume that the systematic component of the actor's utility for alternative k is determined by some linear combination of alternative-specific and individual-specific variables,

$$V_{ik} = X_{ik}\beta + Z_i\gamma_k, \tag{10.2}$$

where X_{ik} are the alternative-specific variables (subscripted by both i and k), and Z_i are the individual-specific variables (subscripted by just i). The coefficients β on the alternative-specific variables are not subscripted, meaning that we estimate just one coefficient for our alternative-specific variables. In contrast, the coefficients γ_k on the individual-specific variables are subscripted by k, meaning that individual-specific variables like gender or race can have differing effects across the choice alternatives. Plugging Eq. 10.2 into Eq. 10.1, we have

$$U_{ik} = X_{ik}\beta + Z_i\gamma_k + \epsilon_{ik}. \tag{10.3}$$

Actor i chooses alternative k if and only if the utility associated with alternative k is larger than or equal to the utility associated with all of the other available alternatives,

$$U_{ik} \geq U_{ij} \quad \forall j \neq k. \tag{10.4}$$

The probability that actor i chooses alternative k is

$$\begin{aligned} P_{ik} = Pr(Y = k) &= Pr(U_{ik} \geq U_{ij} \quad \forall j \neq k) \\ &= Pr(V_{ik} + \epsilon_{ik} \geq V_{ij} + \epsilon_{ij} \quad \forall j \neq k) \\ &= Pr(\epsilon_{ij} - \epsilon_{ik} \leq V_{ik} - V_{ij} \quad \forall j \neq k). \end{aligned} \tag{10.5}$$

To calculate this cumulative probability, we just need to choose an appropriate distribution for our error terms. If we assume that each error term for a choice alternative is independently and identically distributed extreme value,

$$f(\epsilon_{ik}) = e^{-\epsilon_{ik} - e^{-\epsilon_{ik}}}, \tag{10.6}$$

then we get the multinomial logit model.

The probability that individual i chooses alternative k in a multinomial logit model is

$$P_{ik} = Pr(Y = k) = \frac{e^{V_{ik}}}{\sum_{j=1}^{J} e^{V_{ij}}} = \frac{e^{X_{ik}\beta + Z_i\gamma_k}}{\sum_{j=1}^{J} e^{X_{ij}\beta + Z_i\gamma_j}}. \tag{10.7}$$

There's an identification issue that arises with the coefficients on the individual-specific variables. The intuition is fairly straightforward. The value of an individual-specific variable such as age is constant across the choice alternatives for a given individual. In effect, individual-specific variables add some constant to the utility of all the available choice alternatives. But adding a constant to the utility of *all* the alternatives doesn't change the alternative with the highest utility and hence the choice made by individual *i*. Only differences in utility matter for choices, not the absolute level of utility (Train, 2009, 19–20). Because only differences in utility matter, we can't estimate the effect of individual-specific variables whose values are constant across the choice alternatives. The standard solution to this is to constrain one set of the individual-specific coefficients that corresponds to a particular choice alternative to 0. This choice alternative then acts as a "baseline" or "reference" category against which we can evaluate the effect of the individual-specific variables on the other choice alternatives. With this identification constraint, the estimated coefficients γ_k tell us the effect of individual-specific variables such as age on the choice of alternative k *relative to the choice of the selected baseline category*. For example, a positive coefficient on an individual-specific variable for, say, choice alternative 2, tells us that the individual-specific variable increases the probability of choice alternative 2 relative to whatever choice alternative is specified as the baseline category. There are no similar identification constraints needed to estimate the effects of the alternative-specific variables.

To explicitly recognize the identification constraint on the individual-specific variables, we can rewrite Eq. 10.7 as

$$P_{ik} = Pr(Y = k) = \frac{e^{V_{ik}}}{\sum_{j=1}^{J} e^{V_{ij}}} = \frac{e^{X_{ik}\beta + Z_i\gamma_k}}{\sum_{j=1}^{J} e^{X_{ij}\beta + Z_i\gamma_j}} \quad \text{where } \gamma_1 = 0,$$

(10.8)

where we've arbitrarily selected choice alternative 1 to be the baseline category. With the constraint that $\gamma_1 = 0$, we can also break Eq. 10.8 into two equations,

$$P_{i1} = Pr(Y = 1) = \frac{e^{V_{i1}}}{\sum_{j=1}^{J} e^{V_{ij}}} = \frac{e^{X_{i1}\beta}}{e^{X_{i1}\beta} + \sum_{j=2}^{J} e^{X_{ij}\beta + Z_i\gamma_j}}$$

(10.9)

and

$$P_{ik} = Pr(Y = k) = \frac{e^{V_{ik}}}{\sum_{j=1}^{J} e^{V_{ij}}} = \frac{e^{X_{ik}\beta + Z_i\gamma_k}}{e^{X_{i1}\beta} + \sum_{j=2}^{J} e^{X_{ij}\beta + Z_i\gamma_j}} \quad \text{for } k > 1.$$

(10.10)

To illustrate, if we have $J = 3$ choice alternatives and alternative 1 is treated as the baseline category, then the probabilities that individual *i* chooses each of the available alternatives are

$$P_{i1} = Pr(Y = 1) = \frac{e^{X_{i1}\beta + Z_i\gamma_1}}{\sum_{j=1}^{3} e^{X_{ij}\beta + Z_i\gamma_j}} = \frac{e^{X_{i1}\beta + Z_i\gamma_1}}{e^{X_{i1}\beta + Z_i\gamma_1} + e^{X_{i2}\beta + Z_i\gamma_2} + e^{X_{i3}\beta + Z_i\gamma_3}}$$

$$= \frac{e^{X_{i1}\beta}}{e^{X_{i1}\beta} + e^{X_{i2}\beta + Z_i\gamma_2} + e^{X_{i3}\beta + Z_i\gamma_3}}, \tag{10.11}$$

$$P_{i2} = Pr(Y = 2) = \frac{e^{X_{i2}\beta + Z_i\gamma_2}}{\sum_{j=1}^{3} e^{X_{ij}\beta + Z_i\gamma_j}} = \frac{e^{X_{i2}\beta + Z_i\gamma_2}}{e^{X_{i1}\beta + Z_i\gamma_1} + e^{X_{i2}\beta + Z_i\gamma_2} + e^{X_{i3}\beta + Z_i\gamma_3}}$$

$$= \frac{e^{X_{i2}\beta + Z_i\gamma_2}}{e^{X_{i1}\beta} + e^{X_{i2}\beta + Z_i\gamma_2} + e^{X_{i3}\beta + Z_i\gamma_3}}, \tag{10.12}$$

$$P_{i3} = Pr(Y = 3) = \frac{e^{X_{i3}\beta + Z_i\gamma_3}}{\sum_{j=1}^{3} e^{X_{ij}\beta + Z_i\gamma_j}} = \frac{e^{X_{i3}\beta + Z_i\gamma_3}}{e^{X_{i1}\beta + Z_i\gamma_1} + e^{X_{i2}\beta + Z_i\gamma_2} + e^{X_{i3}\beta + Z_i\gamma_3}}$$

$$= \frac{e^{X_{i3}\beta + Z_i\gamma_3}}{e^{X_{i1}\beta} + e^{X_{i2}\beta + Z_i\gamma_2} + e^{X_{i3}\beta + Z_i\gamma_3}}. \tag{10.13}$$

10.2 DATA STRUCTURE AND INTERACTIONS

When it comes to unordered choice models, it's often informative to think about the structure of the underlying data to see what's going on. Suppose we have a choice scenario where individuals are choosing between $J = 3$ alternatives, and we have one individual-specific variable Z_i and one alternative-specific variable X_{ij} that each have an additive effect on an individual's utility. In this setup, the utility that individual i obtains from choice alternative k is

$$U_{ik} = \gamma_{0k} + \gamma_{1k}Z_i + \beta_1 X_{ik} + \epsilon_{ik}. \tag{10.14}$$

In Table 10.1, we show one way in which we might think to store data on the first two individuals and choice opportunities in our sample. There are as many rows for each individual as there are available choice alternatives. The choice alternative associated with each row is indicated in the second column. Our dichotomous dependent variable Y indicating the alternative chosen by the individual is shown in the third column. Although there are six rows in Table 10.1, there are only two "observations" as the rows provide information on just two choice opportunities. The values for our two independent variables appear in the last two columns. The value of the alternative-specific variable varies across the choice alternatives. For example, X_{ij} for individual 1 is 1 for the first alternative, 0 for the second alternative, and 7 for the third alternative.[2] In contrast, the value of the

[2] The value of the alternative-specific variable X_{ij} for a given choice alternative also varies across the individuals. For example, the value of the first alternative is 1 for individual 1 but 2 for individual 2.

Table 10.1 Incomplete data structure for a multinomial logit model				
Individual i	**Alternative j**	**Choice Y**	X_{ij}	Z_i
1	1	0	$X_{11} = 1$	$Z_1 = 1$
1	2	1	$X_{12} = 0$	$Z_1 = 1$
1	3	0	$X_{13} = 7$	$Z_1 = 1$
2	1	1	$X_{21} = 2$	$Z_2 = 4$
2	2	0	$X_{22} = 1$	$Z_2 = 4$
2	3	0	$X_{23} = 6$	$Z_2 = 4$

individual-specific variable is constant across the choice alternatives. As an example, Z_i for individual 1 is always 1 irrespective of the choice alternative.

Although the data structure shown in Table 10.1 contains all of our information, we need to make a change in order to actually estimate an unordered choice model. The problem has to do with the fact that the value of the individual-specific variable Z_i is constant across the choice alternatives. We can't estimate the effect of a variable whose values in a choice scenario never vary over the available alternatives; it's not identified. To solve this issue, we use an "interaction trick." There are two steps to the trick. In the first step, we break up the discrete variable indicating the available choice alternatives in the second column of Table 10.1 into a set of $J = 3$ logically exhaustive and mutually exclusive dichotomous variables that each correspond to one of our three choice alternatives. We'll label these dichotomous variables as A_1, A_2, and A_3 to indicate the associated choice alternative. In the second step, we multiply each of these dichotomous variables by Z_i to create $J = 3$ interaction terms Z_iA_1, Z_iA_2, and Z_iA_3. This two-step process is graphically illustrated in Table 10.2. We now have variables that capture information about Z_i and whose values vary across the choice alternatives.

When estimating our unordered choice model, we'll include $J - 1$ of the interaction terms as well as $J - 1$ of the dichotomous variables that indicate the choice alternatives. The omitted variables in each case should be associated with the same choice alternative. Thus, if we omit the interaction term Z_iA_1, then we'll also omit the dichotomous variable A_1. By omitting one of the interaction terms, we essentially constrain the coefficient on Z_i for one of the choice alternatives to be 0. This is how we satisfy the identification constraint we mentioned earlier from a practical perspective. The choice alternative associated with the omitted variables

Table 10.2 Example data structure for a multinomial logit model I

Individual i	Alternative j	Choice Y	X_{ij}	Z_i	Alternative Dummies A_1	A_2	A_3	$Z_i \times A_1$ Z_iA_1	$Z_i \times A_2$ Z_iA_2	$Z_i \times A_3$ Z_iA_3
1	1	0	$X_{11} = 1$	$Z_1 = 1$	1	0	0	1	0	0
1	2	1	$X_{12} = 0$	$Z_1 = 1$	0	1	0	0	1	0
1	3	0	$X_{13} = 7$	$Z_1 = 1$	0	0	1	0	0	1
2	1	1	$X_{21} = 2$	$Z_2 = 4$	1	0	0	4	0	0
2	2	0	$X_{22} = 1$	$Z_2 = 4$	0	1	0	0	4	0
2	3	0	$X_{23} = 6$	$Z_2 = 4$	0	0	1	0	0	4

becomes the baseline or reference category against which we can interpret the effects of the included variables. Thus, if we omit A_1 and Z_iA_1, then we're making alternative 1 our baseline category. In this case, the estimated coefficient on Z_iA_2 tells us how an increase in Z_i affects the probability of choosing alternative 2 relative to alternative 1. Similarly, the estimated coefficient on Z_iA_3 tells us how an increase in Z_i affects the probability of choosing alternative 3 relative to alternative 1.[3] We can see from this that even though the utility function in Eq. 10.14 is additive, our unordered choice model will have interaction terms in it to account for the inclusion of the individual-specific variable Z_i. The way in which variables like Z_iA_1, Z_iA_2, and Z_iA_3 are named in substantive applications often means that the fact that they're interaction terms is "hidden" from the reader. The coefficients on the dichotomous variables indicating the available choice alternatives act as alternative-specific constants γ_{0j} and tell us the average effect of all the factors that aren't included in the model on the utility of the various alternatives relative to the baseline category. As always, we should make sure to include these alternative-specific constants as they're constitutive components of the interaction terms Z_iA_2 and Z_iA_3.[4] In sum, if we make choice alternative 1 our baseline category, our unordered choice model will include Y as our dependent variable and A_2, A_3, Z_iA_2, Z_iA_3, and X_{ij} as our independent variables; Z_i, as well as A_1 and Z_iA_1, will not, indeed cannot, appear in the model.

Suppose now that our theory implies a conditional claim about how two or more variables interact to influence an individual's choice. To provide some substantive context, we'll assume that we have a vote choice scenario where voters must choose between $J = 3$ candidates. For illustrative purposes, we'll assume that we have two individual-specific variables Z_1 and Z_2 and two alternative-specific variables X_1 and X_2. If each of these variables have an *additive* effect on an individual's utility, then the utility that voter i obtains from candidate k is

$$U_{ik} = \gamma_{0k} + \gamma_{1k}Z_{1i} + \gamma_{2k}Z_{2i} + \beta_1 X_{1ik} + \beta_2 X_{2ik} + \epsilon_{ik}. \tag{10.15}$$

The data structure necessary to test this "additive" specification is illustrated for two voters in panel (a) of Table 10.3. If candidate 1 is our baseline candidate, then our "additive" model would include Y as our

---

[3] The observant among you will have realized that this "interaction trick" is really just an application of the alternative interaction model specification that we discussed earlier in Chapter 3 when one of the modifying variables is discrete.

[4] It often makes sense to include these alternative-specific constants even in choice scenarios where there are no individual-specific variables. This is because estimating a model without them is equivalent to estimating a model without a constant term.

dependent variable and A_2, A_3, $Z_1 A_2$, $Z_1 A_3$, $Z_2 A_2$, $Z_2 A_3$, X_1, and X_2 as our independent variables.

Scholars can make four types of conditional claim about variable-specific interaction in an unordered choice model. The first type involves interaction between alternative-specific variables. Suppose that one of our alternative-specific variables measures the ideological distance between the voter and each of the candidates and that the other indicates whether the voter identifies with the same racial group as the candidate. In line with some of our previous substantive applications, our theory might indicate that voters punish candidates that are ideologically distant from them but less so if they belong to the same racial group. In order to test this conditional claim, we need to create an interaction term $X_1 X_2$ by multiplying together our two alternative-specific variables X_1 and X_2. The values for this interaction term are shown in the "Type 1" column of panel (b) in Table 10.3. The utility that voter i obtains from candidate k is now

$$U_{ik} = \gamma_{0k} + \gamma_{1k} Z_{1i} + \gamma_{2k} Z_{2i} + \beta_1 X_{1ik} + \beta_2 X_{2ik} + \beta_3 X_{1ik} X_{2ik} + \epsilon_{ik}.$$
(10.16)

Continuing with candidate 1 as the baseline candidate, our unordered choice model will include Y as the dependent variable and A_2, A_3, $Z_1 A_2$, $Z_1 A_3$, $Z_2 A_2$, $Z_2 A_3$, X_1, X_2, and $X_1 X_2$ as the independent variables.

The second type of conditional claim we might make about variable-specific interaction involves interaction between individual-specific variables. Suppose that our two individual-specific variables capture the sex (male/female) and race (White/Black) of the voter. We may have an intersectional theory predicting that the effect of a voter's sex (race) on candidate choice depends on their race (sex). In order to test this conditional claim, the utility that voter i obtains from candidate k in Eq. 10.15 would now include the interaction term $Z_1 Z_2$:

$$U_{ik} = \gamma_{0k} + \gamma_{1k} Z_{1i} + \gamma_{2k} Z_{2i} + \gamma_{3k} Z_{1i} Z_{2i} + \beta_1 X_{1ik} + \beta_2 X_{2ik} + \epsilon_{ik}.$$
(10.17)

Note that we have the same identification issue with the individual-specific interaction term $Z_1 Z_2$ as we do with any individual-specific variable – its values don't vary over the candidates in a choice scenario and, as a result, we can't include $Z_1 Z_2$ in our empirical unordered choice model. Just as we did with Z_1 and Z_2 separately, we have to instead multiply $Z_1 Z_2$ by each of the three dummy variables that indicate the different candidates. By doing this we end up with three new interaction terms $Z_1 Z_2 A_1$, $Z_1 Z_2 A_2$,

Table 10.3 Example data structure for a multinomial logit model II

(a) Additive Baseline Model

Individual i	Alternative j	Choice Y	X₁	X₂	Z₁	Z₂	A₁	A₂	A₃	Z₁A₁	Z₁A₂	Z₁A₃	Z₂A₁	Z₂A₂	Z₂A₃
1	1	0	1	2	1	3	1	0	0	1	0	0	3	0	0
1	2	1	0	4	1	3	0	1	0	0	1	0	0	3	0
1	3	0	7	1	1	3	0	0	1	0	0	1	0	0	3
2	1	1	2	3	4	4	1	0	0	4	0	0	4	0	0
2	2	0	1	2	4	4	0	1	0	0	4	0	0	4	0
2	3	0	6	4	4	4	0	0	1	0	0	4	0	0	4

(b) Interaction Terms Needed to Evaluate Different Types of Conditional Claim

Individual i	Alternative j	Choice Y	Type 1	Type 2			Type 3		Type 4		
			X_1X_2	$Z_1Z_2A_1$	$Z_1Z_2A_2$	$Z_1Z_2A_3$	X_1Z_1	$X_1Z_1A_1$	$X_1Z_1A_1$	$X_1Z_1A_2$	$X_1Z_1A_3$
1	1	0	2	3	0	0	1	1	1	0	0
1	2	1	0	0	3	0	0	0	0	0	0
1	3	0	7	0	0	3	7	0	0	0	7
2	1	1	6	16	0	0	8	8	8	0	0
2	2	0	2	0	16	0	4	0	0	4	0
2	3	0	24	0	0	16	24	0	0	0	24

Note: Panel (a) shows the data structure for an unordered choice model with two alternative-specific variables X_1 and X_2 and two individual-specific variables Z_1 and Z_2 that each additively affect an individual's utility. Panel (b) shows various interaction terms that would be needed to evaluate different types of conditional claims about the effects of the independent variables. The first type of conditional claim involves an interaction between the two alternative-specific variables. The second type involves an interaction between the two individual-specific variables. The third type involves an interaction between alternative-specific variable X_1 and individual-specific variable Z_1, where the nature of the interaction is not expected to vary across the choice alternatives. The fourth type is the same as the third type except that the nature of the interaction is expected to vary across the choice alternatives.

442

and $Z_1 Z_2 A_3$.[5] The values for these interaction terms are shown in the three "Type 2" columns in panel (b) of Table 10.3. As before, we can include only $J - 1$ of these interaction terms in our model, and we interpret the effects of these interaction terms on the choice of the candidates relative to the baseline candidate. Continuing with candidate 1 as the baseline candidate, our unordered choice model will include Y as our dependent variable and A_2, A_3, $Z_1 A_2$, $Z_1 A_3$, $Z_2 A_2$, $Z_2 A_3$, $Z_1 Z_2 A_2$, $Z_1 Z_2 A_3$, X_1, and X_2 as our independent variables.

The last two types of conditional claim we might make about variable-specific interaction involve interaction between alternative-specific and individual-specific variables. They differ in terms of whether or not they predict that the nature of the interaction varies across the candidates. A large literature suggests that some policy issues, such as welfare, healthcare, and education, may be more important or salient to women than men (Alexander and Andersen, 1986; Shapiro and Mahajan, 1986; Huddy and Terkildsen, 1993; Swers, 2002). We might think of these particular issues as "women's issues." To the extent that this is true, we'd expect female voters to punish candidates who are more ideologically distant from them on women's issues than male voters. We could evaluate a conditional claim like this if one of our alternative-specific variables, say X_1, measures the ideological distance between a voter and each of the candidates on a "women's policy" issue such as welfare, and one of our individual-specific variables, say Z_1, indicates whether the voter is female as opposed to male.

Before we can proceed, we need to know whether our theory predicts that the conditional way that men and women punish candidates for being ideologically distant with respect to welfare policy varies across the candidates. If our theory predicts that the interaction between ideological distance and gender doesn't vary across the candidates, then we just need to create the interaction term $X_1 Z_1$ by multiplying together X_1 and Z_1. The values for this interaction term are shown in the "Type 3" column of panel (b) in Table 10.3. This time the utility that voter i obtains from candidate k is

$$U_{ik} = \gamma_{0k} + \gamma_{1k} Z_{1i} + \gamma_{2k} Z_{2i} + \beta_1 X_{1ik} + \beta_2 X_{2ik} + \beta_3 X_{1ik} Z_{1i} + \epsilon_{ik}.$$

$$(10.18)$$

Note that while Z_1 doesn't vary across the candidates, the interaction term $X_1 Z_1$ does. This means that we can include it in our empirical model and estimate its effect. Although Z_1 is a constitutive component of the interaction term $X_1 Z_1$, we obviously can't include it in the empirical model because its values don't vary across the candidates. Instead, and assuming that candidate 1 is the baseline candidate, we should be sure to

---

[5] We can also think about creating these three interaction terms by multiplying $Z_1 A_1$ and $Z_2 A_1$ to get $Z_1 Z_2 A_1$, $Z_1 A_2$ and $Z_2 A_2$ to get $Z_1 Z_2 A_2$, and $Z_1 A_3$ and $Z_2 A_3$ to get $Z_1 Z_2 A_3$.

include $Z_1 A_2$ and $Z_1 A_3$. Our unordered choice model will include Y as our dependent variable and A_2, A_3, $Z_1 A_2$, $Z_1 A_3$, $Z_2 A_2$, $Z_2 A_3$, X_1, X_2, and $X_1 Z_1$ as our independent variables. Note that in this setup, we're essentially treating the interaction term $X_1 Z_1$ as an alternative-specific variable and therefore estimating just one corresponding coefficient β_3.

In contrast, our theory may predict that the interaction between voter gender and voter–candidate ideological distance with respect to welfare policy *does* vary across the candidates. Now the utility that voter i obtains from candidate k is

$$U_{ik} = \gamma_{0k} + \gamma_{1k} Z_{1i} + \gamma_{2k} Z_{2i} + \gamma_{3k} X_{1ik} Z_{1i} + \beta_1 X_{1ik} + \beta_2 X_{2ik} + \epsilon_{ik}. \qquad (10.19)$$

In this setup, we're essentially treating the interaction term $X_1 Z_1$ as an individual-specific variable and allowing its effect to vary across the candidates. To allow the conditionality between voter gender and voter–candidate ideological distance with respect to welfare policy to vary across the candidates, we need to multiply the interaction term $X_1 Z_1$ by each of the three dummy variables that indicate the different candidates. This gives us three new interaction terms $X_1 Z_1 A_1$, $X_1 Z_1 A_2$, and $X_1 Z_1 A_3$. The values of these new interaction terms are shown in the three "Type 4" columns in panel (b) of Table 10.3. As before, we can include only $J - 1$ of these new interaction terms in our model. When specifying their unordered choice model, scholars should make sure to include all of the constitutive components associated with each of their interaction terms where possible. In the current example, our unordered choice model will include Y as our dependent variable and A_2, A_3, $Z_1 A_2$, $Z_1 A_3$, $Z_2 A_2$, $Z_2 A_3$, X_1, X_2, $X_1 Z_1$, $X_1 Z_1 A_2$, and $X_1 Z_1 A_3$ as our independent variables.

As our discussion of these examples indicates, scholars need to think carefully about the precise nature of the interaction(s) implied by their conditional claims and what this implies for the structure of their data and how they estimate their unordered choice model.

> ⚠ **Important:** Scholars can make four types of conditional claim about variable-specific interaction in an unordered choice model. The first type involves interaction between alternative-specific variables. The second type involves interaction between individual-specific variables. The third and fourth types involve interaction between alternative-specific and individual-specific variables where the nature of the interaction either does or doesn't vary across the choice alternatives.

10.3 INTERPRETATION

In what follows, we'll look at different ways to interpret the results from an "interactive" unordered choice model. We'll continue with our vote choice scenario where voters are choosing between $J = 3$ candidates, and we have two individual-specific variables and two alternative-specific variables. We'll focus on the situation where we have a conditional claim involving variable-specific interaction between one of the alternative-specific variables and one of the individual-specific variables. And we'll assume that the nature of the interaction doesn't vary across the candidates.[6] In this setup, the utility that voter i obtains from candidate k is

$$U_{ik} = \gamma_{0k} + \gamma_{1k}Z_1 + \gamma_{2k}Z_2 + \beta_1 X_{1k} + \beta_2 X_{2k} + \beta_3 X_{1k}Z_1 + \epsilon_k, \quad (10.20)$$

where we've dropped the subscripts for voter i on the variables. When estimating our model, we'll treat candidate 1 as the baseline candidate.

10.3.1 Coefficients

If we focus on the utility associated with a particular candidate, we can interpret the coefficients in largely the same way as we've seen in previous interaction models with a continuous dependent variable. The marginal effect of the alternative-specific variable X_{1k} on the utility associated with candidate k is

$$\frac{\partial U_{ik}}{\partial X_{1k}} = \beta_1 + \beta_3 Z_1. \quad (10.21)$$

We see that β_1 tells us the effect of a one-unit increase in X_{1k} on the voter's utility for candidate k when the individual-specific variable $Z_1 = 0$. The marginal effect of X_{1k} increases by β_3 for each unit increase in Z_1. In other words, the modifying effect of Z_1 is β_3. This means that we can test for a statistically significant interaction between X_{1k} and Z_1 on a voter's utility for a candidate by conducting a z-test that $\beta_3 = 0$. The marginal effect of individual-specific variable Z_1 on the utility associated with candidate k is

$$\frac{\partial U_{ik}}{\partial Z_1} = \gamma_{1k} + \beta_3 X_{1k}. \quad (10.22)$$

We see that γ_{1k} tells us the effect of a one-unit increase in Z_1 on the voter's utility for candidate k relative to the baseline candidate when the alternative-specific variable $X_{1k} = 0$. Due to the symmetry of interactions, the marginal effect of Z_1 on the voter's utility for candidate k increases by β_3 for each unit increase in X_{1k}. We can use the model coefficients to produce the same types of marginal effect plots or tables we've seen in previous chapters to show how the effect of X_{1k} (Z_1) on the voter's utility for candidate k changes across the observed range of values for Z_1 (X_{1k}). There's nothing really new here if we focus our interpretation on the utility associated with candidate k.

---

[6] We examine other types of variable-specific interaction in the exercises at the end of the chapter.

What do the coefficients tell us about the *probability* of choosing particular candidates? When the marginal effect of an alternative-specific variable like X_{1k} on voter utility is positive, we know that increasing X_{1k} increases the probability that candidate k is chosen. And when the marginal effect on voter utility is negative, we know that the probability of choosing candidate k decreases. As the value of X_{1k} increases, the probability that candidate k is chosen uniformly increases if the marginal effect for X_{1k} on voter utility is positive and uniformly decreases if the marginal effect on voter utility is negative. The probabilities that the other candidates are chosen uniformly move in the opposite direction (Glasgow, 2022, 32). Thus, if the marginal effect of X_{1k} on voter utility is positive (negative) and we increase X_{1k} for, say, candidate 2, then the probabilities that the voter chooses candidate 1 or candidate 3 decrease (increase).

Things are a little more complicated when it comes to an individual-specific variable like Z_1. The marginal effect of Z_1 on voter utility varies across the candidates. When the marginal effect of Z_1 on the voter's utility for candidate k is positive, we know that the probability of choosing candidate k increases *relative to the baseline candidate*. And when the marginal effect on voter utility is negative, we know that the probability of choosing candidate k decreases *relative to the baseline candidate*. As the value of Z_1 increases, the probability that the candidate for whom the marginal effect on voter utility is largest uniformly increases, and the probability that the candidate for whom the marginal effect on voter utility is smallest uniformly decreases. The probabilities that the other candidates are chosen may not uniformly increase or decrease as the value of Z_1 increases (Glasgow, 2022, 31).

As before, we can't make any inferences about the precise magnitude of any change in the probability of choosing a particular candidate from the magnitude of the marginal effect of an independent variable on voter utility. This is one reason why we often choose to move beyond the coefficients and marginal effects with respect to voter utility when substantively interpreting the results from an unordered choice model. And as with the other discrete choice models that we've examined, we can't infer anything from the sign and significance of the coefficient on the interaction term about the sign and significance of the interaction effect of X_{1k} and Z_1 on the choice probabilities of the various candidates.

⚠ **Important:** When the marginal effect of an alternative-specific variable, say X_{1k}, on individual i's utility is positive (negative), then the probability that they choose alternative k increases (decreases). The probabilities of choosing the other alternatives move in the opposite direction. When the marginal effect of an individual-specific variable, say Z_1, on individual i's utility is positive (negative), then the probability of choosing alternative k *relative to the baseline alternative* increases (decreases). An increase in an individual-specific variable such as Z_1 increases (decreases) the probability of choosing the alternative for which the marginal effect of Z_1 on individual i's utility is largest (smallest); the probabilities of choosing the other alternatives do not uniformly increase or decrease with Z_1.

10.3.2 Marginal Effects on Probabilities

As with an ordered choice model, we can calculate the marginal effect of an independent variable on the probability of choosing each of the choice alternatives in an unordered choice model. In other words, there'll be a separate marginal effect when it comes to the choice probabilities for each of the J choice alternatives. We can calculate marginal effects for both individual-specific and alternative-specific variables. As we'll see, we can calculate two types of marginal effect for alternative-specific variables.

10.3.2.1 Individual-Specific Variables

Suppose that we're interested in the marginal effects of an individual-specific variable like Z_2 in Eq. 10.20 that's included *additively* in a multinomial logit model. Recall from Eq. 10.8 that the probability that individual i chooses candidate k is

$$P_{ik} = Pr(Y = k) = \frac{e^{V_{ik}}}{\sum_{j=1}^{J} e^{V_{ij}}}. \tag{10.23}$$

We can find the marginal effect of Z_2 on the probability of choosing candidate k by taking the derivative of this with respect to Z_2,

$$
\begin{aligned}
\frac{\partial P_{ik}}{\partial Z_2} = \frac{\partial Pr(Y = k)}{\partial Z_2} &= \frac{\partial \left(\frac{e^{V_{ik}}}{\sum_{j=1}^{J} e^{V_{ij}}} \right)}{\partial Z_2} \\
&= \frac{e^{V_{ik}} \cdot \gamma_{2k} \cdot \sum_{j=1}^{J} e^{V_{ij}} - e^{V_{ik}} \cdot \sum_{j=1}^{J} e^{V_{ij}} \cdot \gamma_{2j}}{\left(\sum_{j=1}^{J} e^{V_{ij}} \right)^2} \\
&= \frac{e^{V_{ik}} \cdot \gamma_{2k} \cdot \sum_{j=1}^{J} e^{V_{ij}}}{\left(\sum_{j=1}^{J} e^{V_{ij}} \right)^2} - \frac{e^{V_{ik}} \cdot \sum_{j=1}^{J} e^{V_{ij}} \cdot \gamma_{2j}}{\left(\sum_{j=1}^{J} e^{V_{ij}} \right)^2}
\end{aligned}
$$

$$= \frac{e^{V_{ik}}}{\sum_{j=1}^{J} e^{V_{ij}}} \cdot \gamma_{2k} - \frac{e^{V_{ik}}}{\sum_{j=1}^{J} e^{V_{ij}}} \cdot \sum_{j=1}^{J} \frac{e^{V_{ij}}}{\sum_{j=1}^{J} e^{V_{ij}}} \cdot \gamma_{2j}$$

$$= P_{ik} \cdot \gamma_{2k} - P_{ik} \cdot \sum_{j=1}^{J} P_{ij} \cdot \gamma_{2j}$$

$$= P_{ik} \left[\gamma_{2k} - \sum_{j=1}^{J} P_{ij} \cdot \gamma_{2j} \right]. \tag{10.24}$$

Note that γ_{2k} is the marginal effect of Z_2 on the voter's utility for candidate k. The marginal effect of Z_2 on the choice probability shown in Eq. 10.24 will always be positive for the choice alternative (candidate) with the largest coefficient and negative for the choice alternative with the smallest coefficient. It will not be uniformly positive or negative for the other choice alternatives. This means that we must be careful in drawing inferences about the sign of the marginal effect of an individual-specific variable from the sign of its coefficient. By construction, the marginal effects for an individual-specific variable like Z_2 will sum to 0 across the choice alternatives (Glasgow, 2022, 36).

As with the other discrete choice models we've examined, the marginal effect of an individual-specific variable on the various choice probabilities depends on the value of V_{ik} at which the increase takes place. In other words, the effect of an individual-specific variable on the choice probabilities always depends on the values of *all* of the variables in the model, including its own. This tells us that the effects of the individual-specific variables on the choice probabilities are always conditional when we estimate a multinomial logit model, even when the variable of interest is included additively. This conditionality arises from the compression that occurs when we take the unbounded utility values and map them onto a probability space that's bounded between 0 and 1.

Suppose now that we're interested in the marginal effect of an individual-specific variable like Z_1 in Eq. 10.20 that's included *interactively* in a multinomial logit model. We again find the marginal effect on the probability of choosing candidate k by taking the derivative of P_{ik} in Eq. 10.23 with respect to Z_1:

$$\frac{\partial P_{ik}}{\partial Z_1} = \frac{\partial Pr(Y=k)}{\partial Z_1} = \frac{\partial \left(\frac{e^{V_{ik}}}{\sum_{j=1}^{J} e^{V_{ij}}} \right)}{\partial Z_1}$$

$$= P_{ik} \left[\gamma_{1k} + \beta_3 X_{1k} - \sum_{j=1}^{J} P_{ij} \cdot \left(\gamma_{1j} + \beta_3 X_{1j} \right) \right]. \tag{10.25}$$

The same basic insights apply here as with an additive individual-specific variable. The marginal effect of Z_1 on the choice probability shown in Eq. 10.25 will always be positive for the choice alternative (candidate) with the largest marginal effect on voter utility and negative for the choice alternative with the smallest marginal effect on voter utility. It will not be uniformly positive or negative for the other choice alternatives. As before, the marginal effects for an individual-specific variable like Z_1 will sum to 0 across the choice alternatives.

> ⚠ **Important:** The marginal effects of an individual-specific variable such as Z_1 on the probability of choosing each of the alternatives sum to 0 across the choice alternatives.

Scholars with a hypothesis about how the effect of Z_1 on the probability of choosing candidate k varies with the value of X_{1k} might wish to calculate the marginal effect of Z_1 across the observed range of values for X_{1k} while keeping the other independent variables constant at their chosen baseline values. One way to present this type of information is in the form of a marginal effect plot.

10.3.2.2 Alternative-Specific Variables

Suppose that we're interested in the marginal effect of an alternative-specific variable on the probability of choosing candidate k. Unlike individual-specific variables, the values of alternative-specific variables vary over the choice alternatives. This means that we can calculate two types of marginal effects for alternative-specific variables (Glasgow, 2022). We can calculate the marginal effect of an alternative-specific variable on the probability of choosing candidate k when we change the value of this variable for candidate k. We refer to this marginal effect as the *direct marginal effect*. Alternatively, we can calculate the marginal effect of an alternative-specific variable on the probability of choosing candidate k when we change the value of this variable for some other candidate. We refer to this marginal effect as the *cross-marginal effect*. To provide some substantive context, suppose that the alternative-specific variable captures the ideological distance between the voter and each of the candidates on the left-right policy dimension. The direct marginal effect tells us how an increase in the ideological distance between the voter and candidate k affects the probability that the voter chooses candidate k. In contrast, the cross-marginal effect tells us how an increase in the ideological distance between the voter and some other candidate, say candidate m, affects the probability that the voter chooses candidate k.

We'll start by looking at the *direct marginal effect* of an alternative-specific variable. Suppose that we're interested in the direct marginal effect on the probability of choosing candidate k of an alternative-specific variable like X_{2k} in Eq. 10.20 that's included *additively* in a multinomial logit model. We can find this by taking the derivative of P_{ik} in Eq. 10.23 with respect to X_{2k}:

$$\frac{\partial P_{ik}}{\partial X_{2k}} = \frac{\partial Pr(Y = k)}{\partial X_{2k}} = \frac{\partial \left(\frac{e^{V_{ik}}}{\sum_{j=1}^{J} e^{V_{ij}}} \right)}{\partial X_{2k}}$$

$$= \frac{e^{V_{ik}} \cdot \beta_2 \cdot \sum_{j=1}^{J} e^{V_{ij}} - e^{V_{ik}} \cdot e^{V_{ik}} \cdot \beta_2}{\left(\sum_{j=1}^{J} e^{V_{ij}} \right)^2}$$

$$= \frac{e^{V_{ik}} \cdot \beta_2 \cdot \sum_{j=1}^{J} e^{V_{ij}}}{\left(\sum_{j=1}^{J} e^{V_{ij}} \right)^2} - \frac{e^{V_{ik}} \cdot e^{V_{ik}} \cdot \beta_2}{\left(\sum_{j=1}^{J} e^{V_{ij}} \right)^2}$$

$$= \frac{e^{V_{ik}}}{\sum_{j=1}^{J} e^{V_{ij}}} \cdot \beta_2 - \frac{e^{V_{ik}}}{\sum_{j=1}^{J} e^{V_{ij}}} \cdot \frac{e^{V_{ik}}}{\sum_{j=1}^{J} e^{V_{ij}}} \cdot \beta_2$$

$$= P_{ik} \cdot \beta_2 - P_{ik} \cdot P_{ik} \cdot \beta_2$$

$$= P_{ik} (1 - P_{ik}) \cdot \beta_2. \tag{10.26}$$

Note that β_2 is the marginal effect of X_{2k} on voter utility. Because P_{ik} and $1 - P_{ik}$ are both probabilities, the sign of the direct marginal effect shown in Eq. 10.26 will be the same as the sign of the marginal effect of the alternative-specific variable on voter utility. We again see that the effect of a variable like X_{2k} on the choice probabilities is conditional and depends on the values of *all* of the variables in the model even though it's included additively.

⚠ **Important:** We can calculate two types of marginal effect for alternative-specific variables. A *direct marginal effect* tells us how the probability of choosing alternative k changes when we increase the value of the alternative-specific variable for alternative k. A *cross-marginal effect* tells us how the probability of choosing alternative k changes when we increase the value of the alternative-specific variable for some alternative that is not k.

Suppose that we're now interested in the direct marginal effect on the probability of choosing candidate k of an alternative-specific variable like X_{1k} in Eq. 10.20 that's included *interactively* in a multinomial logit model.

We can find this by taking the derivative of P_{ik} in Eq. 10.23 with respect to X_{1k}:

$$\frac{\partial P_{ik}}{\partial X_{1k}} = \frac{\partial Pr(Y=k)}{\partial X_{1k}} = \frac{\partial\left(\frac{e^{V_{ik}}}{\sum_{j=1}^{J} e^{V_{ij}}}\right)}{\partial X_{1k}} = P_{ik}\,(1-P_{ik})\cdot(\beta_1+\beta_3 Z_1).$$

$$(10.27)$$

The sign of this direct marginal effect will again be the same as the sign of the marginal effect of the alternative-specific variable on voter utility, $\beta_1+\beta_3 Z_1$.

We now turn to examine the *cross-marginal effect* of an alternative-specific variable. Suppose that we're interested in the cross-marginal effect on the probability of choosing candidate k of an alternative-specific variable like X_2 in Eq. 10.20 that's included *additively* in a multinomial logit model. To calculate this marginal effect, we need to specify the candidate other than candidate k for whom we're increasing X_2. In this regard, we'll assume that we're interested in how an increase in X_2 for candidate m affects the probability of choosing candidate k. To calculate this particular cross-marginal effect, we take the derivative of P_{ik} in Eq. 10.23 with respect to X_{2m}, where $m \neq k$:

$$\frac{\partial P_{ik}}{\partial X_{2m}} = \frac{\partial Pr(Y=k)}{\partial X_{2m}} = \frac{\partial\left(\frac{e^{V_{ik}}}{\sum_{j=1}^{J} e^{V_{ij}}}\right)}{\partial X_{2m}} = \frac{-e^{V_{ik}}\cdot e^{V_{im}}\cdot\beta_2}{\left(\sum_{j=1}^{J} e^{V_{ij}}\right)^2}$$

$$= \frac{-e^{V_{ik}}\cdot e^{V_{im}}\cdot\beta_2}{\sum_{j=1}^{J} e^{V_{ij}}\cdot\sum_{j=1}^{J} e^{V_{ij}}} = -P_{ik}\cdot P_{im}\cdot\beta_2. \quad (10.28)$$

We can, of course, calculate the cross-marginal effect of X_2 for any of the $J-1$ candidates who aren't candidate k.

Suppose that we're now interested in the cross-marginal effect on the probability of choosing candidate k of an alternative-specific variable like X_1 in Eq. 10.20 that's included *interactively* in a multinomial logit model. As before, we'll assume that we're increasing the value of X_1 for candidate m. We can find our desired cross-marginal effect by taking the derivative of P_{ik} in Eq. 10.23 with respect to X_{1m}, where $m \neq k$:

$$\frac{\partial P_{ik}}{\partial X_{1m}} = \frac{\partial Pr(Y=k)}{\partial X_{1m}} = \frac{\partial\left(\frac{e^{V_{ik}}}{\sum_{j=1}^{J} e^{V_{ij}}}\right)}{\partial X_{1m}} = -P_{ik}\cdot P_{im}\cdot(\beta_1+\beta_3 Z_1).$$

$$(10.29)$$

Because P_{ik} and P_{im} are both probabilities, the sign of the cross-marginal effects shown in Eq. 10.28 and Eq. 10.29 will be the opposite of the sign of the marginal effect of the alternative-specific variable on voter utility. The direct marginal effect and associated cross-marginal effects for an alternative-specific variable for a given candidate k will necessarily sum to zero.

> ⚠ **Important:** The direct marginal effect and associated cross-marginal effects for an alternative-specific variable on the probability of choosing some alternative k sum to 0.

Scholars with a hypothesis about how the direct marginal and cross-marginal effects of X_1 on the probability of choosing candidate k vary with the value of Z_1 might wish to calculate these marginal effects across the observed range of values for Z_1 while keeping the other independent variables constant at their chosen baseline values. One way to present this type of information is in the form of a marginal effect plot.

10.3.3 Differences in Predicted Probabilities

We can also report differences in each of the predicted choice probabilities when presenting the effects of the independent variables in a multinomial logit model. In what follows, we focus on the effect of those independent variables that are included interactively.

10.3.3.1 Alternative-Specific Variables

The effect of increasing an alternative-specific variable like X_{1k} in Eq. 10.20 from some baseline value X_{1kb} to some counterfactual value X_{1kc} on the probability of choosing candidate k when the other independent variables are each set at their baseline values is just the difference in the probability of choosing candidate k in the counterfactual scenario c as opposed to in the baseline scenario b,

$$\Delta P_{ik} = \Delta Pr(Y = k) = \frac{e^{V_{ikc}}}{\sum_{j=1}^{J} e^{V_{ijc}}} - \frac{e^{V_{ikb}}}{\sum_{j=1}^{J} e^{V_{ijb}}}. \quad (10.30)$$

In the counterfactual scenario, we have

$$V_{ikc} = \gamma_{0k} + \gamma_{1k}Z_{1b} + \gamma_{2k}Z_{2b} + \beta_1 X_{1kc} + \beta_2 X_{2kb} + \beta_3 X_{1kc}Z_{1b} \quad (10.31)$$

and

$$V_{ijc} = \gamma_{0j} + \gamma_{1j}Z_{1b} + \gamma_{2j}Z_{2b} + \beta_1 X_{1jb} + \beta_2 X_{2jb} + \beta_3 X_{1jb}Z_{1b} \quad (10.32)$$

for $j \neq k$. And in the baseline scenario, we have

$$V_{ijb} = \gamma_{0j} + \gamma_{1j}Z_{1b} + \gamma_{2j}Z_{2b} + \beta_1 X_{1jb} + \beta_2 X_{2jb} + \beta_3 X_{1jb}Z_{1b} \quad (10.33)$$

for all j, including $j = k$. In this example, we've shown how to calculate a "direct difference" in the predicted probability of choosing candidate k. We can, of course, adopt the same basic approach with some minor modifications to calculate a "cross-difference" in the predicted probability of choosing candidate k, in which we examine the effect of increasing X_1 for some candidate $j \neq k$ on the probability that the voter selects candidate k.

10.3.3.2 Individual-Specific Variables

We can also use Eq. 10.30 to calculate the effect of increasing an individual-specific variable like Z_1 in Eq. 10.20 from some baseline value Z_{1b} to some counterfactual value Z_{1c} on the probability of choosing candidate k when the other independent variables are each set at their baseline values. Now in the counterfactual scenario, we have

$$V_{ijc} = \gamma_{0j} + \gamma_{1j}Z_{1c} + \gamma_{2j}Z_{2b} + \beta_1 X_{1jb} + \beta_2 X_{2jb} + \beta_3 X_{1jb}Z_{1c} \quad (10.34)$$

for all j, including $j = k$. And in the baseline scenario, we have

$$V_{ijb} = \gamma_{0j} + \gamma_{1j}Z_{1b} + \gamma_{2j}Z_{2b} + \beta_1 X_{1jb} + \beta_2 X_{2jb} + \beta_3 X_{1jb}Z_{1b} \quad (10.35)$$

for all j, including $j = k$. Rather than calculate the absolute change in these choice probabilities, it's often preferable, as discussed previously, to calculate the percentage change.

Scholars with a hypothesis about how the effect of X_{1k} (Z_1) on the probability of choosing candidate k varies with the value of Z_1 (X_{1k}) might wish to calculate the (percentage) change in the probability of choosing candidate k associated with the specified increase in X_{1k} (Z_1) across the observed range of values for Z_1 (X_{1k}) while keeping the other independent variables constant. One way to present this type of information is in the form of a "difference" plot.

10.3.4 Odds Ratios

We can also present the effects of the independent variables in terms of odds ratios. In other words, we can calculate how an increase in the independent variables affects the odds of observing one choice alternative versus another. If our choice scenario has J choice alternatives, there'll be $\frac{J(J-1)}{2}$ pairwise comparisons of choice alternatives that we could calculate. Thus, in our example where we have three candidates, we could make $\frac{3(2)}{2} = 3$ pairwise comparisons. Building on the probability that voter i chooses candidate k in Eq. 10.23, the odds of choosing candidate k versus some other candidate m are

$$\Omega_{k \text{ vs } m} = \frac{P_{ik}}{P_{im}} = \frac{\frac{e^{V_{ik}}}{\sum_{j=1}^{J} e^{V_{ij}}}}{\frac{e^{V_{im}}}{\sum_{j=1}^{J} e^{V_{ij}}}} = \frac{e^{V_{ik}}}{e^{V_{im}}} = e^{V_{ik} - V_{im}}. \tag{10.36}$$

One thing to note here is that the final formulation of Eq. 10.36 only makes reference to candidates k and m. In other words, the relative odds of choosing candidate k over candidate m don't depend on which other candidates are available in the choice set or the characteristics of any other candidate. This property, which always holds in a multinomial logit model, is known as the *independence of irrelevant alternatives*.[7] We can examine the effects of both alternative-specific and individual-specific variables in terms of odds ratios.

10.3.4.1 Alternative-Specific Variables

Suppose that we're interested in evaluating the effect of a one-unit increase in an alternative-specific variable like X_{2k} in Eq. 10.20 that's included *additively*. One way to evaluate this effect is to compare the odds of choosing candidate k versus some other candidate m when the value of X_{2k} is one unit higher than some baseline value $(X_{2k} + 1)$ with the odds when the value of X_{2k} is at its baseline value (X_{2k}), while holding the values of the other independent variables constant,

$$\frac{\Omega_{k \text{ vs } m} \mid X_{2k} + 1}{\Omega_{k \text{ vs } m} \mid X_{2k}} = \frac{\frac{e^{V_{ik} \mid X_{2k}+1}}{e^{V_{im}}}}{\frac{e^{V_{ik} \mid X_{2k}}}{e^{V_{im}}}} = \frac{e^{V_{ik} \mid X_{2k}+1}}{e^{V_{ik} \mid X_{2k}}} = \frac{e^{\beta_2(X_{2k}+1)}}{e^{\beta_2 X_{2k}}}$$

$$= e^{\beta_2(X_{2k}+1-X_{2k})} = e^{\beta_2}. \tag{10.37}$$

From this, we see that a one-unit increase in X_{2k} leads to a multiplicative increase in the odds of choosing candidate k versus candidate m of e^{β_2}.[8]

---

[7] The assumption underpinning the independence of irrelevant alternatives (IIA) property is violated when the error terms aren't independent across the choice alternatives. This will often be the case when two or more choice alternatives are viewed as close substitutes. It will also occur when our model omits an independent variable that's common to two or more choice alternatives. This is because the omitted variable is captured in the error terms, making them appear correlated. This last point highlights that satisfying the IIA assumption has to do with our model specification and isn't some abstract property of the decision-maker or choice scenario. Our estimated coefficients will be inconsistent if the IIA assumption is violated. Those who are concerned about the IIA assumption may wish to switch to a mixed logit model (Train, 1998; McFadden and Train, 2000; Glasgow, 2001, 2022).

[8] We could also examine the effect of increasing X_{2m} by one on the odds of choosing candidate k versus candidate m,

$$\frac{\Omega_{k \text{ vs } m} \mid X_{2m} + 1}{\Omega_{k \text{ vs } m} \mid X_{2m}} = e^{-\beta_2}. \tag{10.38}$$

Importantly, we see that this multiplicative increase doesn't depend on which two candidates we're considering or the values of any of the independent variables. To illustrate, if say, $\beta_2 = 0.2$, then the odds of choosing candidate k versus candidate m changes by a factor of $e^{0.2} = 1.22$ when we increase X_{2k} by one. Since the odds ratio is greater than one, we see that a one-unit increase in X_{2k} is associated with an increase in the odds of choosing candidate k versus candidate m. If we increase X_{2k} by some amount δ, then the odds of choosing candidate k versus candidate m changes by a factor of $e^{\beta_2 \times \delta}$. Thus, if we increase X_{2k} by two units, the odds of choosing candidate k versus candidate m changes by a factor of $e^{\beta_2 \times 2} = e^{0.4} = 1.49$. The percentage change in the odds of choosing candidate k versus candidate m is $\left[e^{\beta_2 \times \delta} - 1 \right] \times 100$. Thus, a two-unit increase in X_{2k} is associated with an $\left[e^{0.4} - 1 \right] \times 100 = 49\%$ increase in the odds of choosing candidate k versus candidate m.

Suppose now that we're interested in evaluating the effect of a one-unit increase in an alternative-specific variable like X_{1k} in Eq. 10.20 that's included *interactively*. The odds ratio comparing the odds of choosing candidate k versus candidate m in the scenario where X_{1k} is one-unit higher than some baseline value $(X_{1k} + 1)$ to the scenario where X_{1k} is at its baseline value (X_{1k}), while holding the values of the other independent variables constant, is

$$\frac{\Omega_{k \text{ vs } m} \mid X_{1k} + 1}{\Omega_{k \text{ vs } m} \mid X_{1k}} = \frac{\frac{e^{V_{ik} \mid X_{1k}+1}}{e^{V_{im}}}}{\frac{e^{V_{ik} \mid X_{1k}}}{e^{V_{im}}}} = \frac{e^{V_{ik} \mid X_{1k}+1}}{e^{V_{ik} \mid X_{1k}}} = \frac{e^{\beta_1(X_{1k}+1)+\beta_3((X_{1k}+1)Z_1)}}{e^{\beta_1 X_{1k}+\beta_3 X_{1k}Z_1}}$$

$$= e^{\beta_1(X_{1k}+1-X_{1k})+\beta_3(X_{1k}+1-X_{1k})Z_1} = e^{\beta_1+\beta_3 Z_1}.$$

(10.39)

From this, we see that the effect of a one-unit increase in X_{1k} on the odds of choosing candidate k versus candidate m depends on the value of Z_1. To illustrate, if, say, $\beta_1 = 0.4$ and $\beta_3 = -0.2$, then the odds of choosing candidate k versus candidate m changes by a factor of $e^{0.4-0.2\times0} = e^{0.4} = 1.49$ when $Z_1 = 0$; it changes by a factor of $e^{0.4-0.2\times1} = e^{0.2} = 1.22$ when $Z_1 = 1$; it changes by a factor of $e^{0.4-0.2\times2} = e^0 = 1$ when $Z_1 = 2$; it changes by a factor of $e^{0.4-0.2\times3} = e^{-0.2} = 0.82$ when $Z_1 = 3$, and so on. If we increase X_{1k} by some amount δ, then the odds of choosing candidate k versus candidate m changes by a factor of $e^{(\beta_1+\beta_3 Z_1)\times\delta}$. And the percentage change in the odds of choosing candidate k versus candidate m is calculated as $\left[e^{(\beta_1+\beta_3 Z_1)\times\delta} - 1 \right] \times 100$. Scholars with hypotheses about how the effect of X_{1k} varies with the value of Z_1 might wish to plot the

The effect of increasing X_{2m} by one on the odds of choosing candidate m versus candidate k is, of course, e^{β_2} by Eq. 10.37.

factor or percentage change in the odds of choosing candidate k versus candidate m across the observed range of values for Z_1.

10.3.4.2 Individual-Specific Variables

Suppose that we're interested in evaluating the effect of a one-unit increase in an individual-specific variable like Z_2 in Eq. 10.20 that's included *additively*. One way to evaluate this effect is to compare the odds of choosing candidate k versus some other candidate m when the value of Z_2 is one unit higher than some baseline value $(Z_2 + 1)$ with the odds when the value of Z_2 is at its baseline value (Z_2), while holding the values of the other independent variables constant,

$$
\frac{\Omega_{k \text{ vs } m} \mid Z_2 + 1}{\Omega_{k \text{ vs } m} \mid Z_2} = \frac{\frac{e^{V_{ik}\mid Z_2+1}}{e^{V_{im}\mid Z_2+1}}}{\frac{e^{V_{ik}\mid Z_2}}{e^{V_{im}\mid Z_2}}} = \frac{e^{V_{ik}\mid Z_2+1}}{e^{V_{im}\mid Z_2+1}} \cdot \frac{e^{V_{im}\mid Z_2}}{e^{V_{ik}\mid Z_2}} = \frac{e^{\gamma_{2k}(Z_2+1)}}{e^{\gamma_{2m}(Z_2+1)}} \cdot \frac{e^{\gamma_{2m}Z_2}}{e^{\gamma_{2k}Z_2}}
$$

$$
= \frac{e^{\gamma_{2k}Z_2+\gamma_{2k}}}{e^{\gamma_{2m}Z_2+\gamma_{2m}}} \cdot \frac{e^{\gamma_{2m}Z_2}}{e^{\gamma_{2k}Z_2}} = \frac{e^{\gamma_{2k}}}{e^{\gamma_{2m}}} = e^{\gamma_{2k}-\gamma_{2m}}. \quad (10.40)
$$

The important point to note here is that, unlike with alternative-specific variables, the effect of an individual-specific variable in terms of odds ratios depends on which two candidates are being compared.[9] This is a consequence of the fact that the coefficients on the individual-specific variables vary across the choice alternatives. This is why odds ratios for individual-specific variables are sometimes referred to as *conditional odds ratios*. As Eq. 10.40 indicates, we obtain the change in the odds of choosing candidate k versus candidate m associated with a one-unit increase in an "additive" individual-specific variable by taking the *difference* in the coefficients on this variable for candidates k and m and exponentiating it. Recall that one of the candidates in the choice set is treated as the baseline candidate, and we assume that the coefficients on the individual-specific variables for this candidate are 0. This means that if candidate m in the odds ratio in Eq. 10.40 is the baseline candidate, then $\gamma_{2m} = 0$, and our odds ratio simplifies to $e^{\gamma_{2k}}$. To illustrate, if say, $\gamma_{2k} = 0.4$ and $\gamma_{2m} = 0.1$, then the odds of choosing candidate k versus candidate m changes by a factor of $e^{0.4-0.1} = e^{0.3} = 1.35$ when we increase Z_2 by one unit. If we increase Z_2 by some amount δ, then the odds of choosing candidate k versus candidate m changes by a factor of $e^{(\gamma_{2k}-\gamma_{2m})\times\delta}$. Thus, if we increase Z_2 by two

---

[9] As with alternative-specific variables, though, the effects of *additive* individual-specific variables in terms of odds ratios don't depend on the values of any of the independent variables.

units, the odds of choosing candidate k versus candidate m changes by a factor of $e^{(0.4-0.1)\times 2} = e^{0.6} = 1.82$. The percentage change in the odds of choosing candidate k versus candidate m is $\left[e^{(\gamma_{2k}-\gamma_{2m})\times\delta} - 1\right] \times 100$. Thus, a two-unit increase in Z_2 is associated with an $\left[e^{0.6} - 1\right] \times 100 = 82\%$ increase in the odds of choosing candidate k versus candidate m.

Suppose now that we're interested in evaluating the effect of a one-unit increase in an individual-specific variable like Z_1 in Eq. 10.20 that's included *interactively*. The odds ratio comparing the odds of choosing candidate k versus candidate m in the scenario where Z_1 is one unit higher than some baseline value $(Z_1 + 1)$ to the scenario where Z_1 is at its baseline value (Z_1), while holding the values of the other independent variable constant, is

$$
\frac{\Omega_{k \text{ vs } m} \mid Z_1 + 1}{\Omega_{k \text{ vs } m} \mid Z_1} = \frac{\frac{e^{V_{ik}|Z_1+1}}{e^{V_{im}|Z_1+1}}}{\frac{e^{V_{ik}|Z_1}}{e^{V_{im}|Z_1}}} = \frac{e^{V_{ik}|Z_1+1}}{e^{V_{im}|Z_1+1}} \cdot \frac{e^{V_{im}|Z_1}}{e^{V_{ik}|Z_1}}
$$

$$
= \frac{e^{\gamma_{1k}(Z_1+1)+\beta_3 X_{1k}(Z_1+1)}}{e^{\gamma_{1m}(Z_1+1)+\beta_3 X_{1m}(Z_1+1)}} \cdot \frac{e^{\gamma_{1m}Z_1+\beta_3 X_{1m}Z_1}}{e^{\gamma_{1k}Z_1+\beta_3 X_{1k}Z_1}}
$$

$$
= \frac{e^{\gamma_{1k}Z_1+\gamma_{1k}+\beta_3 X_{1k}Z_1+\beta_3 X_{1k}}}{e^{\gamma_{1m}Z_1+\gamma_{1m}+\beta_3 X_{1m}Z_1+\beta_3 X_{1m}}} \cdot \frac{e^{\gamma_{1m}Z_1+\beta_3 X_{1m}Z_1}}{e^{\gamma_{1k}Z_1+\beta_3 X_{1k}Z_1}}
$$

$$
= \frac{e^{\gamma_{1k}+\beta_3 X_{1k}}}{e^{\gamma_{1m}+\beta_3 X_{1m}}} = e^{\gamma_{1k}-\gamma_{1m}+\beta_3(X_{1k}-X_{1m})}. \tag{10.41}
$$

From this, we see that the effect of a one-unit increase in Z_1 on the odds of choosing candidate k versus candidate m depends on the *difference* in the values of X_1 for candidates k and m. To illustrate, if, say, $\gamma_{1k} = 0.3$, $\gamma_{1m} = 0.1$, and $\beta_3 = -0.2$, then the odds of choosing candidate k versus candidate m changes by a factor of $e^{0.3-0.1-0.2\times 0} = e^{0.2} = 1.22$ when $X_{1k} - X_{1m} = 0$; it changes by a factor of $e^{0.3-0.1-0.2\times 1} = e^0 = 1$ when $X_{1k} - X_{1m} = 1$; it changes by a factor of $e^{0.3-0.1-0.2\times 2} = e^{-0.2} = 0.82$ when $X_{1k} - X_{1m} = 2$; it changes by a factor of $e^{0.3-0.1-0.2\times 3} = e^{-0.4} = 0.67$ when $X_{1k} - X_{1m} = 3$, and so on. If we increase Z_1 by some amount δ, then the odds of choosing candidate k versus candidate m changes by a factor of $e^{[\gamma_{1k}-\gamma_{1m}+\beta_3(X_{1k}-X_{1m})]\times\delta}$. And the percentage change in the odds of choosing candidate k versus candidate m is calculated as $\left[e^{[\gamma_{1k}-\gamma_{1m}+\beta_3(X_{1k}-X_{1m})]\times\delta} - 1\right] \times 100$. Scholars with hypotheses about how the effect of Z_1 varies with the value of X_1 might wish to plot the factor or percentage change in the odds of choosing candidate k versus candidate m across the observed range of values for $X_{1k} - X_m$.

10.3.5 Interaction Effects

As with the other discrete choice models that we've examined in this part of the book, we're generally interested in two broad types of interaction effect when we have an interactive multinomial logit model. The first is the effect of variable-specific interaction on the *utility* associated with the choice alternatives (candidates). The second is the effect of variable-specific interaction on the *probability* of choosing each of the choice alternatives (voting for each of the candidates). Scholars should think very carefully about which type of interaction effect is most relevant for testing their theoretical predictions. In what follows, we look at interaction effects in the context of the same underlying model as before where we have variable-specific interaction between an individual-specific variable Z_1 and an alternative-specific variable X_{1j} and where the utility that voter i obtains from candidate k is

$$U_{ik} = \gamma_{0k} + \gamma_{1k}Z_1 + \gamma_{2k}Z_2 + \beta_1 X_{1k} + \beta_2 X_{2k} + \beta_3 X_{1k}Z_1 + \epsilon_k. \quad (10.42)$$

10.3.5.1 Interaction Effect on U_{ik}

Suppose that our theory makes a prediction about how Z_1 modifies the effect of X_{1k} on the voter's utility for candidate k. Such a prediction speaks to the interaction effect between X_{1k} and Z_1 on U_{ik}. The marginal effect of X_{1k} on the utility associated with candidate k is

$$\frac{\partial U_{ik}}{\partial X_{1k}} = \beta_1 + \beta_3 Z_1. \quad (10.43)$$

The interaction effect between X_{1k} and Z_1 is just the slope relationship between this marginal effect and Z_1,

$$\frac{\partial \left(\frac{\partial U_{ik}}{\partial X_{1k}} \right)}{\partial Z_1} = \frac{\partial (\beta_1 + \beta_3 Z_1)}{\partial Z_1} = \beta_3. \quad (10.44)$$

This interaction effect tells us how a one-unit increase in Z_1 modifies the effect of a one-unit increase in X_{1k} on the voter's utility for candidate k. Due to the symmetry of interaction, it also tells us how a one-unit increase in X_{1k} modifies the effect of a one-unit increase in Z_1 on the voter's utility for candidate k. As we can see, the sign, magnitude, and statistical significance of the interaction effect between X_{1k} and Z_1 on the utility associated with candidate k can be identified simply by looking at the coefficient on the interaction term β_3. The interaction effect with respect to U_{ik} doesn't depend on the value of any of the independent variables and is therefore unconditional.

10.3.5.2 Interaction Effect on P_{ik}

Suppose now that our theory makes a prediction about the variable-specific interaction effect between X_{1k} and Z_1 on the probability of voting for candidate k. This interaction effect refers to both how Z_1 modifies the "direct" effect of an increase in X_1 for candidate k on the probability of voting for candidate k and how X_{1k} modifies the effect of an increase in Z_1 on the probability of voting for candidate k.[10] Once again, we can think about this type of interaction effect in terms of marginal effects or in terms of "differences in differences."

We'll start with marginal effects. Based on Eq. 10.27, the "direct" marginal effect of X_{1k} on the probability of voting for candidate k is

$$\frac{\partial P_{ik}}{\partial X_{1k}} = \frac{\partial Pr(Y = k)}{\partial X_{1k}} = \frac{\partial \left(\frac{e^{V_{ik}}}{\sum_{j=1}^{J} e^{V_{ij}}} \right)}{\partial X_{1k}} = P_{ik} (1 - P_{ik}) \cdot (\beta_1 + \beta_3 Z_1).$$

(10.45)

This tells us how an increase in X_1 on the part of candidate k affects the probability of voting for candidate k. We can rewrite this direct marginal effect as

$$\frac{\partial P_{ik}}{\partial X_{1k}} = P_{ik} \cdot (\beta_1 + \beta_3 Z_1) - P_{ik}^2 \cdot (\beta_1 + \beta_3 Z_1).$$

(10.46)

To calculate the interaction effect between X_{1k} and Z_1 on P_{ik}, we take the derivative of Eq. 10.46 with respect to Z_1:

$$\frac{\partial \left(\frac{\partial P_{ik}}{\partial X_{1k}} \right)}{\partial Z_1} = \frac{\partial \left[P_{ik} \cdot (\beta_1 + \beta_3 Z_1) - P_{ik}^2 \cdot (\beta_1 + \beta_3 Z_1) \right]}{\partial Z_1}$$

$$= \frac{\partial P_{ik}}{\partial Z_1} \cdot (\beta_1 + \beta_3 Z_1) + \beta_3 \cdot P_{ik} - 2 P_{ik} \cdot \frac{\partial P_{ik}}{\partial Z_1} \cdot (\beta_1 + \beta_3 Z_1) - \beta_3 P_{ik}^2$$

$$= P_{ik} (1 - P_{ik}) \cdot \beta_3 + (1 - 2 P_{ik}) \cdot (\beta_1 + \beta_3 Z_1) \cdot \frac{\partial P_{ik}}{\partial Z_1}.$$

(10.47)

We've already calculated $\frac{\partial P_{ik}}{\partial Z_1}$ in Eq. 10.25. Substituting this into Eq. 10.47, we have

$$\frac{\partial \left(\frac{\partial P_{ik}}{\partial X_{1k}} \right)}{\partial Z_1} = P_{ik} (1 - P_{ik}) \cdot \beta_3$$

$$+ (1 - 2 P_{ik}) \cdot (\beta_1 + \beta_3 Z_1) \cdot P_{ik} \left[\gamma_{1k} + \beta_3 X_{1k} - \sum_{j=1}^{J} P_{ij} \cdot (\gamma_{1j} + \beta_3 X_{1j}) \right].$$

(10.48)

[10] While we focus on this particular interaction effect here, we remind readers that we can also think about the interaction effect between X_{1m} and Z_1 on the probability of voting for candidate k. Such an interaction effect speaks to how Z_1 modifies the "cross" effect of an increase in X_1 for candidate m on the probability of choosing candidate k and how X_{1m} modifies the effect of Z_1 on the probability of choosing candidate k.

Due to the symmetry of interactions, this interaction effect tells us both how Z_1 modifies the marginal effect of X_{1k} on P_{ik} *and* how X_{1k} modifies the marginal effect of Z_1 on P_{ik}.

We immediately see that there's no single interaction effect between X_{1k} and Z_1 when it comes to P_{ik}. The interaction effect is conditional and depends not only on the choice of candidate k but also on the values of all of the independent variables in the model. The interaction effect can be non-zero even if the coefficient on the interaction term β_3 is 0. This means that it's not possible to infer whether there's a statistically significant interaction effect between X_{1k} and Z_1 on P_{ik} by conducting a z-test on the interaction term coefficient. The sign of the interaction effect varies depending on the choice of candidate and the values of the independent variables. In sum, just like with the other non-linear discrete choice models that we've examined, we can't typically draw valid inferences about the magnitude, sign, and statistical significance of the interaction effect between X_{1k} and Z_1 on P_{ik} from the magnitude, sign, and statistical significance of the interaction term coefficient.

> ⚠ **Important:** As with other discrete choice models, we cannot typically draw valid inferences about the magnitude, sign, and statistical significance of the interaction effect between variables on P_{ik} in an unordered choice model from the magnitude, sign, and statistical significance of the interaction term coefficient.

We offer the same advice when it comes to interaction effects in unordered choice models as we did when we discussed interaction effects in other types of discrete choice model. We suspect that most theories in the social sciences are only strong enough to make predictions about the existence and the sign of the interaction effect between variables such as X_{1k} and Z_1 on U_{ik} and not about the existence and sign of the interaction effects between these variables on P_{ik}. To the extent that this is the case, we recommend a two-step process when it comes to evaluating interaction in an interactive unordered choice model. In the first step, we should examine whether we observe the effects predicted by our theory on U_{ik}. In the second step, we should examine whether the magnitudes of these effects are substantively meaningful by looking at their impact on P_{ik}.

So far, we've looked at interaction effects with respect to P_{ik} in terms of marginal effects. However, the same basic insights apply if we think about them in terms of "differences-in-differences." Suppose that we're interested in knowing how a discrete change in Z_1 modifies the effect of a discrete change in X_{1k} on P_{ik} for a particular scenario. To identify this interaction effect, we must again calculate three "differences." The first difference captures the *change* in P_{ik} associated with a discrete increase in X_{1k} from some baseline value X_{1kb} to some counterfactual value X_{1kc} when Z_1 and all of the other independent variables are set at their own baseline values,

$$\Delta P_{ik} | Z_{1b} = \frac{e^{V_{ikc}}}{\sum_{j=1}^{J} e^{V_{ijc}}} - \frac{e^{V_{ikb}}}{\sum_{j=1}^{J} e^{V_{ijb}}}. \tag{10.49}$$

In the counterfactual scenario associated with this first difference, we have

$$V_{ikc} = \gamma_{0k} + \gamma_{1k}Z_{1b} + \gamma_{2k}Z_{2b} + \beta_1 X_{1kc} + \beta_2 X_{2kb} + \beta_3 X_{1kc}Z_{1b} \tag{10.50}$$

and

$$V_{ijc} = \gamma_{0j} + \gamma_{1j}Z_{1b} + \gamma_{2j}Z_{2b} + \beta_1 X_{1jb} + \beta_2 X_{2jb} + \beta_3 X_{1jb}Z_{1b} \tag{10.51}$$

for $j \neq k$. And in the baseline scenario for this first difference, we have

$$V_{ijb} = \gamma_{0j} + \gamma_{1j}Z_{1b} + \gamma_{2j}Z_{2b} + \beta_1 X_{1jb} + \beta_2 X_{2jb} + \beta_3 X_{1jb}Z_{1b} \tag{10.52}$$

for all j, including $j = k$.

The second difference captures the *change* in P_{ik} associated with the same discrete increase in X_{1k} when Z_1 is at its counterfactual value Z_{1c}, and all of the other independent variables remain at their own baseline values,

$$\Delta P_{ik} | Z_{1c} = \frac{e^{V'_{ikc}}}{\sum_{j=1}^{J} e^{V'_{ijc}}} - \frac{e^{V'_{ikb}}}{\sum_{j=1}^{J} e^{V'_{ijb}}}. \tag{10.53}$$

In the counterfactual scenario associated with this second difference, we now have

$$V'_{ikc} = \gamma_{0k} + \gamma_{1k}Z_{1c} + \gamma_{2k}Z_{2b} + \beta_1 X_{1kc} + \beta_2 X_{2kb} + \beta_3 X_{1kc}Z_{1c} \tag{10.54}$$

and

$$V'_{ijc} = \gamma_{0j} + \gamma_{1j}Z_{1c} + \gamma_{2j}Z_{2b} + \beta_1 X_{1jb} + \beta_2 X_{2jb} + \beta_3 X_{1jb}Z_{1c} \tag{10.55}$$

for $j \neq k$. And in the baseline scenario for this second difference, we have

$$V'_{ijb} = \gamma_{0j} + \gamma_{1j}Z_{1c} + \gamma_{2j}Z_{2b} + \beta_1 X_{1jb} + \beta_2 X_{2jb} + \beta_3 X_{1jb}Z_{1c} \tag{10.56}$$

for all j, including $j = k$.

The third difference, which is the interaction effect telling us how the discrete increase in Z_1 modifies the effect of the discrete increase in X_{1k} on P_{ik}, is the difference in these two differences,

$$\textit{Interaction Effect}_{X_{1k}Z_1} = \Delta P_{ik} | Z_{1c} - \Delta P_{ik} | Z_{1b}. \tag{10.57}$$

10.4 SUBSTANTIVE APPLICATION: POLICY PREFERENCES, GENDER, AND PARTY SUPPORT IN THE 1992 UK ELECTIONS

How do policy preferences and gender affect support for political parties in the United Kingdom? In this application, we examine how an individual's policy preferences and gender affected support for the Conservative Party, Labour Party, and Liberal Democrats in the 1992 British general elections. Spatial theories of politics predict that individuals will exhibit more support for parties that share their preferred policy position than those that don't (Downs, 1957; Black, 1958). In effect, voters are expected to "punish" parties that are ideologically distant from them. However, the extent to which individuals punish parties for being ideologically distant is likely to depend on the salience of the particular policy issues under consideration. Individuals are only going to punish parties for being ideologically distant from them if the ideological distance occurs on policy issues that they actually care about and thus give weight to when making their vote choice. Moreover, we'd expect that the magnitude of the punishment imposed by voters on a party that's ideologically distant will increase with the perceived salience of the policy issue. According to this theoretical account, an individual's vote choice is shaped *jointly* by (i) the extent to which the proposed policies of the competing parties deviate from the voter's preferred policy position and (ii) the salience that voters attach to the different policy issues under consideration. Empirical analyses of vote choice, as well as other political phenomena such as citizen-elite ideological congruence (Golder, 2010), should take both of these theoretically distinct components (position and salience) of individual policy preferences into account (Alvarez and Nagler, 1998; Ferland and Golder, 2021).

Importantly, a large literature suggests that men and women hold different policy preferences (Sapiro, 1981; Alexander and Andersen, 1986; Shapiro and Mahajan, 1986; Huddy and Terkildsen, 1993; Chaney, Alvarez and Nagler, 1998) and, as a result, we should expect to see gendered differences in vote choice at election time. Policy issues are gendered if men and women hold different positions with respect to them or they attribute different levels of salience to them (Kaufmann and Petrocik, 1999; Inglehart and Norris, 2000; Reingold, 2000; Iversen and Rosenbluth, 2006; Gottlieb, Grossman and Robinson, 2018).[11] We might think to label policy issues that are particularly salient to women as *women's issues* and policy

---

[11] They can also be considered gendered if they disproportionately impact one group more than the other (Reingold, 2000; Swers, 2002).

issues that are particularly salient to men as *men's issues*.[12] Our upcoming application examines gendered differences in party vote choice. In doing so, it provides a potentially useful way to empirically construct a *behavioral* measure of women's and men's issues by identifying those policy issues on which women and men place disproportionate weight when voting. This is arguably better than asking women and men to state what issues are most important to them as there's often a disconnect between someone's reported attitudes and their actual behavior (Gross and Niman, 1975; Schuman and Johnson, 1976; Fazio and Roskos-Ewoldsen, 1994; Jerolmack and Khan, 2014).

In our empirical analysis, we focus on three distinct policy issues. The first is defense policy.[13] Existing research suggests that women and men hold different positions with respect to defense policy. Although results are somewhat mixed at the elite level (Norris and Lovenduski, 1989; Norris, 1996b; Caprioli, 2000; Regan and Paskeviciute, 2003; Koch and Fulton, 2011), survey research at the mass level typically shows that women are less likely to support increases in defense spending and the use of force than men (Conover and Sapiro, 1993; Jelen, Thomas and Wilcox, 1994; Eichenberg, 2003). With the exception of Chaney, Alvarez and Nagler (1998), few studies examine whether defense policy is more salient for women or men. Here we argue that men will care more about defense policy than women because security and protection are closely tied to stereotypical masculinity and because men, who have historically made up the vast majority of the armed forces, are more likely to be directly affected by decisions to go to war. To the extent that this is true, we should expect both women and men to punish parties that are ideologically distant from them on defense

---

[12] There's considerable debate in the literature about exactly how to go about identifying women's and men's issues, the extent to which these issues are connected to women's and men's *interests*, and whether it even makes sense to talk about women's and men's issues given the heterogeneity that exists within each of these groups (Reingold, 2000; Reingold and Swers, 2011; Weldon, 2011; Celis et al., 2014).

[13] Defense policy is often considered a male or masculine issue. The basis for this builds on role congruence theory and the idea that people tend to associate particular positions or expertise with gendered stereotypes and personality traits (Eagly, 1987; Eagly, Wood and Diekman, 2000; Eagly and Karau, 2002; Koenig et al., 2011; Ellemers, 2018; Valdini, 2019; Taylor-Robinson and Geva, Forthcoming). Whereas women are often viewed as caring, communal, and compassionate, men tend to be seen as aggressive and decisive. The claim is that male characteristics are considered more "congruent" with making defense policy, with the result that voters are more likely to support male candidates as opposed to female candidates when defense and security issues are salient (Lawless, 2004; Falk and Kenski, 2006; Holman, Merolla and Zechmeister, 2016). This line of reasoning doesn't directly address whether women and men hold different positions with respect to defense policy or whether they differ in how salient they perceive defense policy to be when it comes to their voting behavior.

policy but for men to do so more. This leads us to our *Defense Distance Hypothesis*.

Defense Distance Hypothesis: An increase in the ideological distance between voters and a political party with respect to defense policy reduces a voter's utility for that party but less so for women than men.

Our second policy issue has to do with the desired level of government intervention in the economy. Existing studies suggest that women and men hold different policy positions with respect to the appropriate size of government, with women preferring a larger and more activist role for the government, especially in the areas of education, health, and social welfare (Chaney, Alvarez and Nagler, 1998; Wängnerud, 2000). At least two potential explanations have been proposed for this (Campbell, 2004). The first is that women, as opposed to men, are driven by an "ethics of care" that causes them to be more altruistic and demand government action to tackle inequality (Gilligan, 1982). The second, which can be derived from either feminist standpoint theory or rational choice theories, argues that women have stronger preferences for a larger government size for more selfish reasons related to their specific experiences as women as well as the fact that their relative poverty means that they benefit disproportionately from government programs and progressive taxation. Along these lines, several scholars find that women's participation in the labor market is associated with increased government size (Iversen and Rosenbluth, 2006; Cavalcanti and Tavares, 2011). As women enter the labor force, they have less time to perform domestic activities such as childcare that have traditionally fallen on their shoulders. This increases their demand for the government to step in and help in areas related to things like childcare, education, and healthcare. Concerns about female poverty arising from increasing rates of divorce and single motherhood have also been linked with increased demand from women for a stronger social welfare net (Edlund and Pande, 2002).[14] But what about gendered differences in the *salience* of this policy dimension? The fact that women can expect to disproportionately benefit from a larger government size, perhaps because more extensive government services help them to challenge traditional norms regarding the gendered division of labor or role of women in society, suggests that the appropriate level of government intervention in the economy will be a more salient issue for women than men. To the extent

---

[14] In line with this general reasoning, studies find that the extension of women's suffrage in the United States and Western Europe was associated with increases in the size and scope of the government's role (Lott and Kenny, 1999; Aidt and Dallal, 2008). Bravo-Ortega, Eterovic and Paredes (2018), though, find no such positive relationship in a more global sample.

that this is true, we should expect both women and men to punish parties that are ideologically distant from them on policy related to the appropriate size of the government but for women to do so more. This leads us to our *Government Intervention Distance Hypothesis*.

Government Intervention Distance Hypothesis: An increase in the ideological distance between voters and a political party with respect to the appropriate level of government intervention in the economy reduces a voter's utility for that party but more so for women than men.

Our third policy issue has to do with women's rights. Perhaps surprisingly, empirical studies generally find that women and men hold similar positions with respect to feminism and women's rights (Simon and Landis, 1989; Banaszak and Plutzer, 1993; Hayes, 1997; Mayeri et al., 2008), especially when these rights are broadly conceived (Jelen, Thomas and Wilcox, 1994).[15] As Rhodebeck (1996, 387) notes in the United States context, "gender is a poor predictor of support for the Equal Rights Amendment (Mansbridge, 1985), equitable treatment of women in the workplace (Klein, 1985), or liberal abortion laws (Cook, Jelen and Wilcox, 1992)." In general, there's evidence that women's and men's positions toward women's rights have both liberalized and increasingly converged since the 1970s, although men have fairly consistently held more liberal attitudes than women when it comes to abortion rights (Bolzendahl and Myers, 2004). Part of the reason why preferences for women's rights don't vary substantially by gender is that opposition to equal rights for women, reproductive freedom, and non-traditional family relations has historically been common among more "traditional" women. But what about gendered differences in the *salience* of this policy dimension? While men and women may hold similar attitudes with respect to women's rights, we might suspect that this issue is more salient for women than it is for men because policy change in this area has a more direct effect on their lives. In line with this, a 2019 survey in 34 countries by the Pew Research Center finds that in 20 countries, a greater percentage of women than men said that gender equality is very important; in 13 countries, there were no differences across gender (Horowitz and Fetterolf, 2020). To the extent that this is true, we should expect both women and men to punish parties that are ideologically distant from them on policy related to women's rights but for women to do so more. This leads us to our *Women's Rights Distance Hypothesis*.

---

[15] While early research suggested that support for feminism and women's rights develops differently for women and men (Klein, 1984), later research has found few significant gender differences in the development of feminist consciousness (Bolzendahl and Myers, 2004).

Women's Rights Distance Hypothesis: An increase in the ideological distance between voters and a political party on the issue of women's rights reduces a voter's utility for that party but more so for women than men.

If gender modifies the effect of ideological distance on party support, then it automatically follows that ideological distance must also modify the effect of gender on party support. Due to the symmetry of interactions, we already have predictions about exactly how ideological distance modifies the effect of gender on party support. From the *Defense Distance Hypothesis*, we know that an increase in the ideological distance to a party on defense policy is expected to have a positive modifying effect on how being female as opposed to male affects the utility associated with the party. And from the *Government Intervention Distance Hypothesis* and the *Women's Rights Distance Hypothesis*, we know that an increase in the ideological distance to a party with respect to the appropriate size of the government or women's rights is expected to have a negative modifying effect on how being female affects the utility associated with the party. What we don't know from our theoretical discussion so far is the predicted sign of the effect of being female as opposed to male on the utility associated with a party. So what can we say about how gender affects party vote choice?

According to the existing literature, women have historically been more conservative and, hence, more likely to vote for right-wing parties than men (Duverger, 1955; Norris, 1996b; Inglehart and Norris, 2000). Women's support for right-wing parties was often attributed to the greater emphasis that these parties place on religion and family values than left-wing parties. And men's lower support for right-wing parties was often attributed to their position in the workforce and the strong connection between trade unions and left-wing parties. However, a realignment has occurred over the last three or four decades such that women, at least in post-industrial societies, are now less likely to vote for right-wing parties than men. One explanation for this realignment has to do with the cultural and structural changes that have accompanied economic development (Inglehart and Norris, 2000, 2003; Edlund and Pande, 2002). In particular, women's increased labor force participation has led them to develop a stronger preference for the type of more interventionist government typically favored by left-wing parties (Iversen and Rosenbluth, 2006; Cavalcanti and Tavares, 2011). The precise timing of the cultural and structural changes that have led to this political realignment has varied across countries. While there's evidence that women have been less likely to vote for right-wing parties than men since about 1980 in the United States (Norris, 2001; Box-Steffensmeier, De Boef and Lin, 2004), this new version of the "gender gap" in voting only started to emerge in Western European countries during the 1990s. With respect to the United Kingdom, Norris (1996a, 339) argues that British women in the early 1990s (the country and period of our substantive application) remained some of the most conservative in Europe.

Studies that examine the impact of gender on party vote choice, like those cited above, tend to focus on the fact that women and men hold different policy positions. It's important to recognize, though, that our upcoming analysis incorporates information about the policy positions of voters. It does so by taking account of the ideological distance between voters and parties on defense policy, the appropriate size of the government, and women's issues. This means that any effect of gender that we find on party vote choice will be the effect of gender for reasons *other than the ideological distances to the parties on these three dimensions*. What might these "other reasons" be? We can think of at least three possibilities. First, there might be other policy issues beyond those in our analysis that individuals care about when making their vote choice.[16] Second, individuals might take account of valence considerations, such as leader charisma or competence, when making their vote choice (Ansolabehere and Snyder, 2000; Schofield, 2003; Adams, Merrill and Grofman, 2005).[17] Third, individuals may consider how salient particular policy issues seem to be for the parties and not just the positions that parties take on these issues (Budge and Farlie, 1983; Budge et al., 2001). In other words, individuals may take into account the difference in how much parties talk about particular policy issues compared to how much they'd like the parties to talk about these issues. It's plausible that gendered differences might exist with respect to any of these three things. Importantly, it's difficult to know with any certainty which of the three things we've raised will be behind any gender effect that we find in our upcoming empirical analysis of party vote choice in the 1992 British elections. Given this, we don't believe that our theory is sufficiently precise after taking account of voter and party policy positions to make a clear prediction about the effect of being female on party vote choice beyond the prediction that gender will modify the effect of ideological distance in a manner consistent with our three *Distance*

---

[16] Note, though, that any effect of gender that we find would be picking up gender differences in the effect of ideological distance on other policy issues only to the extent that the ideological distance between individuals and the parties on these other policy issues are uncorrelated with the ideological distances on the three policy issues already in our model.

[17] Note, though, that voters may interpret competence and other valence characteristics in a way that's correlated with policy differences. For example, voters often perceive parties that share their policy positions as more competent than parties that don't.

hypotheses.[18] We can, of course, still empirically evaluate the effect of being female based on our results.

We test our hypotheses using survey data from the 1992 British General Election Study (Heath et al., 1993).[19] We restrict our sample to those respondents who voted for the Conservative Party, Labour Party, or Liberal Democrats in England or Wales.[20] Our dependent variable, *Party Choice*, is an unordered discrete variable with $J = 3$ party choice alternatives that equals 1 if the respondent voted for the Conservative Party, 2 if they voted for the Labour Party, and 3 if they voted for the Liberal Democrats. With respect to their ideological positions, the Conservative Party is a right-wing party, the Labour Party is a left-wing party, and the Liberal Democrats are a centrist party. In terms of our key alternative-specific variables, we have three policy distance variables. *Defense Distance$_k$*, *Government Intervention Distance$_k$*, and *Women's Rights Distance$_k$* are measured on 0–6 scales and capture the absolute distance between a respondent's self-placement on the named policy dimension and their placement of party k on the same policy dimension.[21] In terms of our key individual-specific variables, *Female* is a dichotomous variable that equals 1 if the respondent identifies as female and 0 if they identify as male. *Defense Distance$_k$ × Female*, *Government Intervention Distance$_k$ × Female*, and *Women's Rights Distance$_k$ × Female* are included to evaluate the claim that the salience of the three policy distance measures for vote choice vary with someone's gender. In terms of control variables, we include individual-specific variables capturing a

[18] When discussing the symmetry of interactions in Chapter 2, we encouraged scholars, *when possible*, to derive and test hypotheses about the effects of each of the variables involved in an interaction. We did so because this provides a stronger test of the underlying theory. As the application in this chapter demonstrates, our theories aren't always precise enough to do this.

[19] In order to test the conditional implications of our theory, we need information on how individuals vote and how they place themselves and the competing parties on different types of policy dimensions. The 1992 British General Election Study is one of the few available election studies to provide all of the necessary information.

[20] Of those respondents who indicated that they voted in England and Wales, only 18 voted for a different party; all of these respondents were removed from the analysis.

[21] Respondents were asked to place themselves and each of the three parties on several 1–7 policy dimensions. The defense policy dimension is coded such that 1 indicates that we should spend much less money on defence and 7 indicates that we should spend much more. The government intervention policy dimension is coded such that 1 indicates that the government should just let each person get ahead on their own and 7 indicates that the government should see to it that every person has a job and a good standard of living. The women's rights policy dimension is coded such that 1 indicates that a woman's place is in the home and 7 indicates that women should have an equal role with men in running business, industry, and government. While the concept of policy distance is continuous, its operationalization here technically leads to a discrete variable with values 0, 1, 2, 3, 4, 5, and 6. For the purposes of this application, we treat our three policy distance variables "as if" they were continuous.

respondent's household income, education, class, trade union membership, and age.[22]

Given the unordered and discrete nature of our dependent variable, we test our hypotheses with a multinomial logit model where the utility associated with party k for individual i is

$$U_{ik} = \beta_1 Defense\ Distance_k + \beta_2 Defense\ Distance_k \times Female$$
$$+ \beta_3 Government\ Intervention\ Distance_k$$
$$+ \beta_4 Government\ Intervention\ Distance_k \times Female$$
$$+ \beta_5 Women's\ Rights\ Distance_k$$
$$+ \beta_6 Women's\ Rights\ Distance_k \times Female$$
$$+ \gamma_{0k} + \gamma_{1k} Female + \gamma_k Controls + \epsilon_k. \qquad (10.58)$$

To identify the effects of the individual-specific variables, we treat the left-wing Labour Party as the baseline or reference category.

The marginal effect of *Defense Distance$_k$* on the utility of party k is

$$\frac{\partial U_{ik}}{\partial Defense\ Distance_k} = \beta_1 + \beta_2 Female. \qquad (10.59)$$

The magnitude of this effect can be interpreted as a behavioral measure of the *salience* of defense policy for an individual's vote choice. According to the *Defense Distance Hypothesis*, an increase in the ideological distance between an individual and party k on defense policy should reduce the utility associated with party k for both men and women. Thus, $\beta_1 + \beta_2 Female$ should be negative for all values of *Female*. It follows that β_1, the effect of increased ideological distance with respect to defense policy for men, should be negative. The negative effect of increased ideological distance on this policy dimension should be smaller for women because we expect women to view defense policy as less salient to their vote choice than men. As a result, β_2 should be positive. For $\beta_1 + \beta_2 Female$ to always be negative, it must be the case that $|\beta_2| < |\beta_1|$.

The marginal effect of *Government Intervention Distance$_k$* on the utility of party k is

$$\frac{\partial U_{ik}}{\partial Government\ Intervention\ Distance_k} = \beta_3 + \beta_4 Female. \qquad (10.60)$$

---

[22] *Income* measures household income in quartiles (1–4) with higher values indicating higher income, *Education* measures an individual's highest level of education (0 = no qualifications, 1 = O-level or CSE, 2 = A-level, 3 = higher education below degree, 4 = degree), *Working Class* is a dichotomous variable that equals 1 if the respondent identifies as working class and 0 if they identify as middle class, *Trade Union* is a dichotomous variable that equals 1 if the respondent is a trade union member and 0 otherwise, and *Age* is the respondent's age in years.

The magnitude of this effect can be interpreted as a behavioral measure of the *salience* of policy regarding the appropriate size of government for an individual's vote choice. According to the *Government Intervention Distance Hypothesis*, an increase in the ideological distance between an individual and party k with respect to policy regarding the appropriate level of government intervention should reduce the utility associated with party k for both men and women. Thus, $\beta_3 + \beta_4 Female$ should be negative for all values of *Female*. It follows that β_3, the effect of increased ideological distance with respect to the appropriate size of government for men, should be negative. The negative effect of increased ideological distance on this policy dimension should be larger for women because we expect women to view the appropriate size of government as more salient to their vote choice than men. As a result, β_4 should be negative.

The marginal effect of *Women's Rights Distance$_k$* on the utility of party k is

$$\frac{\partial U_{ik}}{\partial \, Women's \ Rights \ Distance_k} = \beta_5 + \beta_6 Female. \tag{10.61}$$

The magnitude of this effect can be interpreted as a behavioral measure of the *salience* of women's rights policy for an individual's vote choice. According to the *Women's Rights Distance Hypothesis*, an increase in the ideological distance between an individual and party k on women's rights policy should reduce the utility associated with party k for both men and women. Thus, $\beta_5 + \beta_6 Female$ should be negative for all values of *Female*. It follows that β_5, the effect of increased ideological distance on women's rights policy for men, should be negative. The negative effect of increased ideological distance on this policy dimension should be larger for women because we expect women to view women's rights as more salient to their vote choice than men. As a result, β_6 should be negative.

The effect of a one-unit increase in *Female* on the utility of party k is

$$\frac{\partial U_{ik}}{\partial Female} = \gamma_{1k} + \beta_2 Defense \ Distance_k$$
$$+ \beta_4 Government \ Intervention \ Distance_k$$
$$+ \beta_6 Women's \ Rights \ Distance_k. \tag{10.62}$$

The fact that the coefficient γ_{1k} is subscripted by k means that the effect of being female shown in Eq. 10.62 varies across the different parties. This particular coefficient tells us the effect of being female for an individual who holds the exact same positions on policies related to defense, the appropriate size of government, and women's rights as party k. Our theory wasn't sufficiently precise to make predictions about the sign of the effect of being female on party vote choice for any set of values for our three ideological distance variables. However, we do know from our three *Distance* hypotheses that the effect of being female should increase with *Defense Distance$_k$* ($\beta_2 < 0$) and decrease with *Government Intervention Distance$_k$* ($\beta_4 < 0$) and *Women's Rights Distance$_k$* ($\beta_6 < 0$).

Before we look at the results from our multinomial logit model, we briefly examine whether women and men in the 1992 UK elections held different positions with respect to our three policy issues. Recall that all of the policy issues are measured on a 1–7 scale, where higher numbers indicate increased support for defence spending, government intervention, or women's rights. Women (3.63) indicate slightly higher support for defense spending than men (3.53). While this difference (0.11) is statistically significant ($p < 0.05$, two-tailed), it's substantively small and represents only 7% of one standard deviation in people's preferences for defense spending. When it comes to government intervention, women (5.36) desire a larger government than men (5.19). This difference (0.17) is again statistically significant ($p < 0.01$) but substantively very small. Similar results hold for attitudes toward women's rights. Women (5.92) hold slightly more favorable attitudes toward women's rights than men (5.82). This difference (0.10) is statistically significant ($p < 0.10$) but is again substantively very small. Overall, there appear to be no large or substantively meaningful differences between women and men when it comes to their reported preferences on our three different policy issues. The absence of significant differences in the policy positions of women and men, though, doesn't necessarily imply that there are no gender differences in party vote choice. Maybe there are differences between women and men when it comes to the salience of these policy issues for their actual voting behavior? Our upcoming empirical analysis examines this possibility.

The results from our multinomial logit model are shown in Table 10.4. As predicted, the coefficients on *Defense Distance$_k$*, *Government Intervention Distance$_k$*, and *Women's Rights Distance$_k$* are all negative and statistically significant, indicating that men receive less utility from a party, and are therefore less likely to vote for a party, when that party becomes more ideologically distant from them on each of our three policy issues. Specifically, a one-unit increase in the ideological distance between an individual and a party on defense policy is associated with a 0.70 [0.45, 0.95] unit reduction in the utility that men obtain from the party. The reductions in utility that men experience when the ideological distance between them and a party increases by one unit on the government intervention and the women's rights policy dimensions are 0.42 [0.24, 0.60] and 0.54 [0.26, 0.81], respectively. Since our three policy dimensions are measured on the same 1–7 scale, we can compare the magnitudes of the coefficients on *Defense Distance$_k$*, *Government Intervention Distance$_k$*,

Table 10.4 Determinants of party vote choice in the 1992 British general election I

Dependent Variable: *Party Choice: Conservatives (1), Labour (2), Liberal Democrats (3)*

	Multinomial Logit Model	
Alternative-Specific Variables		
Defense Distance$_k$	−0.70***	
	(0.13)	
Defense Distance$_k$ × Female	0.18	
	(0.17)	
Government Intervention Distance$_k$	−0.42***	
	(0.09)	
Government Intervention Distance$_k$ × Female	−0.26*	
	(0.13)	
Women's Rights Distance$_k$	−0.54***	
	(0.14)	
Women's Rights Distance$_k$ × Female	0.11	
	(0.19)	
Individual-Specific Variables		
	Conservative Party	*Liberal Democrats*
Female	0.84**	0.53
	(0.35)	(0.34)
Income	0.66***	0.40**
	(0.17)	(0.18)
Education	0.43***	0.30*
	(0.16)	(0.16)
Working Class	−0.77**	−0.49
	(0.35)	(0.38)
Trade Union	−0.56*	−0.19
	(0.33)	(0.34)
Age	0.02	0.02
	(0.01)	(0.01)
Constant	−1.62*	−2.57***
	(0.87)	(0.93)
Number of Individuals	480	
Number of Party Alternatives	1,440	
Log Likelihood	−284.55	

Standard errors in parentheses. *$p < 0.10$; **$p < 0.05$; ***$p < 0.01$ (two-tailed)

Note: The data include voters in England and Wales who voted for either the Conservative Party, the Labour Party, or the Liberal Democrats in the 1992 British general election. The Labour Party is treated as the baseline category for the individual-specific variables.

and *Women's Rights Distance$_k$* to evaluate the relative salience of these policy issues for the vote choice of men. In this regard, *Defense Distance$_k$* has the largest negative coefficient, suggesting that defense policy was more salient than policies related to government size and women's rights when men decided for whom to vote in the 1992 British elections. Substantively, the negative coefficient on *Defense Distance$_k$* is 66.6% larger than the negative coefficient on *Government Intervention Distance$_k$* and 30.3% larger than the negative coefficient on *Women's Rights Distance$_k$*. The difference in the magnitudes of the *Defense Distance$_k$* and *Government Intervention Distance$_k$* coefficients is statistically significant ($p < 0.08$, two-tailed); however, this isn't the case for the comparison with the *Women's Rights Distance$_k$* coefficient ($p < 0.39$).

What about the modifying effect of gender? Do women weight the importance of our three policy issues differently to men when making their vote choice? As predicted, the coefficient on *Defense Distance$_k$ × Female* is positive, suggesting that defense policy is less salient for women than men when it comes to their vote choice. To be specific, a one-unit increase in the ideological distance to a party on defense policy is associated with a 0.52 [0.30, 0.74] unit reduction in the utility that women obtain from that party. This negative effect is 25.7% smaller than it is for men. The difference in the magnitude of the effects between women and men isn't statistically significant, though, raising doubts about whether it's appropriate to infer that the salience of defense policy for someone's vote choice varies with their gender. Also as predicted, the coefficient on *Government Intervention Distance$_k$ × Female* is negative and statistically significant ($p < 0.057$, two-tailed), indicating that policy about the appropriate size of government is more salient for women than men. To be specific, a one-unit increase in the ideological distance to a party on the government intervention dimension is associated with a 0.68 [0.48, 0.87] unit reduction in the utility that women obtain from that party. This negative effect is 61.3% larger than it is for men. Contrary to our predictions, the coefficient on *Women's Rights Distance$_k$ × Female* is positive, suggesting that women care less about a party's policy position on women's rights than men when choosing for whom to vote. More specifically, a one-unit increase in the ideological distance to a party on the women's rights dimension is associated with a 0.42 [0.18, 0.67] unit reduction in the utility that women obtain from that party. This negative effect is 20.8% smaller than it is for men. The difference in the magnitude of the effects between women and men isn't statistically significant, though, suggesting that we might not want to rule out the possibility that women and men don't really differ in how they weight policy on women's rights in their vote calculus. The effects of increasing the ideological distance to a party on each of our three policy issues by one

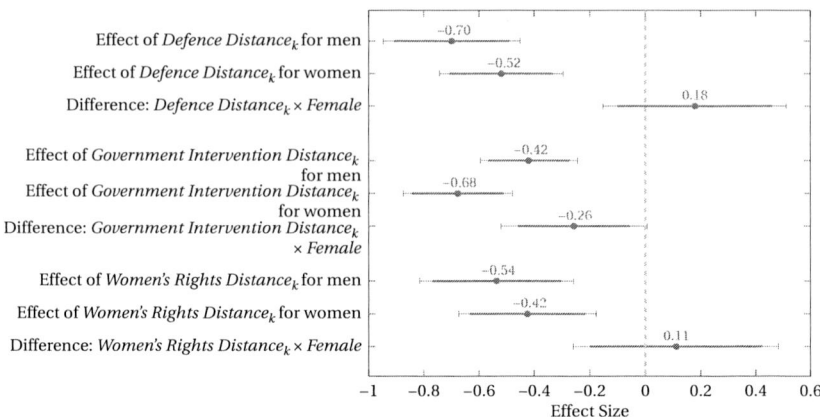

Figure 10.1 The conditional effect of ideological distance on voter utility by gender in the 1992 legislative elections in the United Kingdom

Note: Figure 10.1 shows the effect of a one-unit increase in the ideological distance to a party on three policy issues on voter utility for that party for men and women along with the associated difference (interaction effect). The various effects are shown as small circles along with their corresponding two-tailed 95% (thin) and 90% (thick) confidence intervals.

unit on the utility of male and female voters are presented graphically in Figure 10.1. The various effects are shown as small circles along with their corresponding 95% (thin) and 90% (thick) confidence intervals.

Although we don't have a specific theoretical prediction about the effect of gender on voter utility, we can still examine this effect. As the results in the *Conservative Party* column in Table 10.4 indicate, the coefficient on *Female* is positive and statistically significant for the case where we're comparing the right-wing Conservative Party to the baseline left-wing Labour Party. This indicates that women prefer the Conservative Party over the Labour Party more than men when the Conservative Party shares the same policy position as the voter on all three policy issues. Put differently, ideologically conservative British women in 1992 favored the Conservative Party over the Labour Party more than ideologically conservative British men.[23] The coefficients on *Defense Distance$_k$* × *Female*

---

[23] Norris (1996*a*, 1999) has suggested that there may have been a "gender generation gap" in the United Kingdom in the early 1990s, where older women born prior to World War II were more likely to vote for right-wing parties than men, and younger women born afterward were more likely to vote for left-wing parties than men. We tested this claim by interacting *Female* with a dichotomous variable indicating whether someone was born before or after World War II. While the coefficient on the new individual-specific interaction term had the predicted negative sign for the Conservative–Labour comparison, suggesting that the positive gender gap in support for the Conservative Party between women and men was smaller for those born in

and *Women's Rights Distance$_k$* × *Female* are both statistically insignificant, indicating that the positive gender gap between women and men in favor of the Conservative Party doesn't vary as the Conservative Party becomes more ideologically distant in terms of its policies on defense spending or women's rights. In contrast, the coefficient on *Government Intervention Distance$_k$* × *Female* is negative and statistically significant, indicating that the extent to which women favor the Conservative Party more than men declines as the Conservative Party becomes more ideologically distant in terms of its policy regarding the appropriate size of government. We saw earlier that women view policy related to the appropriate size of government as more salient to their vote choice than men. Thus, what we're seeing here is that the positive gender gap where women favor the Conservative Party over men declines as the party becomes more ideologically distant on a policy that's particularly salient to women.

Recall from Eq. 10.62 that the effect of being female on the utility associated with party k is

$$\frac{\partial U_{ik}}{\partial Female} = \gamma_{1k} + \beta_2 Defense\ Distance_k$$

$$+ \beta_4 Government\ Intervention\ Distance_k$$

$$+ \beta_6 Women's\ Rights\ Distance_k. \tag{10.63}$$

It's not possible to fully visualize this effect with either a 2-D or 3-D "marginal effect" plot because its magnitude varies with the values of *three* policy distance variables. One possible strategy in the current application is to plot the effect of being female on voter utility for the Conservative Party across the observed range of values for *Government Intervention Distance$_k$*, while setting *Defense Distance$_k$* (1.50) and *Women's Rights Distance$_k$* (1.45) to their mean observed values for the Conservative Party. This strategy seems reasonable here because we've already seen that, unlike *Government Intervention Distance$_k$*, *Defense Distance$_k$* and *Women's Rights Distance$_k$* don't significantly modify the effect of being female. Such a plot is shown in Figure 10.2. As the plot indicates, women receive statistically significantly more utility than men from the Conservative Party (relative to the Labour Party) only when the Conservative Party is less than one unit away from them with respect to its policy on government size. The histogram suggests that we're only likely to see a pro-Conservative gender gap between women and men among about 25% of voters.[24] These results

---

the post-war period, it wasn't statistically significant. In sum, we didn't find compelling evidence for a gender generation gap in the 1992 UK elections.

[24] The histogram doesn't distinguish between women and men because there were only trivial differences in the percentage of women and men at different values of *Government Intervention Distance$_k$* for the Conservative Party.

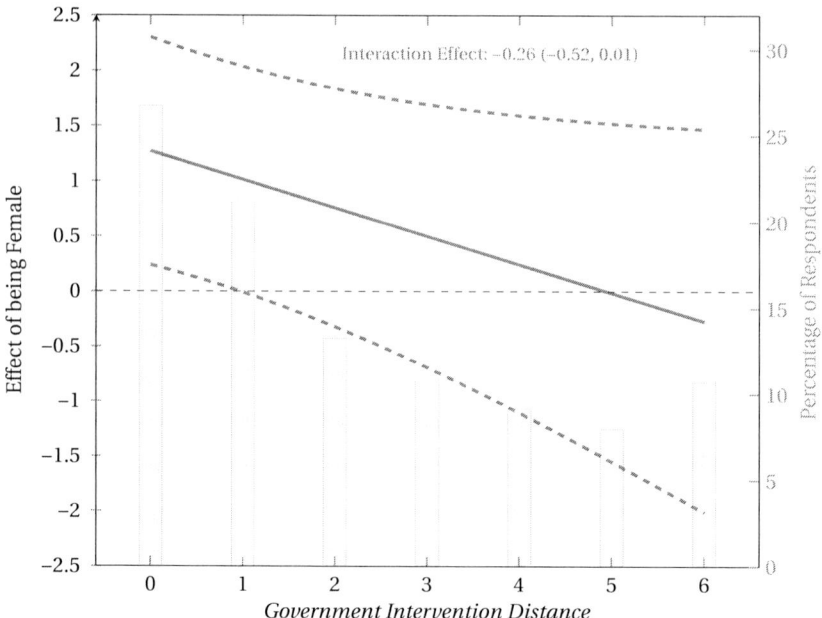

Figure 10.2 The conditional effect of being female on voter utility for the Conservative Party during the 1992 legislative elections in the United Kingdom I

Note: Figure 10.2 shows how the effect of being female instead of male affects the utility someone receives from the Conservative Party (as opposed to the Labour Party) when we vary the Conservative Party's value of *Government Intervention Distance*$_k$ across its observed range while holding *Defense Distance*$_k$ (1.50) and *Women's Rights Distance*$_k$ (1.45) at their observed means for the Conservative Party. The curved dashed lines represent two-tailed 95% confidence intervals. We also show the interaction effect between *Female* and *Government Intervention Distance*$_k$. The histogram shows the percentage of respondents at different values of *Government Intervention Distance*$_k$ for the Conservative Party.

suggest that the pro-Conservative gender gap between women and men in the 1992 United Kingdom elections wasn't unconditional and instead depended on the Conservative Party not being too ideologically distant on issues that were particularly salient to women, such as the appropriate size of the government.

Of course, the plot in Figure 10.2 only shows the effect of *Female* for one particular combination of values for *Defense Distance*$_k$ and *Women's Rights Distance*$_k$. Another possible strategy is to produce a series of 3-D "marginal effect" plots where we show the effect of *Female* across the observed range of values for two of our policy distance variables for different values of our third policy distance variable. In Figure 10.3, we show one potential way of doing this. In panel (a), the meshed surface shows

(a) The Conditional Effect of being Female when *Women's Rights Distance*$_k$ = 0

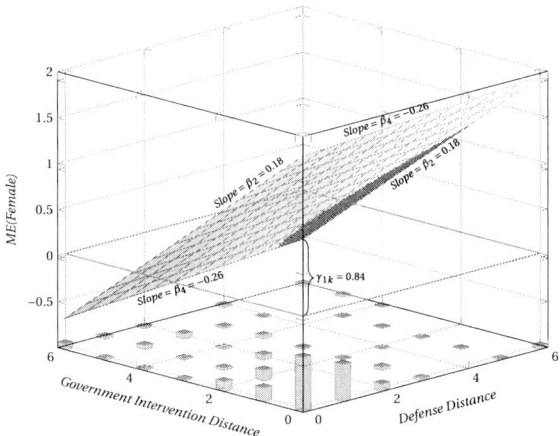

(b) The Conditional Effect of being Female when *Women's Rights Distance*$_k$ = 1

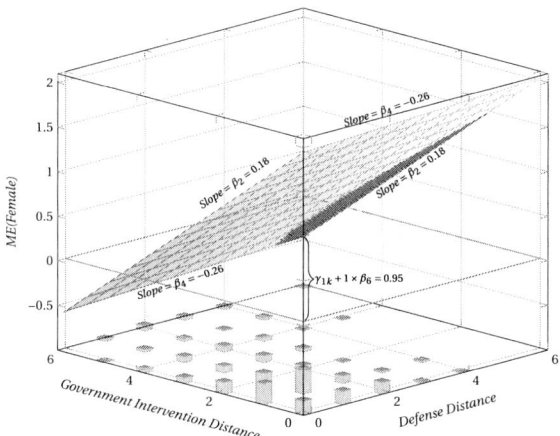

(c) *Defense Distance*$_k$ × *Female*: 0.18 (−0.15, 0.51)
Government Intervention Distance$_k$ × *Female*: −0.26 (−0.52, 0.01)
Women's Rights Distance$_k$ × *Female*: 0.11 (−0.26, 0.48).

Figure 10.3 The conditional effect of being female on voter utility for the Conservative Party during the 1992 legislative elections in the United Kingdom II

Note: The meshed surfaces show how the effect of being female on voter utility for the Conservative Party varies with *Defense Distance*$_k$ and *Government Intervention Distance*$_k$ when *Women's Rights Distance*$_k$ is at its modal value of 0 (a) and when *Women's Rights Distance*$_k$ is at its median value of 1 (b). The surfaces are shaded darker when the effect is statistically significant ($p < 0.05$, two-tailed). The "zero surfaces" help to identify when the effect is positive or negative. The 3-D histograms on the base of the "cubes" indicate the percentage of sample observations at the different values of *Defense Distance*$_k$, *Government Intervention Distance*$_k$, and *Women's Rights Distance*$_k$. In panel (c), we report three interaction effects indicating how *Defense Distance*$_k$, *Government Intervention Distance*$_k$, and *Women's Rights Distance*$_k$ modify the effect of being female.

how the effect of being female on voter utility for the Conservative Party varies with the values of *Defense Distance$_k$* and *Government Intervention Distance$_k$* when *Women's Rights Distance$_k$* is at its modal value of 0. The slope of the surface upward to the right indicates the positive modifying effect of *Defense Distance$_k$*, while the slope downward to the left indicates the negative modifying effect of *Government Intervention Distance$_k$*. The meshed surface is shaded darker when the effect is statistically significant ($p < 0.05$, two-tailed). The horizontal "zero surface" helps us to see when the effect of being female is positive, zero, or negative. To help readers evaluate the substantive importance of the information contained in the marginal effect plot, we provide information about the distribution of the values of the two modifying variables *Defense Distance$_k$* and *Government Intervention Distance$_k$* when *Women's Rights Distance$_k$* = 0 in the form of a 3-D histogram on the base of the "cube" housing the plot. While it's difficult to include a specific axis for the 3-D histogram in the plot, the relative heights of the "columns" do give a sense of where the observations are located in the available space and are comparable across panels. In panel (b), the meshed surface provides the same information but for the scenario where *Women's Rights Distance$_k$* is at its median value of 1. The slopes of the meshed surface in panel (b) are identical to those in panel (a) because we've assumed that the modifying effects of the policy distance variables are all independent or unconditional. We can see the positive modifying effect of *Women's Rights Distance$_k$* by the fact that the marginal effect surface in panel (b) is now higher than that in panel (a). In panel (c), we report three interaction effects indicating how *Defense Distance$_k$*, *Government Intervention Distance$_k$*, and *Women's Rights Distance$_k$* modify the effect of being female. The first two interaction effects speak to the magnitude and statistical significance of the slopes of the meshed surfaces in panels (a) and (b) as we vary the values of *Defense Distance$_k$* and *Government Intervention Distance$_k$*. The third interaction effect speaks to the magnitude and statistical significance of the change in the height of the meshed surface as we move from panel (a) to panel (b). Depending on the goal of the substantive application, scholars might wish to increase the number of plots they show by choosing additional values for the third modifying variable *Women's Rights Distance$_k$* or by choosing different values for it.

The main insight from Figure 10.3 is that the effect of being female on voter utility from the Conservative Party is only positive and statistically significant when the Conservative Party isn't too ideologically distant. We can see this by the fact that the darker shaded surface is located in the front corner where *Defense Distance$_k$* and *Government Intervention Distance$_k$*

are both zero.[25] We also see that the darker surface disappears more quickly as we increase *Government Intervention Distance$_k$* than as we increase *Defense Distance$_k$*. This fits with our claim that women care more about policy with respect to the appropriate size of the government than they do with respect to defense policy.

The information that we've provided so far speaks to the directional nature of the conditional effects of ideological distance and gender. What we haven't examined yet is whether the magnitude of these conditional effects are substantively meaningful. This is because there's no meaningful scale for voter utility. To examine whether our results are substantively significant, we need to calculate effects in terms of quantities of interest that have a meaningful scale.

Suppose that we're interested in the substantive effect of our ideological distance variables. For illustrative purposes, we'll focus on the substantive effect of ideological distance with respect to the appropriate size of government. As we've seen, there are different ways of doing this. One way is to look at the marginal effect of an increase in *Government Intervention Distance* on the probability of voting for each of our three parties. To do this, we must first choose which party is going to become more ideologically distant on this policy dimension. In our example, we'll assume that it's the Conservative Party that becomes more ideologically distant. Given the non-linear nature of the multinomial logit model, we also need to specify the baseline scenario in which the increase in *Government Intervention Distance* occurs. In our example, we'll assume that we have a 47 year old middle class voter who isn't a trade union member and who has median income (*Income* = 6, £12,000−£14,999) and median education (*Education* = 1, O-levels). In the baseline scenario, our voter is also assumed to have median ideological distances to each of the parties on all three issue dimensions. This means that they're one unit away from both the Conservatives and Liberal Democrats and two units away from Labour on *Defense Distance*; one unit away from both Labour and the Liberal Democrats and two units way from the Conservatives on *Government Intervention Distance*; and one unit away from all three parties on *Women's Rights Distance*. We'll calculate marginal effects for the case where our "representative" voter is male and for the case where they're female.

The marginal effect of increasing the ideological distance between a voter and the Conservative Party with respect to policy on the appropriate

---

[25] Although we don't show it in Figure 10.3, the effect of being female on voter utility from the Conservative Party is never statistically significant – there's no darker surface at all – once *Women's Rights Distance$_k$* is greater than 3.1.

size of government on the probability of voting for each of our three parties is shown in Table 10.5 for both men and women. The effect on the probability of voting for the Conservative Party is a *direct marginal effect*, whereas the effects on the probabilities of voting for the other two parties are *cross-marginal effects*. Direct marginal effects are calculated based on Eq. 10.27, while cross-marginal effects are calculated based on Eq. 10.29. Two-tailed 95% confidence intervals, calculated by the method of simulated coefficients, are shown in brackets.[26] We've also reported the baseline probability in gray that our "representative" individual votes for each of the three parties in order to provide some context for evaluating the substantive magnitude of the marginal effects. In the "Total" column, we've summed the marginal effects and probabilities across the three parties. We've done this simply to confirm our earlier claim that the direct and cross-marginal effects across the choice alternatives in an unordered choice model always sum to 0 and to see that the predicted probabilities of voting for the three parties correctly sum to 1. At the bottom of Table 10.5, we report how being female instead of male modifies the marginal effect of increasing the ideological distance to the Conservative Party with respect to policy regarding the appropriate size of government on the probability of voting for each party. This is simply the difference in the marginal effects calculated for women and men by party.

Table 10.5 indicates that the predicted probability that our hypothetical voter, whether male or female, votes for the Conservative Party in the baseline scenario is over 0.86 and is close to 0 for the Labour Party. As expected, the marginal effect of increasing the ideological distance to the Conservative Party with respect to policy on the appropriate size of government reduces the probability that our hypothetical individual votes for the Conservative Party. The drop in support for the Conservative Party is, as predicted, larger (62.2%) for women (-0.07) than men (-0.05). This gender difference, though, isn't quite statistically significant at the two-tailed 95% level in the particular scenario under consideration here.

Rather than focus on marginal effects, we again recommend that scholars calculate differences in predicted probabilities when substantively interpreting the results from a non-linear model like an unordered choice model. In Table 10.6, we show how the predicted probabilities of voting for each of the three parties change for men and women when we

---

[26] One thing to note is that the confidence intervals aren't symmetric. This is especially noticeable for the baseline probabilities, which are close to 1 and 0. As you'll recall from our discussion about measures of uncertainty in Chapter 8, this is an advantage of the method of simulated coefficients over the delta method, which always produces symmetric confidence intervals.

Table 10.5 The marginal effect of increasing *Government Intervention Distance* for the Conservative Party on the probability of voting for different parties in the 1992 United Kingdom legislative elections

| | Direct Marginal Effects | Cross-Marginal Effects | | |
	Conservatives	Labour	Liberal Democrats	Total
Men				
Marginal Effect	−0.05 [−0.1, −0.02]	0.01 [0.002, 0.03]	0.04 [0.01, 0.09]	0
Baseline Probability	0.86 [0.69, 0.96]	0.02 [0.01, 0.07]	0.11 [0.03, 0.28]	1
Women				
Marginal Effect	−0.07 [−0.14, −0.02]	0.01 [0.002, 0.03]	0.06 [0.02, 0.13]	0
Baseline Probability	0.87 [0.71, 0.96]	0.02 [0.004, 0.05]	0.11 [0.03, 0.27]	1
The Modifying Effect of being Female	−0.02 [−0.08, 0.02]	0.001 [−0.01, 0.01]	0.02 [−0.01, 0.07]	0

Note: The quantities reported in Table 10.5 are based on a baseline scenario where we have a **47** year old middle class voter who isn't a trade union member and who has median income and education levels; they also have median levels of ideological distance to each of the three parties on all three policy issues. Two-tailed **95%** confidence intervals are shown in brackets. The "Total" column sums either the direct marginal and cross-marginal effects or the baseline probabilities across the three parties. The modifying effect of being female instead of male captures the difference in the marginal effects for women and men for each party.

Table 10.6 The effect of a two-unit increase in *Government Intervention Distance* for the Conservative Party on the probability of voting for different parties in the 1992 United Kingdom legislative elections

	Conservatives	Labour	Liberal Democrats
Men			
Baseline Scenario	0.86 [0.69, 0.96]	0.02 [0.01, 0.07]	0.11 [0.03, 0.28]
Counterfactual Scenario	0.74 [0.47, 0.91]	0.05 [0.01, 0.13]	0.22 [0.06, 0.47]
Difference	−0.13 [−0.24, −0.04]	0.02 [0.005, 0.07]	0.10 [0.03, 0.21]
Percentage Difference	−15.0% [−32.6, 4.2]	97.7% [46.8, 165.3]	97.7% [46.8, 165.3]
Women			
Baseline Scenario	0.87 [0.71, 0.96]	0.02 [0.004, 0.05]	0.11 [0.03, 0.27]
Counterfactual Scenario	0.66 [0.37, 0.87]	0.05 [0.01, 0.13]	0.30 [0.10, 0.57]
Difference	−0.22 [−0.36, −0.09]	0.03 [0.01, 0.08]	0.19 [0.07, 0.33]
Percentage Difference	−25.4% [−48.1, −9.1]	190.3% [95.2, 319.3]	190.3% [95.2, 319.3]
The Modifying Effect of being Female			
Difference	**−0.09** [−0.23, 0.03]	**0.01** [−0.02, 0.04]	**0.08** [−0.02, 0.21]
Percentage Difference	**−94.7%** [−323.0, 17.7]	**50.5%** [−46.1, 260.8]	**107.8%** [−14.7, 363.7]

Note: The baseline scenario is a 47 year old middle class voter who isn't a trade union member and who has median income and education levels; they also have median levels of ideological distance to each of the parties on all three policy issues. The counterfactual scenario sees a two-unit increase in *Government Intervention Distance* for the Conservative Party. Two-tailed **95%** confidence intervals are shown in brackets. The modifying effect of being female captures the difference and percentage difference in the differences between women and men reported higher up the table.

increase the ideological distance between the same hypothetical voter as before and the Conservative Party by two units with respect to policy regarding the appropriate size of government. This change in ideological distance is basically equivalent to a one-standard deviation increase (2.04) in *Government Intervention Distance* for the Conservative Party. For each gender group, the first two rows report the predicted probabilities of voting for each of the parties in the baseline and counterfactual scenarios. The next two rows report the difference and percentage difference in the predicted probabilities across the baseline and counterfactual scenarios. At the bottom of Table 10.6, we show in bold how being female instead of male modifies the effect of the two-unit increase in *Government Intervention Distance* for the Conservative Party on the probability of voting for each of the parties. This is simply the difference or percentage difference in the differences reported higher up the table. Two-tailed 95% confidence intervals calculated using the method of simulated coefficients are shown in brackets.

As predicted, increasing the ideological distance between our hypothetical voter and the Conservative Party with respect to policy regarding the appropriate size of government reduces the probability of voting for the Conservative Party for both women and men, with the effect being larger for women (25.4%) than men (15.0%). The percentage difference in the size of the effect across women and men is substantively large (94.7%) but isn't quite statistically significant, at least at the two-tailed 95% level. It might appear that the Liberal Democrats are the primary beneficiary when the ideological distance between our voter and the Conservative Party increases. This is because the probability of voting Liberal Democrat increases by 0.10 for men and 0.30 for women in the counterfactual scenario, whereas the probability of voting Labour increases by just 0.02 for men and 0.05 for women. However, this inference is unwarranted. When we look at the *percentage* difference in the probabilities of voting for the Liberal Democrats and Labour in the counterfactual scenario, we see that they're the same for both the Liberal Democrats and Labour: 97.7% for men and 190.3% for women. This will always be the case due to the independence of irrelevant alternatives assumption in the multinomial logit model. In our hypothetical scenario, we've changed the ideological distance between the voter and the Conservative Party. This changes how likely someone is to vote for the Conservative Party relative to the other parties. However, the ideological distances between the voter and the other two parties haven't changed, and so the independence of irrelevant alternatives assumption means that the probability of voting for the Liberal Democrats relative to the Labour Party in the counterfactual scenario remains unchanged from the baseline scenario. This means, as we've seen, that the percentage changes in the probabilities of voting for these two parties will be identical.[27]

In addition to marginal effects and differences, scholars can also use odds ratios to evaluate the substantive importance of the effects of their variables. Based on Eq. 10.39, we see that a one-unit increase in the ideological distance between our voter and the Conservative Party with respect to policy regarding the appropriate size of government reduces the odds that someone votes for the Conservatives as opposed to either Labour or the Liberal Democrats by 33.9% [21.6, 44.9] if they're a man and by 48.8% [38.1, 58.1] if they're a woman. The magnitude of this negative effect on voting Conservative is 48.9% [−0.9, 133.3] larger for women than it is for men. This substantively large negative modifying effect of being female fits with the idea that women care more about policy related to the appropriate size of government than men and therefore punish parties that are more ideologically distant on this policy dimension more than men.

So far, we've examined the substantive effect of the alternative-specific variable *Government Intervention Distance_k* on someone's vote choice in the 1992 United Kingdom elections and how this varies with their gender. We now switch things around to look at the substantive effect of the individual-specific variable *Female* and how this varies with the ideological distance between a voter and a party with regard to policy on the appropriate size of government. For illustrative purposes, we'll focus on how the effect of being female on someone's party vote choice depends on their ideological distance to the Conservative Party on this policy dimension. Where necessary, we'll assume that we're dealing with the same baseline scenario as before. Given the discrete nature of *Female*, it doesn't make sense to examine the substantive effect of gender on party vote choice in terms of marginal effects. As a result, we'll focus on differences and odds ratios.

In Figure 10.4, we show the change (left column) and percentage change (right column) in the probabilities of voting for the Conservatives (top row), Labour (middle row), and the Liberal Democrats (bottom row) associated with being female rather than male as we alter the ideological distance between our hypothetical voter and the Conservative Party with respect to policy regarding the appropriate size of government. Scholars such as Norris (1996a) have argued that women in the 1992 British general elections were more likely than men to vote for the right-wing Conservatives and less likely than men to vote for Labour. The results

---

[27] Those who wish to relax this particular assumption to allow for non-proportional substitution patterns across the choice alternatives can do this by adopting a mixed logit model (Train, 1998; McFadden and Train, 2000; Glasgow, 2001, 2022).

presented in Figure 10.4 suggest that things are more complicated than this. The plots in panels (a) and (b) provide some evidence that women are more likely to vote for the Conservative Party than men, but only when the Conservative Party isn't too ideologically distant from them with respect to the appropriate size of government. We see this from the fact that the solid lines in both plots are positive only when *Government Intervention Distance (Conservative Party)* < 2.4 (61.3%). The number in parentheses indicates the percentage of respondents that satisfy the stated condition. Importantly, though, this pro-Conservative bias among women is substantively small and never statistically significant. At its greatest magnitude, the probability of voting for the Conservatives is only 0.02 or 2.7% higher for women than men. The plots in panels (c) and (d) provide some evidence that women are less likely to vote for Labour than men, but only when the Conservative Party isn't too ideologically distant with respect to policy on the appropriate size of government. We see this from the fact that the solid lines in both plots are negative when *Government Intervention Distance (Conservative Party)* < 4.1 (81.3%) and statistically significant when *Government Intervention Distance (Conservative Party)* < 0.5 (26.8%). We might think that this negative effect is substantively large in that women have a 58.9% lower probability of voting for Labour than men when *Government Intervention Distance (Conservative Party)* = 0. However, we need to be cautious here as the probability of voting for Labour in this particular scenario is very low for both men (0.009) and women (0.004). Overall, the plots in Figure 10.4 provide limited evidence that gender played a substantively meaningful role in determining vote choice in these particular elections once we incorporate information about the policy positions of voters and parties.[28]

Rather than focus on differences, we can also examine the substantive effect of gender in terms of odds ratios. In Figure 10.5, we show how the odds (left column) and percentage change in the odds (right column) of voting for the Conservatives rather than Labour change as we alter the ideological distance between our voter and the Conservative Party with respect to policy regarding the appropriate size of government.[29]

---

[28] The solid lines shown in panels (e) and (f) suggest that women are less likely than men to vote for the Liberal Democrats when *Government Intervention Distance (Conservative Party)* < 2.2 (61.3%) but more likely when when *Government Intervention Distance (Conservative Party)* is larger than this. Importantly, though, the effect of being female is never statistically significant.

[29] Recall that the effects of individual-specific variables such as *Female* vary across the parties. This means that we can't examine how these variables change the *unconditional* odds of voting for a particular party. Instead, we can only examine how these variables affect the *conditional* odds of voting for a particular party such as the Conservatives versus some other party such as Labour.

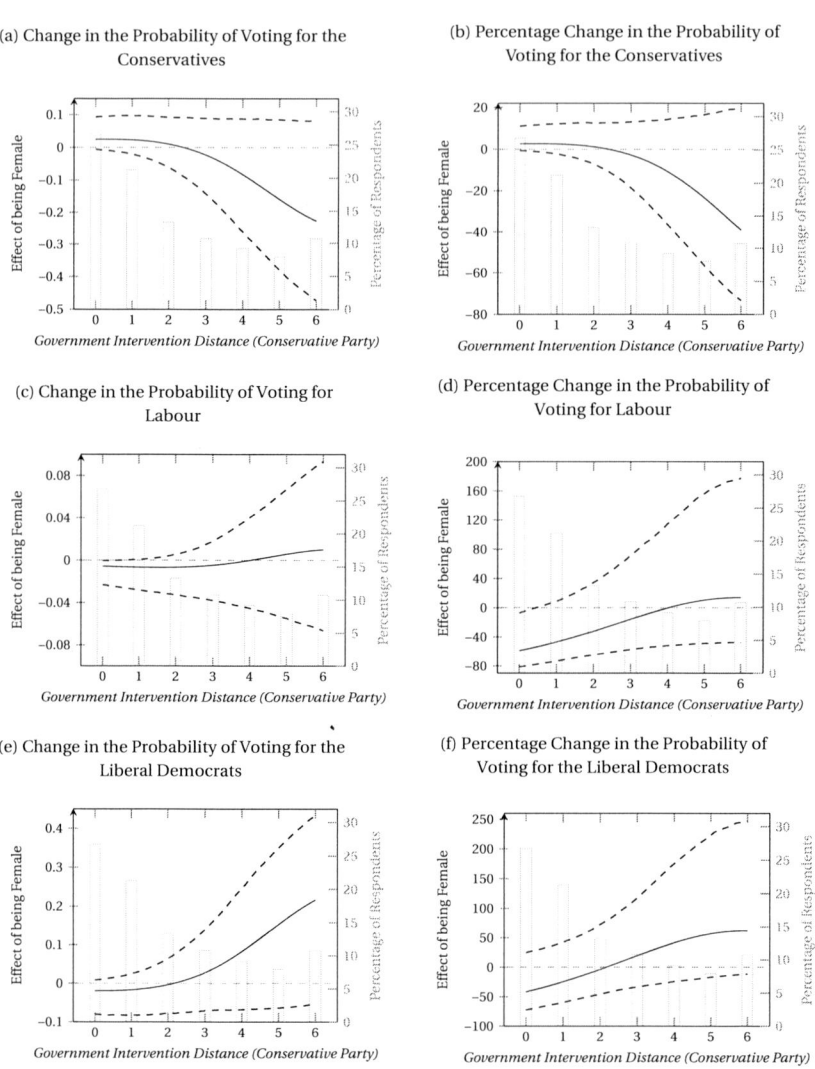

(a) Change in the Probability of Voting for the Conservatives

(b) Percentage Change in the Probability of Voting for the Conservatives

(c) Change in the Probability of Voting for Labour

(d) Percentage Change in the Probability of Voting for Labour

(e) Change in the Probability of Voting for the Liberal Democrats

(f) Percentage Change in the Probability of Voting for the Liberal Democrats

Figure 10.4 (a)–(f) The effect of being female on the probability of voting for the various parties in the 1992 British general elections

Note: Figure 10.4 shows how the effect of being female instead of male affects the change (left column) or percentage change (right column) in the probability of voting for the various parties in the 1992 British general elections when we vary the the ideological distance between a voter and the Conservative Party with respect to policy regarding the appropriate size of government across its observed range. The plots are all drawn for the scenario where we have a 47 year old middle class voter who isn't a trade union member and who has median income and education levels; they also have median levels of ideological distance to Labour and the Liberal Democrats on all three issue dimensions. The curved dashed lines represent two-tailed 95% confidence intervals. The histograms show the percentage of respondents at different values of *Government Intervention Distance$_k$* for the Conservative Party.

(a) Change in the Odds of Voting for the Conservatives rather than Labour

(b) Percentage Change in the Odds of Voting for the Conservatives rather than Labour

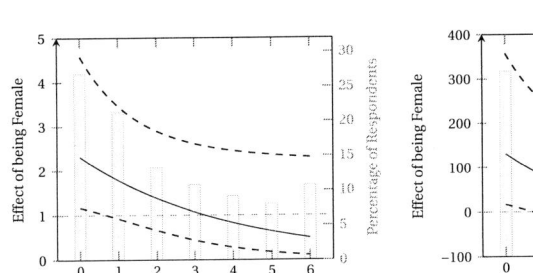
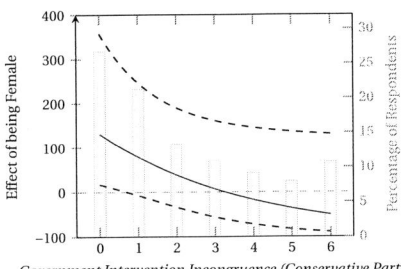

Figure 10.5 The effect of being female on the odds of voting for the Conservatives rather than Labour in the 1992 British general elections

Note: Figure 10.5 shows how being female instead of male affects the odds (a) or percentage change in the odds (b) of voting for the Conservatives rather than Labour in the 1992 British general elections when we vary the the ideological incongruence of the Conservative Party on the government intervention dimension. The curved dashed lines represent two-tailed 95% confidence intervals. The histograms show the percentage of respondents at different values of *Government Intervention Incongruence$_k$* for the Conservative Party.

The results suggest that women are more likely to vote for the right-wing Conservative Party rather than the left-wing Labour Party so long as the ideological distance between the voter and the Conservative Party isn't too great. Specifically, we see that women are more likely than men to vote Conservative as opposed to Labour when *Government Intervention Distance (Conservative Party)* < 3.3 (72.1%). This difference is statistically significant whenever *Government Intervention Distance (Conservative Party)* < 0.8 (26.8%). This pro-Conservative gender gap is substantively large when the voter holds the same policy position as the Conservative Party. To be specific, the odds of voting for the Conservatives rather than Labour are 2.31 times or 131% larger for women than men in this scenario.

We now briefly summarize the main results of our empirical analysis. On the whole, women and men in the 1992 British elections held substantively similar positions when it came to policy related to defense, the appropriate size of the government, and women's rights. In addition, all three policy issues were salient to both women and men when it came to their vote choice. In this regard, we found evidence that both women and men punished political parties that were ideologically distant from them on each of the three policy dimensions. This supports the idea that individuals cared about policy positions and voted "spatially" at election time. There was compelling evidence that women saw policy related to the appropriate size of government as *more* salient to their vote choice than men. Although men also care about this particular policy dimension, this gender difference in salience suggests that we might usefully consider the appropriate size of government to be a "women's issue." While scholars sometimes consider defense policy to be a "men's issue," we found only weak evidence that men actually place more weight on this policy dimension when it comes to their voting behavior than women. There was no evidence that women and men differ in how much weight they place on policy related to women's rights in their vote calculus. This fits with previous research suggesting that there may be fewer gender differences than we might think when it comes to women's rights, at least when these rights are viewed in general terms (Hayes, 1997; Bolzendahl and Myers, 2004; Mayeri et al., 2008). Finally, our empirical analysis suggests that gender played only a limited role when it came to party vote choice after taking account of the policy positions of voters and parties. There was some evidence that women were more likely than similarly-situated men to vote for the right-wing Conservative Party as opposed to the left-wing Labour Party but only so long as the Conservative Party wasn't too ideologically distant on issues that were particularly salient to women, such as the appropriate size of the government. Overall, the analysis suggests that spatial considerations did interact with gender to influence party vote choice in the 1992 British general elections.

In Part III of the book, we've extended our discussion of theories positing interaction to scenarios in which we have some kind of limited dependent variable. We've focused, in particular, on those cases where the conditional implications of our theories can be evaluated with an interactive discrete choice model. Recall that a discrete choice model is one in which a choice is made from a finite set of mutually exclusive and collectively exhaustive choice alternatives. In Chapter 8, we looked at situations in which our choice set has just two choice alternatives, and so our dependent variable was binary or dichotomous. This led us to examine in some detail how to evaluate conditional claims using a logit or probit model. In Chapter 9, we looked at situations in which our choice set has more than two *ordered* choices. This led us to examine how to evaluate conditional claims using an ordered logit or ordered probit model. In the current chapter, we looked at situations in which our choice set has more than two *unordered* choices. And this led us to examine how to evaluate conditional claims using a multinomial logit model.

While there are differences between these discrete choice models, we've seen that we can always evaluate the quantities of interest necessary to test the conditional implications of our theory in terms of marginal effects and differences.[30] This will also be the case for models that deal with other types of limited dependent variables such as counts (Cameron and Trivedi, 2004) or durations (Box-Steffensmeier and Jones, 2004). The nature of the desired quantities of interest in these other models will obviously be different. In a count model, for example, the quantity of interest will usually be an expected count of some event; and in a duration model, the quantity of interest will typically be the amount of time until some event occurs or something similar. Whatever the nature of the desired quantity of interest, though, we can always evaluate it in terms of marginal effects or differences. This means looking at how the desired quantity of interest responds to a small or marginal change in a variable of interest by taking a derivative or by examining how its value changes in response to a discrete change in the variable of interest. As a result, the core insights about testing the implications of theories positing interaction that we've developed in the particular context of discrete choice models easily extends to cases where we're dealing with some other kind of limited dependent variable.

10.5 EXERCISES

1. Our substantive application in the chapter looked at how policy preferences and gender affected party vote choice in the 1992 British general elections. Here, we revisit this application in order to practice how to calculate various quantities of interest. We modify our analysis somewhat by including only one ideological distance variable, $Government\ Intervention\ Distance_k$. Thus, the utility associated with party k for individual i is now

$$U_{ik} = \beta_1 Government\ Intervention\ Distance_k$$
$$+ \beta_2 Government\ Intervention\ Distance_k \times Female$$
$$+ \gamma_{0k} + \gamma_{1k} Female + \gamma_k Controls + \epsilon_k. \quad (10.64)$$

The results from our slightly modified analysis are shown in Table 10.7.

a. Calculate the marginal effect of $Government\ Intervention\ Distance_k$ on the utility associated with party k for men and women. What is the modifying effect of being female instead of male on this marginal effect?

b. Calculate the direct marginal effects of $Government\ Intervention\ Distance_k$ on the probability of voting for the Conservative Party

---

[30] In the case of logit-based models, we've also see that we can evaluate them in terms of odds ratios.

Table 10.7 Determinants of party vote choice in the 1992 British general election II

Dependent Variable: *Party Choice: Conservatives (1), Labour (2), Liberal Democrats (3)*

	Multinomial Logit Model	
Alternative-Specific Variables		
Government Intervention Distance$_k$	−0.482***	
	(0.073)	
Government Intervention Distance$_k$ × Female	−0.227**	
	(0.110)	
Individual-Specific Variables		
	Conservative Party	Liberal Democrats
Female	0.488*	0.435
	(0.280)	(0.300)
Income	0.755***	0.408***
	(0.139)	(0.152)
Education	0.175	0.272*
	(0.130)	(0.142)
Working Class	−1.120***	−0.550*
	(0.287)	(0.330)
Trade Union	−0.560**	−0.168
	(0.269)	(0.298)
Age	0.018**	0.012
	(0.008)	(0.009)
Constant	−1.257*	−2.228***
	(0.720)	(0.810)
Number of Individuals	547	
Number of Party Alternatives	1,641	
Log Likelihood	−387.63	

Standard errors in parentheses. *$p < 0.10$; **$p < 0.05$; ***$p < 0.01$ (two-tailed)

Note: The data include voters in England and Wales who voted for either the Conservative Party, the Labour Party, or the Liberal Democrats in the 1992 British general election. The Labour Party is treated as the baseline category for the individual-specific variables.

for men and women for the scenario where our individuals aren't trade union members, don't belong to the working class, are 30 years old, have median income (*Income* = 6) and education

($Education = 1$), and are located 2 units away from the Conservative Party and 1 unit away from the other parties with respect to policy on the appropriate size of the government. Based on your answers, what is the modifying effect of being female instead of male on the marginal effect of *Government Intervention Distance$_k$* on the probability of voting for the Conservative Party in our chosen scenario?

c. Calculate the cross-marginal effects of increasing *Government Intervention Distance* for the Labour Party on the probability of voting for the Conservative Party for men and women in the same scenario as before.

d. Without doing any complicated calculations, what must the cross-marginal effects of increasing *Government Intervention Distance* for the Liberal Democrats on the probability of voting for the Conservative Party be for men and women in the same scenario as before? Explain.

e. Calculate the change in the probability of voting for the Conservative Party for men and women when we increase *Government Intervention Distance$_k$* by 2 units in the same scenario as before. Based on your answers, what is the modifying effect of being female instead of male on the effect of a 2-unit increase in *Government Intervention Distance$_k$* on the probability of voting for the Conservative Party in our chosen scenario?

f. Calculate the percentage change in the odds of voting for the Conservative Party as opposed to some other party for both men and women if we increase the ideological distance to the Conservative Party with respect to policy on the appropriate size of government by 2 units. What is the modifying effect of gender?

2. In the chapter, we looked at a vote choice scenario where voters are choosing between $J = 3$ candidates and we have two individual-specific variables and two alternative-specific variables. We assumed at the time that we had a conditional claim involving variable-specific interaction between one of the alternative-specific variables X_{1k} and one of the individual-specific variables Z_1. To be specific, the utility that voter i obtains from candidate k in our setup was

$$U_{ik} = \gamma_{0k} + \gamma_{1k}Z_1 + \gamma_{2k}Z_2 + \beta_1 X_{1k} + \beta_2 X_{2k} + \beta_3 X_{1k}Z_1 + \epsilon_k. \quad (10.65)$$

Let's now switch things up and assume that our conditional claim involves variable-specific interaction between the two alternative-specific variables X_{1k} and X_{2k} such that the utility that voter i obtains from candidate k is

$$U_{ik} = \gamma_{0k} + \gamma_{1k}Z_1 + \gamma_{2k}Z_2 + \beta_1 X_{1k} + \beta_2 X_{2k} + \beta_3 X_{1k}X_{2k} + \epsilon_k.$$
$$(10.66)$$

a. Calculate the marginal effect of X_{1k} on the utility that voter i obtains from candidate k. How do we interpret the coefficient β_1?

b. Calculate the *direct marginal effect* of X_{1k} on the probability of voting for candidate k. Explain in words what this marginal effect captures?

c. Calculate the *cross-marginal effect* of X_1 on the probability of voting for candidate k. Assume that we're increasing X_1 for candidate m. Explain in words what this marginal effect captures?

d. What is the effect of a one-unit increase in X_{1k} on the odds of choosing candidate k versus candidate m?

e. What is the interaction effect of X_{1k} and X_{2k} on the utility that voter i obtains from candidate k?

f. What is the interaction effect of X_{1k} and X_{2k} on the probability of voting for candidate k?

3. Let's switch things up again and assume that our conditional claim involves variable-specific interaction between the two individual-specific variables Z_1 and Z_2 such that the utility that voter i obtains from candidate k is

$$U_{ik} = \gamma_{0k} + \gamma_{1k}Z_1 + \gamma_{2k}Z_2 + \gamma_{3k}Z_1 Z_2 + \beta_1 X_{1k} + \beta_2 X_{2k} + \epsilon_k. \quad (10.67)$$

a. Calculate the marginal effect of Z_1 on the utility that voter i obtains from candidate k. How do we interpret the coefficient γ_{1k}?

b. Calculate the marginal effect of Z_1 on the probability of voting for candidate k.

c. What is the effect of a one-unit increase in Z_1 on the odds of choosing candidate k versus candidate m?

d. What is the interaction effect of Z_1 and Z_2 on the utility that voter i obtains from candidate k?

e. What is the interaction effect of Z_1 and Z_2 on the probability of voting for candidate k?

Appendix A

Basic Properties of Variances

In Chapter 4, we discuss how to calculate measures of uncertainty for various quantities of interest in linear-interactive models. This typically involves calculating a variance. In this appendix, we present some of the basic properties of variances.

The variance of a constant is zero,

$$\text{var}(a) = 0. \tag{A.1}$$

Adding a constant to all values of a random variable doesn't change its variance,

$$\text{var}(X + a) = \text{var}(X). \tag{A.2}$$

Scaling all values of a random variable by a constant scales the variance by the square of the constant,

$$\text{var}(aX) = a^2 \text{var}(X). \tag{A.3}$$

The variance of a sum of two random variables is

$$\text{var}(aX + bY) = a^2 \text{var}(X) + b^2 \text{var}(Y) + 2ab \times \text{cov}(X, Y), \tag{A.4}$$

and

$$\text{var}(aX - bY) = a^2 \text{var}(X) + b^2 \text{var}(Y) - 2ab \times \text{cov}(X, Y). \tag{A.5}$$

The variance of a linear combination of random variables is

$$
\begin{aligned}
\text{var}\left(\sum_{i=1}^{N} a_i X_i\right) &= \sum_{i,j=1}^{N} a_i a_j \times \text{cov}(X_i, X_j) \\
&= \sum_{i=1}^{N} a_i^2 \text{var}(X_i) + \sum_{i \neq j} a_i a_j \times \text{cov}(X_i, X_j) \\
&= \sum_{i=1}^{N} a_i^2 \text{var}(X_i) + 2 \sum_{1 < i < j \leq N} a_i a_j \times \text{cov}(X_i, X_j).
\end{aligned} \tag{A.6}
$$

Note that when applying these properties to calculate variances for quantities of interest derived from an interaction model, we treat the coefficients as the random variables and the independent variables as the constants. Thus, the variance for $\beta_1 + \beta_3 Z$ is

$$\mathrm{var}\,(\beta_1 + \beta_3 Z) = \mathrm{var}\,(\beta_1) + Z^2 \mathrm{var}\,(\beta_3) + 2Z\mathrm{cov}\,(\beta_1, \beta_3). \tag{A.7}$$

Appendix B

Marginal Effects and Variances for Various Linear-Interactive Models

In Tables B.1 and B.2, we provide information on the marginal effects and variances for various commonly-used linear-interactive model specifications (Aiken and West, 1991). The model specifications in Table B.1 are linear in X (and Z), while those in Table B.2 include quadratic terms for X.

Table B.1 Marginal effects and variances for various linear-interactive models

Case	Equation	Marginal Effect	Variance
1a	$Y = \beta_0 + \beta_1 X + \beta_2 Z + \beta_3 XZ + \epsilon$	$\frac{\partial Y}{\partial X} = \beta_1 + \beta_3 Z$	$\text{var}\left(\frac{\partial Y}{\partial X}\right) = \text{var}(\beta_1) + Z^2\text{var}(\beta_3) + 2Z\text{cov}(\beta_1,\beta_3)$
1b	$Y = \beta_0 + \beta_1 X + \beta_2 Z + \beta_3 XZ + \epsilon$	$\frac{\partial Y}{\partial Z} = \beta_2 + \beta_3 X$	$\text{var}\left(\frac{\partial Y}{\partial Z}\right) = \text{var}(\beta_2) + X^2\text{var}(\beta_3) + 2X\text{cov}(\beta_2,\beta_3)$
2	$Y = \beta_0 + \beta_1 X + \beta_2 Z + \beta_3 W + \epsilon$ $+\beta_4 XZ + \beta_5 ZW$	$\frac{\partial Y}{\partial X} = \beta_1 + \beta_4 Z$	$\text{var}\left(\frac{\partial Y}{\partial X}\right) = \text{var}(\beta_1) + Z^2\text{var}(\beta_4) + 2Z\text{cov}(\beta_1,\beta_4)$
3	$Y = \beta_0 + \beta_1 X + \beta_2 Z + \beta_3 W$ $+\beta_4 XZ + \beta_5 XW + \beta_6 ZW + \epsilon$	$\frac{\partial Y}{\partial X} = \beta_1 + \beta_4 Z + \beta_5 W$	$\text{var}\left(\frac{\partial Y}{\partial X}\right) = \text{var}(\beta_1) + Z^2\text{var}(\beta_4) + W^2\text{var}(\beta_5)$ $+2Z\text{cov}(\beta_1,\beta_4) + 2W\text{cov}(\beta_1,\beta_5)$ $+2ZW\text{cov}(\beta_4,\beta_5)$
4	$Y = \beta_0 + \beta_1 X + \beta_2 Z + \beta_3 W$ $+\beta_4 XZ + \beta_5 XW + \beta_6 ZW$ $+\beta_7 XZW + \epsilon$	$\frac{\partial Y}{\partial X} = \beta_1 + \beta_4 Z + \beta_5 W + \beta_7 ZW$	$\text{var}\left(\frac{\partial Y}{\partial X}\right) = \text{var}(\beta_1) + Z^2\text{var}(\beta_4) + W^2\text{var}(\beta_5)$ $+Z^2 W^2\text{var}(\beta_7)$ $+2Z\text{cov}(\beta_1,\beta_4) + 2W\text{cov}(\beta_1,\beta_5)$ $+2ZW\text{cov}(\beta_1,\beta_7) + 2ZW\text{cov}(\beta_4,\beta_5)$ $+2WZ^2\text{cov}(\beta_4,\beta_7) + 2ZW^2\text{cov}(\beta_5,\beta_7)$

Note: Information from Aiken and West (1991).

Table B.2 Marginal effects and variances for various linear-interactive models (quadratic terms)

Case	Equation	Marginal Effect	Variance
1	$Y = \beta_0 + \beta_1 X + \beta_2 X^2 + \epsilon$	$\frac{\partial Y}{\partial X} = \beta_1 + 2\beta_2 X$	$\text{var}(\frac{\partial Y}{\partial X}) = \text{var}(\beta_1) + 4X^2\text{var}(\beta_2) + 4X\text{cov}(\beta_1, \beta_2)$
2	$Y = \beta_0 + \beta_1 X + \beta_2 X^2$ $+ \beta_3 Z + \epsilon$	$\frac{\partial Y}{\partial X} = \beta_1 + 2\beta_2 X$	$\text{var}(\frac{\partial Y}{\partial X}) = \text{var}(\beta_1) + 4X^2\text{var}(\beta_2) + 4X\text{cov}(\beta_1, \beta_2)$
3a	$Y = \beta_0 + \beta_1 X + \beta_2 X^2$ $+ \beta_3 Z + \beta_4 XZ + \epsilon$	$\frac{\partial Y}{\partial X} = \beta_1 + 2\beta_2 X + \beta_4 Z$	$\text{var}(\frac{\partial Y}{\partial X}) = \text{var}(\beta_1) + 4X^2\text{var}(\beta_2) + Z^2\text{var}(\beta_4)$ $+ 4X\text{cov}(\beta_1, \beta_2) + 2Z\text{cov}(\beta_1, \beta_4)$ $+ 4XZ\text{cov}(\beta_2, \beta_4)$
3b	$Y = \beta_0 + \beta_1 X + \beta_2 X^2$ $+ \beta_3 Z + \beta_4 XZ + \epsilon$	$\frac{\partial Y}{\partial Z} = \beta_3 + \beta_4 X$	$\text{var}(\frac{\partial Y}{\partial Z}) = \text{var}(\beta_3) + X^2\text{var}(\beta_4) + 2X\text{cov}(\beta_3, \beta_4)$
4a	$Y = \beta_0 + \beta_1 X + \beta_2 X^2$ $+ \beta_3 Z + \beta_4 XZ + \beta_5 X^2 Z + \epsilon$	$\frac{\partial Y}{\partial X} = \beta_1 + 2\beta_2 X + \beta_4 Z + 2\beta_5 XZ$	$\text{var}(\frac{\partial Y}{\partial X}) = \text{var}(\beta_1) + 4X^2\text{var}(\beta_2) + Z^2\text{var}(\beta_4)$ $+ 4X^2 Z^2\text{var}(\beta_5)$ $+ 4X\text{cov}(\beta_1, \beta_2) + 2Z\text{cov}(\beta_1, \beta_4)$ $+ 4XZ\text{cov}(\beta_2, \beta_4) + 4XZ\text{cov}(\beta_1, \beta_5)$ $+ 8X^2\text{cov}(\beta_2, \beta_5) + 4XZ^2\text{cov}(\beta_4, \beta_5)$
4b	$Y = \beta_0 + \beta_1 X + \beta_2 X^2$ $+ \beta_3 Z + \beta_4 XZ + \beta_5 X^2 Z + \epsilon$	$\frac{\partial Y}{\partial Z} = \beta_3 + \beta_4 X + \beta_5 X^2$	$\text{var}(\frac{\partial Y}{\partial Z}) = \text{var}(\beta_3) + X^2\text{var}(\beta_4) + X^4\text{var}(\beta_5)$ $+ 2X\text{cov}(\beta_3, \beta_4) + 2X^2\text{cov}(\beta_3, \beta_5)$ $+ 2X^3\text{cov}(\beta_4, \beta_5)$

Note: Information from Aiken and West (1991).

Appendix C

Calculating the Smallest Standard Error for the Marginal Effect of X on Y

Suppose that we estimate the following linear-interactive specification:

$$Y = \beta_0 + \beta_1 X + \beta_2 Z + \beta_3 XZ + \epsilon. \tag{C.1}$$

The marginal effect of X on Y is

$$\frac{\partial Y}{\partial X} = \beta_1 + \beta_3 Z. \tag{C.2}$$

The estimated standard error associated with this effect is calculated as

$$se\left(\frac{\partial Y}{\partial X}\right) = se\left(\beta_1 + \beta_3 Z\right) = \sqrt{var\left(\beta_1\right) + Z^2 var\left(\beta_3\right) + 2Zcov\left(\beta_1, \beta_3\right)}$$

$$= \left[var\left(\beta_1\right) + Z^2 var\left(\beta_3\right) + 2Zcov\left(\beta_1, \beta_3\right)\right]^{\frac{1}{2}}. \tag{C.3}$$

As we can see, the standard error associated with the marginal effect of X on Y varies with the value of Z. To calculate the value of Z that minimizes the standard error, we take the derivative of the standard error with respect to Z, set it equal to 0, and solve for Z. Using the power rule and the product rule, we see that

$$\frac{\partial\left[se\left(\frac{\partial Y}{\partial X}\right)\right]}{\partial Z} = \frac{1}{2}\left[var\left(\beta_1\right) + Z^2 var\left(\beta_3\right) + 2Zcov\left(\beta_1, \beta_3\right)\right]^{-\frac{1}{2}}$$
$$\times \left[2Zvar\left(\beta_3\right) + 2cov\left(\beta_1, \beta_3\right)\right]$$

$$= \frac{2Zvar\left(\beta_3\right) + 2cov\left(\beta_1, \beta_3\right)}{2\sqrt{var\left(\beta_1\right) + Z^2 var\left(\beta_3\right) + 2Zcov\left(\beta_1, \beta_3\right)}}$$

$$= \frac{Zvar\left(\beta_3\right) + cov\left(\beta_1, \beta_3\right)}{se\left(\frac{\partial Y}{\partial X}\right)}. \tag{C.4}$$

Setting this equal to 0 and multiplying through by $se\left(\frac{\partial Y}{\partial X}\right)$, we have

$$\frac{\partial\left[se\left(\frac{\partial Y}{\partial X}\right)\right]}{\partial Z} = \frac{Z\text{var}\,(\beta_3) + \text{cov}\,(\beta_1,\beta_3)}{se\left(\frac{\partial Y}{\partial X}\right)} = 0$$

$$Z\text{var}\,(\beta_3) + \text{cov}\,(\beta_1,\beta_3) = 0. \qquad (C.5)$$

Solving for Z, we have

$$Z^* = -\frac{\text{cov}\,(\beta_1,\beta_3)}{\text{var}\,(\beta_3)}. \qquad (C.6)$$

This tells us that Z^* is the value of Z that minimizes the standard error associated with the marginal effect of X on Y. It follows that the confidence interval around the marginal effect of X on Y is narrowest when $Z = Z^*$. Symmetrically, we see that the value of X that minimizes the standard error associated with the marginal effect of Z is

$$X^* = -\frac{\text{cov}\,(\beta_2,\beta_3)}{\text{var}\,(\beta_3)}. \qquad (C.7)$$

Appendix D

Calculating the Values of the Modifying Variable Z at which the Bounds of the Confidence Interval for the Marginal Effect of X Equal 0

Suppose we have a theory positing interaction between X and Z on a continuous dependent variable Y. Marginal effect plots showing how the effect of the independent variables on Y vary with the value of the other modifying variable can be very useful. Whether the effects of the independent variables are statistically significant at a particular value of the modifying variable depends on whether the upper and lower bounds of the confidence interval for the marginal effect are on the same side of the horizontal zero line at this value; that is, whether they contain 0. It's often the case that the marginal effect of an independent variable is statistically significant for some values of the modifying variable but not others. To identify the range of values at which the effect is statistically significant requires identifying when the upper or lower bounds of the confidence intervals cut the horizontal zero line. We now show how to calculate these cut-points.

Let's assume that we have the following linear-interactive specification,

$$Y = \beta_0 + \beta_1 X + \beta_2 Z + \beta_3 XZ + \epsilon, \tag{D.1}$$

and that we're interested in the marginal effect of X,

$$\frac{\partial Y}{\partial X} = \beta_1 + \beta_3 Z. \tag{D.2}$$

The standard error for the marginal effect of X is

$$se\left(\frac{\partial Y}{\partial X}\right) = se(\beta_1 + \beta_3 Z) = \sqrt{var(\beta_1) + Z^2 var(\beta_3) + 2Z cov(\beta_1, \beta_3)}, \tag{D.3}$$

and so the confidence interval for the marginal effect of X is

$$\beta_1 + \beta_3 Z \pm t \times \sqrt{var(\beta_1) + Z^2 var(\beta_3) + 2Z cov(\beta_1, \beta_3)}, \tag{D.4}$$

where t is the critical value from a t-distribution for a particular α level of significance and $n - k$ degrees of freedom. To identify where the upper and lower bounds of the confidence interval cut the horizontal zero line, we simply need to set the confidence interval in Eq. D.4 equal to 0,

$$\beta_1 + \beta_3 Z \pm t \times \sqrt{\mathrm{var}(\beta_1) + Z^2 \mathrm{var}(\beta_3) + 2Z\mathrm{cov}(\beta_1, \beta_3)} = 0, \quad (D.5)$$

and solve for Z.

Moving the product term involving the square root to the right, we can rewrite Eq. D.5 as

$$\beta_1 + \beta_3 Z = \pm t \times \sqrt{\mathrm{var}(\beta_1) + Z^2 \mathrm{var}(\beta_3) + 2Z\mathrm{cov}(\beta_1, \beta_3)}. \quad (D.6)$$

Squaring both sides gives us

$$\beta_1^2 + 2\beta_1\beta_3 Z + \beta_3^2 Z^2 = t^2 \times \left[\mathrm{var}(\beta_1) + Z^2 \mathrm{var}(\beta_3) + 2Z\mathrm{cov}(\beta_1, \beta_3) \right]$$
$$= t^2 \mathrm{var}(\beta_1) + t^2 Z^2 \mathrm{var}(\beta_3) + 2t^2 Z\mathrm{cov}(\beta_1, \beta_3). \quad (D.7)$$

Note that we don't have to take account of the signs of the terms on each side of Eq. D.6 when squaring them because of the $\pm$ sign. Moving everything to the left and grouping terms involving Z^2 and Z together, we have

$$\left[\beta_3^2 - t^2 \mathrm{var}(\beta_3) \right] Z^2 + \left[2\beta_1\beta_3 - 2t^2 \mathrm{cov}(\beta_1, \beta_3) \right] Z + \left[\beta_1^2 - t^2 \mathrm{var}(\beta_1) \right] = 0. \quad (D.8)$$

You'll see that we now have a quadratic equation. This tells us that finding the values of Z at which the upper and lower bounds of the confidence interval for the marginal effect of X are equal to 0 is equivalent to solving the quadratic equation in Eq. D.8 for its roots. We can find the roots using the quadratic formula.

Suppose that we have the following generic quadratic equation:

$$aX^2 + bX + c = 0. \quad (D.9)$$

The quadratic formula tells us that the values of X that solve this equation (the roots of the equation) are

$$X^* = \frac{-b \pm \sqrt{b^2 - 4ac}}{2a}. \quad (D.10)$$

The term $b^2 - 4ac$ is known as the discriminant. If the discriminant is positive, then we have two roots. In our particular context, this means that both the upper and lower bounds of the confidence interval cross the horizontal zero line. This tells us that there's a range of values for the modifying variable Z between the two roots in which the confidence interval for the marginal effect of X contains the horizontal zero line. In other words, there's a range of values for Z between the two roots at which the marginal effect of X isn't statistically significant. It also follows that there must be a range of values for Z where the marginal effect of X is negative and statistically significant and a range of values for Z where it's positive and statistically significant. Of course, these particular ranges for the values of Z may not always be within the range of observed or theoretically possible values for Z in our given application. If the discriminant is negative, then there are no real roots. In our context, this means that neither the upper nor the lower bound of the confidence interval cuts the horizontal zero line. This tells us that the marginal effect of X is never statistically significant.[1]

Translating the terms of our quadratic equation in Eq. D.8 into the terms of the generic quadratic equation in Eq. D.9, we have

$$a = \beta_3^2 - t^2 \text{var}(\beta_3),$$
$$b = 2\beta_1\beta_3 - 2t^2 \text{cov}(\beta_1, \beta_3),$$
$$c = \beta_1^2 - t^2 \text{var}(\beta_1).$$

Using the quadratic formula, we see that the values of Z at which the upper and lower bounds of the confidence interval for the marginal effect of X are 0 are

$$Z^* = \frac{-\left[2\beta_1\beta_3 - 2t^2\text{cov}(\beta_1,\beta_3)\right] \pm \sqrt{\left[2\beta_1\beta_3 - 2t^2\text{cov}(\beta_1,\beta_3)\right]^2 - 4\left[\beta_3^2 - t^2\text{var}(\beta_3)\right]\left[\beta_1^2 - t^2\text{var}(\beta_1)\right]}}{2\left[\beta_3^2 - t^2\text{var}(\beta_3)\right]}.$$

(D.11)

D.1 SUBSTANTIVE EXAMPLE: MARGINAL EFFECT PLOT FOR *DEMAND*

In Chapter 5, we used one of our substantive analyses to examine how demand-side factors and supply-side factors interact to determine the level of women's legislative representation. We estimated the following linear-interactive model specification:

$$
\begin{aligned}
\text{Women's Representation} = \beta_0 &+ \beta_1 Demand + \beta_2 Supply \\
&+ \beta_3 Demand \times Supply \\
&+ \beta_4 Effective\ Gender\ Quota \\
&+ \beta_5 Democracy + \epsilon.
\end{aligned}
$$
(D.12)

Using the results in Table 5.6 and the variance-covariance matrix for the coefficients shown in Eq. 5.52, we produced a marginal effect plot for

---

[1] If the discriminant is zero, then there's a double root. In our context, this would imply that the upper and lower bounds of the confidence interval cross the horizontal zero line at the same value of Z. This would occur only if the conditional standard error for the marginal effect of X is exactly 0 at this point. Practically-speaking, this is never going to be the case.

(a) The Conditional Effect of *Demand*

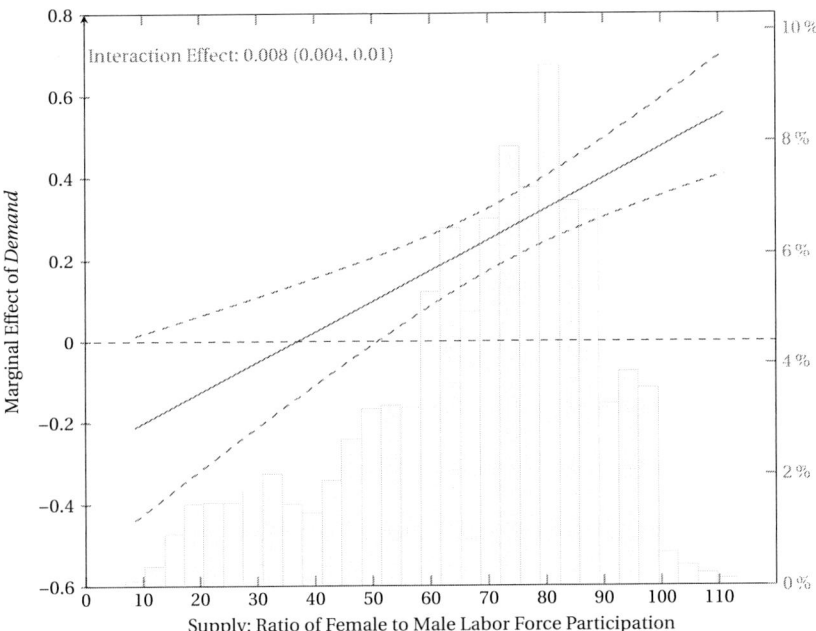

Figure D.1 The conditional effects of *Demand* on the level of women's legislative representation

Note: The plot shows the effect of a one-unit increase in *Demand* on women's legislative representation across the observed range of *Supply*. It also shows the interaction effect between *Demand* and *Supply*. The curved dashed lines indicate two-tailed 95% confidence intervals. The histogram shows the percentage of observations at different values of *Supply*. The plot is based on the results from the standard interaction model shown in the first column of Table 5.6.

Demand in panel (a) of Figure 5.8. We've reproduced that marginal effect plot in Figure D.1.

Based on Figure D.1, we can see that the marginal effect of *Demand* becomes positive and statistically significant once the value of *Supply* is larger than about 50. This is the point at which the upper and lower bounds of the confidence interval both start being above the horizontal zero line. It's difficult to know precisely when the effect of *Demand* becomes positive and statistically significant from simply eye-balling the plot. However, we can use the information in this appendix to identify precisely when this happens.

The confidence interval for the marginal effect of *Demand* is

$$\beta_1 + \beta_3 Supply \pm t \times \sqrt{\text{var}(\beta_1) + (Supply)^2 \text{var}(\beta_3) + 2 \times Supply \times \text{cov}(\beta_1, \beta_3)}.$$
(D.13)

We want to know the values of *Supply* at which the upper and lower bounds of this confidence interval are 0. Based on Eq. D.11, we know that this occurs when

$$Supply^* = \frac{-[2\beta_1\beta_3 - 2t^2\text{cov}(\beta_1,\beta_3)] \pm \sqrt{[2\beta_1\beta_3 - 2t^2\text{cov}(\beta_1,\beta_3)]^2 - 4[\beta_3^2 - t^2\text{var}(\beta_3)][\beta_1^2 - t^2\text{var}(\beta_1)]}}{2[\beta_3^2 - t^2\text{var}(\beta_3)]}.$$
(D.14)

From Table 5.6, we know that $\beta_1 = -0.28$ and $\beta_3 = 0.008$. From the variance-covariance matrix in Eq. 5.52, we know that $\text{var}(\beta_1) = 0.016526$, $\text{var}(\beta_3) = 0.000003$, and $\text{cov}(\beta_1, \beta_3) = -0.000207$. We have $n - k = 186 - 6 = 180$ degrees of freedom and an $\alpha = 0.05$ level of significance because we want two-tailed 95% confidence intervals. It follows that $t = t_{n-k,\alpha/2} = t_{180,0.025} = 1.97323$. We just need to plug these values into Eq. D.14 and solve for *Supply**.

Recall that the terms in square brackets in Eq. D.14 refer to the terms a, b, and c in the generic quadratic equation shown in Eq. D.9,

$$a = \beta_3^2 - t^2\text{var}(\beta_3) = 0.008^2 - 1.97323^2 \times 0.000003 = 0.00004544,$$
$$b = 2\beta_1\beta_3 - 2t^2\text{cov}(\beta_1,\beta_3)$$
$$= 2 \times -0.28 \times 0.008 - 2 \times 1.97323^2 \times -0.000207 = -0.002565,$$
$$c = \beta_1^2 - t^2\text{var}(\beta_1) = (-0.28)^2 - 1.97323^2 \times 0.016526 = 0.012835.$$

From this, we can calculate the discriminant as

$$b^2 - 4ac = (-0.002565)^2 - 4 \times 0.00004544 \times 0.012835 = 0.000004244$$

That the discriminant is positive tells us that we're going to have two roots for the quadratic equation in Eq. D.14. In other words, both the upper and lower bounds of our confidence interval for the marginal effect of X are going to cross the horizontal zero line. Plugging all of the values that we've calculated into the generic quadratic formula gives us

$$Supply^* = \frac{-b \pm \sqrt{b^2 - 4ac}}{2a} = \frac{-(-0.002565) \pm \sqrt{0.000004244}}{2.00004544}$$

$$= \frac{0.002565 \pm 0.00206}{0.00009089}.$$
(D.15)

From this, we see that the bounds of our confidence interval are 0 when

$$Supply^* = \frac{0.002565 + 0.00206}{0.00009089} = 50.884.$$
(D.16)

or

$$Supply^* = \frac{0.002565 - 0.00206}{0.00009089} = 5.550.$$
(D.17)

By looking at the marginal effect plot in Figure D.1, it's easy to see that it's the lower bound of the confidence interval that crosses the horizontal zero line at $Supply = 50.884$. Because the upper and lower bounds of the confidence interval are both above the horizontal zero line when $Supply > 50.884$, we know that the marginal effect of X on Y is positive and statistically significant when $Supply > 50.884$. This is what we reported when discussing this substantive application in Chapter 5. Our analysis here reveals that our quadratic equation has a second root at $Supply = 5.550$. This is where the upper bound of the confidence interval would cut the horizontal zero line.[2] This isn't actually shown in Figure D.1 as, following our recommendations, we only plotted the marginal effect of X and its confidence interval across the observed range of values for $Supply$; the minimum observed value for $Supply$ is 8.61. We know that the marginal effect of X isn't statistically significant when $Supply$ takes on values between the two roots; that is, $5.550 < Supply < 50.884$.

---

[2] Because the upper and lower bounds of the confidence interval are both below the horizontal zero line when $Supply < 5.550$, we know that the marginal effect of X on Y is negative and statistically significant when $Supply < 5.550$.

References

Achen, Christopher H. 1982. *Interpreting and Using Regression*. London: Sage.

Adams, James F., Samuel Merrill and Bernard Grofman. 2005. *A Unified Theory of Party Competition: A Cross-National Analysis Integrating Spatial and Behavioral Factors*. New York: Cambridge University Press.

Adams, James, Michael Clark, Lawrence Ezrow and Garrett Glasgow. 2006. "Are niche parties fundamentally different from mainstream parties? The causes and the electoral consequences of western European parties' policy shifts, 1976–1998." *American Journal of Political Science* 50(3):513–529.

Ai, Chunrong and Edward C. Norton. 2003. "Interaction terms in logit and probit models." *Economic Letters* 80(1):123–129.

Aidt, Toke S. and Bianca Dallal. 2008. "Female voting power: The contribution of women's suffrage to the growth of social spending in Western Europe (1869–1960)." *Public Choice* 134(3–4):132–150.

Aiken, Leona and Stephen West. 1991. *Multiple Regression: Testing and Interpreting Interactions*. London: Sage.

Ajrouch, Kristine J., Alysia Y. Blandon and Toni C. Antonucci. 2005. "Social networks among men and women: The effects of age and socioeconomic status." *Journal of Gerontology* 60B(6):S313–S317.

Albrecht, Stan L., Bruce A. Chadwick and David S. Alcorn. 1977. "Religiosity and deviance: Application of an attitude-behavior contingent consistent model." *Journal for the Scientific Study of Religion* 16(3):263–274.

Alexander, Deborah and Kristi Andersen. 1986. "Gender as a factor in the attribution of leadership traits." *Political Research Quarterly* 46(3):527–545.

Alilunas, Peter. 2011. "The (in)visible people in the room: Men in women's studies." *Men and Masculinities* 14(2):210–229.

Alvarez, R. Michael and Jonathan Nagler. 1998. "When politics and models collide: Estimating models of multiparty elections." *American Journal of Political Science* 42(1): 55–96.

American National Election Studies. 2014. "User's Guide and Codebook for the ANES 2012 Time Series Study." Release Date: May 2016. Ann Arbor, MI and Palo Alto, CA: The University of Michigan and Stanford University.

American National Election Studies. 2019. "The American National Election Study (ANES) 2016 Time Series Study." Release Date: September 2019. Version 20190904. Ann Arbor, MI: American National Election Studies.

Amorim Neto, Octavio and Gary W. Cox. 1997. "Electoral institutions, cleavage structures, and the number of parties." *American Journal of Political Science* 41:149–174.

Anderson, Carol. 2018. *One Person, No Vote: How Voter Suppression Is Destroying Our Democracy*. New York: Bloomsbury Publishing.

Ansolabehere, Stephen and James M. Snyder. 2000. "Valence politics and equilibrium in spatial election models." *Public Choice* 103:327–336.

Aristotle (C. D. C. Reeve, Trans.). 1998. *Politics*. Indianapolis: Hackett.

Banaszak, Lee Ann and Eric Plutzer. 1993. "Contextual determinants of feminist attitudes: National and subnational influences in Western Europe." *American Political Science Review* 87(1):145–157.

Banks, Antoine J. and Nicholas A. Valentino. 2012. "Emotional substrates of white racial attitudes." *American Journal of Political Science* 56(2):286–297.

Baron, Raoul M. and David A. Kenny. 1986. "The moderator–mediator variable distinction in social psychological research: Conceptual, strategic, and statistical considerations." *Journal of Personality and Social Psychology* 105(6):1173–1182.

Beck, Nathaniel and Simon Jackman. 1998. "Beyond linearity by default: Generalized additive models." *American Journal of Political Science* 42(2):596–627.

Beck, Nathaniel, Gary King and Langche Zeng. 2000. "Improving quantitative studies of international conflict: A conjecture." *American Political Science Review* 94(1):21–35.

Beckwith, Karen and Lisa Baldez. 2007. "Intersectionality." *Politics & Gender* 3(2):229–232.

Bernard, Jessie. 1974. "Age, sex, and feminism." *The Annals of the American Academy of Political and Social Science* 415:120–137.

Berry, William D., Jacqueline H. R. DeMeritt and Justin Esarey. 2010. "Testing for interaction in binary logit and probit models: Is a product term essential?" *American Journal of Political Science* 54(1):248–266.

Berry, William D., Matt Golder and Daniel Milton. 2012. "Improving tests of theories positing interaction." *Journal of Politics* 74(3):653–671.

Black, Duncan. 1958. *The Theory of Committees and Elections*. New York: Cambridge University Press.

Block, Ray and Matt Golder. 2023. "Is there a substantive–descriptive representation trade-off?" Unpublished manuscript, Pennsylvania State University.

Block, Ray, Matt Golder and Sona Golder. 2023. "Evaluating claims of intersectionality." *Journal of Politics* 85(3):795–811.

Bock, Jarrod, Jennifer Byrd-Craven and Melissa Burkley. 2017. "The role of sexism in voting in the 2016 presidential election." *Personality and Individual Differences* 119:189–193.

Bolzendahl, Catherine and Daniel J. Myers. 2004. "Feminist attitudes and support for gender equality: Opinion change in women and men, 1974–1998." *Social Forces* 83(2):759–790.

Bowen, Harry P. 2012. "Testing moderating hypotheses in limited dependent variable and other nonlinear models: Secondary versus total interactions." *Journal of Management* 38(3):860–889.

Box-Steffensmeier, Janet M. and Bradford S. Jones. 2004. *Event History Modeling: A Guide for Social Scientists*. New York: Cambridge University Press.

Box-Steffensmeier, Janet M., Suzanna De Boef and Tse-Min Lin. 2004. "The dynamics of the partisan gender gap." *American Political Science Review* 98(3):515–528.

Brambor, Thomas, William Roberts Clark and Matt Golder. 2006. "Understanding interaction models: Improving empirical analyses." *Political Analysis* 14(1):63–82.

Braumoeller, Bear. 2004. "Hypothesis testing and multiplicative interaction terms." *International Organization* 58:807–820.

Bravo-Ortega, Claudio, Nicolas A. Eterovic and Valentina Paredes. 2018. "What do women want? Female suffrage and the size of government." *Economic Systems* 42:132–150.

Budge, Ian and Dennis J. Farlie. 1983. *Explaining and Predicting Elections*. London: George Allen and Unwin.

Budge, Ian, Hans-Dieter Klingemann, Andrea Volkens, Judith Bara and Eric Tanenbaum. 2001. *Mapping Policy Preferences. Estimates for Parties, Electors, and Governments 1945–1998*. New York: Oxford University Press.

Burkett, Stephen R. and Mervin White. 1974. "Hellfire and delinquency: Another look." *Journal for the Scientific Study of Religion* 13:455–462.

Burnside, Craig and David Dollar. 2000. "Aid, policies, and growth." *American Economic Review* 90(4):847–868.

Cai, Weiyi and Scott Clement. 2016. "What Americans think of feminism today." *The Washington Post*, January 27.

Calasanti, Toni and Kathleen F. Slevin, eds. 2006. *Age Matters: Realigning Feminist Thinking*. New York: Routledge.

Calasanti, Toni M. and Kathleen F. Slevin. 2001. *Gender, Social Inequalities, and Aging*. Walnut Creek, CA: Alta Mira Press.

Calasanti, Toni, Kathleen F. Slevin and Neal King. 2006. "Ageism and feminism: From 'et cetera' to center." *NWSA Journal* 18(1):13–30.

Cameron, A. Colin and Pravin K. Trivedi. 2004. *Regression Analysis of Count Data*. New York: Cambridge University Press.

Cameron, A. Colin and Pravin K. Trivedi. 2005. *Microeconometrics: Methods and Applications*. New York: Cambridge University Press.

Campbell, Rosie. 2004. "Gender, ideology, and issue preference: Is there such a thing as a political women's interest in Britain?" *British Journal of Politics and International Relations* 6:20–44.

Caprioli, Mary. 2000. "Gendered conflict." *Journal of Peace Research* 37(1):51–68.

Cassese, Erin C. 2017. "Why Donald Trump never really had a 'woman' problem among Republican voters." *LSE US Centre [blog]*, January 26.

Cassese, Erin C. and Tiffany D. Barnes. 2019. "Reconciling sexism and women's support for Republican candidates: A look at gender, class, and Whiteness in the 2012 and 2016 presidential races." *Political Behavior* 41(3):677–700.

Cattaneo, Matias D., Nicolás Idrobo and Rocío Titiunik. 2020. *A Practical Introduction to Regression Discontinuity Designs*. New York: Cambridge University Press.

Cavalcanti, Tiago V. De V. and José Tavares. 2011. "Women prefer larger governments: Growth, structural transformation, and government size." *Economic Inquiry* 49(1):155–171.

Celis, Karen, Sarah Childs, Johanna Kantola and Mona Lena Krook. 2014. "Constituting women's interests through representative claims." *Politics & Gender* 10:149–174.

Chaney, Carole Kennedy, R. Michael Alvarez and Jonathan Nagler. 1998. "Explaining the gender gap in US presidential elections 1980–1992." *Political Research Quarterly* 51(2):311–339.

Cho, Sumi, Kimberlé Crenshaw and Leslie McCall. 2013. "Toward a field of intersectionality studies: Theory, applications, and praxis." *Signs: Journal of Women in Culture and Society* 38(4):785–810.

Clark, William Roberts. 2003. *Capitalism, Not Globalism: Capital Mobility, Central Bank Independence, and the Political Control of the Economy*. Ann Arbor: University of Michigan Press.

Clark, William Roberts and Matt Golder. 2006. "Rehabilitating Duverger's theory: Testing the mechanical and strategic modifying effects of electoral laws." *Comparative Political Studies* 39:163–192.

Clark, William Roberts, Michael Gilligan and Matt Golder. 2006. "A simple multivariate test for asymmetric hypotheses." *Political Analysis* 14(3):311–331.

Cohen, Dov, Faith Shin, Xi Liu, Peter Ondish and Michael W. Kraus. 2017. "Defining social class across time and between groups." *Personality and Social Psychology Bulletin* 43(11):1530–1545.

Conover, Pamela J. and Virginia Sapiro. 1993. "Gender, feminist consciousness, and war." *American Journal of Political Science* 37(4): 1079–1099.

Cook, Elizabeth A., Ted G. Jelen and Clyde Wilcox. 1992. *Between Two Absolutes: Public Opinion and the Politics of Abortion.* Boulder, CO: Westview.

Cox, Gary W. 1997. *Making Votes Count: Strategic Coordination in the World's Electoral Systems.* New York: Cambridge University Press.

Crabtree, Charles, Matt Golder, Thomas Gschwend and Indriði H. Indriðason. 2020. "It's not only what you say, it's also how you say it: The strategic use of campaign sentiment." *Journal of Politics* 82(3): 1044–1060.

Crenshaw, Kimberlé. 1989. "Demarginalizing the intersection of race and sex: A black feminist critique of antidiscrimination doctrine, feminist theory, and antiracist policies." *University of Chicago Legal Forum* 1:139–167.

Crenshaw, Kimberlé. 1991. "Mapping the margins: Intersectionality, identity politics, and violence against women of color." *Stanford Law Review* 43(6):1241–1300.

Davenport, Lauren. 2018. *Politics Beyond Black and White: Biracial Identity and Attitudes in America.* New York: Cambridge University Press.

Davenport, Lauren. 2020. "The fluidity of racial classifications." *Annual Review of Political Science* 23:221–240.

Davis, Kathy and Dubravka Zarkov. 2017. "*EJWS* retrospective on intersectionality." *European Journal of Women's Studies* 24(4):313–320.

Davis, Nancy J. and Robert V. Robinson. 1991. "Men's and women's consciousness of gender inequality: Austria, West Germany, Great Britain, and the United States." *American Sociological Review* 56(1):72–84.

Dawson, Michael C. 1995. *Behind the Mule: Race and Class in African-American Politics.* Princeton, NJ: Princeton University Press.

Dawson, Michael C. 2001. *Black Visions: The Roots of Contemporary African-American Political Ideologies*. Chicago: University of Chicago Press.

Dhima, Kostanca. 2022. "Reexamining the supply and demand framework for explaining women's legislative representation." Unpublished manuscript, Texas A&M University.

Dhima, Kostanca. 2022. "Women's legislative representation: Supply, mass and elite demand, and regime type." Unpublished manuscript, Texas A&M University.

Dittmar, Kelly. 2016. "Watching election 2016 with a gender lens." *PS: Political Science & Politics* 49(4):807–812.

D'Orazio, Vito, Steven T. Landis, Glenn Palmer and Philip Schrodt. 2014. "Separating the wheat from the chaff: Applications of automated document classification using support vector machines." *Political Analysis* 22(2):224–242.

Döring, Holger and Philip Manow. 2015. "Parliaments and governments database (ParlGov): Information on parties, elections, and cabinets in modern democracies." www.parlgov.org/.

Downs, Anthony. 1957. *An Economic Theory of Democracy*. New York: Harper & Row.

Durkheim, Emile (Edited by Steven Lukes). 2003/1895. *The Rules of Sociological Method and Selected Texts on Sociology and Its Method*. New York: Free Press.

Duverger, Maurice. 1954. *Political Parties: Their Organization and Activity in the Modern State*. New York: John Wiley & Sons, Inc.

Duverger, Maurice. 1955. *The Political Role of Women*. New York: UNESCO.

Eagly, Alice H. 1987. *Sex Differences in Social Behavior: A Social-Role Interpretation*. Hillsdale, NJ: Erlbaum.

Eagly, Alice H. and Steven J. Karau. 2002. "Role congruity theory of prejudice toward female leaders." *Psychological Review* 109(3): 573–598.

Eagly, Alice H., Wendy Wood and Amanda B. Diekman. 2000. Social role theory of sex differences and similarities: A current appraisal. In *The Developmental Social Psychology of Gender*, ed. Thomas Eckes and Hanns M. Trautner. Mahwah, NJ: Erlbaum pp. 123–174.

Edlund, Lena and Rohini Pande. 2002. "Why have women become left-wing? The political gender gap and the decline in marriage." *Quarterly Journal of Economics* CXVII(3):917–961.

Efron, Bradley and Robert J. Tibshirani, eds. 1994. *An Introduction to the Bootstrap*. New York: Chapman & Hall.

Eichenberg, Richard. 2003. "Gender differences in attitudes toward the use of force by the United States, 1990–2003." *International Security* 28(1):10–141.

Ellemers, Naomi. 2018. "Gender stereotypes." *Annual Review of Psychology* 69:275–298.

Epstein, David L., Robert Bates, Jack Goldstone, Ida Kristensen and Sharyn O'Halloran. 2006. "Democratic transitions." *American Journal of Political Science* 50(3):551–569.

Eviatar, Daphne. 2003. "Do aid studies govern policies or reflect them?" *New York Times*, July 26.

EVS. 2015. "European Values Survey Longitudinal Data File 1981–2018 (EVS 1981–2008)." GESIS Data Archive, Cologne. ZA4804 Data File Version 3.0.0.

Falk, Erika and Kate Kenski. 2006. "Issue saliency and gender stereotypes: Support for women as presidents in times of war and terrorism." *Social Science Quarterly* 87(1):1–18.

Fazio, Russell H. and David R. Roskos-Ewoldsen. 1994. Acting as we feel: When and how attitudes guide behavior. In *Persuasion: Psychological Insights and Perspectives*, ed. Sharon Shavitt and Timothy C. Brock. Needham Heights, MA: Allyn & Bacon pp. 71–147.

Ferland, Benjamin. 2018. "The alternative specification of interaction models with a discrete modifying variable." *The Political Methodologist*, February 12. https://polmeth.org

Ferland, Benjamin and Matt Golder. 2021. Citizen representation and electoral systems. In *Oxford Research Encyclopedia of Politics*. Oxford University Press (online only).

Feuersänger, Christian. 2010. "Manual for Package PGFPLOTS." http://pgfplots.sourceforge.net/.

Fleming, J. Marcus. 1962. "Domestic financial policies under fixed and floating exchange rates." *IMF Staff Papers* 9(3):369–380.

Frasure-Yokley, Lorrie. 2018. "Choosing the velvet glove: Women voters, ambivalent sexism, and vote choice in 2016." *Journal of Race, Ethnicity and Politics* 3(1):3–25.

Friedrich, Robert. 1982. "In defense of multiplicative terms in multiple regression equations." *American Journal of Political Science* 26: 797–833.

Gandhi, Jennifer and Adam Przeworski. 2006. "Cooperation, cooptation, and rebellion under dictatorships." *Economics and Politics* 18: 174–204.

Gelman, Andrew and Hal Stern. 2006. "The difference between 'significant' and 'not significant' is not itself statistically significant." *The American Statistician* 60(4):328–331.

Gelman, Andrew and Jennifer Hill. 2006. *Data Analysis Using Regression and Multilevel/Hierarchical Models*. New York: Cambridge University Press.

Gillespie, Andra and Nadia E. Brown. 2019. "#BlackGirlMagic demystified: Black women as voters, partisans and political actors." *Phylon* 56(2):37–58.

Gilligan, Carol, ed. 1982. *In a Different Voice: Psychological Theory and Women's Development*. Cambridge, MA: Harvard University Press.

Glasgow, Garrett. 2001. "Mixed logit models for multiparty elections." *Political Analysis* 9:116–136.

Glasgow, Garrett. 2022. *Interpreting Discrete Choice Models*. New York: Cambridge University Press.

Glasgow, Garrett, Matt Golder and Sona N. Golder. 2011. "Who 'wins'? Determining the party of the prime minister." *American Journal of Political Science* 55(4):937–954.

Glasgow, Garrett, Matt Golder and Sona N. Golder. 2012. "New empirical strategies for the study of parliamentary government formation." *Political Analysis* 20:248–270.

Golder, Matt. 2010. "Ideological congruence and electoral institutions." *American Journal of Political Science* 54(1):90–106.

Golder, Matt. 2016. "Far right parties in Europe." *Annual Review of Political Science* 19:477–497.

Golder, Sona N. 2006*a*. "Pre-electoral coalition formation in parliamentary democracies." *British Journal of Political Science* 36:193–212.

Golder, Sona N. 2006*b*. *The Logic of Pre-Electoral Coalition Formation*. Columbus, OH: Ohio State University Press.

Goldfeld, Stephen M. and Richard E. Quandt. 1973. "The estimation of structural shifts by switching regressions." *Annals of Economic and Social Measurement* 2(4):475–485.

Gottlieb, Jessica, Guy Grossman and Amanda Lea Robinson. 2018. "Do men and women have different policy preferences in Africa? Determinants and implications of gender gaps in policy prioritization." *British Journal of Political Science* 48(3):611–636.

Green, Donald P. and Holger L. Kern. 2012. "Modeling heterogenous treatment effects in survey experiments with Bayesian additive regression trees." *Public Opinion Quarterly* 76(3):491–511.

Greene, William H., ed. 2003. *Econometric Analysis*. Upper Saddle River, NJ: Prentice Hall.

Gross, S. J. and C. M. Niman. 1975. "Attitude behavior consistency: A review." *Public Opinion Quarterly* 39(3):358–368.

Gujarati, Damodar N. 2003. *Basic Econometrics*. New York: McGraw Hill.

Hainmueller, Jens and Chad Hazlett. 2014. "Kernell regularized least squares: Reducing misspecification bias with a flexible and interpretable machine learning approach." *Political Analysis* 22(2): 143–168.

Hainmueller, Jens, Jonathan Mummolo and Yiqing Zu. 2019. "How much should we trust estimates from multiplicative interaction models? Simple tools to improve empirical practice." *Political Analysis* 27(2): 679–708.

Hancock, Ange-Marie. 2004. *The Politics of Disgust: The Public Identity of the Welfare Queen*. New York: New York University Press.

Hancock, Ange-Marie. 2007. "When multiplication doesn't equal quick addition: Examining intersectionality as a research paradigm." *Perspectives on Politics* 5(1):63–79.

Hancock, Ange-Marie. 2016. *Intersectionality: An Intellectual History*. New York: Oxford University Press.

Hanmer, Michael J. and Kerem Ozan Kalkan. 2013. "Behind the curve: Clarifying the best approach to calculating predicted probabilities and marginal effects from limited dependent variable models." *American Journal of Political Science* 57(1):263–277.

Hastie, Trevor and Robert Tibshirani. 1986. "Generalized additive models." *Statistical Science* 1(3):297–318.

Hastie, Trevor, Robert Tibshirani and Jerome Friedman. 2017. *The Elements of Statistical Learning: Data Mining, Inference, and Prediction*. New York: Springer.

Hayes, Bernadette. 1997. "Gender, feminism, and electoral behavior in Britain." *Electoral Studies* 16(2):203–216.

Heath, Anthony Francis, Roger Jowell, John Kevin Curtice, A. Brand and James C. Mitchell. 1993. "British General Election Study 1992." Cross-Section Survey [computer file]. Colchester, Essex: UK Data Archive [distributor], April 1993.

Higgins, Paul C. and Gary L. Albrecht. 1977. "Hellfire and delinquency revisited." *Social Forces* 55(4):952–958.

Hirschi, Travis and Rodney Stark. 1969. "Hellfire and delinquency." *Social Problems* 17(1):202–213.

Holman, Mirya R., Jennifer L. Merolla and Elizabeth J. Zechmeister. 2016. "Terrorist threat, male stereotypes, and candidate evaluations." *Political Research Quarterly* 69(1):134–147.

Horowitz, Juliana Menasce and Janell Fetterolf. 2020. "Worldwide optimism about future of gender equality, even as many see advantages for men." *Pew Research Center*, April 30. www.pewresearch.org/global/2020/04/30/worldwide-optimism-about-future-of-gender-equality-even-as-many-see-advantages-for-men/

Huber, Joan and Glenna Spitze. 1983. *Sex Stratification: Children, Housework, and Jobs.* New York: Academic.

Huber, Lindsay Perez. 2016. "Make America great again: Donald Trump, racist nativism and the virulent adherence to White supremacy amid US demographic change." *Charleston Law Review* 10:215–248.

Huddy, Leonie and Nayda Terkildsen. 1993. "The consequences of gender stereotypes for women candidates at different levels and types of office." *Political Research Quarterly* 46(3):503–525.

Huddy, Leonie, Erin Cassese and Mary-Kate Lizotte. 2012. Gender, public opinion, and political reasoning. In *Political Women and American Democracy*, ed. Christina Wolbrecht, Karen Beckwith and Lisa Baldez. New York: Cambridge University Press pp. 31–49.

Hughes, Melanie M., Pamela Paxton, Amanda Clayton and Pär Zetterberg. 2017. "Quota Adoption and Reform Over Time (QAROT), 1947–2015." Ann Arbor, MI: Inter-university Consortium for Political and Social Research [distributor], 2017-08-16. https://doi.org/10.3886/E100918V1.

Hughes, Melanie M., Pamela Paxton, Amanda Clayton and Pär Zetterberg. 2019. "Global gender quota adoption, implementation, and reform." *Comparative Politics* 51(2):219–238.

Imai, Kosuke, Luke Keele, Dustin Tingley and Teppei Yamamoto. 2011. "Unpacking the black box of causality: Learning about causal mechanisms from experimental and observational studies." *American Political Science Review* 105(4):765–789.

Inglehart, Ronald and Pippa Norris. 2000. "The developmental theory of the gender gap: Women's and men's voting behavior in global perspective." *International Political Science Review* 21(4):441–463.

Inglehart, Ronald and Pippa Norris. 2003. *Rising Tide: Gender Equality and Cultural Change around the World.* New York: Cambridge University Press.

Inter-Parliamentary Union. 2019. "Women in National Parliaments." Data available from World Bank Indicators at https://data.worldbank.org/indicator/SG.GEN.PARL.ZS.

Iversen, Torbern and Frances Rosenbluth. 2006. "The political economy of gender: Explaining cross-national variation in the gender division of labor and the gender voting gap." *American Journal of Political Science* 50(1):1–19.

Jelen, Ted, Sue Thomas and Clyde Wilcox. 1994. "The gender gap in comparative perspective: Gender differences in abstract ideology and concrete issues in Western Europe." *European Journal of Political Research* 25(2):171–186.

Jerolmack, Colin and Shamus Khan. 2014. "Talk is cheap: Ethnography and the attitudinal fallacy." *Sociological Methods & Research* 43(2):178–209.

Jordan-Zachary, Julia. 2003. "The female bogeyman: Political implications of criminalizing black women." *Souls* 5(2):42–62.

Junn, Jane. 2017. "The Trump majority: White womanhood and the making of female voters in the US." *Politics, Groups, and Identities* 5(2):343–352.

Junn, Jane and Natalie Masuoka. 2020. "The gender gap is a race gap: Women voters in US presidential elections." *Perspectives on Politics* 18(4):1135–1145.

Kam, Cindy D. and Robert J. Franzese. 2007. *Modeling and Interpreting Interactive Hypotheses in Regression Analysis.* Ann Arbor, MI: University of Michigan Press.

Kaufman, Robert L., ed. 2019. *Interaction Effects in Linear and Generalized Linear Models: Examples and Applications using Stata.* London: Sage.

Kaufmann, Karen M. and John R. Petrocik. 1999. "The changing politics of American men: Understanding the sources of the gender gap." *American Journal of Political Science* 43(3):864–887.

Keele, Luke. 2008. *Semiparametric Regression for the Social Sciences.* Hoboken, NJ: John Wily & Sons.

King, Gary, ed. 2001. *Unifying Political Methodology: The Likelihood Theory of Statistical Inference.* Ann Arbor, MI: University of Michigan Press.

King, Gary and Langche Zeng. 2006. "The dangers of extreme counterfactuals." *Political Analysis* 14(2):131–159.

King, Gary, Michael Tomz and Jason Wittenberg. 2000. "Making the most of statistical analyses: Improving interpretation and presentation." *American Journal of Political Science* 44(2):341–355.

Klein, Ethel. 1984. *Gender Politics.* Cambridge, MA: Harvard University Press.

Klein, Ethel. 1985. "The gender gap: Different issues, different answers." *The Brookings Review* 3:33–37.

Koch, Michael T. and Sarah A. Fulton. 2011. "In the defense of women: Gender, office holding, and national security policy in established democracies." *Journal of Politics* 73(1):1–16.

Koenig, Anne M., Alice H. Eagly, Abigail A. Mitchell and Riina Ristikari. 2011. "Are leader stereotypes masculine? A meta-analysis of three research paradigms." *Psychological Bulletin* 137(4):616–642.

Kramer, Betty J. and Edward H. Thompson, eds. 2002. *Men as Caregivers: Theory, Research, and Service Implications.* New York: Springer.

Krinsky, Itzhak and A. Leslie Robb. 1986. "On approximating the statistical properties of elasticities." *Review of Economics and Statistics* 68(4):715–719.

Lawless, Jennifer L. 2004. "Women, war, and winning elections: Gender stereotyping in the post-September 11th era." *Political Research Quarterly* 53(3):479–490.

Lewis, Angela K. 2013. *Conservatism in the Black Community: To the Right and Misunderstood*. London: Routledge.

Lijphart, Arend, ed. 1994. *Electoral Systems and Party Systems: A Study of Twenty-Seven Democracies, 1945–1990*. New York: Oxford University Press.

Lindh, Arvid and Leslie McCall. 2020. "Class position and political opinion in rich democracies." *Annual Review of Sociology* 46: 419–441.

Long, J. Scott, ed. 1997. *Regression Models for Categorical and Limited Dependent Variables*. Thousand Oaks, CA: Sage Publications.

Lott, John R. Jr. and Lawrence W. Kenny. 1999. "Did women's suffrage change the size and scope of government?" *Journal of Political Economy* 107(6):1163–1198.

Lowi, Theodore J. 1992. "The party crasher." *New York Times*, August 23.

MacKinnon, David P. 2008. *Introduction to Statistical Mediation Analysis*. New York: Routledge.

Mansbridge, Jane J. 1985. "Myth and reality: The ERA and the gender gap in the 1980 election." *Public Opinion Quarterly* 49:164–178.

Mansbridge, Jane J. 1999. "Should blacks represent blacks and women represent women? A contingent 'yes'." *Journal of Politics* 61:628–657.

Maoz, Zeev. 1998. "Realist and cultural critiques of the democratic peace: A theoretical and empirical re-assessment." *International Interactions* 24(1):3–89.

Marshall, Monty G., Ted Robert Gurr and Keith Jaggers. 2019. "Polity IV Project: Political Regime Characteristics and Transitions, 1800-2018, Dataset Users' Manual." www.systemicpeace.org/inscr/p4manualv2018.pdf.

Matland, Richard E. 2005. Enhancing women's political representation: Legislative recruitment and electoral systems. In *Women in Parliament: Beyond Numbers, A Revised Edition*, ed. Julie Ballington and Azza Karam. Stockholm: International IDEA Publications pp. 93–111.

Mayeri, Serena, Ryan Brown, Nathaniel Persily and Son-Ho Kim. 2008. Gender equality. In *Public Opinion and Constitutional Controversy*, ed. Nathaniel Persily, Jack Citrin and Patrick J. Egan. New York: Oxford University Press pp. 31–49.

McCall, Leslie. 2005. "The complexity of intersectionality." *Signs: Journal of Women in Culture and Society* 30(3):1771–1800.

McFadden, Daniel and Kenneth Train. 2000. "Mixed MNL models of discrete response." *Journal of Applied Econometrics* 15:447–470.

Montgomery, Jacob M. and Santiago Olivella. 2018. "Tree-based models for political science data." *American Journal of Political Science* 62(3):729–744.

Mudde, Cas. 2007. *Populist Radical Right Parties in Europe*. New York: Cambridge University Press.

Mundell, Robert A. 1963. "Capital mobility and stabilization policy under fixed and flexible exchange rates." *Canadian Journal of Economic and Political Science* 29(4):475–485.

Nagler, Jonathan. 1991. "The effect of registration laws and education on US voter turnout." *American Political Science Review* 85(4): 1393–1405.

Nagler, Jonathan. 1994. "Scobit: An alternative estimator to logit and probit." *American Journal of Political Science* 38(1):230–255.

Nelder, J. A. and R. W. M. Wedderburn. 1972. "Generalized linear models." *Journal of the Royal Statistical Society, Series A* 135: 370–384.

Nellis, Ashley. 2016. "The color of justice: Racial and ethnic disparity in state prisons." *The Sentencing Project*, October 13. www .sentencingproject.org/reports/the-color-of-justice-racial-and-ethnic-disparity-in-state-prisons-the-sentencing-project/.

Newman, Benjamin J., Sono Shah and Loren Collingwood. 2018. "Race, place, and building a base: Latino population growth and the nascent Trump campaign for president." *Public Opinion Quarterly* 82(1):122–134.

Newman, Benjamin, Jennifer L. Merolla, Sono Shah, Danielle Casarez Lemi, Loren Collingwood and S. Karthik Ramakrishnan. 2021. "The Trump effect: An experimental investigation of the emboldening effect of racially inflammatory communication." *British Journal of Political Science* 51(3):1138–1159.

Norris, Pippa. 1996a. "Mobilising the 'women's vote': The gender-generation gap in voting behavior." *Parliamentary Affairs* 49(2): 333–342.

Norris, Pippa. 1996b. "Women politicians: Legislative recruitment in advanced democracies." *Parliamentary Affairs* 49(1):89–102.

Norris, Pippa. 1999. Gender: A gender-generation gap? In *Critical Elections: British Parties and Voters in Long-Term Perspective*, ed. Geoffrey Evans and Pippa Norris. London: Sage.

Norris, Pippa. 2001. The gender gap: Old challenges, new approaches. In *Women and American Politics: Agenda Setting for the 21st Century*, ed. Susan J. Carroll. Oxford: Oxford University Press.

Norris, Pippa and Joni Lovenduski. 1989. "Women candidates for parliament: Transforming the agenda?" *British Journal of Political Science* 19(1):106–115.

Norris, Pippa and Joni Lovenduski. 1993. "'If only more candidates came forward': Supply-side explanations of candidate selection in Britain." *British Journal of Political Science* 23(3):373–408.

Norton, Edward C., Hua Wang and Chunrong Ai. 2004. "Computing interaction effects and standard errors in logit and probit models." *The Stata Journal* 4(2):154–167.

Oehlert, Gary W. 1992. "A note on the delta method." *The American Statistician* 46(1):27–29.

Oneal, John R. and Bruce M. Russett. 1997. "The classic liberals were right: Democracy, interdependence, and conflict 1950–1985." *International Studies Quarterly* 41(2):267–294.

Pampel, Fred. 2011. "Cohort changes in the socio-demographic determinants of gender egalitarianism." *Social Forces* 89(3):961–982.

Parker, Christopher. 2016. "Race and politics in the age of Obama." *Annual Review of Sociology* 42:217–230.

Paxton, Pamela, Sheri Kunovich and Melanie M. Hughes. 2007. "Gender in politics." *Annual Review of Political Science* 33:263–84.

Pennebaker, James W., Roger J. Booth and Martha E. Francis. 2007. "Linguistic inquiry and word count: LIWC [Computer software]. Austin, TX: [Operator's manual]."

Phillips, Anne. 1998. *The Politics of Presence: The Political Representation of Gender, Ethnicity, and Race*. New York: Oxford University Press.

Phillips, Anne. 2014. *The Price of the Ticket: Barack Obama and the Rise and Decline of Black Politics*. New York: Oxford University Press.

Pitkin, Hannah Fenichel. 1967. *The Concept of Representation*. Berkeley, CA: University of California Press.

Plutzer, Eric. 1988. "Work life, family life, and women's support of feminism." *American Sociological Review* 853(4):640–649.

Przeworski, Adam, Michael E. Alvarez, José Antonio Cheibub and Fernando Limongi. 2000. *Democracy and Development: Political Institutions and Well-Being in the World, 1950–1990*. New York: Cambridge University Press.

Quinlan, J. R. 1986. "Induction of decision trees." *Machine Learning* 1(1):81–106.

Rainey, Carlisle. 2016. "Compression and conditional effects: A product term is essential when using a logistic regression to test for interaction." *Political Science Research and Methods* 4(3):621–639.

Regan, Patrick M. and Aida Paskeviciute. 2003. "Women's access to politics and peaceful states." *Journal of Peace Research* 40(3): 287–302.

Reingold, Beth. 2000. *Representing Women: Sex, Gender, and Legislative Behavior in Arizona and California*. Chapel Hill, NC: University of North Carolina Press.

Reingold, Beth and Michele Swers. 2011. "An endogenous approach to women's interests: When interests are interesting in and of themselves." *Politics & Gender* 7(3):429–435.

Reingold, Beth, Kerry L. Haynie and Kirsten Widner. 2020. *Race, Gender, and Political Representation: Toward a More Intersectional Approach*. New York: Oxford University Press.

Rhodebeck, Laurie A. 1996. "The structure of men's and women's feminist orientations: Feminist identity and feminist opinion." *Gender & Society* 10(4):386–403.

Rhodes, Albert Lewis and Albert J. Reiss. 1970. "The 'religious factor' and delinquent behavior." *Journal of Research in Crime and Delinquency* 7:83–98.

Rigueur, Leah Wright. 2014. *The Loneliness of the Black Republican: Pragmatic Politics and the Pursuit of Power*. Princeton, NJ: Princeton University Press.

Riker, William. 1982. "The two-party system and Duverger's law: An essay on the history of political science." *American Political Science Review* 76:752–766.

Saperstein, Aliya and Andrew M. Penner. 2012. "Racial fluidity and inequality in the United States." *American Journal of Sociology* 118(3):676–727.

Sapiro, Virginia. 1981. "Research frontier essay: When are interests interesting? The problem of political representation of women." *American Political Science Review* 75:701–716.

Schaffner, Brian F., Matthew MacWilliams and Tatishe Nteta. 2018. "Understanding White polarization in the 2016 vote for president: The sobering role of racism and sexism." *Political Science Quarterly* 133(1):9–34.

Schenker, Nathaniel and Jane F. Gentleman. 2001. "On judging the significance of differences by examining the overlap between confidence intervals." *The American Statistician* 55:182–186.

Schofield, Norman J. 2003. "Valence competition in the spatial stochastic model." *Journal of Theoretical Politics* 15(4):371–383.

Schuman, Howard and Michael P. Johnson. 1976. "Attitudes and behavior." *Annual Review of Sociology* 2:161–207.

Schwarz, Hermann. 1873. "Communication." *Archives des Sciences Physiques et Naturelles* 48:38–44.

Shapiro, Robert Y. and Harpreet Mahajan. 1986. "Gender differences in policy preferences: A summary of trends from the 1960s to the 1980." *Public Opinion Quarterly* 50(1):42–61.

Simon, Rita J. and Jean M. Landis. 1989. "A report: Women's and men's attitudes about a woman's role and place." *Public Opinion Quarterly* 53(2):265–276.

Smith, Preston H. 2018. "Self-help," Black conservatives, and the reemergence of Black privatism. In *Without Justice for All: The New Liberalism and Our Retreat from Racial Equality*, ed. Adolph Reed. London: Routledge pp. 257–289.

Stark, Rodney. 1996. "Religion as context: Hellfire and delinquency one more time." *Sociology of Religion* 57(2):163–173.

Stark, Rodney, Lori Kent and Daniel P. Doyle. 1982. "Religion and delinquency: The ecology of a 'lost' relationship." *Journal of Research in Crime and Delinquency* 19(1):4–24.

Stockemer, Daniel. 2011. "Women's parliamentary representation in Africa: The impact of democracy and corruption on the number of female deputies in national parliaments." *Political Studies* 59: 693–712.

Stolper, W. F. and Paul A. Samuelson. 1941. "Protection and real wages." *Review of Economic Studies* 9(1):58–73.

Subramanian, Ram, Kristi Riley and Chris Mai. 2018. "Divided Justice: Trends in Black and White jail incarceration 1990–2013." Vera Institute of Justice.

Swain, Randall D. 2018. "Negative Black stereotypes, support for excessive use of force by police, and voter preference for Donald Trump during the 2016 presidential primary election cycle." *Journal of African American Studies* 22:109–124.

Swers, Michele L., ed. 2002. *The Difference Women Make: The Policy Impact of Women in Congress*. Chicago, IL: University of Chicago Press.

Taagepera, Rein. 1998a. "Effective magnitude and effective threshold." *Electoral Studies* 17:393–404.

Taagepera, Rein. 1998b. "Nationwide inclusion and exclusion thresholds of representation." *Electoral Studies* 17:405–417.

Taylor-Robinson, Michelle M. and Nehemia Geva. Forthcoming. Mental templates of leaders – do they include women? In *Women in Government: An Experimental Study of Attitudes about Governing Ability,* ed. Michelle M. Taylor-Robinson and Nehemia Geva. New York: Oxford University Press.

Thompson, Edward H. 2000. Gendered caregiving of husbands and sons. In *Intersections of Aging: Readings in Social Gerontology*, ed. Elizabeth W. Markson and Lisa A. Hollis-Sawyer. Los Angeles, CA: Roxbury Publishing Company pp. 333–344.

Tillery, Alvin B. 2019. Obama's legacy for race relations. In *The Obama Legacy*, ed. Bert A. Rockman and Andrew Rudalevige. London: Routledge pp. 71–90.

Tomlinson, Barbara. 2013. "To tell the truth and not get trapped: Desire, distance, and intersectionality at the scene of argument." *Signs: Journal of Women in Culture and Society* 38(4):993–1017.

Train, Kenneth E. 1998. "Recreation demand models with taste differences over people." *Land Economics* 74:230–239.

Train, Kenneth E. 2009. *Discrete Choice Methods with Simulation (Second Edition)*. New York: Cambridge University Press.

Tripp, Aili Mari and Alice Kang. 2008. "The global impact of quotas on the fast track to increased female legislative representation." *Comparative Political Studies* 41(3):338–361.

Valdini, Melody Eiils. 2019. *The Inclusion Calculation: Why Men Appropriate Women's Representation*. New York: Oxford University Press.

Valdini, Melody Ellis. 2012. "A deterrent to diversity: The conditional effect of electoral rules on the nomination of women candidates." *Electoral Studies* 31:740–749.

Vapnik, Vladimir N. 1995. *The Nature of Statistical Learning Theory*. New York: Springer.

Vapnik, Vladimir N. 1998. *Statistical Learning Theory*. New York: John Wiley & Sons.

Ver Hoef, Jay M. 2012. "Who invented the delta method?" *The American Statistician* 66(2):124–127.

Wängnerud, Lena. 2000. "Testing the politics of presence: Women's representation in the Swedish Riksdag." *Scandinavian Political Studies* 23(1):67–91.

Warner, Leah H. 2007. "A best practices guide to intersectional approaches in psychological research." *Sex Roles* 59:454–463.

Weaver, Vesla M. 2010. "Political consequences of the carceral state." *American Political Science Review* 104(4):817–833.

Weldon, S. Laurel. 2006. "The structure of intersectionality: A comparative politics of gender." *Sex Roles* 2(2):235–248.

Weldon, S. Laurel. 2011. "Perspective against interests: Sketch of a feminist political theory of women." *Politic & Gender* 7(3):441–446.

Williams, Vanessa. 2017. "What's wrong with White women voters? Here's the problem with that question." *Washington Post*, December 22.

World Bank. 2017. "World Development Indicators." Washington, DC: The World Bank, https://datacatalog.worldbank.org/dataset/world-development-indicators.

Wright, Gerald. 1976. "Linear models for evaluating conditional relationships." *American Journal of Political Science* 2:349–373.

WVS. 2015. "World Values Survey 1981–2015 Official Aggregate v.20150418." World Values Survey Association (www.worldvaluessurvey.org). Aggregate File Producer: JDSystems, Madrid.

Yoon, Mi Yung. 2004. "Explaining women's legislative representation in sub-Saharan Africa." *Legislative Studies Quarterly* 29(3):447–468.

Young, Iris Marion. 2002. *Inclusion and Democracy*. New York: Oxford University Press.

Zeng, Langche. 1999. "Prediction and classification with neural network models." *Sociological Methods and Research* 27(4):499–524.

Zuberi, Tukufu. 2000. "Deracializing social statistics: Problems in the quantification of race." *The Annals of the American Academy of Political and Social Science* 568:172–185.

Solutions

EXERCISES FROM CHAPTER 2

1. • *Education, Income, and Spending:* Mediation.
   • *Drugs, Exercise, and Illness:* Moderation.
   • *Development, Culture, and Democracy:* Mediation.
   • *Federalism, Ethnic Diversity, and Conflict:* Moderation.
   • *Incumbency, Deterrence, and Electoral Advantage:* Mediation.
   • *Ethnic Cues, Anxiety, Support for Immigration:* Mediation.
   • *Inequality, Asset Mobility, and Democracy:* Moderation.

2. *Gender Party Gap Theory*

   $P_{Female \times Black}$: The marginal effect of each of *Female* and *Black* is negatively related to the other variable.

   $P_{Female|White}$: The marginal effect of *Female* on *Wages* is negative when $Black = 0$.

   $P_{Female|Black}$: The marginal effect of *Female* on *Wages* is negative when $Black = 1$.

   $P_{Black|Male}$: The marginal effect of *Black* on *Wages* is negative when $Female = 0$.

   $P_{Black|Female}$: The marginal effect of *Black* on *Wages* is negative when $Female = 1$.

   Female Hypothesis: Women earn less money than men. The negative effect of being a woman is greater if you're Black as opposed to White.

   Black Hypothesis: Blacks earn less money than Whites. The negative effect of being Black is greater if you're female as opposed to male.

   Party System Size Theory

   $P_{Social\ Diversity \times Electoral\ Permissiveness}$: The marginal effect of each of *Social Diversity* and *Electoral Permissiveness* is positively related to the other variable.

   $P_{Social\ Diversity|Electoral\ Permissiveness_{min}}$: The marginal effect of *Social Diversity* on *Parties* is positive (or zero) when *Electoral Permissiveness* is at its *lowest* value.

   $P_{Social\ Diversity|Electoral\ Permissiveness_{max}}$: The marginal effect of *Social Diversity* on *Parties* is positive when *Electoral Permissiveness* is at its *highest* value.

$P_{Electoral\ Permissiveness|Social\ Diversity_{min}}$: The marginal effect of *Electoral Permissiveness* on *Parties* is zero when *Social Diversity* is at its *lowest* value.

$P_{Electoral\ Permissiveness|Social\ Diversity_{max}}$: The marginal effect of *Electoral Permissiveness* on *Parties* is positive when *Social Diversity* is at its *highest* value.

Social Diversity Hypothesis: Social diversity has a positive (or zero) effect on the number of parties when the electoral system is not permissive. This positive effect is greater when electoral system permissiveness is high.

Electoral Permissiveness Hypothesis: Electoral system permissiveness has no effect on the number of parties when there's no social diversity. Electoral system permissiveness has a positive and growing effect on the number of parties as social diversity increases.

Democratization Theory

$P_{X \times Middle\ Class}$: The marginal effect of each of X and *Middle Class* is positively related to the other variable.

A variety of answers are possible at this point. The key is that X can only have an effect on *Democratization* when *Middle Class* $= 1$.

$P_{X|Middle\ Class_{min}}$: The marginal effect of X on *Democratization* is zero when *Middle Class* $= 0$.

$P_{X|Middle\ Class_{max}}$: The marginal effect of X on *Democratization* is positive when *Middle Class* $= 1$.

$P_{Middle\ Class|X_{min}}$: The marginal effect of *Middle Class* on *Democratization* is positive when X is at its *lowest* value.

$P_{Middle\ Class|X_{max}}$: The marginal effect of *Middle Class* on *Democratization* is positive when X is at its *highest* value.

X Hypothesis: X has no effect on democratization when there's no middle class but has a positive effect when there is a middle class.

Middle Class Hypothesis: The existence of a middle class always has a positive effect on democratization. This positive effect increases with X.

Clinical Trial Theory

$P_{Drug \times Male}$: The marginal effect of each of *Drug* and *Male* is positively related to the other variable.

$P_{Drug|Female}$: The marginal effect of *Drug* on *Depression* is negative when *Male* $= 0$.

$P_{Drug|Male}$: The marginal effect of *Drug* on *Depression* is negative when *Male* $= 1$.

$P_{Male|Drug_{min}}$: The marginal effect of *Male* on *Depression* is negative when $Drug = 0$.

$P_{Male|Drug_{max}}$: The marginal effect of *Male* on *Depression* is [negative, zero, positive] when $Male = 1$.

Drug Hypothesis: Taking the drug always lowers the level of depression. This negative effect is smaller for men than for women.

Male Hypothesis: Men have lower levels of depression than women when no drug is being used. Depending on one's theory, this negative effect declines, disappears, or even reverses when the drug is being used.

3. All interactions are symmetric. As a result, it's necessarily the case that the sign of the slope relationship between *ME(X|Z)* and *Z* is the same as the sign of the slope relationship between *ME(Z|X)* and *X*. According to $H_{X|Z}$, the slope of the relationship between *ME(X|Z)* and *Z* is *negative*; that is, the positive effect of *X* on *Y* gets smaller with *Z*. According to $H_{Z|X}$, the slope of the relationship between *ME(Z|X)* and *X* is *positive*; that is, the negative effect of *Z* on *Y* first gets smaller with increases in *X*, then disappears, and eventually becomes positive. In effect, the two hypotheses make opposite predictions about the sign of the coefficient on the interaction term. As a result, something is wrong with the hypotheses.

4. The marginal effect of *X* is

$$\frac{\partial Y}{\partial X} = \beta_1 + \beta_3 Z. \tag{S.1}$$

From this, we see that β_1 tells us the marginal effect of *X* on *Y* when $Z = 0$. The marginal effect of *Z* is

$$\frac{\partial Y}{\partial Z} = \beta_2 + \beta_3 X. \tag{S.2}$$

From this, we see that β_2 tells us the marginal effect of *Z* on *Y* when $X = 0$. From both marginal effects, we see that β_3 indicates how the slope of the relationship between *X* and *Y* changes with *Z* and how the slope of the relationship between *Z* and *Y* changes with *X*. In other words, β_3 captures the "interaction effect." The coefficient on the constant term, β_0, tells us the predicted level of *Y* when *X* and *Z* are both zero.

5. • *X* and *Z* are both important causal variables in determining *Y*: How exactly do *X* and *Z* influence *Y*? What does "important" mean?

   • The interaction between *X* and *Z* is positively related to *Y*: How exactly do *X* and *Z* influence *Y*?

- X has less effect on Y when Z is high: How exactly does X influence Y?
- The negative effect of X on Y can be diminished through Z: Is this a mediating or moderating relationship?
- The effect of X on Y is greater when Z is absent than when Z is present: Is the effect of X on Y positive or negative?
- X should be a stronger predictor of Y when Z is high: Is the effect of X on Y positive or negative? What does "stronger predictor" mean?
- The effect of X on Y is contingent on Z: Is the effect of X on Y positive or negative? How exactly does the effect of X on Y depend on Z?
- X has a negative effect on Y when Z is present: What's the effect of X on Y when Z is absent?
- Y should be viewed as a process, not an event: This is not a hypothesis – there is no relational claim between at least two variables.
- X has an insignificant effect on Y: Is the effect of X on Y positive or negative? Is the theory really strong enough to make a claim about statistical significance for a given sample size?
- X increases Y independent of Z: Is this a mediating or moderating relationship? Should Z be included in the model in an additive or interactive manner?
- The impact of X on Y is less when Z is present than when Z is absent: Is the effect of X on Y positive or negative?
- X and Z interact to determine Y: How exactly do X and Z interact to influence Y? What's the effect of X on Y?
- X has the most influence on Y: What's the effect of X on Y? What does "most influence" mean?

6. $P_{Consumer\ Sentiment \times Recession}$: The marginal effect of each of *Consumer Spending* and *Recession* is positively related to the other variable.

   $P_{Consumer\ Sentiment | Recession_{min}}$: The marginal effect of *Consumer Sentiment* on *Consumer Spending* is positive when *Recession* $= 0$.

   $P_{Consumer\ Sentiment | Recession_{max}}$: The marginal effect of *Consumer Sentiment* on *Consumer Spending* is positive when *Recession* $= 1$.

   $P_{Recession | Consumer\ Sentiment_{min}}$: The marginal effect of *Recession* on *Consumer Spending* is negative when *Consumer Spending* is at its *lowest* value.

   $P_{Recession | Consumer\ Sentiment_{max}}$: The marginal effect of *Recession* on *Consumer Spending* is negative when *Consumer Spending* is at its *highest* value.

Consumer Sentiment Hypothesis: An increase in consumer sentiment always increases consumer spending. This positive effect is greater in recessions.

Recession Hypothesis: Recessions always reduce consumer spending. This negative effect is less when consumer sentiment is high.

(a) The marginal effect of *Consumer Sentiment* on *Consumer Spending* is

$$\frac{\partial Consumer\ Spending}{\partial Consumer\ Sentiment} = \beta_1 + \beta_3 Recession. \tag{S.3}$$

The marginal effect of *Consumer Sentiment* when we're not in a recession is β_1. The marginal effect of *Consumer Sentiment* when we're in a recession is $\beta_1 + \beta_3$. According to the *Consumer Sentiment Hypothesis*, we should find that $\beta_1 > 0$ and $\beta_3 > 0$. The predicted marginal effect plot for *Consumer Sentiment* shows two points, one for the effect of *Consumer Sentiment* when we're not in a recession and one for the effect of *Consumer Sentiment* when we're in a recession. The point indicating the effect of *Consumer Sentiment* when we're in a recession is higher than the point indicating the effect of *Consumer Sentiment* when we're not in a recession. The difference in the height of the two points is equal to the coefficient on the interaction term, β_3, and is also the same as the slope of the line that would be depicted in a marginal effect plot for *Recession*.

(b) The effect of a one-unit discrete change in *Recession* can be calculated as

$$\frac{\partial Consumer\ Spending}{\partial Recession} = \beta_2 + \beta_3 Consumer\ Sentiment. \tag{S.4}$$

According to the *Recession Hypothesis*, we should find that $\beta_2 < 0$, $\beta_3 > 0$, and $\beta_2 + \beta_3 Consumer\ Sentiment < 0$ for all values of *Consumer Sentiment*. The predicted plot depicting the effect of a one-unit discrete change in *Recession* on *Consumer Spending* shows a line with a negative intercept sloping up to the right but always remaining below zero. The slope of the line is equal to the coefficient on the interaction term, β_3, and is also the same as the difference in the height of the two points that would be depicted in a marginal effect plot for *Consumer Sentiment*.

EXERCISES FROM CHAPTER 3

1. a. We let our new rescaled variable be X^*, where $X = X^* + 5$. We can now rewrite our original interactive model specification as

$$Y = \beta_0 + \beta_1(X^* + 5) + \beta_2 Z + \beta_3(X^* + 5)Z + \epsilon. \tag{S.5}$$

Multiplying through, we have

$$Y = \beta_0 + \beta_1 X^* + 5\beta_1 + \beta_2 Z + \beta_3 X^* Z + 5\beta_3 Z + \epsilon. \tag{S.6}$$

Collecting terms, we have

$$Y = (\beta_0 + 5\beta_1) + \beta_1 X^* + (\beta_2 + 5\beta_3)Z + \beta_3 X^* Z + \epsilon. \tag{S.7}$$

From this, we can see that the constant for the model with the rescaled variable X^* is $1.00 + 5 \times 0.20 = 2.00$; we don't have enough information to calculate the standard error. The coefficient and standard error on X^* is the same as in the original model, 0.20 (0.08). The coefficient on Z is $-2.00 + 5 \times 0.40 = 0$; we don't have enough information to calculate the standard error. The coefficient and standard error on the interaction term is the same as in the original model, 0.40 (0.13).

b. It's not possible to have an *a priori* prediction that the marginal effect of Z on Y is zero when $X = 0$ unless the X variable is measured with a natural zero. This is because, as we just saw, it's possible to arbitrarily manipulate the scale of the X variable when it doesn't have a natural zero to make the coefficient on Z be zero (or any other value). Since our theories don't typically tell us which scale to use for our variables, we can't have an *a priori* prediction that the marginal effect of Z on Y is zero when $X = 0$.

2. a. The coefficient on the constitutive term *Social Divisions*$_c$ tells us that the marginal effect of *Social Divisions* is positive when *Electoral Permissiveness* is at its mean value. Specifically, a one-unit increase in *Social Divisions* increases the number of parties by 1.11 when *Electoral Permissiveness* is 1.53.

b. The coefficient on the constitutive term *Electoral Permissiveness*$_c$ tells us that the marginal effect of *Electoral Permissiveness* is positive when *Social Divisions* is at its mean value. Specifically, a one-unit increase in *Electoral Permissiveness* increases the number of parties by 0.61 when *Social Divisions* is 1.57.

c. The marginal effect of *Social Divisions* when *Electoral Permissiveness* is zero isn't shown in the regression output. However, we can calculate it from the information provided. Using Eq. 3.9 from Chapter 3, we see that the marginal effect of

Table S.1 Social divisions, electoral system permissiveness, and the number of parties		
Dependent Variable: *Number of Political Parties*		
	Mean-centered	**Uncentered**
Social Divisions	1.11***	−0.16
	(0.38)	(?)
Electoral Permissiveness	0.61***	−0.69
	(0.16)	(?)
Social Divisions × *Electoral Permissiveness*	0.83***	0.83***
	(0.29)	(0.29)
Constant	3.69***	3.01
	(0.22)	(?)
Observations	51	51
R^2	0.30	0.30
Variance Inflation Factor	1.19	6.00

Standard errors in parentheses. *$p < 0.10$; **$p < 0.05$; ***$p < 0.01$ (two-tailed)

Social Divisions when *Electoral Permissiveness* is zero is $\delta_1 - \delta_3\overline{Z} = 1.11 - (0.83 \times 1.53) = -0.16$. We don't have enough information to calculate the standard error.

d. The marginal effect of *Electoral Permissiveness* when *Social Divisions* is zero is also not shown in the regression output. However, we can calculate it from the information provided. Using Eq. 3.9 from Chapter 3, we see that the marginal effect of *Electoral Permissiveness* when *Social Divisions* is zero is $\delta_2 - \delta_3\overline{X} = 0.61 - (0.83 \times 1.57) = -0.69$. We don't have enough information to calculate the standard error.

e. Variance inflation factor scores should never be used to justify the omission of constitutive terms in an interaction model. As we saw in the chapter, the centered model is exactly equivalent to the uncentered model. Since centering doesn't provide us with any new information, it can't reduce multicollinearity. The fact that the coefficients on the constitutive terms are statistically significant in the centered model but statistically insignificant in the uncentered model isn't a sign that the centered model has lower levels of multicollinearity, just that the coefficients on these constitutive terms capture completely different quantities of interest in the two models.

f. The second column of Table S.1 includes as much information as we can get from the results of the centered model about the coefficients, standard errors, and R^2 that we would get if we were to estimate an uncentered model.

3. a. The fact that the model specification shown in Eq. 3.47 doesn't include all of the constitutive elements of the interaction terms isn't problematic. As we saw in Chapter 3, this model specification is exactly equivalent to a more standard interactive model that includes *Niche*, *Public Opinion Shift*, *Niche×Public Opinion Shift*, and *Controls* as independent variables. It's impossible to include *Public Opinion Shift* as a separate variable as there's perfect multicollinearity with the two interaction terms. Given that the model includes a constant, it's not possible to include both *Niche* and *Mainstream* in the model specification.

b. The coefficient on the interaction term *Public Opinion Shift × Mainstream* tells us the marginal effect of *Public Opinion Shift* on *Party Position Shift* when the party is a mainstream party. Specifically, a one-unit increase in *Public Opinion Shift* is associated with a shift in the party's position of 0.97 units in the same direction if the party is a mainstream party.

c. The coefficient on the interaction term *Public Opinion Shift × Niche* tells us the marginal effect of *Public Opinion Shift* on *Party Position Shift* when the party is a niche party. Specifically, a one-unit increase in *Public Opinion Shift* is associated with a shift in the party's position of 0.55 units in the opposite direction if the party is a niche party.

d. The coefficient on *Niche* tells us the effect of being a niche party as opposed to a mainstream party on *Party Position Shift* when *Public Opinion Shift* = 0. Specifically, niche parties move their party position by 0.02 units compared to mainstream parties when there's no change in public opinion; this difference isn't statistically significant.

e. It would be appropriate to omit *Niche* and replace it with *Mainstream*. The important thing given that the model specification includes a constant is that one of *Niche* or *Mainstream* is included in the model. Using *Mainstream* instead of *Niche* wouldn't change how we interpret the coefficients on the interaction terms.

f. The second column of Table S.2 includes as much information as we can get from the results of the alternative interactive model

Table S.2 Political responsiveness of niche and mainstream parties		

Dependent Variable: Party Position Shift

	Alternative Specification	Standard Specification
Niche	0.02	0.02
	(0.12)	(0.12)
Public Opinion Shift × Mainstream	0.97***	
	(0.19)	
Public Opinion Shift × Niche	−0.55**	−1.52
	(0.26)	(?)
Public Opinion Shift		0.97***
		(0.19)
Constant	0.09	0.09
	(0.18)	(0.18)
Controls	Yes	Yes
Observations	158	158
R^2	0.33	0.33

Standard errors in parentheses. $^*p < 0.10$; $^{**}p < 0.05$; $^{***}p < 0.01$ (two-tailed)

specification in the first column about the coefficients, standard errors, and R^2 that we would get if we were to estimate a standard interactive model.

g. The coefficient on *Public Opinion Shift* tells us the marginal effect of *Public Opinion Shift* on *Party Position Shift* when the party is a mainstream party (*Niche* $= 0$). Specifically, a one-unit change in public opinion leads to a 0.97 unit shift in the party's position in the same direction when the party is a mainstream party.

h. The coefficient on *Niche* tells us the effect of being a niche party as opposed to a mainstream party on *Party Position Shift* when *Public Opinion Shift* $= 0$. Specifically, niche parties move their party position by 0.02 units compared to mainstream parties when there's no change in public opinion; this difference isn't statistically significant.

i. The coefficient on the interaction term *Public Opinion Shift* × *Niche* tells us how the type of party – mainstream or niche –

Table S.3 Symbolic racism, anger, and opposition to race-conscious policies I

	Dependent Variable: *Oppose Race Policies*			
	Full Sample		**Split Sample**	
	Standard Specification	**Alternative Specification**	**Calm**	**Angry**
Symbolic Racism	0.52***		0.52	0.78
	(0.05)		(?)	(?)
Anger	−0.11	−0.11		
	(0.10)	(0.10)		
Symbolic Racism × Calm		0.52***		
		(0.05)		
Symbolic Racism × Anger	0.26**	0.78		
	(0.12)	(?)		
Constant	0.27***	0.27***	0.27***	0.16
	(0.04)	(0.04)	(0.04)	(?)
Controls	No	No	No	No
Observations	182	182	142	40
R^2	0.46	0.46	?	?

Standard errors in parentheses. *$p < 0.10$; **$p < 0.05$; ***$p < 0.01$ (two-tailed)

modifies the slope of the relationship between changes in public opinion and changes in party position and how changes in public opinion modify the how mainstream and niche parties change their party positions.

4. a. The second column of Table S.3 includes as much information as we can get from the results of the standard interactive model specification in the first column about the coefficients, standard errors, and R^2 that we would get if we were to estimate the alternative interactive model.

   b. The third and fourth columns of Table S.3 include as much information as we can get from the results of the standard interactive model specification in the first column about the coefficients, standard errors, and R^2 that we would get if we were to adopt a split sample strategy.

Table S.4 Gender, race, and support for the Republican Party in the United States

Dependent Variable: *Like Republican Party, 0–10*

	Alternative Specification I	Alternative Specification II	Standard Specification
White Female	−0.04 (0.12)	2.53 (?)	
Black Male	−1.50*** (0.27)	1.07 (?)	
Black Female	−2.57*** (0.22)		
White Male		2.57*** (0.22)	
Female			−0.04 (0.12)
Black			−1.50*** (0.27)
Female×Black			−1.03 (?)
Age	0.02*** (0.003)	0.02*** (0.003)	0.02*** (0.003)
Constant	4.47*** (0.18)	1.90 (?)	4.47*** (0.18)
Observations	2,858	2,858	2,858
R^2	0.07	0.07	0.07

Standard errors in parentheses. *$p < 0.10$; **$p < 0.05$; ***$p < 0.01$ (two-tailed)

c. If the results in the first column of Table S.3 were based on a model that includes control variables, then we wouldn't be able to identify any information about the coefficients, standard errors, and R^2 that we'd get if we were to adopt the split sample strategy. This is because the split sample strategy allows the effects of *all* of the variables to vary across the different values of *Anger*. We have no way of knowing what the coefficients, standard errors, and R^2

would be from the split sample strategy without actually estimating our model on the two sub-samples.

5. a. Yes, the model shown in Eq. 3.47 is an interaction model. We know this because the variables *White Female, Black Male,* and *Black Female* are dichotomous interaction terms and because the model specification is exactly equivalent to the standard interactive model specification shown in Eq. 3.48.

b. The coefficient on *White Female* indicates that support for the Republican Party is 0.04 units lower among White women than White men. Put differently, it indicates that the effect of being female as opposed to male for White people is −0.04. This effect is not statistically significant. The coefficient on *Black Male* indicates that support for the Republican Party is 1.50 units lower among Black men than White men. Put differently, it indicates that the effect of being Black rather than White for men is −1.50. The coefficient on *Black Female* indicates that support for the Republican Party is 2.57 points lower among Black women than White men. Put differently, it indicates that the effect of jointly changing the gender and race of White men is −2.57.

c. The interaction effect of gender and race on support for the Republican Party is calculated as
$\gamma_3 - \gamma_1 - \gamma_2 = -2.57 - (-0.04) - (1.50) = -1.03$.

d. The second column of Table S.4 includes as much information as we can get from the results in the first column about the coefficients, standard errors, and R^2 that we'd obtain if we reestimated our model with Black women as the baseline category.

e. The third column of Table S.4 includes as much information as we can get from the results in the first column about the coefficients, standard errors, and R^2 that we'd obtain if we estimated the standard interactive model in Eq. 3.48.

EXERCISES FROM CHAPTER 4

1. a. The variances for the coefficients in Table 4.1 are shown in the diagonal elements, top left to bottom right, of the variance-covariance matrix in Eq. 4.44. If we square root these elements, we obtain the standard errors for the coefficients in Table 4.1. As an example, if we square root the element in the second row and second column, $\sqrt{0.00915827} = 0.09570$, we obtain the standard error for the coefficient on *Anger*. If we square root the element in the fourth row and fourth column, $\sqrt{0.00155169} = 0.03939$, we obtain the standard error for the

Table S.5 Symbolic racism, anger, and opposition to race-conscious policies II

Dependent Variable: *Oppose Race Policies*

	Model 1
Symbolic Racism	0.52126***
	(0.05464)
Anger	−0.11196
	(0.09570)
Symbolic Racism × *Anger*	0.25792**
	(0.12454)
Constant	0.26545***
	(0.03939)
Observations	182
R^2	0.46

Standard errors in parentheses. *$p < 0.10$; **$p < 0.05$; ***$p < 0.01$ (two-tailed)

   coefficient on the constant term. The remaining standard errors are shown in Table S.5.

b. The covariances are shown in the off-diagonals of the variance-covariance matrix in Eq. 4.44. Since the variance-covariance matrix is symmetric, $\text{cov}(\beta_1, \beta_3) = \text{cov}(\beta_3, \beta_1) = -0.00298512$, and $\text{cov}(\beta_2, \beta_3) = \text{cov}(\beta_3, \beta_2) = -0.01125366$.

c. The marginal effect of *Symbolic Racism* on *Oppose Race Policies* when someone isn't angry is given by the coefficient on *Symbolic Racism*, 0.52126. In other words, a one-unit increase in *Symbolic Racism* is associated with a 0.52 increase in *Oppose Race Policies* for someone who isn't angry. The variance associated with this effect is given by the element in the first row and first column of the variance-covariance matrix in Eq. 4.44, 0.00298512. The standard error is $\sqrt{0.00298512} = 0.054636$. The t-statistic is $0.52126/0.054636 = 9.54$. The p-value associated with a test statistic of this size and $n - k = 182 - 4 = 178$ degrees of freedom is $p < 0.00001$. The critical t-value for an $\alpha = 0.05$ level of

significance and $n - k = 178$ degrees of freedom is 1.973.
Thus, the two-tailed 95% confidence interval is
$0.52 \pm 1.973 \times 0.054636 = 0.52 \pm 0.107797$. Thus, the marginal
effect is 0.52 (0.413, 0.629).

d. The marginal effect of *Symbolic Racism* on *Oppose Race Policies*
when someone is angry is given by
$\beta_1 + \beta_3 \times 1 = 0.52126 + 0.25792 = 0.77918$. In other words, a
one-unit increase in *Symbolic Racism* is associated with a 0.78
increase in *Oppose Race Policies* for someone who's angry. The
variance associated with this effect is given by

$$\text{var}(\beta_1 + \beta_3) = \text{var}(\beta_1) + 1^2 \times \text{var}(\beta_3) + 2 \times 1 \times \text{cov}(\beta_1, \beta_3)$$
$$= 0.00298512 + 0.01550988 + 2 \times -0.00298512$$
$$= 0.012525. \tag{S.8}$$

The standard error associated with the effect is $\sqrt{0.012525}$
$= 0.11191$. The t-statistic is $0.77918/0.11191 = 6.96$. The p-value
associated with a test statistic of this size and $n - k = 182 - 4$
$= 178$ degrees of freedom is $p < 0.00001$. The critical t-value for
an $\alpha = 0.05$ level of significance and $n - k = 178$ degrees of
freedom is 1.973. Thus, the two-tailed 95% confidence interval is
$0.78 \pm 1.973 \times 0.11191 = 0.78 \pm 0.2208$. Thus, the marginal effect
is 0.78 (0.558, 1.000).

e. The effect of being angry on *Oppose Race Policies* when an
individual's level of racial resentment is 0 is given by the coefficient
on *Anger*, -0.11196. In other words, the effect of being angry as
opposed to calm is associated with a 0.11 reduction in *Oppose
Race Policies* for someone who has no racial resentment. The
variance associated with this effect is given by the element in the
second row and second column of the variance-covariance matrix
in Eq. 4.44, 0.00915827. The standard error is $\sqrt{0.00915827}$
$= 0.095699$. The t-statistic is $-0.11196/0.095699 = -1.17$. The
p-value associated with a test statistic of this size and $n - k$
$= 182 - 4 = 178$ degrees of freedom is $p = 0.244$. The critical
t-value for an $\alpha = 0.05$ level of significance and $n - k = 178$
degrees of freedom is 1.973. Thus, the two-tailed 95% confidence
interval is $-0.11 \pm 1.973 \times 0.095699 = -0.11 \pm 0.1888$. Thus, the
effect is -0.11 (-0.301, 0.077).

f. The effect of being angry on *Oppose Race Policies* when an
individual's level of racial resentment is 0.5 is given by $\beta_2 + \beta_3$
$\times 0.5 = -0.11196 + 0.25792 \times 0.5 = 0.017$. In other words, being

angry is associated with a 0.017 increase in *Oppose Race Policies* for someone whose level of racial resentment is 0.5. The variance associated with this effect is given by

$$\text{var}(\beta_2 + \beta_3 \times 0.5)$$
$$= \text{var}(\beta_2) + 0.5^2 \times \text{var}(\beta_3) + 2 \times 0.5 \times \text{cov}(\beta_2, \beta_3)$$
$$= 0.00915827 + 0.25 \times 0.01550988 + 1 \times -0.01125366$$
$$= 0.00178. \tag{S.9}$$

The standard error associated with the effect is $\sqrt{0.00178} = 0.0422$. The t-statistic is $0.017/0.0422 = 0.40$. The p-value associated with a test statistic of this size and $n - k = 182 - 4 = 178$ degrees of freedom is $p = 0.69$. The critical t-value for an $\alpha = 0.05$ level of significance and $n - k = 178$ degrees of freedom is 1.973. Thus, the two-tailed 95% confidence interval is $0.017 \pm 1.973 \times 0.0422 = 0.017 \pm 0.083$. Thus, the effect is 0.017 (−0.066, 0.100).

g. The effect of being angry on *Oppose Race Policies* when an individual's level of racial resentment is 1 is given by $\beta_2 + \beta_3 \times 1 = -0.11196 + 0.25792 \times 1 = 0.14596$. In other words, being angry is associated with a 0.15 increase in *Oppose Race Policies* for someone whose level of racial resentment is 1. The variance associated with this effect is given by

$$\text{var}(\beta_2 + \beta_3 \times 1)$$
$$= \text{var}(\beta_2) + 1^2 \times \text{var}(\beta_3) + 2 \times 1 \times \text{cov}(\beta_2, \beta_3)$$
$$= 0.00915827 + 0.01550988 + 2 \times -0.01125366$$
$$= 0.00216. \tag{S.10}$$

The standard error associated with the effect is $\sqrt{0.00216} = 0.04648$. The t-statistic is $0.14596/0.04648 = 3.14$. The p-value associated with a test statistic of this size and $n - k = 182 - 4 = 178$ degrees of freedom is $p = 0.002$. The critical t-value for an $\alpha = 0.05$ level of significance and $n - k = 178$ degrees of freedom is 1.973. Thus, the two-tailed 95% confidence interval is $0.15 \pm 1.973 \times 0.04648 = 0.15 \pm 0.092$. Thus, the effect is 0.15 (0.054, 0.238).

h. The predicted value of *Oppose Race Policies* in the baseline scenario where an individual is calm and has a level of racial resentment equal to 0.5 is

$$\hat{Y}_b = \beta_0 + \beta_1 \times 0.5 = 0.26545 + 0.52126 \times 0.5 = 0.52608. \tag{S.11}$$

The variance for this predicted value is

$$\text{var}\left(\hat{Y}_b\right) = \text{var}\left(\beta_0 + \beta_1 \times 0.5\right)$$

$$= \text{var}(\beta_0) + 0.5^2 \times \text{var}(\beta_1) + 2 \times 0.5 \times \text{cov}(\beta_0, \beta_1)$$

$$= 0.00155169 + 0.25 \times 0.00298512 - 0.00199709$$

$$= 0.000301. \tag{S.12}$$

The standard error is $\sqrt{0.000301} = 0.017346$. The two-tailed 95% confidence interval for the predicted value is $0.53 \pm 1.973 \times 0.017346 = 0.53 \pm 0.034$. Thus, the predicted value is 0.53 (0.492, 0.560).

i. The predicted value of *Oppose Race Policies* in the counterfactual scenario where an individual is angry and has a level of racial resentment equal to 0.5 is

$$\hat{Y}_c = \beta_0 + \beta_1 \times 0.5 + \beta_2 \times 1 + \beta_3 \times 0.5 \times 1$$

$$= 0.26545 + 0.52126 \times 0.5 - 0.11196 + 0.26545 \times 0.5$$

$$= 0.54308. \tag{S.13}$$

The variance for this predicted value is

$$\text{var}\left(\hat{Y}_c\right) = \text{var}\left(\beta_0 + \beta_1 \times 0.5 + \beta_2 \times 1 + \beta_3 \times 0.5 \times 1\right)$$

$$= \text{var}(\beta_0) + 0.5^2 \times \text{var}(\beta_1) + 1^2 \times \text{var}(\beta_2) + (0.5 \times 1)^2 \times \text{var}(\beta_3)$$

$$+ 2 \times 0.5 \times \text{cov}(\beta_0, \beta_1) + 2 \times 1 \times \text{cov}(\beta_0, \beta_2) + 2 \times (0.5 \times 1) \times \text{cov}(\beta_0, \beta_3)$$

$$+ 2 \times 0.5 \times 1 \times \text{cov}(\beta_1, \beta_2) + 2 \times 0.5 \times (0.5 \times 1) \times \text{cov}(\beta_1, \beta_3)$$

$$+ 2 \times 1 \times (0.5 \times 1) \times \text{cov}(\beta_2, \beta_3). \tag{S.14}$$

Simplifying, we have

$$\text{var}\left(\hat{Y}_c\right) = \text{var}(\beta_0) + 0.25 \times \text{var}(\beta_1) + \text{var}(\beta_2) + 0.25 \times \text{var}(\beta_3)$$

$$+ \text{cov}(\beta_0, \beta_1) + 2 \times \text{cov}(\beta_0, \beta_2) + \text{cov}(\beta_0, \beta_3)$$

$$+ \text{cov}(\beta_1, \beta_2) + 0.5 \times \text{cov}(\beta_1, \beta_3)$$

$$+ \text{cov}(\beta_2, \beta_3). \tag{S.15}$$

Substituting the appropriate values from the variance-covariance matrix in Eq. 4.44, we have

$$\text{var}\left(\hat{Y}_c\right) = 0.00155169 + 0.25 \times 0.00298512 + 0.00915827 + 0.25 \times 0.01550988$$

$$- 0.00199709 + 2 \times -0.00155169 + 0.00199709$$

$$+ 0.00199709 + 0.5 \times -0.00298512$$

$$- 0.01125366$$

$$= 0.0014812. \tag{S.16}$$

The standard error is $\sqrt{0.0014812} = 0.0384863$. The two-tailed 95% confidence interval for the predicted value is $0.54 \pm 1.973 \times 0.0384863 = 0.54 \pm 0.076$. Thus, the predicted value is 0.54 (0.467, 0.619).

j. The quick way to answer this question is to recognize that the difference in the predicted value of *Oppose Race Policies* between an individual who has a level of racial resentment equal to 0.5 and is angry and an equivalent individual who's calm is equivalent to the effect of being angry on *Oppose Race Policies* when an individual's level of resentment is 0.5. This is something that we already calculated in part f. The difference in predicted value is therefore 0.017 (−0.066, 0.100). We can, of course, calculate the difference,

$$\hat{Y}_c - \hat{Y}_b = 0.54308 - 0.52608 = 0.017. \tag{S.17}$$

Building on Eq. 4.38 earlier in the chapter, the variance for this difference is

$$\begin{aligned}
\mathrm{var}\left(\hat{Y}_c - \hat{Y}_b\right) &= \mathrm{var}\left[(1-0)(\beta_2 + \beta_3 \times 0.5)\right] \\
&= (1-0)^2 \mathrm{var}(\beta_2 + \beta_3 \times 0.5) \\
&= 1^2\left[\mathrm{var}(\beta_2) + 0.5^2 \times \mathrm{var}(\beta_3) + 2 \times 0.5 \times \mathrm{cov}(\beta_2, \beta_3)\right] \\
&= \mathrm{var}(\beta_2) + 0.5^2 \times \mathrm{var}(\beta_3) + 2 \times 0.5 \times \mathrm{cov}(\beta_2, \beta_3). \tag{S.18}
\end{aligned}$$

Substituting the appropriate values from the variance-covariance matrix in Eq. 4.44, we have

$$\mathrm{var}\left(\hat{Y}_c - \hat{Y}_b\right) = 0.00915827 + 0.25 \times 0.01550988 - 0.01125366$$

$$= 0.00178. \tag{S.19}$$

The standard error is $\sqrt{0.00178} = 0.0422$. This means that the two-tailed 95% confidence interval is $0.017 \pm 1.973 \times 0.0422 = 0.017 \pm 0.083$. Thus, the difference in predicted values is 0.017 (−0.066, 0.100). This is exactly what we got before.

EXERCISES FROM CHAPTER 5

1. a. The confidence intervals aren't linear in marginal effect plots in which the interaction is linear because the magnitude of the standard error associated with the marginal effect varies with the value of the modifying variable shown on the horizontal axis. There is a value of the modifying variable MV^* at which the standard error, and hence the distance between the upper and

lower bounds of the confidence intervals, is minimized. The standard error, and thus the distance between the confidence intervals, increases at an increasing rate as the value of the modifying variable increases or decreases from MV^*. It increases at an increasing rate, thereby producing the hour-glass shape, due to the squared term for the modifying variable in the equation for the standard error associated with the marginal effect.

b. Using information from the estimated variance-covariance matrix for the coefficients in Eq. 5.52, we see that the value of *Supply* at which the confidence interval for the marginal effect of *Demand* is narrowest is

$$Supply^* = -\frac{\text{cov}\,(\beta_1, \beta_3)}{\text{var}\,(\beta_3)} = -\frac{-0.000207}{0.000003} = 69. \tag{S.20}$$

This is different from the mean value of *Supply*, which is 67.88.

c. Using information from the estimated variance-covariance matrix for the coefficients, we see that the value of *Demand* at which the confidence interval for the marginal effect of *Supply* is narrowest is

$$Demand^* = -\frac{\text{cov}\,(\beta_2, \beta_3)}{\text{var}\,(\beta_3)} = -\frac{-0.000107}{0.000003} = 35.67. \tag{S.21}$$

This is different from the mean value of *Demand*, which is 52.01.

d. The confidence interval for the marginal effect of *Supply* is

$$\beta_2 + \beta_3 Demand \pm t \times \sqrt{\text{var}\,(\beta_2) + (Demand)^2\,\text{var}\,(\beta_3) + 2 \times Demand \times \text{cov}\,(\beta_2, \beta_3)}. \tag{S.22}$$

We want to know the values of *Demand* at which the upper and lower bounds of this confidence interval are 0. Based on Eq. D.11, we know that this occurs when

$$Demand^* = \frac{-[2\beta_2\beta_3 - 2t^2\text{cov}\,(\beta_2,\beta_3)] \pm \sqrt{[2\beta_2\beta_3 - 2t^2\text{cov}\,(\beta_2,\beta_3)]^2 - 4[\beta_3^2 - t^2\text{var}\,(\beta_3)][\beta_2^2 - t^2\text{var}\,(\beta_2)]}}{2[\beta_3^2 - t^2\text{var}\,(\beta_3)]}. \tag{S.23}$$

From Table 5.6, we know that $\beta_2 = -0.12$ and $\beta_3 = 0.008$. From the variance-covariance matrix in Eq. 5.52, we know that var $(\beta_2) = 0.005190$, var $(\beta_3) = 0.000003$, and cov $(\beta_2, \beta_3) = -0.000107$. We have $n - k = 186 - 6 = 180$ degrees of freedom and an $\alpha = 0.05$ level of significance. It follows that $t = t_{n-k,\alpha/2} = t_{180,0.025} = 1.97323$. If we plug these values into Eq. S.23, we obtain

$$
\begin{aligned}
Demand^* &= \frac{-[-0.00109] \pm \sqrt{[-0.00109]^2 - 4[0.000052][-0.00581]}}{2[-0.00581]} \\
&= \frac{0.00109 \pm 0.00155}{0.000105}.
\end{aligned}
\tag{S.24}
$$

From this, we have *Demand* * = 25.18 or −4.41. This tells us that the lower bound of the confidence interval is 0 when *Demand* is 25.18 and that the upper bound is 0 when *Demand* is −4.41.[3]

2. a. It's relatively easy to sketch a marginal effect plot for *Symbolic Racism* as we only need to display three quantities of interest: (1) the marginal effect of *Symbolic Racism* when *Anger* = 0, (2) the marginal effect of *Symbolic Racism* when *Anger* = 1, and (3) the interaction effect. Two of these quantities are provided directly in the regression output. The interaction effect is given by the coefficient on the interaction term *Symbolic Racism*×*Anger* in Table 5.7. We can use the standard error associated with this coefficient, along with the degrees of freedom in the model, to calculate a two-tailed 95% confidence interval. The interaction effect is 0.26 (0.01, 0.50). The marginal effect of *Symbolic Racism* when someone's calm is given by the coefficient on *Symbolic Racism*. We can use the standard error associated with this coefficient to calculate a confidence interval. The marginal effect of *Symbolic Racism* when someone's calm is 0.52 (0.41, 0.63). The marginal effect of *Symbolic Racism* when someone's angry is not directly provided in the regression output. However, it's easily calculated as $\beta_1 + \beta_3 = 0.78$. We can calculate the associated confidence interval using the information in the variance-covariance matrix for the coefficients in Eq. 5.57. As we saw in the solutions to Exercise 1 at the end of Chapter 4, the marginal effect of *Symbolic Racism* when someone's angry is 0.78 (0.56, 1.00). Putting these three quantities together, the marginal effect plot should look like the one shown in Figure S.1.

    b. The standard error associated with the effect of *Anger* on *Oppose Race Policies* is minimized when the value of *Symbolic Racism* is

$$Symbolic\ Racism^* = -\frac{\text{cov}(\beta_2, \beta_3)}{\text{var}(\beta_3)} = -\frac{-0.01125366}{0.01550988} = 0.73.$$

(S.25)

    c. The marginal effect plot for *Anger* is shown in Figure S.2. To construct this plot, we start with the straight line that shows how the effect of *Anger* varies across the observed range of *Symbolic Racism*, 0 − 1. The coefficient on *Anger*, −0.11, tells us the effect of *Anger* when *Symbolic Racism* = 0. The marginal effect of *Anger* when *Symbolic Racism* = 1 is

---

[3] Using more precise values for the coefficients, variances, and covariances than we provided here, *Demand** is 26.52 and −4.22; these are the values we reported in Chapter 5.

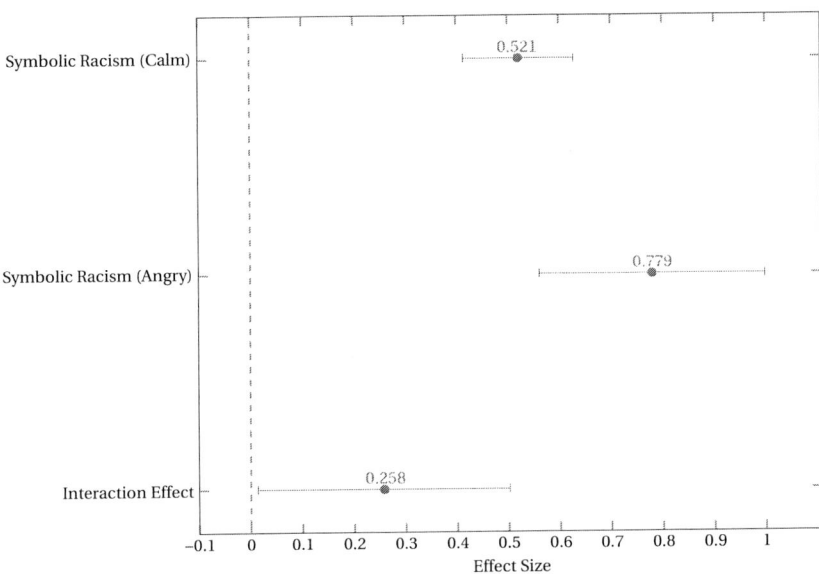

Figure S.1 The effect of *Symbolic Racism* on *Oppose Race Policies*

$\beta_2 + \beta_3 \times 1 = -0.11 + 0.26 \times 1 = 0.15$. Since the interaction effect is linear, we simply join these two points up with a straight line to obtain the desired marginal effect line for *Anger*.

Without a computer to automate the process, it can be tedious to draw the confidence intervals for the marginal effect of *Anger* as accurately as they are shown in Figure S.2. However, we can sketch them using some of the information that we have at hand. We start by noting that the standard error associated with the coefficient on *Anger* in Table 5.7 can be used, along with the degrees of freedom in the model, to easily calculate the confidence interval when *Symbolic Racism* = 0. We've indicated this particular confidence interval in Figure S.2 with the vertical black error bar at *Symbolic Racism* = 0. We then take advantage of the fact that we calculated the confidence intervals for the marginal effect of *Anger* when *Symbolic Racism* = 0.5 and when *Symbolic Racism* = 1 when completing Exercise 1 at the end of Chapter 4. These other confidence intervals are also shown as black vertical error bars in Figure S.2. We also know that the confidence interval is narrowest when *Symbolic Racism* = 0.73; this is indicated by the dashed vertical line in Figure S.2.

Finally, we can calculate where the upper and lower bounds of the confidence interval equal 0 and therefore cut the horizontal

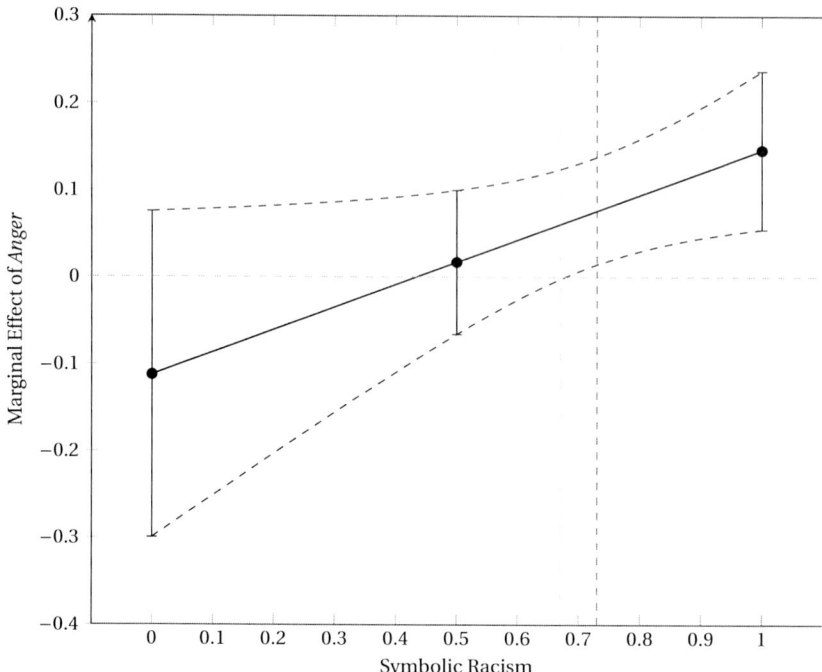

Figure S.2 The effect of *Anger* on *Oppose Race Policies*

Note: Figure S.2 shows the effect of *Anger* on *Oppose Race Policies* across the observed range of *Symbolic Racism*. The curved dashed lines indicate two-tailed 95% confidence intervals. The three vertical black error bars indicate the confidence intervals for the effect of *Anger* when *Symbolic Racism* is 0, 0.5, and 1. All three of these confidence intervals were calculated when completing Exercise 1 at the end of Chapter 4. The dashed vertical line indicates the value of *Symbolic Racism* at which the confidence interval for the effect of *Anger* is narrowest. The vertical gray line indicates the value of *Symbolic Racism* at which the lower bound of the confidence interval is 0.

zero line. The confidence interval for the marginal effect of *Anger* is

$$\beta_2 + \beta_3 Symbolic\ Racism \pm t$$
$$\times \sqrt{\operatorname{var}(\beta_2) + (Symbolic\ Racism)^2 \operatorname{var}(\beta_3) + 2 \times Symbolic\ Racism \times \operatorname{cov}(\beta_2, \beta_3)}.$$
$$(S.26)$$

Based on Eq. D.11, we know that the values of *Symbolic Racism* at which the bounds of this confidence interval are 0 are

Symbolic Racism*

$$= \frac{-\left[2\beta_2\beta_3 - 2t^2 \operatorname{cov}(\beta_2, \beta_3)\right] \pm \sqrt{\left[2\beta_2\beta_3 - 2t^2 \operatorname{cov}(\beta_2, \beta_3)\right]^2 - 4\left[\beta_3^2 - t^2 \operatorname{var}(\beta_3)\right]\left[\beta_2^2 - t^2 \operatorname{var}(\beta_2)\right]}}{2\left[\beta_3^2 - t^2 \operatorname{var}(\beta_3)\right]}.$$
$$(S.27)$$

From Table 5.7, we know that $\beta_2 = -0.11$ and $\beta_3 = 0.26$. From the variance-covariance matrix in Eq. 5.57, we know that var $(\beta_2) = 0.00915827$, var $(\beta_3) = 0.01550988$, and that cov $(\beta_2, \beta_3) = -0.0112537$. We have $n - k = 182 - 4 = 178$ degrees of freedom and an $\alpha = 0.05$ level of significance. It follows that $t = t_{n-k,\alpha/2} = t_{178,0.025} = 1.97338$. If we plug these values into Eq. S.27, we obtain

$$Symbolic\ Racism^* = \frac{-[0.03045] \pm \sqrt{[0.03045]^2 - 4[0.00720][-0.02356]}}{2[0.00720]}$$

$$= \frac{-0.03045 \pm 0.04007}{0.0144}. \tag{S.28}$$

From this, we have *Symbolic Racism** = 0.668 or −4.897. This tells us that the lower bound of the confidence interval is 0 when *Symbolic Racism* is 0.668 and that the upper bound is 0 when *Symbolic Racism* is −4.897. Since the theoretical range of *Symbolic Racism* is 0–1, we know that the upper bound of the confidence is always above the horizontal line for all of the possible values of *Symbolic Racism*. The fact that the lower bound of the confidence interval cuts the horizontal zero line when *Symbolic Racism* = 0.668 is indicated by the vertical gray line in Figure S.2.

Putting all of this information together allows us to sketch out the general shape of the confidence intervals shown in Figure S.2.

d. It would be good to know the frequency distribution of *Symbolic Racism* so that we could include a histogram for the modifying variable on the horizontal axis.

EXERCISES FROM CHAPTER 6

1. a. The variances for the coefficients in Table 6.4 are shown in the diagonal elements, top left to bottom right, of the variance-covariance matrix in Eq. 6.55. If we square root these elements, we obtain the standard errors for our coefficients. The "missing" standard errors appear in Table S.6.

   b. Based on Eq. 6.18, the marginal effect of being female rather than male for a 30 year old college graduate is $\beta_1 + \beta_4 \times 1 + \beta_5 \times 30 = 13.97506 + 3.460115 \times 1 - 0.1782375 \times 30 = 12.08805$. In other words, a 30 year old female college graduate exhibits 12.09 units more support for feminism than a 30 year old male college graduate. Based on Eq. 6.6, the variance associated with this effect is given by

Table S.6 Gender, educational attainment, age, and support for feminism III

Dependent Variable: *Feminism*, 0–100

	Model 1
Female	13.97506***
	(2.66595)
Education	2.411783**
	(1.214694)
Age	0.2223337***
	(.0350729)
Female × *Education*	3.460115**
	(1.693807)
Female × *Age*	−0.1782375***
	(0.0483869)
Constant	74.79999
	(2.181709)
Controls	Yes
Observations	2,710
R^2	0.31

Standard errors in parentheses. $^*p < 0.10$; $^{**}p < 0.05$; $^{***}p < 0.01$ (two-tailed)
Note: *Controls* refers to the seven additional control variables that are shown in Table 6.1.

$$\begin{aligned}
\text{var}(\beta_1 + \beta_4 + 30\beta_5) &= \text{var}(\beta_1) + 1^2\text{var}(\beta_4) + 30^2\text{var}(\beta_5) \\
&\quad + 2 \times 1 \times \text{cov}(\beta_1, \beta_4) + 2 \times 30 \times \text{cov}(\beta_1, \beta_5) \\
&\quad + 2 \times 1 \times 30 \times \text{cov}(\beta_4, \beta_5) \\
&= 7.1072878 + 1^2 \times 2.8689833 + 30^2 \times 0.0023413 \\
&\quad + 2 \times 1 \times -1.2959764 + 2 \times 30 \times -0.11614581 \\
&\quad + 2 \times 1 \times 30 \times -0.0005735. \\
&= 2.48833. \quad\quad\quad\quad\quad\quad\quad\quad (\text{S.29})
\end{aligned}$$

The standard error associated with the effect is $\sqrt{2.47684} = 1.5774$. The t-statistic is $12.08805/1.5774 = 7.66$. We have $n - k = 2710 - 13 = 2,697$ degrees of freedom and an

Table S.7 Demand, supply, regime type, and women's legislative representation III

Dependent Variable: *Women's Representation, 0–100*

	Model 1
Demand	0.2586052
	(0.43248)
Supply	0.4157726**
	(0.202021)
Democracy	3.08445***
	(0.952859)
Demand×Supply	−0.0012633
	(0.005955)
Demand×Democracy	−0.0543243**
	(0.026859)
Supply×Democracy	−0.0505661***
	(0.014246)
Demand×Supply×Democracy	0.0008223**
	(0.000363)

Standard errors in parentheses. $^*p < 0.10$; $^{**}p < 0.05$; $^{***}p < 0.01$ (two-tailed)

$\alpha = 0.05$ level of significance. It follows that $t = t_{n-k,\alpha/2}$ $= t_{2,697,0.025} = 1.9608$. As a result, the two-tailed 95% confidence interval is $12.09 \pm 1.9608 \times 1.5774 = 12.09 \pm 3.09$. Thus, the effect of being female for a 30 year old college graduate is 12.06 (9.00,15.18).

a. The variances for the coefficients in Table 6.5 are shown in the diagonal elements, top left to bottom right, of the variance-covariance matrix in Eq. 6.57. Taking the square root of these elements provides the standard errors for our coefficients. The "missing" standard errors appear in Table S.7.

b. Based on Eq. 6.53, the effect of a 20 unit increase in *Demand* when *Supply* = 80 and *Democracy* = 15 is $(\beta_1 + \beta_4 \times 80 + \beta_5 \times 15 + \beta_7 \times 80 \times 15) \times 20 = 6.5887$. Based on Eq. 6.34, the variance associated with this effect is given by

$$\text{var}\,(20\beta_1 + 1600\beta_4 + 300\beta_5 + 24000\beta_7)$$
$$= 20^2\text{var}\,(\beta_1) + 1600^2\text{var}\,(\beta_4) + 300^2\text{var}\,(\beta_5) + 24000^2\text{var}\,(\beta_7)$$
$$+ 2 \times 20 \times 1600 \times \text{cov}\,(\beta_1, \beta_4) + 2 \times 20 \times 300 \times \text{cov}\,(\beta_1, \beta_5)$$
$$+ 2 \times 20 \times 24000 \times \text{cov}\,(\beta_1, \beta_7) + 2 \times 1600 \times 300 \times \text{cov}\,(\beta_4, \beta_5)$$
$$+ 2 \times 1600 \times 24000\text{cov}\,(\beta_4, \beta_7) + 2 \times 300 \times 24000 \times \text{cov}\,(\beta_5, \beta_7).$$
$$\text{(S.30)}$$

From the variance-covariance matrix in 6.57, we know that
$\text{var}\,(\beta_1) = 0.18704159$, $\text{var}\,(\beta_4) = 0.00003546$,
$\text{var}\,(\beta_5) = 0.0007214$, $\text{var}\,(\beta_7) = 0.0000001318$,
$\text{cov}\,(\beta_1, \beta_4) = -0.00243877$, $\text{cov}\,(\beta_1, \beta_5) = -0.01087325$,
$\text{cov}\,(\beta_4, \beta_5) = 0.00014096$, $\text{cov}\,(\beta_4, \beta_7) = -0.000002027$,
$\text{cov}\,(\beta_5, \beta_7) = -0.000009343$. Substituting these values into Eq.
S.30, we have

$$\text{var}\,(20\beta_1 + 1600\beta_4 + 300\beta_5 + 24000\beta_7) = 0.87. \qquad \text{(S.31)}$$

The standard error associated with this effect is $\sqrt{0.87} = 0.93$.
The t-statistic is $6.5887/0.93 = 7.1$. We have
$n - k = 186 - 9 = 177$ degrees of freedom and an $\alpha = 0.05$ level of
significance. It follows that $t = t_{n-k,\alpha/2} = t_{177,0.025} = 1.9735$. As a
result, the two-tailed 95% confidence interval is
$6.59 \pm 1.9735 \times 0.93 = 6.59 \pm 1.84$. Thus, the effect of a 20 unit
increase in *Demand* when *Supply* $= 80$ and *Democracy* $= 15$ is
6.59 (4.75, 8.43).

EXERCISES FROM CHAPTER 7

1. Consider the following cubic regression model:

$$Y = \beta_0 + \beta_1 X + \beta_2 X^2 + \beta_3 X^3 + \epsilon. \qquad \text{(S.32)}$$

a. The coefficient on the X^3 term tells us whether the cubic
relationship between X and Y moves from negative infinity when
X is low to positive infinity when X is high or from positive
infinity when X is low to negative infinity when X is large.
Loosely-speaking, it indicates whether the non-linear relationship
between X and Y eventually goes up to the right or down to the
right when X gets large. The left panel in Figure S.3 shows a
predicted value plot from a cubic model where β_3 is positive,
while the right panel shows a predicted value plot where β_3 is
negative. The intuition behind this relationship with the sign of β_3
comes from the fact that the cubic term will tend to dominate the
relationship between X and Y when the magnitude of X gets large.
This tells us that Y will go toward either positive or negative

infinity as the absolute magnitude of X increases. If β_3 is positive, larger negative values of X when cubed will move Y toward negative infinity, while larger positive values of X when cubed will move Y toward positive infinity (left panel). If β_3 is negative, larger negative values of X when cubed will move Y toward positive infinity, while larger positive values of X when cubed will move Y toward negative infinity (right panel).

b. The marginal effect of X on Y is

$$\frac{\partial Y}{\partial X} = \beta_1 + 2\beta_2 X + 3\beta_3 X^2. \tag{S.33}$$

This tells us about the slope of the relationship between X and Y. Depending on the value of X, the slope may be positive, negative, or zero. From this, we see that β_1 indicates the marginal effect of X on Y when X is 0; it tells us the slope of the relationship between X and Y when $X = 0$. To illustrate, β_1 is negative (-144) in the cubic model shown in the left panel of Figure S.3, and, as expected, the slope of the relationship between X and Y is negative when $X = 0$. In contrast, β_1 is positive (144) in the cubic model shown in the right panel, and, as expected, the slope of the relationship between X and Y is now positive when $X = 0$.

c. The modifying effect of X on the marginal effect of X is

$$\frac{\partial\left(\frac{\partial Y}{\partial X}\right)}{\partial X} = \frac{\partial\left(\beta_1 + 2\beta_2 X + 3\beta_3 X^2\right)}{\partial X} = 2\beta_2 + 6\beta_3 X. \tag{S.34}$$

This tells us about the *curvature* of the relationship between X and Y. In other words, it tells us how the slope of the relationship between X and Y changes with a marginal increase in X. From

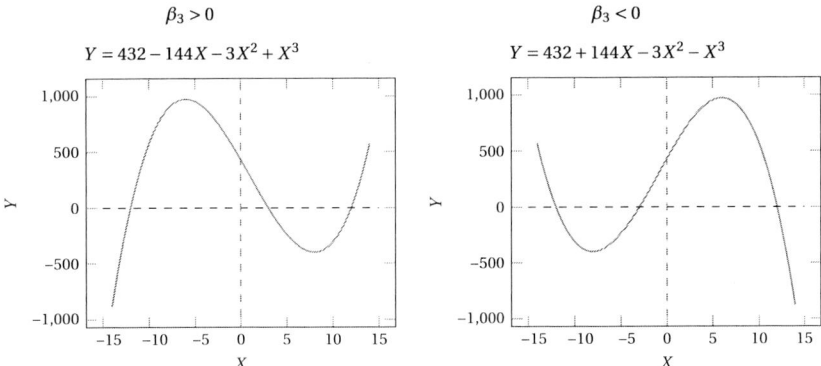

Figure S.3 The sign of the coefficient on X^3

this, we see that β_2 tells us how the slope of the relationship between X and Y is changing at $X = 0$. The precise interpretation of the sign of β_2 will depend on the sign of β_1. Suppose that the slope of the relationship between X and Y is negative at $X = 0$ ($\beta_1 < 0$). In this scenario, a negative β_2 indicates that the negative slope at $X = 0$ is in the process of becoming steeper. In contrast, a positive β_2 indicates that the negative slope at $X = 0$ is in the process of becoming less steep. Now suppose that the slope of the relationship between X and Y is positive at $X = 0$ ($\beta_1 > 0$). In this scenario, a negative β_2 indicates that the positive slope at $X = 0$ is in the process of becoming less steep. In contrast, a positive β_2 indicates that the positive slope at $X = 0$ is in the process of becoming steeper. To illustrate, the negative coefficient on X^2 in the left panel of Figure S.3 ($\beta_1 < 0$) indicates that the negative slope at $X = 0$ is in the process of becoming steeper; that is, the negative effect of X is becoming larger. The negative coefficient on X^2 in the right panel ($\beta_1 > 0$) indicates that the positive slope at $X = 0$ is in the process of becoming less steep; that is, the positive effect of X is becoming smaller.

d. If they exist, the turning points in a cubic relationship between X and Y occur when the slope of the cubic function is 0. The points of a cubic function where the slope is 0 are often referred to as the *critical points* or the *stationary points*. To find these critical points, we set the marginal effect of X equal to 0 and solve for X,

$$\frac{\partial Y}{\partial X} = \beta_1 + 2\beta_2 X + 3\beta_3 X^2 = 0. \tag{S.35}$$

Using the quadratic formula on Eq. S.35, we find that

$$X^* = \frac{-\beta_2 \pm \sqrt{\beta_2^2 - 3\beta_1\beta_3}}{3\beta_3}. \tag{S.36}$$

This tells us the values of X at which the slope of the cubic function is 0.

Although the slope of a cubic function is 0 at X^*, this does not necessarily mean that X^* always indicate a turning point where the slope of the relationship between X and Y changes sign. This is because the slope of a cubic function can be 0 at *an inflection point* rather than a turning point. The term in the square root of Eq. S.36, $\beta_2^2 - 3\beta_1\beta_3$, is known as the *discriminant*. If the discriminant is positive, then there are two critical points that are both turning points, one a local maximum and the other a local minimum.[4] If the discriminant is equal to 0, then we have only one critical point, and this represents an *inflection point* rather than a turning point. With an inflection point, the curvature (the rate of change) of the slope of the non-linear relationship between X and Y changes sign from positive to negative or negative to positive but not the slope of the relationship itself. And, for the sake of completeness, if the discriminant is negative, then there are no (real) critical values of X. Thus, in the case where $\beta_2^2 - 3\beta_1\beta_3 \leq 0$, the non-linear relationship between X and Y is monotonic, and its slope doesn't change sign; there are no turning points. This is why we reported earlier that a polynomial of degree three (a cubic model) has *at most* two turning points.

To illustrate, we'll find the critical values of X associated with the turning points in the cubic relationship shown in the right panel of Figure S.3. We see that $\beta_1 = 144$, $\beta_2 = -3$, and $\beta_3 = -1$. In this case, the discriminant is $(-3)^2 - 3(144)(-1) = 441$, which is positive. This tells us that we have two turning points, a local maximum and a local minimum. The critical values of X at which the marginal effect of X is 0 are

$$
\begin{aligned}
X^* &= \frac{-(-3) \pm \sqrt{441}}{-3} \\
&= \frac{3 \pm 21}{-3} \\
&= -1 \pm 7.
\end{aligned}
\tag{S.37}
$$

From this, we see that the critical values of X are -8 and 6. To find out which critical value is associated with the local maximum and which is associated with the local minimum, we take the derivative of the marginal effect of X with respect to X, which we saw earlier was $2\beta_2 + 6\beta_3 X$. When we plug in $X = -8$, we have $2(-3) + 6(-1)(-8) = 42$. Since this is positive, $X = -8$ is associated with the local minimum. When we plug in $X = 6$, we have $2(-3) + 6(-1)(6) = -42$. Since this is negative, $X = 6$ is associated with a local maximum.

---

[4] To determine which of the critical values of X is associated with the local maximum and which is associated with the local minimum, we can take the derivative of the marginal effect of X with respect to X to see how the curvature of the slope relationship between X and Y is changing. If this second derivative is negative when we plug in our critical value of X, then the slope of the relationship between X and Y is changing from positive to negative at this point, and we have a local maximum. If it's positive when we plug in our critical value of X, then the slope of the relationship between X and Y is changing from negative to positive at this point, and we have a local minimum.

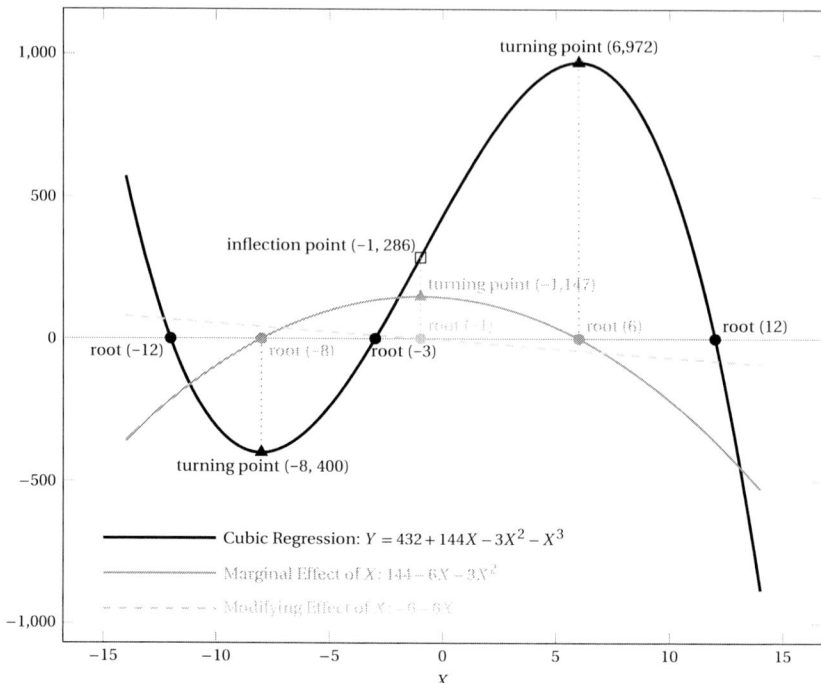

Figure S.4 Cubic regression model, the marginal effect of X, and the modifying effect of X

e. The solid black line in Figure S.4 shows the cubic relationship between X and Y based on the results from the following cubic regression model:

$$Y = 432 + 144X - 3X^2 - X^3.$$
(S.38)

The "roots" of the cubic function indicate the values of X at which the cubic function is 0 and are shown as black circles, $X = \{-12, -3, 12\}$.
The solid dark gray line shows the marginal effect of X on Y,

$$\frac{\partial Y}{\partial X} = 144 - 6X - 3X^2.$$
(S.39)

This is a quadratic equation that tells us the slope of the relationship between X and Y. As expected, the dark gray line is below the horizontal zero line when the slope of the relationship between X and Y is negative and above it when the slope of the relationship between X and Y is positive. The "roots" of the marginal effect (quadratic) function indicate the values of X at which the marginal effect of X is 0 and are shown as dark gray circles, $X = \{-8, 6\}$. These circles tell us the stationary or critical points of the cubic function; that is, they tell us when the slope of the relationship between X and Y is 0. As a result, they correspond to the values of X associated with the turning points of the cubic function; that is, the local minimum and maximum. The dashed light gray line shows the modifying effect of X on the marginal effect of X,

$$\frac{\partial Y}{\partial X} = -6 - 6X. \tag{S.40}$$

This is a linear equation and tells us the curvature of the relationship between X and Y. Put differently, it tells us how the slope relationship between X and Y (the marginal effect of X shown in dark gray) is changing with X. The "root" of the modifying effect (linear) function indicates the value of X at which the modifying effect of X is 0 and is shown as a light gray circle ($X = -1$). This circle tells us when the slope of the marginal effect of X is 0, or equivalently the inflection point in the cubic relationship between X and Y at which the *change in the slope* of the cubic function switches from positive to negative. As expected, the dashed light gray line is positive to the left of the inflection point (negative slopes of the cubic function are becoming less steep and positive slopes are becoming more steep) and negative to the right (positive slopes of the cubic function are becoming less steep and negative slopes are becoming more steep).

f. The marginal effect of X on Y is

$$\frac{\partial Y}{\partial X} = \beta_1 + 2\beta_2 X + 3\beta_3 X^2. \tag{S.41}$$

The variance of the marginal effect of X is

$$\mathrm{var}\left(\beta_1 + 2\beta_2 X + 3\beta_3 X^2\right) = \mathrm{var}\left(\beta_1\right) + 4X^2 \mathrm{var}\left(\beta_2\right) + 9X^4 \mathrm{var}\left(\beta_3\right)$$
$$+ 4X \mathrm{cov}\left(\beta_1, \beta_4\right) + 6X^2 \mathrm{cov}\left(\beta_1, \beta_3\right)$$
$$+ 12X^3 \mathrm{cov}\left(\beta_2, \beta_3\right). \tag{S.42}$$

g. The modifying effect of X on Y is

$$\frac{\partial\left(\frac{\partial Y}{\partial X}\right)}{\partial Y} = 2\beta_2 + 6\beta_3 X. \tag{S.43}$$

The variance of the modifying effect of X is

$$\mathrm{var}\left(2\beta_2 + 6\beta_3 X\right) = 4\mathrm{var}\left(\beta_2\right) + 36X^2 \mathrm{var}\left(\beta_3\right) + 24X \mathrm{cov}\left(\beta_1, \beta_3\right). \tag{S.44}$$

h. The effect of increasing X in the cubic regression model shown in Eq. S.32 from some baseline value X_b to some counterfactual value X_c is calculated as

$$\hat{Y}_{X_c} - \hat{Y}_{X_b} = \left(\beta_0 + \beta_1 X_c + \beta_2 X_c^2 + \beta_3 X_c^3\right) - \left(\beta_0 + \beta_1 X_b + \beta_2 X_b^2 + \beta_3 X_b^3\right)$$
$$= \beta_1 (X_c - X_b) + \beta_2 \left(X_c^2 - X_b^2\right) + \beta_3 \left(X_c^3 - X_b^3\right). \tag{S.45}$$

If we think in terms of a one-unit increase in X such that $X_c = X_b + 1$, then this difference becomes

$$\hat{Y}_{X_b+1} - \hat{Y}_{X_b} = \beta_1 (X_b + 1 - X_b) + \beta_2 \left((X_b + 1)^2 - X_b^2\right) + \beta_3 \left((X_b + 1)^3 - X_b^3\right)$$
$$= \beta_1 + \beta_2 + \beta_3 + 2\beta_2 X_b + 3\beta_3 X_b + 3\beta_3 X_b^2$$
$$= \beta_1 + (1 + 2X_b)\beta_2 + \left(1 + 3X_b + 3X_b^2\right)\beta_3. \tag{S.46}$$

The variance associated with the effect of a one-unit increase in X is

$$\text{var}\left[\beta_1 + (1 + 2X_b)\beta_2 + \left(1 + 3X_b + 3X_b^2\right)\beta_3\right] =$$
$$\text{var}(\beta_1) + (1 + 2X_b)^2 \text{var}(\beta_2) + \left(1 + 3X_b + 3X_b^2\right)^2 \text{var}(\beta_3)$$
$$+ 2(1 + 2X_b)\text{cov}(\beta_1, \beta_2) + 2\left(1 + 3X_b + 3X_b^2\right)\text{cov}(\beta_1, \beta_3)$$
$$+ 2(1 + 2X_b)\left(1 + 3X_b + 3X_b^2\right)\text{cov}(\beta_2, \beta_3). \tag{S.47}$$

i. With a cubic regression model, we're interested in the effect of one variable, X. However, X is interacted with two other variables that happen to also be X; this is, after all, how we obtain X^3. Thus, we're dealing with a theoretical situation similar to the one discussed earlier in Chapter 6 where we had a *fully interactive relationship* between three variables X, Z, and W. The only difference is that the three interacting variables in a cubic regression model are the same variable, X. This means that we have two levels of interaction: (1) how X modifies the marginal effect of X on Y and (ii) how X modifies the modifying effect of X on the marginal effect of X. We recommend that scholars make predictions about the marginal effect of X and both levels of interaction.

Usually, scholars can make five key predictions from a theory predicting that there's a cubic relationship between X and Y. The first two predictions relate to the marginal effect of X on Y when X is at its lowest and highest values ($P_{X|X_{\min}}$, $P_{X|X_{\max}}$). The second two predictions relate to the modifying effect of X on the marginal effect of X when X is at its lowest and highest values ($P_{XX|X_{min}}$, $P_{XX|X_{max}}$). The last prediction relates to the second-order interaction effect and how X modifies the modifying effect of X on the marginal effect of X (P_{XXX}).

To evaluate these five predictions, scholars can usefully employ two plots. The first plot would show the marginal effect of X on Y, $\beta_1 + 2\beta_2 X + 3\beta_3 X^2$, across the observed range of X. The second plot would show the modifying effect of X on the marginal effect of X, $2\beta_2 + 6\beta_3 X$, across the observed range of X. Somewhere in these plots, scholars should also report the modifying effect of X on the modifying effect of X on the marginal effect of X, $6\beta_3$. Scholars may also wish to present a predicted value plot showing how the predicted value of Y changes with X. However, while such a plot nicely illustrates the non-linear relationship between X and Y, it, like a predicted value plot from a quadratic regression, provides little leverage when it comes to evaluating the implications from a theory predicting a cubic relationship between X and Y.

2. a. The marginal effect of X is

$$\frac{\partial Y}{\partial X} = \beta_1 + \beta_3 Z + \beta_5 Z^2. \tag{S.48}$$

The variance of the marginal effect of X is

$$\text{var}\left(\beta_1 + \beta_3 Z + \beta_5 Z^2\right) = \text{var}(\beta_1) + Z^2 \text{var}(\beta_3) + Z^4 \text{var}(\beta_5)$$
$$+ 2Z\text{cov}(\beta_1, \beta_3) + 2Z^2\text{cov}(\beta_1, \beta_5) + 2Z^3\text{cov}(\beta_3, \beta_5). \tag{S.49}$$

The interaction effect between X and Z is

$$\frac{\partial\left(\frac{\partial Y}{\partial X}\right)}{\partial Z} = \beta_3 + 2\beta_5 Z. \tag{S.50}$$

The variance of the interaction effect between X and Z is

$$\text{var}(\beta_3 + 2\beta_5 Z) = \text{var}(\beta_3) + 4Z^2 \text{var}(\beta_5) + 4Z\text{cov}(\beta_3, \beta_5). \tag{S.51}$$

b. The marginal effect of X is

$$\frac{\partial Y}{\partial X} = \beta_1 + \beta_3 Z + \beta_5 Z^2 + \beta_7 Z^3. \tag{S.52}$$

The variance of the marginal effect of X is

$$\text{var}\left(\beta_1 + \beta_3 Z + \beta_5 Z^2 + \beta_7 Z^3\right) = \text{var}(\beta_1) + Z^2\text{var}(\beta_3) + Z^4\text{var}(\beta_5) + Z^6\text{var}(\beta_7)$$
$$+ 2Z\text{cov}(\beta_1, \beta_3) + 2Z^2\text{cov}(\beta_1, \beta_5) + 2Z^3\text{cov}(\beta_1, \beta_7)$$
$$+ 2Z^3\text{cov}(\beta_3, \beta_5) + 2Z^4\text{cov}(\beta_3, \beta_5) + 2Z^5\text{cov}(\beta_5, \beta_7). \tag{S.53}$$

The interaction effect between X and Z is

$$\frac{\partial \left(\frac{\partial Y}{\partial X} \right)}{\partial Z} = \beta_3 + 2\beta_5 Z + 3\beta_7 Z^2. \tag{S.54}$$

The variance of the interaction effect is

$$\mathrm{var}\left(\beta_3 + 2\beta_5 Z + 3\beta_7 Z^2\right) = \mathrm{var}\left(\beta_3\right) + 4Z^2 \mathrm{var}\left(\beta_5\right) + 9Z^4 \mathrm{var}\left(\beta_7\right)$$
$$+ 4Z\mathrm{cov}\left(\beta_3, \beta_5\right) + 6Z^2 \mathrm{cov}\left(\beta_3, \beta_7\right) + 12Z^3 \mathrm{cov}\left(\beta_5, \beta_7\right). \tag{S.55}$$

EXERCISES FROM CHAPTER 8

1. a. The marginal effect of *Disproportionality* on the propensity of pre-electoral coalition formation is

$$\frac{\partial PEC^*}{\partial Disproportionality} = \beta_1 + \beta_3 Polarization. \tag{S.56}$$

Plugging in the relevant coefficients from Table 8.3, we see that the marginal effect is $0.0280 + 0.0005 \times Polarization$. The marginal effect of *Polarization* on the propensity of pre-electoral coalition formation is

$$\frac{\partial PEC^*}{\partial Polarization} = \beta_1 + \beta_3 Disproportionality. \tag{S.57}$$

Plugging in the relevant coefficients from Table 8.3, we see that the marginal effect is $0.0037 + 0.0005 \times Polarization$.

   b. We can say that an increase in *Disproportionality* has a positive effect on the probability of pre-electoral coalition formation across all values of *Polarization*. We know that this positive effect is statistically significant when *Polarization* is 0. We can't determine from the information in Table 8.3 for what other values of *Polarization* the positive effect of *Disproportionality* is statistically significant. While we can say that *Polarization* has a statistically significant positive modifying effect on the effect of *Disproportionality* on the *propensity* to form pre-electoral coalitions, we can make no inference about the sign or statistical significance of this modifying effect on the effect of *Disproportionality* on the *probability* of pre-electoral coalition formation.

We can say that an increase in *Polarization* has a positive effect on the probability of pre-electoral coalition formation across all values of *Disproportionality*. We know that this positive effect isn't statistically significant when $Disproportionality = 0$. We can't determine from the information in Table 8.3 if and when the positive effect of *Polarization* becomes statistically significant. While we can say that *Disproportionality* has a statistically significant positive modifying effect on the effect of *Polarization* on the *propensity* to form pre-electoral coalitions, we can make no inference about the sign or statistical significance of this modifying effect on the effect of *Polarization* on the *probability* of pre-electoral coalition formation.

c. From Eq. 8.6, we know that the probability of pre-electoral coalition formation in a logit model is

$$Pr(PEC = 1) = \Lambda(X\beta) = \frac{e^{X\beta}}{1 + e^{X\beta}}, \tag{S.58}$$

where $\Lambda(\cdot)$ represents the cumulative standard logistic distribution. From Eqs. 8.21 and 8.22, we know that the marginal effect of *Ideological Incompatibility* is

$$\frac{\partial PEC}{\partial Ideological\ Incompatibility} = \lambda(X\beta) \cdot \beta_4 = \Lambda(X\beta) \cdot [1 - \Lambda(X\beta)] \cdot \beta_4. \tag{S.59}$$

From Table 8.3, we have $\beta_4 = -0.0105$. For our chosen scenario, $X\beta = -3.2367 + 0.0280 \times 5 + 0.0037 \times 24.68 + 0.0005 \times 5 \times 24.68 - 0.0105 \times 19.51 = -3.149$. From this, we have $\Lambda(-3.149) = 0.041$ and $1 - \Lambda(-3.149) = 0.959$. And so the marginal effect of *Ideological Incompatibility* on the probability of pre-electoral coalition formation in the chosen scenario is $0.041 \times 0.959 \times -0.0105 = -0.0004$. From footnote 4, we see that the largest marginal effect of *Ideological Incompatibility* is $0.5 \times 0.5 \times \beta_4 = 0.5 \times 0.5 \times -0.0105 = -0.0026$.

d. From Eq. 8.56, we know that the marginal effect of *Disproportionality* is

$$\frac{\partial PEC}{\partial Disproportionality} = \Lambda(X\beta) \cdot [1 - \Lambda(X\beta)] \cdot [\beta_1 + \beta_3 Polarization]. \tag{S.60}$$

Given our chosen scenario where *Disproportionality* = 5, *Polarization* = 24.68, and *Ideological Incompatibility* = 19.51, we have $X\beta = -3.149$, $\Lambda(X\beta) = \Lambda(-3.149) = 0.041$, $1 - \Lambda(X\beta) = 1 - \Lambda(-3.149) = 0.959$, and $\beta_1 + \beta_3 Polarization = 0.0280 + 0.0005 \times 24.68 = 0.0403$. This means that the marginal effect of *Disproportionality* in the chosen scenario is $0.041 \times 0.959 \times 0.0403 = 0.0016$.

The marginal effect of *Polarization* is

$$\frac{\partial PEC}{\partial Polarization} = \Lambda(X\beta) \cdot \left[1 - \Lambda(X\beta)\right] \cdot \left[\beta_2 + \beta_3 Disproportionality\right].$$

(S.61)

Given our chosen scenario, we have $\beta_2 + \beta_3 Disproportionality$ $= 0.0037 + 0.0005 \times 5 = 0.0062$. This means that the marginal effect of *Disproportionality* in the chosen scenario is $0.041 \times 0.959 \times 0.0062 = 0.0002$.

e. From Eq. 8.72, the interaction effect of *Disproportionality* and *Polarization* on the probability of pre-electoral coalition formation is

$$\frac{\partial \left(\frac{\partial Pr(PEC)}{\partial Disproportionality} \right)}{\partial Polarization} = \Lambda(X\beta)\left[1 - \Lambda(X\beta)\right] \cdot \beta_3$$
$$+ \Lambda(X\beta)\left[1 - \Lambda(X\beta)\right] \cdot \left[1 - 2\Lambda(X\beta)\right]$$
$$\cdot (\beta_1 + \beta_3 Polarization) \cdot (\beta_2 + \beta_3 Disproportionality).$$

(S.62)

Given our chosen scenario, we have $\Lambda(X\beta) = 0.041$, $1 - \Lambda(X\beta) = 0.959$, $\beta_3 = 0.0005$, $1 - 2\Lambda(X\beta) = 0.9177$, $\beta_1 + \beta_3 Polarization = 0.0403$, and $\beta_2 + \beta_3 Disproportionality = 0.0062$. Thus, the interaction effect of *Disproportionality* and *Polarization* on the probability of pre-electoral coalition formation in the chosen scenario is $0.041 \times 0.959 \times 0.0005 + 0.041 \times 0.959 \times 0.9177 \times 0.0403 \times 0.0062 = 0.00003$.

f. For our first chosen baseline scenario, where *Disproportionality* $= 2$, we have $X\beta = -3.2367 + 0.0280 \times 2 + 0.0037 \times 15 + 0.0005 \times 2 \times 15 - 0.0105 \times 24 = -3.3622$. This means that the probability of pre-electoral coalition formation in our baseline scenario is $\Lambda(-3.3622) = 0.0335$.

For our first chosen counterfactual scenario, we have $X\beta = -3.2367 + 0.0280 \times 2 + 0.0037 \times 25 + 0.0005 \times 2 \times 25 - 0.0105 \times 24 = -3.3152$. This means that the probability of pre-electoral coalition formation in our counterfactual scenario is $\Lambda(-3.3152) = 0.0351$.

The change in the probability of pre-electoral coalition formation associated with a 10 unit increase in *Polarization* in this particular baseline scenario is $0.0351 - 0.0335 = 0.0016$. The percentage change in the probability of pre-electoral coalition formation is $(0.0016/0.0335) \times 100 = 4.64\%$.

For our second chosen baseline scenario, where
Disproportionality $= 10$, we have $X\beta = -3.2367 + 0.0280 \times 10 +$
$0.0037 \times 15 + 0.0005 \times 10 \times 15 - 0.0105 \times 24 = -3.0782$. This
means that the probability of pre-electoral coalition formation in
our second baseline scenario is $\Lambda(-3.0782) = 0.0440$.

For our second chosen counterfactual scenario, we have
$X\beta = -3.2367 + 0.0280 \times 10 + 0.0037 \times 25 + 0.0005 \times 10 \times 25 -$
$0.0105 \times 24 = -2.9912$. This means that the probability of
pre-electoral coalition formation in our counterfactual scenario is
$\Lambda(-2.9912) = 0.0478$.

The change in the probability of pre-electoral coalition formation
associated with a 10 unit increase in *Polarization* in the second
baseline scenario is $0.0478 - 0.0440 = 0.0038$. The percentage
change in the probability of pre-electoral coalition formation is
$(0.0016/0.0335) \times 100 = 8.65\%$.

Thus, the change in the probability of pre-electoral coalition
formation associated with a 10 unit increase in *Polarization* from
15 to 25 in our chosen scenario is
$[(0.0038 - 0.0016)/0.0016] \times 100 = 145\%$ larger when
Disproportionality is 10 rather than 2.

g. The odds ratio for *Ideological Incompatibility* indicates that a
one-unit increase in *Ideological Incompatibility* is associated with
a 0.9896 factor change or 1.04% decrease in the odds of
pre-electoral coalition formation.

The odds ratio for *Disproportionality* indicates that a one-unit
increase in *Disproportionality* is associated with a 1.028 factor
change or 2.8% increase in the odds of pre-electoral coalition
formation *when Polarization* $= 0$.

The odds ratio for *Polarization* indicates that a one-unit increase
in *Polarization* is associated with a 1.0037 factor change or 0.37%
increase in the odds of pre-electoral coalition formation *when*
Disproportionality $= 0$.

h. Building on Eq. 8.66, the effect of increasing *Disproportionality*
by five-units on the odds of pre-electoral coalition formation is
$e^{(\beta_1 + \beta_3 Polarization)) \times 5}$. Given that $\beta_1 = 0.0280$, $\beta_3 = 0.0005$, and
Polarization $= 10$ in our first scenario, we see that a five-unit
increase in *Disproportionality* increases the odds of pre-electoral
coalition formation by a factor of $e^{(0.0280 + 0.0005 \times 10)) \times 5} = 1.179$ or
by 17.9%.

In the second scenario *Polarization* $= 30$, and so a five-unit
increase in *Disproportionality* increases the odds of pre-electoral

coalition formation by a factor of $e^{(0.0280+0.0005\times30))\times5} = 1.240$ or by 24%.

Thus, the factor change in the odds of pre-electoral coalition formation associated with a five-unit increase in *Disproportionality* is $[(1.240 - 1.179)/1.179] \times 100 = 5.13\%$ larger when *Polarization* is 30 rather than 10.

i. The fact that the probability of pre-electoral coalition formation is very low in most scenarios tells us that $Pr(PEC)$ is likely to be less than 0.5 in most circumstances. The coefficients on *Disproportionality* and *Polarization* in Table 8.4 are both positive. Based on the information in Figure 8.4, we can, therefore, infer that the interaction effect between *Disproportionality* and *Polarization* will likely be positive in most scenarios. In other words, we'll find that an increase in *Disproportionality* (*Polarization*) is associated with an increase in the effect of *Polarization* (*Disproportionality*) on the probability of pre-electoral coalition formation. We can't infer anything about the substantive magnitude or statistical significance of this interaction effect.

The coefficients on *Disproportionality* and *Ideological Incompatibility* in Table 8.4 have different signs. Based on this, we can infer that the interaction effect between *Disproportionality* and *Ideological Incompatibility* will likely be negative in most scenarios. In other words, we'll find that an increase in *Disproportionality* (*Ideological Incompatibility*) is associated with a decrease in the effect of *Ideological Incompatibility* (*Disproportionality*) on the probability of pre-electoral coalition formation. We can't infer anything about the substantive magnitude or statistical significance of this interaction effect. Based on the information in Figure 8.4, we won't find a compression-based interaction between pairs of our independent variables only if (1) either of the coefficients on the two variables are 0 or (2) the probability of pre-electoral coalition formation in our baseline scenario is 0.5.

EXERCISES FROM CHAPTER 9

1. a. *Prejudice Hypothesis: An increase in someone's level of prejudice increases their propensity to openly express prejudice but only in the presence of prejudiced elite speech.*

Prejudiced Speech Hypothesis: Exposure to prejudiced elite speech increases the propensity to openly express prejudice but only among those who are sufficiently prejudiced to begin with.

Given the strength of the underlying theory provided by Newman et al. (2021), and following our recommendations, we state our hypotheses about the effect of *Prejudice* and *Prejudiced Speech* with respect to the underlying latent *propensity* to express prejudice rather than the *probability* to express prejudice. The marginal effect of *Prejudice* on the propensity to express prejudice is

$$\frac{\partial Expressed\ Prejudice^*}{\partial Prejudice} = \beta_1 + \beta_3 Prejudiced\ Speech. \qquad (S.63)$$

According to the *Prejudice Hypothesis*, an increase in someone's level of prejudice should have no effect on their propensity to express prejudice in the absence of prejudiced elite speech. As a result, β_1 should be 0. However, it should have a positive effect in the presence of prejudiced elite speech. As a result, $\beta_1 + \beta_3$ should be positive. It necessarily follows that β_3 should also be positive. The marginal effect of *Prejudiced Speech* on the propensity to express prejudice is

$$\frac{\partial Expressed\ Prejudice^*}{\partial Prejudiced\ Speech} = \beta_2 + \beta_3 Prejudice. \qquad (S.64)$$

According to the *Prejudiced Speech Hypothesis*, exposure to prejudiced elite speech should have no effect on someone's level of expressed prejudice if they're not already prejudiced. As a result, β_2 should be 0. However, it should have a positive and growing effect among those with high levels of prejudice. It follows that $\beta_1 + \beta_3 Prejudice$ should become positive once *Prejudice* is sufficiently large and that β_3 should be positive.

b. As predicted, the marginal effect of *Prejudice* is close to 0 (-0.078) and statistically insignificant for individuals in the control group who weren't exposed to prejudiced elite speech. As predicted, the marginal effect of *Prejudice* is positive ($-0.078 + 3.04 = 2.962$) among those who were exposed to prejudiced elite speech. Although we can't tell from the results provided whether this effect is statistically significant, we do know that it is statistically significantly different from the marginal effect in the control group. This is because the coefficient on the interaction term is statistically significant.

As predicted, the marginal effect of *Prejudiced Speech* is close to 0 (-0.597) and statistically insignificant for individuals who aren't prejudiced at all. We know that the marginal effect of *Prejudiced Speech* becomes positive when *Prejudice* is greater than

$0.597/3.04 = 0.196$ and that it reaches a maximum value of $-0.597 + 3.04 = 2.443$ when *Prejudice* is at its highest value of 1. Based on the results provided, we don't have enough information to know if and when the positive marginal effect of *Prejudiced Speech* on the propensity to express prejudice is statistically significant.

c. We can say that an increase in *Prejudice* has no statistically significant effect on the probability of choosing a higher value of *Expressed Prejudice* when there's no exposure to prejudiced elite speech but that it has a positive effect when there's exposure. We can also say that an increase in *Prejudice* increases the probability of completely agreeing with the discriminatory behavior described in the vignette and reduces the probability of completely disagreeing with it. We can make no inference about how it affects the probability of choosing one of the intermediate values of *Expressed Prejudice*.

We can say that exposure to prejudiced elite speech has no statistically significant effect on the probability of choosing a higher value of *Expressed Prejudice* when individuals aren't prejudiced at all but that it has a positive effect when *Prejudice* > 0.196. We can also say that exposure to prejudiced elite speech increases the probability of completely agreeing with the discriminatory behavior described in the vignette and reduces the probability of completely disagreeing with it when *Prejudice* > 0.196. We can make no inference about how it affects the probability of choosing one of the intermediate values of *Expressed Prejudice*.

d. From Eq. 9.4, we know that the probability that individual i chooses alternative j in an ordered logit model is

$$P_{ij} = \Lambda(\tau_j - X\beta) - \Lambda(\tau_{j-1} - X\beta), \tag{S.65}$$

where $\Lambda(\cdot)$ represents the cumulative standard logistic distribution. And from Eq. 9.11, we know that the marginal effect of *Prejudice* on the probability that individual i picks choice alternative j is

$$\frac{\partial P_{ij}}{\partial Prejudice} = \left[\lambda(\tau_{j-1} - X\beta) - \lambda(\tau_j - X\beta)\right] \cdot (\beta_1 + \beta_3 Prejudiced\ Speech), \tag{S.66}$$

where $\lambda(\cdot)$ is the standard logistic probability density function. Our dependent variable, *Expressed Prejudice*, has five possible values. Following Eq. 9.12, the marginal effect of *Prejudice* on each of our five choice probabilities is

$$\frac{\partial P_{i1}}{\partial Prejudice} = [\lambda(\tau_1 - X\beta)] \cdot - (\beta_1 + \beta_3 Prejudiced\ Speech)$$

$$\frac{\partial P_{i2}}{\partial Prejudice} = [\lambda(\tau_1 - X\beta) - \lambda(\tau_2 - X\beta)] \cdot (\beta_1 + \beta_3 Prejudiced\ Speech)$$

$$\frac{\partial P_{i3}}{\partial Prejudice} = [\lambda(\tau_2 - X\beta) - \lambda(\tau_3 - X\beta)] \cdot (\beta_1 + \beta_3 Prejudiced\ Speech)$$

$$\frac{\partial P_{i4}}{\partial Prejudice} = [\lambda(\tau_3 - X\beta) - \lambda(\tau_4 - X\beta)] \cdot (\beta_1 + \beta_3 Prejudiced\ Speech)$$

$$\frac{\partial P_{i5}}{\partial Prejudice} = [\lambda(\tau_4 - X\beta)] \cdot (\beta_1 + \beta_3 Prejudiced\ Speech). \tag{S.67}$$

Recall from Eq. 8.22 that $\lambda(\cdot)$ evaluated at, say, $\tau_1 - X\beta$ is

$$\lambda(\tau_1 - X\beta) = \frac{e^{\tau_1 - X\beta}}{\left(1 + e^{\tau_1 - X\beta}\right)^2}. \tag{S.68}$$

From Table 9.3, $\tau_1 = 0.094$, $\tau_2 = 2.58$, $\tau_3 = 3.79$, and $\tau_4 = 4.46$. We're now ready to calculate the desired marginal effects for those in the control group and those who received the *Prejudiced Speech* treatment.

Let's start with the control group. Based on the results in Table 9.3, $X\beta = -0.078 \times 0.3 - 0.597 \times 0 + 3.04 \times 0.3 \times 0 = -0.0234$.[5] This means that we have

$$\lambda(\tau_1 - X\beta) = \frac{e^{0.094 + 0.0234}}{\left(1 + e^{0.094 + 0.0234}\right)^2} = \frac{e^{0.1174}}{\left(1 + e^{0.1174}\right)^2} = \frac{1.125}{2.125^2} = 0.249$$

$$\lambda(\tau_2 - X\beta) = \frac{e^{2.58 + 0.0234}}{\left(1 + e^{2.58 + 0.0234}\right)^2} = \frac{e^{2.6034}}{\left(1 + e^{2.6034}\right)^2} = \frac{13.51}{14.51^2} = 0.064$$

$$\lambda(\tau_3 - X\beta) = \frac{e^{3.79 + 0.0234}}{\left(1 + e^{3.79 + 0.0234}\right)^2} = \frac{e^{3.8134}}{\left(1 + e^{3.8134}\right)^2} = \frac{45.30}{46.30^2} = 0.021$$

$$\lambda(\tau_4 - X\beta) = \frac{e^{4.46 + 0.0234}}{\left(1 + e^{4.46 + 0.0234}\right)^2} = \frac{e^{4.4834}}{\left(1 + e^{4.4834}\right)^2} = \frac{88.54}{89.54^2} = 0.011.$$

We also have
$\beta_1 + \beta_3 Prejudiced\ Speech = -0.078 + 3.04 \times 0 = -0.078$.
Substituting all of these values into Eq. S.67, we have

$$\frac{\partial P_{i1}}{\partial Prejudice} = 0.249 \times 0.078 = 0.02$$

$$\frac{\partial P_{i2}}{\partial Prejudice} = [0.249 - 0.064] \times -0.078 = -0.01$$

---

[5] All of the other variables are set to 0 in this scenario, and so we've ignored them in presenting this calculation.

$$\frac{\partial P_{i3}}{\partial Prejudice} = [0.064 - 0.021] \times -0.078 = -0.003$$

$$\frac{\partial P_{i4}}{\partial Prejudice} = [0.021 - 0.011] \times -0.078 = 0.001$$

$$\frac{\partial P_{i5}}{\partial Prejudice} = 0.011 \times -0.078 = 0.001. \tag{S.69}$$

We now turn to the *Prejudiced Speech* treatment group. Based on the results in Table 9.3,
$X\beta = -0.078 \times 0.3 - 0.597 \times 1 + 3.04 \times 0.3 \times 1 = 0.2916.$[6] This means that we have

$$\lambda(\tau_1 - X\beta) = \frac{e^{0.094-0.2916}}{\left(1 + e^{0.094-0.2916}\right)^2} = \frac{e^{-0.1976}}{\left(1 + e^{-0.1976}\right)^2} = \frac{0.821}{1.821^2} = 0.248$$

$$\lambda(\tau_2 - X\beta) = \frac{e^{2.58-0.2916}}{\left(1 + e^{2.58-0.2916}\right)^2} = \frac{e^{2.2884}}{\left(1 + e^{2.2884}\right)^2} = \frac{9.859}{10.859^2} = 0.084$$

$$\lambda(\tau_3 - X\beta) = \frac{e^{3.79-0.2916}}{\left(1 + e^{3.79-0.2916}\right)^2} = \frac{e^{3.4984}}{\left(1 + e^{3.4984}\right)^2} = \frac{33.063}{34.063^2} = 0.028$$

$$\lambda(\tau_4 - X\beta) = \frac{e^{4.46-0.2916}}{\left(1 + e^{4.46-0.2916}\right)^2} = \frac{e^{4.1684}}{\left(1 + e^{4.1684}\right)^2} = \frac{64.612}{65.612^2} = 0.015.$$

We also have
$\beta_1 + \beta_3 Prejudiced\ Speech = -0.078 + 3.04 \times 1 = 2.962.$
Substituting all of these values into Eq. S.67, we have

$$\frac{\partial P_{i1}}{\partial Prejudice} = 0.248 \times -2.962 = -0.73$$

$$\frac{\partial P_{i2}}{\partial Prejudice} = [0.248 - 0.084] \times 2.962 = 0.49$$

$$\frac{\partial P_{i3}}{\partial Prejudice} = [0.084 - 0.028] \times 2.962 = 0.16$$

$$\frac{\partial P_{i4}}{\partial Prejudice} = [0.028 - 0.015] \times 2.962 = 0.04$$

$$\frac{\partial P_{i5}}{\partial Prejudice} = 0.015 \times 2.962 = 0.04. \tag{S.70}$$

We can ascertain how exposure to prejudiced elite speech modifies the marginal effect of *Prejudice* on each of our five choice probabilities in our chosen scenario by finding the difference between the marginal effects from the treatment and control

---

[6] All of the other variables remain set to 0 in this scenario, and so we've ignored them in presenting this calculation.

Table S.8 The marginal effect of *Prejudice* on the probability of different values of *Expressed Prejudice*

	Control Group	Prejudiced Speech Treatment	Difference
1: Completely Unacceptable	0.02	−0.73	−0.75
2: Unacceptable	−0.01	0.49	0.50
3: Neither Good Nor Bad	−0.003	0.16	0.17
4: Acceptable	0.001	0.04	0.04
5: Completely Acceptable	0.001	0.04	0.05

Note: The quantities reported in Table S.8 are calculated for an individual with *Prejudice* $= 0.3$. The "Difference" column indicates the modifying effect of *Prejudiced Speech* and is equal to *Prejudiced Speech* Treatment minus Control Group.

groups for each of our choice categories. We summarize our results in Table S.8. One thing to note is that although the sign of the coefficient on the interaction term β_3 is positive, the sign of the modifying effect of *Prejudiced Speech* on the marginal effect of *Prejudice* on the probability of the choice alternatives isn't always positive in this scenario; it depends on the value of the choice alternative.

e. From Eq. 9.13, we know that the change in the probability from an ordered logit model that individual i chooses alternative j associated with exposure to prejudiced elite speech is

$$\Delta P_{ij} = \left[\Lambda(\tau_j - X_c\beta) - \Lambda(\tau_{j-1} - X_c\beta)\right] \\ - \left[\Lambda(\tau_j - X_b\beta) - \Lambda(\tau_{j-1} - X_b\beta)\right], \tag{S.71}$$

where $X_b\beta$ indicates the value of $X\beta$ when *Prejudiced Speech* $= 0$ and the other independent variables are set at their "baseline" values, $X_c\beta$ indicates the value of $X\beta$ when *Prejudiced Speech* $= 1$ and the other independent variables remain at their "baseline" values, and $\Lambda(\cdot)$ represents the cumulative standard logistic distribution. Following Eq. 9.16, the change in the probability that individual i chooses each of the five alternatives in *Expressed Prejudice* is

$$\Delta P_{i1} = \Lambda(\tau_1 - X_c\beta) - \Lambda(\tau_1 - X_b\beta)$$
$$\Delta P_{i2} = \left[\Lambda(\tau_2 - X_c\beta) - \Lambda(\tau_1 - X_c\beta)\right] - \left[\Lambda(\tau_2 - X_b\beta) - \Lambda(\tau_1 - X_b\beta)\right]$$
$$\Delta P_{i3} = \left[\Lambda(\tau_3 - X_c\beta) - \Lambda(\tau_2 - X_c\beta)\right] - \left[\Lambda(\tau_3 - X_b\beta) - \Lambda(\tau_2 - X_b\beta)\right]$$
$$\Delta P_{i4} = \left[\Lambda(\tau_4 - X_c\beta) - \Lambda(\tau_3 - X_c\beta)\right] - \left[\Lambda(\tau_4 - X_b\beta) - \Lambda(\tau_3 - X_b\beta)\right]$$
$$\Delta P_{i5} = \left[1 - \Lambda(\tau_4 - X_c\beta)\right] - \left[1 - \Lambda(\tau_4 - X_b\beta)\right]. \tag{S.72}$$

Recall that $\Lambda(\cdot)$ evaluated at, say, $\tau_1 - X\beta$ is

$$\Lambda(\tau_1 - X\beta) = \frac{e^{\tau_1 - X\beta}}{1 + e^{\tau_1 - X\beta}}. \tag{S.73}$$

From Table 9.3, $\tau_1 = 0.094$, $\tau_2 = 2.58$, $\tau_3 = 3.79$, and $\tau_4 = 4.46$. We're now ready to calculate the desired change in probabilities for the baseline scenarios where *Prejudice* $= 0.25$ and *Prejudice* $= 0.75$.

Let's start with the baseline scenario where *Prejudice* $= 0.25$. Based on the results in Table 9.3,
$$X_b\beta = -0.078 \times 0.25 - 0.597 \times 0 + 3.04 \times 0.25 \times 0 = -0.0195,$$
and $X_c\beta = -0.078 \times 0.25 - 0.597 \times 1 + 3.04 \times 0.25 \times 1 = 0.1435.^{[7]}$
This means that we have

$$\Lambda(\tau_1 - X_b\beta) = \frac{e^{0.094 + 0.0195}}{\left(1 + e^{0.094 + 0.0195}\right)} = \frac{e^{0.1135}}{\left(1 + e^{0.1135}\right)} = \frac{1.12}{2.12} = 0.528$$

$$\Lambda(\tau_2 - X_b\beta) = \frac{e^{2.58 + 0.0195}}{\left(1 + e^{2.58 + 0.0195}\right)} = \frac{e^{2.5995}}{\left(1 + e^{2.5995}\right)} = \frac{13.457}{14.457} = 0.931$$

$$\Lambda(\tau_3 - X_b\beta) = \frac{e^{3.79 + 0.0195}}{\left(1 + e^{3.79 + 0.0195}\right)} = \frac{e^{3.8095}}{\left(1 + e^{3.8095}\right)} = \frac{45.128}{46.128} = 0.978$$

$$\Lambda(\tau_4 - X_b\beta) = \frac{e^{4.46 + 0.0195}}{\left(1 + e^{4.46 + 0.0195}\right)} = \frac{e^{4.4795}}{\left(1 + e^{4.4795}\right)} = \frac{88.19}{89.19} = 0.989$$

and

$$\Lambda(\tau_1 - X_c\beta) = \frac{e^{0.094 - 0.1435}}{\left(1 + e^{0.094 - 0.1435}\right)} = \frac{e^{-0.0495}}{\left(1 + e^{-0.0495}\right)} = \frac{0.952}{1.952} = 0.488$$

$$\Lambda(\tau_2 - X_c\beta) = \frac{e^{2.58 - 0.1435}}{\left(1 + e^{2.58 - 0.1435}\right)} = \frac{e^{2.4365}}{\left(1 + e^{2.4365}\right)} = \frac{11.433}{12.433} = 0.920$$

$$\Lambda(\tau_3 - X_c\beta) = \frac{e^{3.79 - 0.1435}}{\left(1 + e^{3.79 - 0.1435}\right)} = \frac{e^{3.6465}}{\left(1 + e^{3.6465}\right)} = \frac{38.340}{39.340} = 0.975$$

$$\Lambda(\tau_4 - X_c\beta) = \frac{e^{4.46 - 0.1435}}{\left(1 + e^{4.46 - 0.1435}\right)} = \frac{e^{4.3165}}{\left(1 + e^{4.3165}\right)} = \frac{74.926}{75.926} = 0.987.$$

[7] Again, all of the other variables are set to 0 in this scenario, and so we've ignored them in presenting these calculations.

Substituting all of these values into Eq. S.72, we have

$$\Delta P_{i1} = 0.488 - 0.528 = -0.041$$
$$\Delta P_{i2} = [0.920 - 0.488] - [0.931 - 0.528] = 0.432 - 0.403 = 0.029$$
$$\Delta P_{i3} = [0.975 - 0.920] - [0.978 - 0.931] = 0.055 - 0.047 = 0.008$$
$$\Delta P_{i4} = [0.987 - 0.975] - [0.989 - 0.978] = 0.012 - 0.011 = 0.002$$
$$\Delta P_{i5} = [1 - 0.987] - [1 - 0.989] = 0.013 - 0.011 = 0.002. \quad\quad (S.74)$$

We now turn to the baseline scenario where *Prejudice* $= 0.75$.
Based on the results in Table 9.3,
$X_b\beta = -0.078 \times 0.75 - 0.597 \times 0 + 3.04 \times 0.75 \times 0 = -0.0585$,
and $X_c\beta = -0.078 \times 0.75 - 0.597 \times 1 + 3.04 \times 0.75 \times 1 = 1.6245$.
This means that we have

$$\Lambda(\tau_1 - X_b\beta) = \frac{e^{0.094+0.0585}}{(1+e^{0.094+0.0585})} = \frac{e^{0.1525}}{(1+e^{0.1525})} = \frac{1.165}{2.165} = 0.538$$

$$\Lambda(\tau_2 - X_b\beta) = \frac{e^{2.58+0.0585}}{(1+e^{2.58+0.0585})} = \frac{e^{2.6385}}{(1+e^{2.6385})} = \frac{13.992}{14.992} = 0.933$$

$$\Lambda(\tau_3 - X_b\beta) = \frac{e^{3.79+0.0585}}{(1+e^{3.79+v})} = \frac{e^{3.8485}}{(1+e^{3.8485})} = \frac{46.923}{47.923} = 0.979$$

$$\Lambda(\tau_4 - X_b\beta) = \frac{e^{4.46+0.0585}}{(1+e^{4.46+0.0585})} = \frac{e^{4.5185}}{(1+e^{4.5185})} = \frac{91.698}{92.698} = 0.989$$

and

$$\Lambda(\tau_1 - X_c\beta) = \frac{e^{0.094-1.6245}}{(1+e^{0.094-1.6245})} = \frac{e^{-1.5305}}{(1+e^{-1.5305})} = \frac{0.216}{1.216} = 0.178$$

$$\Lambda(\tau_2 - X_c\beta) = \frac{e^{2.58-1.6245}}{(1+e^{2.58-1.6245})} = \frac{e^{0.9555}}{(1+e^{0.9555})} = \frac{2.60}{3.60} = 0.722$$

$$\Lambda(\tau_3 - X_c\beta) = \frac{e^{3.79-1.6245}}{(1+e^{3.79-1.6245})} = \frac{e^{2.1655}}{(1+e^{2.1655})} = \frac{8.719}{9.719} = 0.897$$

$$\Lambda(\tau_4 - X_c\beta) = \frac{e^{4.46-1.6245}}{(1+e^{4.46-1.6245})} = \frac{e^{2.8355}}{(1+e^{2.8355})} = \frac{17.039}{18.039} = 0.945.$$

Substituting all of these values into Eq. S.72, we have

$$\Delta P_{i1} = 0.178 - 0.538 = -0.36$$
$$\Delta P_{i2} = [0.722 - 0.178] - [0.933 - 0.538] = 0.544 - 0.395 = 0.149$$
$$\Delta P_{i3} = [0.897 - 0.722] - [0.979 - 0.933] = 0.175 - 0.046 = 0.129$$
$$\Delta P_{i4} = [0.945 - 0.897] - [0.989 - 0.979] = 0.048 - 0.01 = 0.037$$
$$\Delta P_{i5} = [1 - 0.945] - [1 - 0.989] = 0.055 - 0.011 = 0.045. \quad\quad (S.75)$$

We can ascertain how increasing *Prejudice* from 0.25 to 0.75
modifies the effect of exposure to prejudiced elite speech on each

Table S.9 The effect of *Prejudiced Speech* on the probability of different values of *Expressed Prejudice*

	Prejudice = 0.25	Prejudice = 0.75	Difference
1: Completely Unacceptable	−0.041	−0.36	−0.32
2: Unacceptable	0.029	0.149	0.12
3: Neither Good Nor Bad	0.008	0.129	0.121
4: Acceptable	0.002	0.037	0.035
5: Completely Acceptable	0.002	0.045	0.043

Note: The "Difference" column indicates the modifying effect of changing *Prejudice* from 0.25 to 0.75.

of our five choice probabilities in our chosen scenario by finding the difference between the effects from the *Prejudice* = 0.75 and *Prejudice* = 0.25 groups for each of our choice categories. We summarize our results in Table S.9.

f. From Eq. 9.21, we know that the effect of a 0.5-unit increase in *Prejudice* on the odds of observing a value of 2 or less versus any higher value for *Expressed Prejudice* is $e^{-(\beta_1 + \beta_3 Prejudiced\ Speech) \times 0.5}$. Plugging in the values from Table 9.3, we have $e^{-(-0.078 + 3.04 Prejudiced\ Speech) \times 0.5}$. When *Prejudiced Speech* = 0, we have $e^{-(-0.078) \times 0.5} = e^{0.039} = 1.04$. In other words, the odds of observing a value of 2 or less (*Completely Unacceptable*, *Unacceptable*) increase by a factor of 1.04 or by 4%. Due to the parallel regression assumption, this factor and percentage change applies to the odds of observing any value k or less, and not just a value of 2 or less. When *Prejudiced Speech* = 1, we have $e^{-(-0.078 + 3.04) \times 0.5} = e^{-1.481} = 0.23$. In other words, the odds of observing a value of 2 or less increase by a factor of 0.23 or, equivalently, decrease by 77.3%.

The odds of observing a value of 3 or more is the same as the odds of observing a value greater than 2. From Eq. 9.21, we know that the effect of receiving the *Prejudiced Speech* treatment on the odds of observing a value of 2 or less is $e^{-(\beta_2 + \beta_3 Prejudice)} = e^{-(-0.597 + 3.04 Prejudice)}$. The effect on the odds of observing a value of 3 or more is therefore just the inverse of this, $\frac{1}{e^{-(-0.597 + 3.04 Prejudice)}}$. When *Prejudice* = 0.25, we have

$\frac{1}{e^{-(-0.597+3.04\times0.25)}} = \frac{1}{e^{-0.163}} = \frac{1}{0.85} = 1.18$. In other words, the odds of observing a value of 3 or more (*Neither Good Nor Bad, Acceptable, Completely Acceptable*) increase by a factor of 1.18 or by 18%. When *Prejudice* = 0.75, we have $\frac{1}{e^{-(-0.597+3.04\times0.75)}} = \frac{1}{e^{-1.683}} = \frac{1}{0.19} = 5.38$. In other words, the odds of observing a value of 3 or more increase by a factor of 5.38 or by 438%.

EXERCISES FROM CHAPTER 10

1. a. Based on 10.64, the marginal effect of *Government Intervention Distance$_k$* on the utility associated with candidate k is $\beta_1 + \beta_2 \times Female = -0.482 - 0.227 \times Female$. Thus, the marginal effect is -0.482 for men and $-0.482 - 0.227 \times 1 = -0.71$ for women. The modifying effect of being female instead of male on the marginal effect of *Government Intervention Distance$_k$* on the utility associated with candidate k is $\beta_2 = -0.227$.

    b. From Eq. 10.27, we know that the direct marginal effect of an alternative-specific variable like *Government Intervention Distance$_k$* on the probability of choosing alternative k is $P_{ik}(1 - P_{ik}) \cdot (\beta_1 + \beta_3 Z_1)$. And from Eq. 10.8, we know that

$$P_{ik} = \frac{e^{V_{ik}}}{\sum_{j=1}^{J} e^{V_{ij}}}. \tag{S.76}$$

In our example, k refers to the Conservative Party. Based on the results in Table 10.7, and for the scenario in which our individual is male, we have

$$V_{iCons} = -0.482 \times 2 + 0.755 \times 6 + 0.175 \times 1 + 0.018 \times 30 - 1.257 = 3.024$$

$$V_{iLab} = -0.482 \times 1 = -0.482$$

$$V_{iLibdem} = -0.482 \times 1 + 0.408 \times 6 + 0.272 \times 1 + 0.012 \times 30 - 2.228 = 0.37.$$

From this, we have

$$P_{iCons} = \frac{e^{V_{iCons}}}{\sum_{j=1}^{J} e^{V_{ij}}} = \frac{e^{V_{iCons}}}{e^{V_{iCons}} + e^{V_{iLab}} + e^{V_{iLibdem}}}$$

$$= \frac{e^{3.024}}{e^{3.024} + e^{-0.482} + e^{0.37}}$$

$$= \frac{20.57}{20.57 + 0.62 + 1.45} = 0.91. \tag{S.77}$$

Thus, the direct marginal effect of *Government Intervention Distance$_k$* on the probability of voting for the Conservative Party *for a man* in our chosen scenario is
$0.91 \times (1 - 0.91) \times (-0.482) = -0.04$.
Based on the results in Table 10.7, and for the scenario in which our individual is female, we have

$$V_{i\text{Cons}} = -0.482 \times 2 - 0.227 \times 2 \times 1 + 0.488 \times 1 + 0.755 \times 6 \\ + 0.175 \times 1 + 0.018 \times 30 - 1.257 = 3.058 \tag{S.78}$$

$$V_{i\text{Lab}} = -0.482 \times 1 - 0.227 \times 1 \times 1 = -0.709$$

$$V_{i\text{Libdem}} = -0.482 \times 1 - 0.227 \times 1 \times 1 + 0.435 \times 1 + 0.408 \times 6 \\ + 0.272 \times 1 + 0.012 \times 30 - 2.228 = 0.578.$$

From this, we have

$$P_{i\text{Cons}} = \frac{e^{V_{i\text{Cons}}}}{\sum_{j=1}^{J} e^{V_{ij}}} = \frac{e^{V_{i\text{Cons}}}}{e^{V_{i\text{Cons}}} + e^{V_{i\text{Lab}}} + e^{V_{i\text{Libdem}}}}$$

$$= \frac{e^{3.058}}{e^{3.058} + e^{-0.709} + e^{0.578}}$$

$$= \frac{21.28}{21.28 + 0.49 + 1.78} = 0.90. \tag{S.79}$$

Thus, the direct marginal effect of *Government Intervention Distance$_k$* on the probability of voting for the Conservative Party *for a woman* in our chosen scenario is
$0.90 \times (1 - 0.90) \times (-0.482 - 0.227 \times 1) = -0.06$.
The modifying effect of being female on the direct marginal effect of *Government Intervention Distance$_k$* on the probability of voting for the Conservative Party in our chosen scenario is the difference in the two direct marginal effects that we've calculated,
$-0.06 - (-0.04) = -0.02$.

c. From Eq. 10.29, we know that the cross-marginal effect of an alternative-specific variable like *Government Intervention Distance$_m$* on the probability of choosing alternative k is $-P_{ik} \cdot P_{im} \cdot (\beta_1 + \beta_3 Z_1)$. In our example, k refers to the Conservative Party and m refers to the Labour Party. Let's start with the case where our voter is male. We already know $P_{i\text{Cons}}$ for the male case from Eq. S.77. Based on our earlier calculations, we know that

$$P_{i\text{Lab}} = \frac{e^{V_{i\text{Lab}}}}{\sum_{j=1}^{J} e^{V_{ij}}} = \frac{e^{V_{i\text{Lab}}}}{e^{V_{i\text{Cons}}} + e^{V_{i\text{Lab}}} + e^{V_{i\text{Libdem}}}}$$

$$= \frac{e^{-0.482}}{e^{3.024} + e^{-0.482} + e^{0.37}}$$

$$= \frac{0.62}{20.57 + 0.62 + 1.45} = 0.03. \qquad (S.80)$$

This means that the cross-marginal effect of increasing *Government Intervention Distance* for the Labour Party on the probability of voting for the Conservative Party *for a man* in our chosen scenario is $-0.91 \times 0.03 \times -0.482 = 0.01$.

Let's switch to the case where our voter is female. We already know $P_{i\text{Cons}}$ for the female case from Eq. S.79. Based on our earlier calculations, we know that

$$P_{i\text{Lab}} = \frac{e^{V_{i\text{Lab}}}}{\sum_{j=1}^{J} e^{V_{ij}}} = \frac{e^{V_{i\text{Lab}}}}{e^{V_{i\text{Cons}}} + e^{V_{i\text{Lab}}} + e^{V_{i\text{Libdem}}}}$$

$$= \frac{e^{0.49}}{e^{3.058} + e^{-0.709} + e^{0.578}}$$

$$= \frac{21.28}{21.28 + 0.49 + 1.78} = 0.02. \qquad (S.81)$$

This means that the cross-marginal effect of increasing *Government Intervention Distance* for the Labour Party on the probability of voting for the Conservative Party *for a woman* in our chosen scenario is $-0.90 \times 0.02 \times (-0.482 - 0.227 \times 1) = 0.01$.

d. The direct marginal effect and associated cross-marginal effects for an alternative-specific variable like *Government Intervention Distance$_k$* for a given choice alternative k will necessarily sum to 0. In the case where our voter is male, we know that the direct marginal effect of *Government Intervention Distance* is -0.04 and that the cross-marginal effect for the Labour Party is 0.01. As a result, the cross-marginal effect for the Liberal Democrats will be $0 - (-0.04) - 0.01 = 0.03$. In the case where our voter is female, we know that the direct marginal effect of *Government Intervention Distance* is -0.06 and that the cross-marginal effect for the Labour Party is 0.01. As a result, the cross-marginal effect for the Liberal Democrats will be $0 - (-0.06) - 0.01 = 0.05$.

e. Let's start with the case where our voter is male. We already know the probability of voting for the Conservatives in the baseline scenario b, $P_{i\text{Cons}_b}$, from Eq. S.77, 0.91. The two-unit increase in

Government Intervention Distance for the Conservative Party in the counterfactual scenario c means that we now have

$$V_{iCons_c} = -0.482 \times 4 + 0.755 \times 6 + 0.175 \times 1 + 0.018 \times 30 - 1.257 = 2.06.$$
(S.82)

This means that the probability of voting for the Conservative Party in the counterfactual scenario is

$$P_{iCons_c} = \frac{e^{V_{iCons_c}}}{\sum_{j=1}^{J} e^{V_{ij}}} = \frac{e^{V_{iCons_c}}}{e^{V_{iCons_c}} + e^{V_{iLab_b}} + e^{V_{iLibdem_b}}}$$

$$= \frac{e^{2.06}}{e^{2.06} + e^{-0.482} + e^{0.37}}$$

$$= \frac{7.85}{7.85 + 0.62 + 1.45} = 0.79.$$
(S.83)

Thus, the change in the probability that a man in our chosen scenario votes for the Conservative Party when we increase the ideological distance to the Conservative Party by two units with respect to policy on the appropriate size of the government is

$$\Delta P_{iCons} = P_{iCons_c} - P_{iCons_b} = 0.79 - 0.91 = -0.12.$$
(S.84)

Let's switch to the case where our voter is female. We already know the probability of voting for the Conservatives in the baseline scenario b, P_{iCons_b}, from Eq. S.79, 0.90. The two-unit increase in *Government Intervention Distance* for the Conservative Party in the counterfactual scenario c means that we now have

$$V_{iCons} = -0.482 \times 4 - 0.227 \times 4 \times 1 + 0.488 \times 1 + 0.755 \times 6$$
$$+ 0.175 \times 1 + 0.018 \times 30 - 1.257 = 1.64.$$
(S.85)

This means that the probability of voting for the Conservative Party in the counterfactual scenario is

$$P_{iCons_c} = \frac{e^{V_{iCons_c}}}{\sum_{j=1}^{J} e^{V_{ij}}} = \frac{e^{V_{iCons_c}}}{e^{V_{iCons_c}} + e^{V_{iLab_b}} + e^{V_{iLibdem_b}}}$$

$$= \frac{e^{1.64}}{e^{1.64} + e^{-0.709} + e^{0.578}}$$

$$= \frac{5.16}{5.16 + 0.49 + 1.78} = 0.69.$$
(S.86)

Thus, the change in the probability that a woman in our chosen scenario votes for the Conservative Party when we increase the ideological distance to the Conservative Party by two units with respect to policy on the appropriate size of the government is

$$\triangle P_{i\mathrm{Cons}} = P_{i\mathrm{Cons}_c} - P_{i\mathrm{Cons}_b} = 0.90 - 0.69 = -0.21. \tag{S.87}$$

It follows that the modifying effect of being female on the effect of a two-unit increase in *Government Intervention Distance* on the part of the Conservative Party on the probability of voting for the Conservative Party in our chosen scenario is $-0.21 - (-0.12) = -0.09$.

f. Based on Eq. 10.64 and Eq. 10.39, the odds ratio comparing the odds of choosing party k versus party m in the scenario where we increase *Government Intervention Distance$_k$* by one unit is

$$\frac{\Omega_{k \text{ vs } m} \mid Government\ Intervention\ Distance_k + 1}{\Omega_{k \text{ vs } m} \mid Government\ Intervention\ Distance_k} = e^{\beta_1 + \beta_2 Female}. \tag{S.88}$$

Using the results in Table 10.7, it follows that the percentage change in the odds that a *man* will vote for the Conservative Party as opposed to another party if we increase the ideological distance to the Conservative Party with respect to policy on the appropriate size of government by two units in our chosen scenario is $\left[e^{-0.482 \times 2} - 1\right] \times 100 = -61.86\%$. And the equivalent percentage change in the odds that a *woman* will vote for the Conservative Party is $\left[e^{(-0.482 - 0.227) \times 2} - 1\right] \times 100 = -75.78\%$. From this, we see that the negative effect of the two-unit increase in the ideological distance to the Conservative Party on the odds of voting for the Conservative Party as opposed to some other party is 13.92 percentage points or 22.5% larger for women than men.

2. a. The marginal effect of X_{1k} on the utility that voter i obtains from candidate k is

$$\frac{\partial U_{ik}}{\partial X_{1k}} = \beta_1 + \beta_3 X_{2k}. \tag{S.89}$$

From this, we see that the coefficient β_1 tells us the effect of a one-unit increase in X_{1k} when $X_{2k} = 0$.

b. The *direct marginal effect* of X_{1k} on the probability of voting for candidate k is

$$\frac{\partial P_{ik}}{\partial X_{1k}} = \frac{\partial \left(\frac{e^{V_{ik}}}{\sum_{j=1}^{J} e^{V_{ij}}} \right)}{\partial X_{1k}}$$

$$= \frac{e^{V_{ik}} \cdot (\beta_1 + \beta_3 X_{2k}) \cdot \sum_{j=1}^{J} e^{V_{ij}} - e^{V_{ik}} \cdot e^{V_{ik}} \cdot (\beta_1 + \beta_3 X_{2k})}{\left(\sum_{j=1}^{J} e^{V_{ij}} \right)^2}$$

$$= \frac{e^{V_{ik}} \cdot (\beta_1 + \beta_3 X_{2k}) \cdot \sum_{j=1}^{J} e^{V_{ij}}}{\left(\sum_{j=1}^{J} e^{V_{ij}} \right)^2} - \frac{e^{V_{ik}} \cdot e^{V_{ik}} \cdot (\beta_1 + \beta_3 X_{2k})}{\left(\sum_{j=1}^{J} e^{V_{ij}} \right)^2}$$

$$= \frac{e^{V_{ik}}}{\sum_{j=1}^{J} e^{V_{ij}}} \cdot (\beta_1 + \beta_3 X_{2k}) - \frac{e^{V_{ik}}}{\sum_{j=1}^{J} e^{V_{ij}}} \cdot \frac{e^{V_{ik}}}{\sum_{j=1}^{J} e^{V_{ij}}} \cdot (\beta_1 + \beta_3 X_{2k})$$

$$= P_{ik} \cdot \beta_2 - P_{ik} \cdot P_{ik} \cdot \beta_2$$

$$= P_{ik} (1 - P_{ik}) \cdot (\beta_1 + \beta_3 X_{2k}). \tag{S.90}$$

This tells us the marginal effect of increasing X_1 for candidate k on the probability that individual i votes for candidate k.

c. The *cross-marginal effect* associated with increasing X_1 for candidate m on the probability of voting for candidate k is

$$\frac{\partial P_{ik}}{\partial X_{1m}} = \frac{\partial \left(\frac{e^{V_{ik}}}{\sum_{j=1}^{J} e^{V_{ij}}} \right)}{\partial X_{1m}} = \frac{-e^{V_{ik}} \cdot e^{V_{im}} \cdot \beta_2}{\left(\sum_{j=1}^{J} e^{V_{ij}} \right)^2}$$

$$= \frac{-e^{V_{ik}} \cdot e^{V_{im}} \cdot (\beta_1 + \beta_3 X_{2k})}{\sum_{j=1}^{J} e^{V_{ij}} \cdot \sum_{j=1}^{J} e^{V_{ij}}}$$

$$= -P_{ik} \cdot P_{im} \cdot (\beta_1 + \beta_3 X_{2k}). \tag{S.91}$$

This tells us the marginal effect of increasing X_1 for candidate m on the probability that individual i votes for candidate k.

d. The effect of a one-unit increase in X_{1k} on the odds of choosing candidate k versus candidate m is

$$\frac{\Omega_{k \text{ vs } m} \mid X_{1k} + 1}{\Omega_{k \text{ vs } m} \mid X_{1k}} = \frac{\frac{e^{V_{ik}\mid X_{1k}+1}}{e^{V_{im}}}}{\frac{e^{V_{ik}\mid X_{1k}}}{e^{V_{im}}}}$$

$$= \frac{e^{V_{ik}\mid X_{1k}+1}}{e^{V_{ik}\mid X_{1k}}} = \frac{e^{\beta_1 (X_{1k}+1)+\beta_3 ((X_{1k}+1)X_{2k})}}{e^{\beta_1 X_{1k}+\beta_3 X_{1k}X_{2k}}}$$

$$= e^{\beta_1 (X_{1k}+1-X_{1k})+\beta_3 (X_{1k}+1-X_{1k})X_{2k}} = e^{\beta_1 + \beta_3 X_{2k}}. \tag{S.92}$$

e. The interaction effect of X_{1k} and X_{2k} on the utility that voter i obtains from candidate k is

$$\frac{\partial\left(\frac{\partial U_{ik}}{\partial X_{1k}}\right)}{\partial X_{2k}} = \frac{\partial\left(\beta_1 + \beta_3 X_{2k}\right)}{\partial X_{2k}} = \beta_3. \tag{S.93}$$

f. The interaction effect of X_{1k} and X_{2k} on the probability of voting for candidate k is

$$\begin{aligned}
\frac{\partial\left(\frac{\partial P_{ik}}{\partial X_{1k}}\right)}{\partial X_{2k}} &= \frac{\partial\left[P_{ik}\left(1-P_{ik}\right)\cdot\left(\beta_1+\beta_3 X_{2k}\right)\right]}{\partial X_{2k}}\\
&= \frac{\partial\left[P_{ik}\cdot\left(\beta_1+\beta_3 X_{2k}\right)-P_{ik}^2\cdot\left(\beta_1+\beta_3 X_{2k}\right)\right]}{\partial X_{2k}}\\
&= \frac{\partial P_{ik}}{\partial X_{2k}}\cdot\left(\beta_1+\beta_3 X_{2k}\right)+\beta_3\cdot P_{ik}-2P_{ik}\cdot\frac{\partial P_{ik}}{\partial X_{2k}}\cdot\left(\beta_1+\beta_3 X_{2k}\right)-\beta_3 P_{ik}^2\\
&= P_{ik}\left(1-P_{ik}\right)\cdot\beta_3+\left(1-2P_{ik}\right)\cdot\left(\beta_1+\beta_3 X_{2k}\right)\cdot\frac{\partial P_{ik}}{\partial X_{2k}}. \tag{S.94}
\end{aligned}$$

Following the logic set out in part b of the exercise, we know that

$$\frac{\partial P_{ik}}{\partial X_{2k}} = P_{ik}\left(1-P_{ik}\right)\cdot\left(\beta_2+\beta_3 X_{1k}\right). \tag{S.95}$$

Substituting this into Eq. S.94, we have

$$\begin{aligned}
\frac{\partial\left(\frac{\partial P_{ik}}{\partial X_{1k}}\right)}{\partial X_{2k}} &= P_{ik}\left(1-P_{ik}\right)\cdot\beta_3\\
&\quad+\left(1-2P_{ik}\right)\cdot\left(\beta_1+\beta_3 X_{2k}\right)\cdot\left[P_{ik}\left(1-P_{ik}\right)\cdot\left(\beta_2+\beta_3 X_{1k}\right)\right]\\
&= P_{ik}\left(1-P_{ik}\right)\cdot\beta_3\\
&\quad+P_{ik}\left(1-P_{ik}\right)\left(1-2P_{ik}\right)\cdot\left(\beta_1+\beta_3 X_{2k}\right)\cdot\left(\beta_2+\beta_3 X_{1k}\right). \tag{S.96}
\end{aligned}$$

Due to the symmetry of interactions, this interaction effect tells us both how X_{2k} modifies the marginal effect of X_{1k} on the probability of voting for candidate k *and* how X_{1k} modifies the marginal effect of X_{2k} on the probability of voting for candidate k.

3. a. The marginal effect of Z_1 on the utility that voter i obtains from candidate k is

$$\frac{\partial U_{ik}}{\partial Z_1} = \gamma_{1k} + \gamma_{3k} Z_2. \tag{S.97}$$

From this, we see that the coefficient γ_{1k} tells us the effect of a one-unit increase in Z_1 on the voter's utility for candidate k relative to the baseline candidate when $Z_2 = 0$.

b. The marginal effect of Z_1 on the probability of voting for candidate k is

$$
\frac{\partial P_{ik}}{\partial Z_1} = \frac{\partial \left(\frac{e^{V_{ik}}}{\sum_{j=1}^{J} e^{V_{ij}}} \right)}{\partial Z_1}
$$

$$
= \frac{e^{V_{ik}} \cdot (\gamma_{1k} + \gamma_{3k} Z_2) \cdot \sum_{j=1}^{J} e^{V_{ij}} - e^{V_{ik}} \cdot \sum_{j=1}^{J} e^{V_{ij}} \cdot (\gamma_{1j} + \gamma_{3j} Z_2)}{\left(\sum_{j=1}^{J} e^{V_{ij}} \right)^2}
$$

$$
= \frac{e^{V_{ik}} \cdot (\gamma_{1k} + \gamma_{3k} Z_2) \cdot \sum_{j=1}^{J} e^{V_{ij}}}{\left(\sum_{j=1}^{J} e^{V_{ij}} \right)^2} - \frac{e^{V_{ik}} \cdot \sum_{j=1}^{J} e^{V_{ij}} \cdot (\gamma_{1j} + \gamma_{3j} Z_2)}{\left(\sum_{j=1}^{J} e^{V_{ij}} \right)^2}
$$

$$
= \frac{e^{V_{ik}}}{\sum_{j=1}^{J} e^{V_{ij}}} \cdot (\gamma_{1k} + \gamma_{3k} Z_2) - \frac{e^{V_{ik}}}{\sum_{j=1}^{J} e^{V_{ij}}} \cdot \sum_{j=1}^{J} \frac{e^{V_{ij}}}{\sum_{j=1}^{J} e^{V_{ij}}} \cdot (\gamma_{1j} + \gamma_{3j} Z_2)
$$

$$
= P_{ik} \cdot (\gamma_{1k} + \gamma_{3k} Z_2) - P_{ik} \cdot \sum_{j=1}^{J} P_{ij} \cdot (\gamma_{1j} + \gamma_{3j} Z_2)
$$

$$
= P_{ik} \left[\gamma_{1k} + \gamma_{3k} Z_2 - \sum_{j=1}^{J} P_{ij} \cdot (\gamma_{1j} + \gamma_{3j} Z_2) \right]. \tag{S.98}
$$

c. The effect of a one-unit increase in Z_1 on the odds of choosing candidate k versus candidate m is

$$
\frac{\Omega_{k \text{ vs } m} \mid Z_1 + 1}{\Omega_{k \text{ vs } m} \mid Z_1} = \frac{\frac{e^{V_{ik} \mid Z_1 + 1}}{e^{V_{im} \mid Z_1 + 1}}}{\frac{e^{V_{ik} \mid Z_1}}{e^{V_{im} \mid Z_1}}} = \frac{e^{V_{ik} \mid Z_1 + 1}}{e^{V_{im} \mid Z_1 + 1}} \cdot \frac{e^{V_{im} \mid Z_1}}{e^{V_{ik} \mid Z_1}}
$$

$$
= \frac{e^{\gamma_{1k}(Z_1+1)+\gamma_{3k}(Z_1+1)Z_2}}{e^{\gamma_{1m}(Z_1+1)+\gamma_{3m}(Z_1+1)Z_2}} \cdot \frac{e^{\gamma_{1m}Z_1+\gamma_{3m}Z_1 Z_2}}{e^{\gamma_{1k}Z_1+\gamma_{3k}Z_1 Z_2}}
$$

$$
= \frac{e^{\gamma_{1k}Z_1+\gamma_{1k}+\gamma_{3k}Z_1 Z_2+\gamma_{3k}Z_2}}{e^{\gamma_{1m}Z_1+\gamma_{1m}+\gamma_{3m}Z_1 Z_2+\gamma_{3m}Z_2}} \cdot \frac{e^{\gamma_{1m}Z_1+\gamma_{3m}Z_1 Z_2}}{e^{\gamma_{1k}Z_1+\gamma_{3k}Z_1 Z_2}}
$$

$$
= \frac{e^{\gamma_{1k}+\gamma_{3k}Z_2}}{e^{\gamma_{1m}+\gamma_{3m}Z_2}} = e^{\gamma_{1k}-\gamma_{1m}+(\gamma_{3k}-\gamma_{3m})Z_2}. \tag{S.99}
$$

d. The interaction effect of Z_1 and Z_2 on the utility that voter i obtains from candidate k is

$$
\frac{\partial \left(\frac{\partial U_{ik}}{\partial Z_1} \right)}{\partial Z_2} = \frac{\partial (\gamma_{1k} + \gamma_{3k} Z_2)}{\partial Z_2} = \gamma_{3k}. \tag{S.100}
$$

e. The interaction effect of Z_1 and Z_2 on the probability of voting for candidate k is

$$
\frac{\partial \left(\frac{\partial P_{ik}}{\partial Z_1} \right)}{\partial Z_2} = \frac{\partial \left(P_{ik} \left[\gamma_{1k} + \gamma_{3k} Z_2 - \sum_{j=1}^{J} P_{ij} \cdot \left(\gamma_{1j} + \gamma_{3j} Z_2 \right) \right] \right)}{\partial Z_2}
$$

$$
= \frac{\partial \left[P_{ik} \left(\gamma_{1k} + \gamma_{3k} Z_2 \right) - P_{ik} \cdot \sum_{j=1}^{J} P_{ij} \cdot \left(\gamma_{1j} + \gamma_{3j} Z_2 \right) \right]}{\partial Z_2}
$$

$$
= \frac{\partial P_{ik}}{\partial Z_2} \cdot \left(\gamma_{1k} + \gamma_{3k} Z_2 \right) + \gamma_{3k} P_{ik}
$$

$$
- \frac{\partial P_{ik}}{\partial Z_2} \cdot \sum_{j=1}^{J} P_{ij} \cdot \left(\gamma_{1k} + \gamma_{3k} Z_2 \right) - P_{ik} \cdot \sum_{j=1}^{J} \left(\frac{\partial P_{ij}}{\partial Z_2} \cdot \left(\gamma_{1k} + \gamma_{3k} Z_2 \right) + \gamma_{3k} P_{ij} \right)
$$

$$
= \frac{\partial P_{ik}}{\partial Z_2} \cdot \left(\gamma_{1k} + \gamma_{3k} Z_2 - \sum_{j=1}^{J} P_{ij} \cdot \left(\gamma_{1k} + \gamma_{3k} Z_2 \right) \right)
$$

$$
+ P_{ik} \cdot \left(\gamma_{3k} - \sum_{j=1}^{J} \left(\frac{\partial P_{ij}}{\partial Z_2} \cdot \left(\gamma_{1k} + \gamma_{3k} Z_2 \right) + \gamma_{3k} P_{ij} \right) \right). \tag{S.101}
$$

Following the logic set out in part b., we know that

$$
\frac{\partial P_{ik}}{\partial Z_2} = P_{ik} \left[\gamma_{2k} + \gamma_{3k} Z_1 - \sum_{j=1}^{J} P_{ij} \cdot \left(\gamma_{2j} + \gamma_{3j} Z_1 \right) \right] \tag{S.102}
$$

and

$$
\frac{\partial P_{ij}}{\partial Z_2} = P_{ij} \left[\gamma_{2j} + \gamma_{3j} Z_1 - \sum_{j=1}^{J} P_{ij} \cdot \left(\gamma_{2j} + \gamma_{3j} Z_1 \right) \right] \tag{S.103}
$$

Substituting these into Eq. S.101, we have

$$
\frac{\partial \left(\frac{\partial P_{ik}}{\partial Z_1} \right)}{\partial Z_2} = P_{ik} \left[\gamma_{2k} + \gamma_{3k} Z_1 - \sum_{j=1}^{J} P_{ij} \cdot \left(\gamma_{2j} + \gamma_{3j} Z_1 \right) \right] \cdot \left(\gamma_{1k} + \gamma_{3k} Z_2 - \sum_{j=1}^{J} P_{ij} \cdot \left(\gamma_{1k} + \gamma_{3k} Z_2 \right) \right)
$$

$$
+ P_{ik} \cdot \left(\gamma_{3k} - \sum_{j=1}^{J} \left(P_{ij} \left[\gamma_{2j} + \gamma_{3j} Z_1 - \sum_{j=1}^{J} P_{ij} \cdot \left(\gamma_{2j} + \gamma_{3j} Z_1 \right) \right] \cdot \left(\gamma_{1k} + \gamma_{3k} Z_2 \right) + \gamma_{3k} P_{ij} \right) \right). \tag{S.104}
$$

Due to the symmetry of interactions, this interaction effect tells us both how Z_2 modifies the marginal effect of Z_1 on the probability of voting for candidate k *and* how Z_1 modifies the marginal effect of Z_2 on the probability of voting for candidate k.

Alphabetical Index